Lecture Notes in Computer Science 4104

Commenced Publication in 1973
Founding and Former Series Editors:
Gerhard Goos, Juris Hartmanis, and Jan van Leeuwen

Thomas Kunz   S.S. Ravi (Eds.)

Ad-Hoc, Mobile, and Wireless Networks

5th International Conference, ADHOC-NOW 2006
Ottawa, Canada, August 17-19, 2006
Proceedings

 Springer

Volume Editors

Thomas Kunz
Carleton University
Department of Systems and Computer Engineering
1125 Colonel By Drive, Ottawa, Ontario, Canada K1S 5B6
E-mail: tkunz@sce.carleton.ca

S.S. Ravi
State University of New York
University at Albany
Department of Computer Science
Albany, NY 12222, USA
E-mail: ravi@cs.albany.edu

Library of Congress Control Number: 2006930271

CR Subject Classification (1998): C.2, D.2, H.4, H.3, I.2.11, K.4.4, K.6.5

LNCS Sublibrary:
SL 5 – Computer Communication Networks and Telecommunications

ISSN   0302-9743
ISBN-10   3-540-37246-6 Springer Berlin Heidelberg New York
ISBN-13   978-3-540-37246-2 Springer Berlin Heidelberg New York

Springer is a part of Springer Science+Business Media

springer.com

© Springer-Verlag Berlin Heidelberg 2006
Printed in Germany

Typesetting: Camera-ready by author, data conversion by Scientific Publishing Services, Chennai, India
Printed on acid-free paper   SPIN: 11814764   06/3142   5 4 3 2 1 0

Preface

The 5th International Conference on Adhoc, Mobile and Wireless Networks (Adhoc-Now 2006) was held during August 17–19, 2006, in Ottawa, Canada. The first four conferences in this series were held in Toronto (2002), Montreal (2003), Vancouver (2004) and Cancun (2005), respectively. The purpose of this conference is to provide a forum for researchers from academia/industry and practitioners to meet and exchange ideas regarding recent developments in the areas of ad hoc wireless networks.

We received more than 170 papers for the conference submitted by authors from the following 25 countries: Australia, Bangladesh, Brazil, Canada, China, France, Germany, Greece, India, Iran, Ireland, Italy, Korea, Malaysia, Mexico, Nepal, Norway, Poland, Spain, Taiwan, The Netherlands, Turkey, UK, USA and Venezuela. Each paper was assigned to three members of the Technical Program Committee (TPC). Since each TPC member was responsible for generating reviews for six to nine papers, we encouraged TPC members to delegate some reviews to other qualified reviewers. The names of all the TPC members and the additional reviewers appear elsewhere in this volume. Based on the reviews, we decided to accept 25 submissions as regular papers and 10 as short papers. Regular papers were given 25 minutes for presentation while short papers were given 15 minutes. All of the accepted papers appear in this volume.

We are indebted to the TPC and the other reviewers for their detailed and timely reviews which enabled us to put together an excellent technical program. We thank Mani B. Srivastava (University of California at Los Angeles, USA) and Victor C. M. Leung (University of British Columbia, Canada) for accepting our invitation to present keynote addresses at the conference. We also thank the Publicity Committee for making sure that the call for papers received the maximum amount of publicity. Ivan Stojmenovic (University of Ottawa, Canada) deserves special thanks for his help with local arrangements. It is also a pleasure to acknowledge a number of student volunteers from Carleton University without whose help the conference could not have run so smoothly. Last but not the least, we thank our sponsors, namely Carleton University (Canada), University at Albany - State University of New York (USA) and the University of Ottawa (Canada).

August 2006

Thomas Kunz
S. S. Ravi
Program Co-chairs
Adhoc-Now 2006

Organization

Organizing Committees

Steering Committee

Michel Barbeau
Evangelos Kranakis
Ioanis Nikolaidis
S. S. Ravi
Violet Syrotiuk

Program Co-chairs

Thomas Kunz and S. S. Ravi

Publicity Committee

Mieso Denko and Pedro M. Ruiz

Local Arrangements

Ivan Stojmenovic

Technical Program Committee

N. Abu-Ghazaleh, SUNY Binghamton
E. Altman, INRIA
M. Barbeau, Carleton Univ.
R. Bazzi, Arizona State Univ.
P. Bose, Carleton Univ.
T. Brown, Univ. of Berne
G. Calinescu, Illinois Inst. of Tech.
E. Chavez, Univ. Michoacana
H. Chen, National Sun Yat-Sen Univ.
J. Cobb, Univ. of Texas Dallas
M. Conti, IIT-CNR
P. Crescenzi, Univ. of Florence
M. Denko, Univ. of Guelph
S. Dobrev, Univ. of Ottawa
M. Dohler, France Telecom
A. Farago, Univ of Texas Dallas
L. Feeney, SICS
S. Fischer, Univ. of Lübeck
A. Gomez-Skarmeta, Univ. of Murcia
A. Hall, ETH
A. Jukan, UIUC
V. King, Univ. of Victoria
E. Kranakis, Carleton Univ.

D. Krizanc, Wesleyan Univ.
S. Krumke, Univ. of Kaiserslautern
T. Kunz, Carleton Univ.
L. Lamont, CRC
E. L. Lloyd, Univ. of Delaware
A. Mielke, LANL
J. Misic, Univ. of Manitoba
P. Morin, Carleton Univ.
L. Narayanan, Concordia Univ.
I. Nikolaidis, Univ. of Alberta
J. Opatrny, Concordia Univ.
M. Papatriantafilou, Chalmers Univ.
P. Penna, Univ. of Salerno
C. Pinotti, Univ. of Perugia
R. Prakash, Univ of Texas Dallas
S. Rajsbaum, UNAM
S. S. Ravi, SUNY Albany
J.-M. Robert, Alcatel
P. M. Ruiz, Univ. of Murcia
C. Schindelhauer, Univ. of Paderborn
A. Scott, Univ. of Lancaster
M. Steenstrup, Stow Research LLC
R. Sundaram, Northeastern Univ.

V. Syrotiuk, Arizona State Univ.
D. Turgut, Univ. of Central FL
A. Vullikanti, VBI and Virginia Tech

P. Ward, Univ. of Waterloo
G. Zaruba, Univ. of Texas Arlington
J. Zhao, ICSI

Adhoc Reviewers

M. Ahmad
I. Ashraf
F. Bian
R. Bruno
S. Commuri
H. De Meer
R. Fonseca
V. Govindaswamy
T. Johansson
V. Kolar
C. Laurendeau
R. Marin
M. Paquette
G. Rossi
T. Staub
A. Tsertou
K. Voulgaris
K. Wang
C. Wu

E. Ancillotti
M. Bechler
P. Boone
C. Cicconetti
M. Couture
J. Diederich
S. Gandham
H. Hellbrück
K. Kang
G. Kuo
S. Lim
A. Martinez
G. Perez
S. Schmidt
H. Tejeda
B. Turgut
M. Waelchli
M. Wang
C. Yeh

K. Anna
T. Bernoulli
E. Borgia
A. Clementi
S. De
B. Ferruccio
A. Gkelias
R. Huang
T. Klie
S. Kuppa
L. Liu
M. Mohsin
A. Pinizziotto
A. Schrader
M. Thoppian
C. Ventre
U. Walther
A. Weyland

List of Sponsors

I. Carleton University, Ottawa, Canada.
   (a) Office of the Vice President (Research and International).
   (b) Dean, Faculty of Engineering and Designs.

II. University at Albany – State University of New York, USA.
   (a) Office of the Vice President for Research.
   (b) College of Computing and Information.
   (c) Department of Computer Science.

III. Dean, Faculty of Engineering, University of Ottawa, Canada.

Table of Contents

Invited Presentations

Sensor Networks: From Smart Dust to Multi-scale, Multi-modal,
Multi-user Observing Systems
Mani B. Srivastava 1

Advances in Wireless Personal Area Networking
Victor C.M. Leung 3

Routing in Sensor Networks

A Comparative Study of Routing Strategies for Wireless Sensor
Networks: Are MANET Protocols Good Fit?
Yasser Gadallah 5

Detecting Disruptive Routers in Wireless Sensor Networks
Steven Cheung, Bruno Dutertre, Ulf Lindqvist 19

Energy-Efficient Data Dissemination in Sensor Networks Using
Distributed Dynamic Tree Management
Kwang-il Hwang, Doo-seop Eom 32

Virtual Coordinates with Backtracking for Void Traversal in Geographic
Routing
Ke Liu, Nael B. Abu-Ghazaleh 46

Routing in MANET

A Distributed Protocol for the Bounded-Hops Converge-Cast
in Ad-Hoc Networks
Andrea E.F. Clementi, Miriam Di Ianni,
Massimo Lauria, Angelo Monti, Gianluca Rossi,
Riccardo Silvestri 60

Discrete Optimization Models for Cooperative Communication
in Ad Hoc Networks
Carlos A.S. Oliveira 73

ROAR: A Multi-rate Opportunistic AODV Routing Protocol
for Wireless Ad-Hoc Networks
  Kwan-Wu Chin, Darryn Lowe   87

Energy-Efficient Face Routing on the Virtual Spanner
  Héctor Tejeda, Edgar Chávez, Juan A. Sanchez, Pedro M. Ruiz   101

Short Papers on Routing

Energy Efficient Multipath Routing Using Network Coding in Wireless
Sensor Networks
  Shan-Shan Li, Pei-Dong Zhu, Xiang-Ke Liao, Wei-Fang Cheng,
  Shao-Liang Peng   114

Formal Verification and Simulation for Performance Analysis
for Probabilistic Broadcast Protocols
  Ansgar Fehnker, Peng Gao   128

Policy-Based Route Optimization for Network Mobility of Next
Generation Wireless Networks
  Moon-Sang Jeong, Yeong-Hun Cho, Jong-Tae Park   142

Geographic Pattern Routing for MANETOR in IVC
  Jiang Hao, Jian-Jin Li, Kun Mean Hou, Chen Lijia   156

Evaluation of the Energy Consumption in MANET
  Géraud Allard, Pascale Minet, Dang-Quan Nguyen,
  Nirisha Shrestha   170

Link Availability at Any Time in MANET
  Jianxin Wang, Xianman Zhu, Jianer Chen   184

Security I

Reputation Based Localized Access Control for Mobile Ad-Hoc Networks
  Sangheethaa Sukumaran, Elijah Blessing   197

Distributively Increasing the Percentage of Similarities Between Strings
with Applications to Key Agreement
  Effie Makri, Yannis C. Stamatiou   211

Key Revocation for Identity-Based Schemes in Mobile Ad Hoc Networks
  Katrin Hoeper, Guang Gong   224

Self-monitoring Security in Ad Hoc Routing
    J.A. Ploskonka, A.R. Hurson 238

Security II

Improved Pairing Protocol for Bluetooth
    Dave Singelée, Bart Preneel 252

Threats to Security in DSRC/WAVE
    Christine Laurendeau, Michel Barbeau 266

LSTOP: A Light-Weight Scalable Truthful Routing Protocol
in MANETs with Selfish Nodes
    Yongwei Wang, Mukesh Singhal 280

Modelling and Analysis of Attacks on the MANET Routing in AODV
    Peter Ebinger, Tobias Bucher 294

Wireless MAC

Initialization for Ad Hoc Radio Networks with Carrier Sensing
and Collision Detection
    Jacek Cichoń, Mirosław Kutyłowski, Marcin Zawada 308

Performance Modeling of a Bottleneck Node in an IEEE 802.11 Ad-Hoc
Network
    Hans van den Berg, Michel Mandjes, Frank Roijers 321

Circularity-Based Medium Access Control in Mobile Ad Hoc Networks
    Mohammad Z. Ahmad, Damla Turgut, R. Bhakthavathsalam 337

Short Papers on Security, QoS and TCP

Improving the ID-Based Key Exchange Protocol in Wireless Mobile
Ad Hoc Networks
    Eun-Jun Yoon, Kee-Young Yoo 349

An Efficient Certificate Management for Mobile Ad-Hoc Network
    Dae-Young Lee, Hyun-Cheol Jeong 355

Performance Improvement of TCP with an Efficient Contention
Window Control Mechanism (ECWC) in IEEE 802.11 Based Multi-hop
Wireless Networks
    Byungjoo Park, In Huh, Haniph Latchman 365

New Service Differentiation Model for End-to-End QoS Provisioning
in Wireless Ad Hoc Networks
  Joo-Sang Youn, Seung-Joon Seok, Chul-Hee Kang 376

Upper Layer Issues I

Authenticated In-Network Programming for Wireless Sensor Networks
  Ioannis Krontiris, Tassos Dimitriou 390

A Congestion Window Adjustment Scheme for Improving TCP
Performance over Mobile Ad-Hoc Networks
  Jung-Hoon Song, Kyung-Hwan Ahn, Dong-Hoon Cho, Ki-Jun Han . . . 404

Predictive Call Admission Control Algorithm for Power-Controlled
Wireless Systems
  Choong Ming Chin, Moh Lim Sim, Sverrir Olafsson 414

Upper Layer Issues II

File System Support for Adjustable Resolution Applications in Sensor
Networks
  Vikram Munishwar, Sameer Tilak, Nael B. Abu-Ghazaleh 428

A Classification and Performance Comparison of Mobility Models
for Ad Hoc Networks
  Emre Atsan, Öznur Özkasap 444

Power-Aware Rate Control for Mobile Multimedia Communications
  *Hye-Soo Kim, Dinh Trieu Duong, Jae-Yun Jeong, Byoung-Kyu Dan,
  Sung-Jea Ko* 458

Author Index 473

Sensor Networks: From Smart Dust to Multi-scale, Multi-modal, Multi-user Observing Systems

Mani B. Srivastava

EE Department, University of California Los Angeles,
Los Angeles, CA 90095-1594

Abstract. Internet and wireless technologies have flattened the world by revolutionizing the exchange of information among individuals and organizations at a global scale over the past decade. Similar technological forces have led to the emergence of embedded networked sensing systems, or sensor networks, that are bringing about the next revolution. This new local revolution is making the world "transparent" by enabling observation of physical, biological, chemical, enterprise, urban, social, and personal processes up close, and at spatial and temporal details that are simply impossible otherwise. Already this technology has led to new science resulting from observation of new phenomena in areas ranging from the investigation of critical microclimate on the scale of a mountain canyon, to distribution of contaminants and their introduction into ground water supplies, to the fine-scale properties of alpine plants. The considerable progress in the past few years have also led to the realization that the early view of sensor networks as "smart dust" - a large and ad hoc but flat and homogeneous single-purpose long-lived collection of static resource-constrained devices - needs to be considerably expanded to a view of these systems as multi-scale, multi-modal, multi-user rapidly-deployable actuated observing systems. The talk will describe how the early technical challenges such as autonomous self-configuration, energy-aware protocols, and efficient embedded software are now giving way to new challenges involving system and data integrity, safety and robustness, software re-configuration, and active sensing. Moreover, as the embedded sensing technology moves from scientific, engineering, defense, and industrial contexts to the wider personal, social and urban contexts, a new class of applications are emerging which draw on sensed information about people, objects, and physical spaces, and integrate with the global Internet and cellular infrastructure. The talk will discuss the privacy and data sharing requirement of these applications, and speculate on their implications on the Internet and cellular network fabric and services.

Biography

Professor Mani Srivastava received both the M.S. and Ph.D. degrees from the University of California, Berkeley, in 1987 and 1992, respectively. His M.S.

T. Kunz and S.S. Ravi (Eds.): ADHOC-NOW 2006, LNCS 4104, pp. 1–2, 2006.

project was on automatic compilation of CMOS bit-slice datapaths as part of the Lager silicon compiler for DSP VLSI, while his Ph.D. dissertation was on hardware-software rapid prototyping and co-design for embedded DSP and control applications. Prior to joining the UCLA Electrical Engineering Department faculty in 1996, Dr. Srivastava worked on mobile and wireless networking at the Networked Computing Research Department at AT&T Bell Labs in Murray Hill, NJ (now Lucent Technologies - Bell Labs Innovations).

Since joining the EE Department at UCLA, Professor Srivastava's research and teaching have focussed on architecture, algorithms, and design optimization techniques for wireless networked and embedded systems, and DSP VLSIs. In recent years his main focus has been on mobile and wireless multimedia systems, and distributed wireless sensor networks, with emphasis on issues such as power-aware computing and communications, and quality of service. He has been the Editor-in-Chief of ACM Sigmobile Mobile Computing and Communications Review since January 2005.

Professor Srivastava holds five patents for: the method for call establishment and rerouting in mobile computing networks; medium access control and air interface subsystems for an indoor wireless ATM network; wireless adapter architecture for mobile computing; scheduling in wireless access protocols based on battery power level; and mobile host roaming in ATM networks. He has published extensively on wireless networking, low-power systems, and embedded system design tools. He is a member of the IEEE and of the ACM.

Advances in Wireless Personal Area Networking

Victor C.M. Leung

Dept. of Electrical and Computer Engineering,
The University of British Columbia, Vancouver, BC, Canada V6T 1Z4

Abstract. Over the last ten years, the emergence of license-free wireless networking technologies has been one of the most exciting developments in the communications area. These technologies are already widely used in wireless local area networks, which provide local area connectivity in computer networks. Wireless personal area networks (WPANs), which communicate using license-free radios over much shorter distances, are emerging to bring ubiquitous network connectivity to consumer electronic devices. First generation WPANs based on the Bluetooth technology are already widely deployed in cellular telephones, audio headsets, and personal digital assistants. In the next few years, high data rate WPANs, especially those employing ultra wideband (UWB) transmissions, will increasingly provide multimedia connectivity to home entertainment systems. This presentation will describe the advances of WPAN technologies from Bluetooth to high data rate WPANs. Research results in Bluetooth and high data rate WPANs accomplished at the University of British Columbia will be highlighted. Future trends and open research problems will be discussed.

Biography

Victor C. M. Leung received the B.A.Sc. (Hons.) and Ph.D. degrees in electrical engineering from the University of British Columbia (U.B.C.) in 1977 and 1981, respectively. From 1981 to 1987, Dr. Leung was a Senior Member of Technical Staff at MPR Teltech Ltd. In 1988, he was a Lecturer in the Department of Electronics at the Chinese University of Hong Kong. He returned to U.B.C. as a faculty member in 1989, where he currently holds the positions of Professor and TELUS Mobility Research Chair the Department of Electrical and Computer Engineering, and is a member of the Institute for Computing, Information and Cognitive Systems. He was a project leader and a member of the Board of Directors in the Canadian Institute for Telecommunications Research, a Network of Centres of Excellence funded by the Canadian Government. His research interests are in the areas of architectural and protocol design and performance analysis for computer and telecommunication networks, with applications in satellite, mobile, personal communications and high speed networks. He has authored or co-authored about 300 publications in refereed international journals and conferences.

The many academic awards that Dr. Leung has received include the APEBC Gold Medal as the head of the graduating class in the Faculty of Applied Science,

T. Kunz and S.S. Ravi (Eds.): ADHOC-NOW 2006, LNCS 4104, pp. 3–4, 2006.
© Springer-Verlag Berlin Heidelberg 2006

UBC, and Natural Sciences and Engineering Research Council Postgraduate
Scholarships. Dr. Leung is a Fellow of IEEE and a voting member of ACM. He
is an editor of the IEEE Transactions on Wireless Communications, an associate
editor of the IEEE Transactions on Vehicular Technology, and an editor of the
International Journal of Sensor Networks. He has served on the committees of numerous international conferences, and was the General Co-chair of IEEE/ACM
MSWiM'05 in Montreal, PQ, and the TPC Vice-chair of IEEE WCNC'05 in
New Orleans, LA. He is the Local Chair of IWCMC 2006 in Vancouver, BC, and
chairs the Next Generation Mobile Networks Symposium in the same conference.

A Comparative Study of Routing Strategies
for Wireless Sensor Networks: Are MANET Protocols
Good Fit?

Yasser Gadallah

Communications Research Centre (CRC) Canada
3701 Carling Avenue, Ottawa, Ontario, Canada K2H 8S2
yasser.gadallah@crc.ca

Abstract. The operation of sensor networks places special requirements on routing algorithms. These requirements stem from the unique nature of these networks. For example, the algorithm needs to be data-oriented and it should impose the smallest possible overhead on the resource-constrained nodes of the network. It has been long claimed that algorithms designed for mobile ad hoc networks (MANET) should not be considered for sensor networks for several reasons. Most important of these reasons is their heavy operation overhead. In this study, we examine this claim by comparing two of the sensor network specific algorithms, directed diffusion and TTDD, with two MANET algorithms, AODV and OLSR which belong to the reactive and proactive routing categories, respectively. We run experiments with realistic sensor network scenarios. We show that MANET protocols perform significantly better than their sensor-specific counterparts. Finally, we offer suggestions for future research based on the results of our study.

Keywords: Sensor Networks, Routing, MANET, Performance Comparison.

1 Introduction

Wireless sensor networks are getting increasing attention due to the recent large advances in hardware and wireless technologies as well as the unlimited applications and possibilities for their use both in the civilian and military fields. These networks have special characteristics that differentiate them from their more general parent class namely, the mobile ad hoc networks (MANET). For example, a wireless sensor network is expected to have, in general, a large number of tiny resource-constrained nodes. The number of nodes can be several orders in magnitude larger than a typical mobile ad hoc network. Also, mobility of this type of nodes is generally limited and they are normally used to sense some phenomena and send their data for processing to more capable nodes either within the same network or via gateways to other networks.

Due to the differences in nature of these networks than that of mobile ad hoc networks, some studies (e.g. [3]) have suggested that the use of routing protocols that have been developed for MANET is not practical for sensor networks and hence

T. Kunz and S.S. Ravi (Eds.): ADHOC-NOW 2006, LNCS 4104, pp. 5–18, 2006.

many new routing protocols (see the survey in [1] for some examples) have been developed specifically for these networks.

In this study, we revisit the assertion of the unsuitability of MANET routing protocols for use in sensor networks, as has been suggested in the past. For this purpose, we first discuss the features of sensor networks and the resulting requirements they place on routing protocols. Then, and with this discussion in mind, we perform direct performance comparisons between protocols that belong to the main MANET routing categories, namely AODV [8] and OLSR [7], and two data dissemination algorithms that have been developed specifically for sensor networks, namely Directed Diffusion [5] and Two-Tier Data Dissemination (TTDD) [10]. In our comparison experiments we use scenarios that would be encountered with the use of wireless sensor networks. Our goal is to find out if the performance difference between the different protocols justifies reconsidering the assertion that MANET protocols are not suitable for use in wireless sensor networks. We also offer suggestions for areas of improvement to the resulting best technique in order to make it more compliant with the routing functionality requirements for sensor networks that we discuss in this study.

This paper is organized as follows. In section 2, we discuss the features of sensor networks and the criteria that should be used to evaluate the suitability of a routing protocol for use in these networks. In section 3, we give an overview of the protocols that we compare in this study. In section 4, we describe our comparison scenarios, metrics and simulation environment. In section 5, we present the results of our experiments. In section 6, we conclude the study and offer suggestions for future work based on this study.

2   Sensor Networks Features and Criteria for Routing

Routing in sensor networks has the main function of ensuring that the information gets transferred successfully from network nodes that "sense" this information from their targets to the end destination where it gets analyzed and made use of. In general, sensor networks are characterized by the following main features:

- The network may be heterogeneous in nature. This means that its nodes may not necessarily have the same computational, storage or mobility capabilities.
- Sensor nodes have limited resources e.g. energy, memory, storage, etc.
- Wireless communication range for sensor nodes is generally limited.
- Network topology is expected to change with time, in general. This could be due to failed nodes, new nodes joining the network, or some node mobility.
- Number of nodes within the network is generally large (could be few thousands).
- There could be multiple data collection points for sensed data.
- Sensor networks are expected to be data-oriented in nature. This means that data of specific attributes is targeted as opposed to node addresses.

Based on these characteristics, we can extract some criteria to use in evaluating the suitability of a routing protocol for use in sensor networks:

- The overhead of the routing protocol specific traffic should not be significant for network bandwidth. Also, impact on node resources should be kept to a minimum.
- The routing protocol should be able to scale with network size and traffic.
- The routing protocol should either be inherently energy-efficient, or adaptable to integration with a suitable energy-efficient algorithm.
- In order to ensure robust operation, the protocol should not be dependent on a node or a subset of network nodes for its functionality.
- The routing protocol should be able to respond to changing network conditions and topology promptly and efficiently without resulting in heavy traffic load.
- The routing protocol should be able to support multi-hop operation due to the limited sensor node wireless ranges.
- The routing algorithm should be able to support the network data-oriented nature.

3 Protocol Overview

In this section, we give an overview of the routing protocols that we compare in this study.

3.1 Directed Diffusion

In the directed diffusion algorithm [5], communication between nodes is done on the basis of one-on-one communications. Therefore, no elaborate routing is done. Sink nodes decide on which data is needed and the intervals at which this data needs to be transmitted. The concept of the gradient, which mainly describes the data, its dissemination frequency and its flow direction, is used to handle data transfers. The process starts by a sink node broadcasting its need for certain data that is described by specific attributes. This request gets propagated through the network until it reaches a node that either possesses this data or knows where to get it. When several responses reach the requesting sink node, it accepts the first one and enforces the continuation of obtaining data from the neighbor from which the fastest response was obtained. The rest of the neighbors continue to receive exploratory events, which are data packets at a reduced rate. This can be a source of wasteful bandwidth and energy consumption. When we consider the criteria that we discussed in the pervious section we note that directed diffusion does not possess inherent energy-efficiency features. Therefore its adaptability to integration with a complementary energy-efficient technique needs to be investigated.

3.2 Two-Tier Data Dissemination (TTDD)

The TTDD design [10] was built on the assumption that network sensor nodes are stationary and location-aware. A source node detects an event and proactively builds a grid structure throughout the sensor field and sets up the forwarding information only at the sensors closest to grid points (dissemination nodes). The cell size, α, is a critical parameter for TTDD. Its choice, according to [10] is dependent on the application at hand. Data sinks need to flood their queries only within their local cells until a dissemination node is reached. The query is then forwarded on the grid upstream towards the source. Each dissemination node along the path keeps information about the downstream node from which it got the query. Data is then sent back to the sink along the same path. TTDD handles sink mobility via assigning changeable *primary* and *immediate* agents to each sink. The primary agent communicates with dissemination nodes. It then sends the data to the immediate agent which relays it to the sink. When a sink node moves out of range of its immediate agent, it acquires a new immediate agent and informs its primary agent. Similarly, the primary agent is replaced when the sink moves away from it. Considering the criteria of section 2, TTDD does not follow an energy-efficiency strategy. Its adaptability to integration with an energy-efficient scheme needs to be examined. It should also be noted that the dependence of TTDD on nodes at grid points only can lead to large energy imbalance. Also, the significant communication overhead resulting from grid construction and maintenance can have a visible effect on energy consumption and network operation.

3.3 AODV

The functionality of the AODV protocol [8], which belongs to the reactive routing category of MANET, is based on maintaining a vector of paths (i.e. routing table) that lead to the different destinations at each node. Route establishment is done on demand. A given node does not have a full knowledge of any of the routes. It only knows the next node along any given route. Each node keeps only one route (the one that has the smallest number of hops, i.e. the shortest route) to any given destination. When a node needs to communicate with another node to which it does not have a route, it sends a route request to its neighbors. This request gets propagated in the network until a node that knows a route is found. It then replies back informing that it knows a route. The classic AODV algorithm also uses periodic HELLO messages to keep neighbors aware of the other nodes in the neighborhood. In more recent versions of AODV, this is done via relying on the link state capabilities of the underlying MAC protocol. This procedure is used for route maintenance. AODV is not a data-oriented algorithm. This is an area that needs to be addressed to make it viable for use in many of the sensor network applications. Also, it does not have an inherent energy-efficiency strategy. However, it has been shown to integrate successfully with energy-efficient algorithms [4]. Also, its dependence on IP addressing can introduce unnecessary overhead to sensor network operations. To overcome this issue, a light IP implementation that is suitable for sensor networks can be used [2].

3.4 OLSR

The OLSR (Optimized Link State Routing) protocol [7] belongs to the proactive routing category in MANET. In this protocol, nodes exchange topology information on a regular basis. Selected multipoint relay (MPR) nodes announce this information periodically to the network. These nodes are also used in determining the route from any network node to a given destination. MPR node selection depends on periodical exchange of HELLO messages between nodes. Also, topology control messages are sent periodically (or upon MPR node changes) to maintain topology information throughout the network. With regard to the criteria discussed in section 2, the aspects that we discussed in the AODV case apply equally to OLSR. In addition, the overhead of the OLSR proactive approach in determining routes to all network nodes can result in unnecessary overhead to sensor network operation even though it can also have a positive impact in lowering packet delivery latency in some scenarios.

4 Comparison Strategy

We perform our comparison experiments based on realistic scenarios that can be encountered in practical sensor network applications. This is to demonstrate the relative abilities and performance of the different algorithms under study in real life situations. For this purpose, we select a topology of specific dimensions and use randomly generated node deployment scenarios.

4.1 Comparison Scenarios

Our goal is to test the performance of each algorithm in situations where sensor nodes are deployed randomly in the field to collect data in a certain area of interest. Examples of this can be in areas with environmental problems such as in the case of forest fires, or in disaster areas to aid search and rescue operations. It can also be in situations such as in a battle field where enemy movements need to be detected. We simulate a heterogeneous network with two types of nodes. The first type is sensor nodes which have limited energy resources and are stationary (no movement). The second type is nodes that are more capable and have significantly more resources, and are generally mobile. This is to simulate the situation of the interactions between sensor nodes that are deployed in the field, and resource-rich nodes which can be either the processing/command centers or gate ways to these centers. In reality they can be mounted on tanks or battle vehicles that move around in the military field, or on the rescue vehicles that are used in the areas of natural disasters and can communicate with deployed sensors to collect data that can guide the operations. We will call this latter type of nodes "sink nodes" throughout the rest of this study. We experiment with the following scenarios:

- Various node populations: to test the scalability with increased network sizes
- Different sink mobility levels: to test the ability to cope with changing topologies
- Different numbers of sink and source (sensor) nodes
- Different data rates (the data rate of sources increases with the increase in the activity of the monitored phenomena)

Throughout our simulations we will assume that all data sinks are interested in data detected by any source sensor. This means that connections need to be established for all source-sink combinations within the network.

4.2 Metrics

We use both network operation performance metrics as well as sensor node energy performance metrics. As far as network operation performance is concerned, we measure the following two aspects:

- Average data delivery ratio: This metric gives an indication of the ability of the algorithm to deliver the data that was detected at source sensors to data sinks. It is measured as the ratio of the number of data packets received at data sinks to the number of data packets detected at source nodes over all source-sink pairs. In our evaluations we will consider that the delivery ratio for a certain source-sink pair is zero for a certain algorithm if the algorithm was unable to enable the pair in question to establish a path and communicate.
- Average data delivery latency: This metric measures the average data delivery delay over all packets that were successfully delivered to data sinks.

As for the energy performance metrics of sensor nodes, we measure the following aspects:

- Average energy consumption: It is the average consumed energy calculated over all sensor nodes within the network. It gives an indication of the effect of the scenario on the lifetime of sensor nodes.
- Energy fairness: This metric is calculated as the standard deviation of remaining sensor node energies. It shows the relative utilization of sensor nodes in routing duties by the routing algorithm. The higher the standard deviation, the lower the energy fairness. Unfair operation leads to faster depletion of some critical nodes which may lead to network partitioning.

4.3 Simulation Environment

We use the network simulator ns-2 [11] with its MAC 802.11 layer implementation in our comparisons. We adjusted the energy parameters to present network node conditions more precisely as we will discuss shortly. The AODV and directed diffusion algorithms are already implemented and included in the regular ns-2 software. We use the INRIA implementation of OLSR [6]. We also use the TTDD code as implemented by the TTDD authors [9]. We run all our simulations in a simulation area of dimensions 1000 × 1000 square meters. Unless otherwise indicated in the discussion of a specific experiment, the simulations parameters default to the following values. We use a total number of network nodes of 100. We use 4 sink nodes, and 4 source nodes. Data are generated at each source node at a rate of 1 packet per second. The size of the data packet is 64 bytes. Sensor nodes are always stationary and data sink nodes are also static unless otherwise indicated. Each sensor node, including source nodes, starts the simulation with limited energy resources. Sink nodes, on the other hand, have plenty of energy resources to simulate a realistic difference between the normally resource restricted sensor nodes and the resource rich data sinks. Energy parameters are given in table 1. For TTDD, the cell size is an important parameter. Since there is no specific method in the original TTDD study to calculate it, we used a cell size of 300 meters based on taking the ratio between the

Table 1. Simulation energy parameters

Initial Sensor Node Energy (J)	200
Rx Sensor Node Power Consumption (W)	0.395
Tx Sensor Node Power Consumption (W)	0.66
Idle Sensor Node Power Consumption (W)	0.035

simulation area size of [10] and that of our study. Each result is an average of ten simulation runs; each of which with a different random node deployment scenario. Each simulation experiment runs for 600 seconds.

5 Evaluation Results

5.1 Performance with Increasing Network Population

In this experiment, we examine the performance of the algorithms with increasing the number of network nodes. In reality, it is expected that many sensor network applications will require deploying a large number of sensor nodes randomly over a certain geographical area of interest. We ran simulation experiments with the default simulation parameters as in Section 4.3 but with varying the number of network nodes (50, 100, 150, 200 and 250 nodes) to check the effect of increasing the network size on protocol performance. Fig. 1 shows the effect on packet delivery ratio. We notice from this figure that the packet delivery ratio for AODV is consistently high at almost 100% regardless of network population for these simulation conditions. Considering directed diffusion, we find that the delivery ratio is comparable to that of AODV at lower node numbers and starts to drop significantly at about 150 nodes. As the number of nodes increases, the number of possible paths between a source-sink pair increases which causes a significant increase in exploratory messages within the network. This in turn leads to increased collisions and packet loss. The same trend is also seen with OLSR but with a higher ratio decline than with directed diffusion. The proactive nature of the OLSR algorithm in trying to establish routes to all network destinations affects its performance with the increase of network nodes due to increased congestion and hence collisions. As for TTDD, we find that the delivery ratio is lower than the other algorithms even at a lower number of nodes due to the fact that some source-sink connections could not be established in some node deployment scenarios hence affecting the overall average. The decline in the case of TTDD as the number of network nodes increases is also significantly steeper than the other cases. The high communication overhead involved in constructing and maintaining the grids in TTDD and the resulting congestion around some critical nodes seems to affect its ability to establish robust source-sink connections as the number of network nodes grows. Considering packet delivery latency, we find that the AODV and TTDD algorithms perform best in this regard, see Fig. 2. Nevertheless, we have to keep in mind the sharp decline of packet delivery as nodes increase in the TTDD case. The increase in the latency with directed diffusion as nodes increase is attributed to the increase in the retransmissions due to the relatively higher congestion with the increased routing and the associated exploratory events overhead with increased path possibilities. In case of OLSR, retransmissions due to the congestion resulting from the topology maintenance overhead cause packet delivery delays as nodes increase. Considering the energy performance, AODV seems to consume the least amount of energy in all four protocols. The amount of consumed energy stays almost the same as the number of network nodes increases. This is due to the reactive routing strategy that it follows which is based on establishing one route per connection and only when needed. The relatively low communication overhead of AODV also helps conserve energy. The TTDD algorithm consumes less energy than directed diffusion and OLSR and performs somewhat close to AODV with small number of network nodes. This changes as the number of nodes increases with the sharp increase in protocol overhead. With OLSR, energy consumption increases sharply as network population increases due to the increase in proactive route maintenance overhead as well as the cost of retransmissions. Fig. 3 shows the energy consumption performance. Fig. 4 shows algorithm relative performances with regard to energy fairness. The figure shows that the standard deviation of remaining sensor node energies, which is the measure for energy fairness, increases only slightly in case of AODV. This means that energy imbalance increases with the increase of network nodes as the focus on specific routes causes some nodes to become more utilized relative to others within the network. In case of directed diffusion, energy imbalance between network nodes starts higher than that of AODV at smaller number of nodes. As the number of nodes increases the imbalance increases only slightly in most cases. In case of TTDD, the imbalance starts at relatively low levels with smaller number of network nodes and increases sharply as the number of nodes increases due to the over utilization of dissemination nodes in routing duties relative to other sensor nodes. At a number of nodes of 150, this trend is reversed and the level of imbalance starts decreasing as more nodes get more utilized through the flooding as the density increases per cell. The large drop in packet delivery at 150 nodes also contributes to this result. OLSR also follows a trend similar to that of TTDD but with less increase in energy imbalance with the increase in network size. As the number of nodes increase beyond a certain number, the imbalance starts to decrease due to the sharp drop in data delivery which reduces the load on MPR nodes.

5.2 Performance with Sink Mobility

In this experiment, we explore the performance of the algorithms under different sink mobility levels. The conditions of this experiment can be encountered in situations such as rescue operations where emergency vehicles (data sinks) join the sensor network to get information from deployed sensor nodes about the conditions in a certain disaster area e.g. presence of live people who need to be rescued. In our experiment, we use the default simulation parameters as described in Section 4.3 and vary the mobility conditions of the sinks by changing the pause time. We use pause times of 100-600 seconds. The sink stops for the pause time and then moves in a random direction at a speed that is uniformly distributed between 0 and 10 m/sec.

Under these conditions, the packet delivery performance of the algorithms is as shown in Fig. 5. We notice that AODV consistently delivers data at a delivery rate of almost 100% regardless of the sink mobility conditions. It is followed closely by

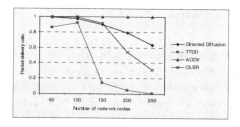

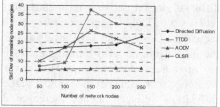

Fig. 1. Effect of network population on packet delivery ratio

Fig. 2. Effect of increasing network popultion on packet delivery latency

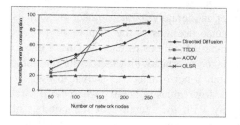

Fig. 3. Effect of network size on energy consumption

Fig. 4. Effect of network size on energy fairness

OLSR. On the other hand, directed diffusion and TTDD show lower delivery ratios with higher mobility. This improves with less mobility (higher pause time). Average packet delivery latency, as shown in Fig. 6, is consistently low with both AODV and OLSR. TTDD shows a slightly higher latency. Directed diffusion shows high delivery latency with high sink mobility. Path establishment between source-sink pairs seems to take longer time with mobility in directed diffusion than all other algorithms. In both the directed diffusion and TTDD cases, the latency decreases with the decrease of mobility as the routes become more stable. As far as energy consumption is concerned, the trend is shown in Fig. 7. All algorithms, with the exception of TTDD, perform in a consistent manner with regard to average power consumption regardless of sink mobility conditions. AODV consumes the least amount of energy, while OLSR comes second followed closely by directed diffusion. TTDD, on the other hand, consumes comparable amount of energy to that of OLSR and directed diffusion at high mobility conditions. Its energy consumption decreases with the decrease of sink mobility due to the corresponding significant decrease of its routing overhead. Energy fairness, as shown by the standard deviation trends in Fig. 8, is generally consistent regardless of mobility in directed diffusion, AODV and OLSR, with AODV being the fairest due to its relatively low routing overhead and hence moderate utilization of routing nodes. In TTDD, energy imbalance between sensor nodes gets generally higher with higher sink mobility. This is attributed to the increased flooding that the sinks perform in their new neighborhoods to establish new routes to sources. The resulting extra energy cost on certain sensor nodes increases the imbalance.

5.3 Performance with Different Numbers of Sources

In this experiment, we explore the effect of increasing the number of sources, and hence the overall data traffic, on protocol performance. When monitoring some phenomenon, several sensor nodes may detect activities pertaining to this phenomenon simultaneously. These sensor nodes will then transmit data to interested sinks as per network configurations. It is important to investigate the effect of such an event on the routing performance. For this purpose we keep the number of sink nodes at 4 and vary the number of sources to 1-6. Otherwise, the other simulation parameters are kept at default values as specified previously. Fig. 9 shows the effect of changing the number of sources on the packet delivery ratio of the four algorithms. We notice from this figure that AODV and OLSR are almost unaffected. Directed diffusion packet delivery ratio starts to suffer moderately as the number of sources reaches 5. TTDD's packet delivery performance declines as the number of sources increases. This reaches high levels as the number of sources reaches 5. This is due to the increase in the grid forming and maintenance overhead with increasing the number of sources. As for the packet delivery latency performance as shown in Fig. 10, we notice that AODV, OLSR and TTDD show low values of latency whereas directed diffusion's delivery latency increases sharply with the increase of the number of sources. This could be due to the increased reliance on common nodes for different sink-source paths as the number of sources increases. This, combined with the exploratory messaging overhead, imposes extra delays on packet transmissions. The energy cost of the different algorithms is shown in Fig. 11. In this figure we notice that energy consumption increases in all cases due to increased traffic. In the cases of AODV and OLSR the rate of increase is low even though the consumption in case of OLSR is much higher due to topology maintenance overhead as we explained in earlier cases. In case of TTDD the rate of increase becomes more visible at larger numbers of sources (starting 4 sources) due to the corresponding increase of protocol overhead. In the case of directed diffusion, the increase in consumption is almost linearly proportional to the increase in the number of sources and it surpasses OLSR's consumption level at 4 sources. This shows that the cost of overhead in case of directed diffusion climbs steadily with the increase in the number of sources. Energy fairness trend, as shown in Fig. 12, shows a decrease in energy fairness in all protocols as the number of sources increases. The rate of increase is lowest in OLSR. This is due to the fact that it proactively searches routes for all destinations in all cases, and the increase is merely a reflection of the increase of the data transmission load on the MPR nodes. With the other protocols, new routes are needed with the increase of sources. This causes an increase of the routing overhead. The overall increase in traffic imposes an additional load on the nodes on the routes. As the overhead of AODV is moderate, the imbalance is the lowest in all algorithms. TTDD shows almost a similar trend like AODV until we reach 5 sources and then the additional overhead pushes the imbalance sharply higher. In the case of directed diffusion, the rate of increased energy imbalance is sharpest in all algorithms with the increase in the number of sources. This is due to the increased overhead, and hence energy costs, in the neighborhoods where sink nodes try to establish communications with sources.

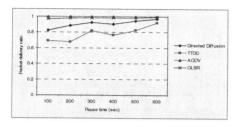

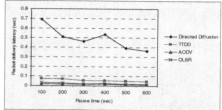

Fig. 5. Effect of Sink mobility on packet delivery ratio

Fig. 6. Effect of sink mobility on packet delivery latency

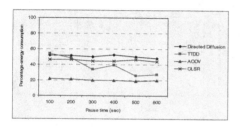

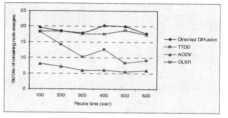

Fig. 7. Effect of sink mobility on energy consumption

Fig. 8. Effect of sink mobility on energy fairness

5.4 Performance with Different Numbers of Sinks

The number of sinks can increase in reality when the demand on a sensor network increases. Examples of this are in situations where new parties (firefighters, soldiers, rescue workers, etc) subscribe to the network from different geographic locations upon the occurrence of certain events in order to obtain sensed data of specific phenomena. Therefore, we experimented with varying the number of sinks to check the effect on the performance of the algorithms. We found the results of this experiment to be similar to the case of varying the number of sources with slight differences. We are not including the figures that show the results due to space constraints.

5.5 Performance with Different Data Transmission Rates

Data transmission rates increase in reality when the activity of the monitored phenomena increases. Sensor nodes then increase data rates to sinks if transmission is triggered by events. In this experiment, we change each source's data rate to 1-5 packets per second. Fig. 13 shows packet delivery ratio trends. We see that there is no visible effect on AODV as data rate increases. The effect on OLSR becomes visible at the relatively high rate of 4 packets/sec. Directed diffusion shows lower delivery ratio earlier. This gets worse as data rate increases. TTDD's delivery ratio drops steeper than the rest of the algorithms at lower data rates. Packet delivery latency, as seen in Fig. 14, is consistently low for AODV, OLSR and TTDD. Directed diffusion, on the other hand, shows an almost linear increase in latency with data rate.

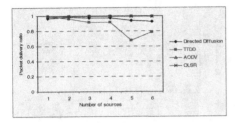

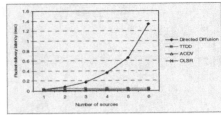

Fig. 9. Effect of number of sources on packet delivery ratio

Fig. 10. Effect of the number of sources on latency

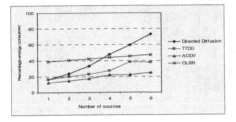

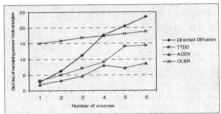

Fig. 11. Effect of the number of sources on energy consumption

Fig. 12. Effect of number of sources on energy fairness

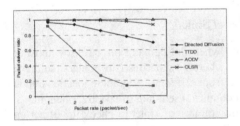

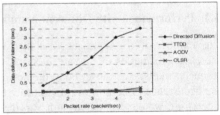

Fig. 13. Effect of data rate on packet delivery ratio

Fig. 14. Effect of data rate on latency

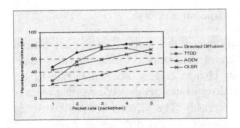

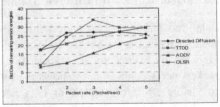

Fig. 15. Effect of data rate on energy consumption

Fig. 16. Effect of data rate on energy fairness

It seems that the increased data rate together with the directed diffusion exploratory message overhead affects the latency. Fig. 15 shows that energy consumption increases as the data rate increases for all algorithms. The increase is close to linear in AODV, OLSR and directed diffusion. In case of TTDD, the increase in consumption is drastic as the rate increases from 1 packet per second up to 3 packets per second and then it becomes almost flat. This is due to the increased congestion with the increased data rate combined with TTDD protocol overhead. The resulting collisions and retransmission attempts cause additional energy waste. Consumption then drops at 4 and 5 packets per second due to the large drop in packet delivery. Energy fairness, as shown by Fig. 16, decreases as the data rate increases in the cases of AODV and OLSR due to the increased utilization of the forwarding nodes with increasing data traffic. This is also the case with TTDD up to a certain data traffic level where energy fairness trend becomes flat due to the drop in its ability to forward traffic. In directed diffusion, since route selection is based on the lowest delay, fairness decreases up to 2 packets per second and then the trend becomes flat as the routing paths become more diversified with the increased overall latency.

5.6 Discussion of Results

Experiment results show that the routing strategies of the different algorithms result in different levels of strength depending on the scenario at hand. The only algorithm that shows consistently stronger performance relative to the other algorithms is the AODV algorithm. Therefore, and from routing strategy and technique perspective, we can conclude that AODV which belongs to the reactive ad hoc routing class of algorithms performs significantly better than the algorithms that were specifically designed for sensor networks for the experiments that we have conducted. However, and as we indicated earlier, there are some issues that have to be addressed for the AODV algorithm in order to make it suitable for the data oriented nature of most of sensor network applications. The strategy to address these issues should not affect the core strengths that are shown by the AODV algorithm. The alternative to this approach would be to resolve the issues that are faced when using sensor network specific algorithms such as the ones we studied in this paper. This alternative could prove challenging as some of these issues are fundamental performance affecting ones.

6 Conclusions

Our goal in this study was to evaluate, from routing strategy perspective, the performance of some routing techniques that were specifically designed for sensor networks versus that of traditional MANET routing techniques. We performed several experiments that reflect realistic scenarios in wireless sensor networks. This included different network node populations, sink mobility levels, number of sources and sinks, and data traffic rates. We compared sensor network-specific techniques namely, directed diffusion and TTDD with the MANET protocols AODV and OLSR.

We found that AODV generally performs better than the other algorithms in all the scenarios that we have experimented with. This means that we can still rely on MANET routing algorithms to deliver reliable performance in realistic sensor networks. Ad hoc routing algorithms, however, lack some of the required features that are needed by sensor networks such as the data-oriented operation. Therefore, we have to make the choice between two strategies for providing the algorithms that best fit the operation of real-life sensor networks. One strategy is to adapt strong ad hoc routing techniques to the needs of sensor networks. The other strategy is to address the performance issues of sensor network specific algorithms to bring their performance to a level comparable to that of their MANET counterparts. Our conclusion from this study is that it is more feasible to align the operation of promising ad hoc routing protocols, such as AODV, with the needs of the sensor network operation. Therefore, for future studies we plan to work on adapting some of the ad hoc network protocols that demonstrate consistently good performance under various conditions to the needs of sensor networks without compromising their robust and proven abilities to cope with changing network conditions and environment.

Acknowledgement

This work was supported by Defence Research and Development Canada.

References

[1] K. Akkaya and M. Younis, "A survey on routing protocols for wireless sensor networks" in Elsevier Journal of Ad Hoc Networks, Volume 3, Issue 3, May 2005, pages 325-349.

[2] A. Dunkles, J. Alonso and T. Voigt, "Making TCP/IP viable for wireless sensor networks", In Proceedings of the First European Workshop on Wireless Sensor Networks (EWSN'04), work-in-progress session, , January 2004, Berlin, Germany.

[3] D. Estrin, R. Govindan, J. Heidemann and S. Kumar, "Next Century Challenges: Scalable Coordination in Sensor Networks", Proceedings of the 5th annual ACM/IEEE international conference on Mobile Computing and Networking (MobiCom '99), Seattle, WA, USA, August 1999, pages 263-269.

[4] Y. Gadallah and T. Kunz, "A protocol independent energy saving technique for mobile ad hoc networks", accepted for the special issue on Mobile and Wireless Networking, International Journal of High Performance Computing and Networking (IJHPCN), Vol 4, Issue 1, 2006.

[5] C. Intanagonwiwat and R. Govindan and D. Estrin and J. Heidemann and F. Silva, "Directed diffusion for wireless sensor networks", ACM/IEEE Transactions on Networking, Volume 11, Issue 1, February 2003, pages 2-16.

[6] INRIA – Project HIPERCOM. http://hipercom.inria.fr/olsr/

[7] P. Jacquet, P. Muhlethaler, T. Clausen, A. Laouiti, A. Qayyum and L. Viennot, " Optimized Link State Routing protocol for Ad Hoc Networks", IEEE INMIC, Pakistan 2001.

[8] C. E. Perkins and E. M. Royer, "Ad-hoc On-Demand Distance Vector Routing", Proceedings of 2nd IEEE workshop on Mobile Computing Systems and Applications, February 1999, pages 90-100.

[9] UCLA Internet Lab – GRAB Project. http://irl.cs.ucla.edu/GRAB/

[10] F.Ye, H. Luo, J. Cheng, S. Lu and L. Zhang , "A Two-Tier Data Dissemination Model for Large-scale Wireless Sensor Networks", Proceedings of the 8th Annual International Conference on Mobile Computing and Networking (ACM MobiCom), September 2002, Atlanta, Georgia, USA, pages 148-159.

[11] The Network Simulator – ns-2. http://www.isi.edu/nsnam/ns/

Detecting Disruptive Routers in Wireless Sensor Networks*

Steven Cheung, Bruno Dutertre, and Ulf Lindqvist

Computer Science Laboratory
SRI International
Menlo Park, CA 94025
{steven.cheung, bruno.dutertre, ulf.lindqvist}@sri.com

Abstract. In wireless sensor networks, sensor nodes are typically resource constrained and used as routers to forward packets for other
nodes. Compromised sensor nodes may be able to cause network-wide
denial of service by dropping or corrupting packets. This paper presents
a novel, lightweight approach for detecting disruptive routers that drop
or corrupt packets.

Keywords: Denial of service, intrusion detection, routing security, sensor networks.

1   Introduction

Wireless sensor networks consist of small, resource-constrained nodes that can
perform sensing, computation, and communication. These networks present a
cost-effective solution to a range of applications such as physical perimeter monitoring, urban warfare defense, monitoring of natural phenomena such as seismic
activities, and wildlife tracking. Some of these applications are security-critical.
Wireless sensor networks have certain characteristics that make them especially
vulnerable to denial-of-service (DoS) attacks:

Resource constraints: Sensor nodes are typically resource constrained. In
particular, sensor nodes may run on batteries. When the energy reserve is
depleted, the sensor nodes may be rendered useless. An attack vector is to
drain the energy reserve of sensor nodes. Also, limited memory, computation power, energy reserve, and network bandwidth of sensor nodes prevent
resource-intensive security techniques from being employed in sensor networks.

Exposure: In a sensor network, nodes may be spread across a large geographic
region, and may be in remote locations that are more likely to be subjected
to physical attacks.

---

* This research is sponsored by DARPA under contract number F30602-02-C-0212 and
National Science Foundation under grant number CNS-0434997. The views herein
are those of the authors and do not necessarily reflect the views of the supporting
agencies.

T. Kunz and S.S. Ravi (Eds.): ADHOC-NOW 2006, LNCS 4104, pp. 19–31, 2006.

Collaborative processing: Typically, the radios of sensor nodes are of short range and can only communicate directly with other nodes in the vicinity. Multihop routing is needed for sensor networks of nontrivial size. Sensor nodes are not only used as data sources, but also used as routers to forward messages from other nodes. A compromised sensor node may thus be able to affect communication of the entire sensor network.

Wood's and Stankovic's paper [15] presents an overview of the DoS threats against sensor networks such as signal jamming and routing attacks.

This paper presents a "hint-based" approach for detecting disruptive nodes that (maliciously) drop packets or corrupt packets in sensor networks, which may cause network-wide DoS. Our approach is based on probabilistically sending "hints", which contain packet digests and local routing information, via different paths (with respect to the ones used for routing the corresponding packets) to reach the destination. A main challenge for disruptive router detection in sensor networks is to address the severe resource constraints for the sensor nodes. The distinguishing characteristics of our scheme include its low runtime overhead for routers—which make it applicable to sensor networks—and its ability to detect various routing misbehavior such as dropping or corrupting packets.

2   Our Model

We use a graph to represent a static sensor network, with vertices representing sensor nodes and the base station and edges representing communication links. Sensor nodes may generate sensor reports (also called data packets) and send them to the base station. Also, sensor nodes may act as routers to forward packets for other sensor nodes.

We make the following network assumptions.

1. *Communication pattern:* The communication pattern involves sensor nodes (sources) sending packets to a base station (sink), called convergecast in the literature (e.g., [5]).
2. *Resource-rich base station:* We assume that the base station is a resource-rich node, which does not have the resource constraints of the sensor nodes. For example, a base station may be a regular notebook computer.
3. *Key distribution:* Every node shares a (secret) symmetric key with the base station[1].
4. *Network connectivity:* The network remains connected after the removal of disruptive routers.

---

[1] A key distribution scheme is as follows: A master key k_m and a keyed one-way hash function G may be used to generate the node key $G_{k_m}(i)$ for node i. The master key is stored in the base station, and the node keys are placed in the corresponding nodes before deployment. The base station may then reconstruct the node keys using G and k_m, when needed.

5. *Shortest-path routing:* A router always chooses a shortest path to route data packets to the base station. Also, the sink knows the distance between any sensor and itself.

6. *Bounded message delay:* The total delay (including propagation, queueing, and processing delays) for a packet to travel from its source to the sink is bounded above by α. (Thus, when the sink receives a data or hint packet, it can compute the time when it can safely discard the packet.)

We define a *disruptive router* as a router that (maliciously) drops or corrupts its transit packets. Moreover, a disruptive router may change its behavior for packets used for routing control and diagnosis to avoid being diagnosed as non-functional or disruptive. In particular, a disruptive router may correctly forward packets pertaining to a diagnosis that reveal its anomalous behavior. We assume that after an adversary compromises a sensor node, she can turn it into a disruptive router (say by reprogramming it), or extracts its secret key and uses an adversary-owned node (which could be a more powerful node) to replace the compromised node. Also, we assume that the adversary cannot obtain the secret key of a sensor node, unless she compromises it. Thus the adversary cannot use her nodes to masquerade as a router that she has not compromised.

3   Our Approach: Hint-Based Detection and Diagnosis

In our approach, as a packet traverses the network from its source to the base station (typically, a significantly more resourceful node than the sensors), the source and every intermediate node probabilistically send a *hint* of the packet to the destination. Moreover, the hint specifies a set of nodes that should not be used for routing the hint. Ganesan et al.'s [4] (braided) multipath routing algorithms may be used for routing hints, for example. The basic idea is that if the next-hop sensor for the sensor packet is N (as determined by the underlying routing protocol), the hint should not be routed through N. If N misbehaves, it may not affect both the sensor packet and the hint. Based on the packets and hints received, the base station may detect misbehaving nodes in the network.

A hint contains (1) a digest of the packet, used by the destination to verify whether the corresponding packet reaches its destination and to verify its integrity, (2) its local routing information such as the next-hop router, and (3) a set of nodes that should not be used for routing the hint, called *exclusion set.* (In the protocol we will describe in next section, the set contains just the next-hop neighbor.) The base station may increase the suspicion that a node X is disruptive if (1) the base station receives a hint but not the corresponding packet, and (2) X is in the exclusion set of the hint. Based on the packets and the hints received, the base station detects and locates disruptive routers.

Figure 1 depicts an example of our approach: Sensor A forwards a packet p to the base station S, and N is the next-hop sensor (as determined by the underlying routing protocol). With a specified probability, A will also send a

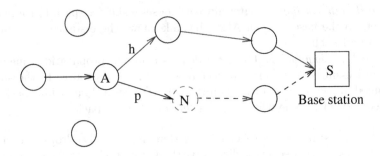

Fig. 1. Hint-based Detection Approach: An Example

hint h, which will not be routed through N. Thus, if the node N misbehaves, it cannot affect both the packet and the hint. If S receives h but not p, it may increase the suspicion that N is a disruptive router.

3.1 Hint Generation

Let node N be the next-hop neighbor for sending a packet from node A to the base station. When node A receives a packet p, A will forward it to N. Moreover, if N is not the base station, then A, with probability δ, will send an authenticated *hint* to a neighbor other than N. The purpose of the hint is to test whether N is a disruptive router. Note that a node does not generate a hint for a hint that it receives. The hint contains the identities of A and N, a sequence number seq_{AN}, a packet identifier sid, and a message authentication code. The structure of the hint is

$$\langle A, N, \{seq_{AN}, sid_p\}_{k_{as}^{enc}}, G_{k_{as}^{mac}}(A, N, \{seq_{AN}, sid_p\}_{k_{as}^{enc}})\rangle,$$

where G is a keyed one-way hash function[2], and k_{as} is a shared symmetric key between A and the base station S, which is split into two subkeys k_{as}^{enc} and k_{as}^{mac}, for encryption and message authentication respectively.

The sequence number seq_{AN} denotes the number of packets forwarded from A to N, and serves as a nonce to counter replay attacks. The packet identifier is used to associate the hint with the packet p. We use $H(p)$ as the packet identifier for packet p, where H is a shared cryptographic hash function. The message authentication code is used to ensure the integrity and authenticity of the hint.

To prevent a malicious node from acting only on packets that do not have an associated hint to avoid detection, we use two techniques to obfuscate the association: encryption and random time delay. When a node sends a hint, it encrypts the packet identifier and the sequence number using a secret key it shares with the base station. Thus, the only nodes that can link the hint and the packet using the content of the hint are the hint-generating node and the

---

[2] A keyed one-way hash function G has the property that, given a data string m and a key k, one can compute $G_k(m)$ efficiently. However, it is computationally infeasible to find k given m and $G_k(m)$.

base station. Moreover, a random delay is introduced between the sending time of a packet and that of the associated hint by storing the hint it generates in a FIFO buffer and sending it at a later time. This is used to prevent a malicious node from inferring the relationship between the packet and the hint by using its send time.

Suppose a node generates a hint with probability δ', when it receives a data packet. (We will discuss the relationship between δ and δ' in the following.) The node then puts the hint into its buffer of size n. If the buffer has room, the hint is appended at the end of the buffer. Otherwise, the node chooses a hint in the buffer at random and drops it, before the new hint is appended to the buffer. After forwarding the data packet, the node may send the hint at the front of its buffer with probability δ', if the buffer is nonempty. One can formulate this buffer system as a Markov chain with state transitions triggered by the arrivals of data packets. Let $P(k)$, where $0 \leq k \leq n$, denote the probability that the buffer has k hints. In steady state, $P(k) = 1/(n+1)$, for $0 \leq k \leq n$. Moreover, the probability that a hint is dropped (because of a full buffer) is equal to $P(n)\delta'(1 - \delta')$. Thus the "effective" hint generation rate, δ, is equal to $\delta' - P(n)\delta'(1-\delta')$. If we choose nine as the buffer size, δ is about 10% less than δ'. We argue that the costs for introducing this buffer scheme to uncouple the send time of a data packet and the corresponding hint—a small constant-sized buffer and generating hints at a slightly higher rate—are acceptable.

3.2   Hint Routing

If a disruptive node can affect both a packet and the corresponding hints, one may not be able to detect it. In our protocol, hints are sent using a special communication primitive $sendx(D, L, p)$, where D is the destination, L is a set of intermediate nodes to avoid in the route, and p is the packet to send. For our hint-based detection protocol, A sends the hint h to the base station S using $sendx(S, \{N\}, h)$. For certain routing protocols, realizing $sendx()$ requires no or minor adaptation of the underlying routing protocol. This section describes several such algorithms and the adaptation needed, if any.

For constraint-based routing [14], one may specify node exclusion using a *route constraint*. When route discovery is performed to find a path from Node A to the base station, the route constraint for excluding node N will be used to guide the search process.

In [4], multipaths are used to increase the robustness of routing. In particular, the authors present heuristics for constructing disjoint paths and braided paths (which are partially disjoint paths) for directed diffusion-type routing schemes. These multipaths may be used to realize $sendx()$ by routing p through N and routing h through an alternate path.

For routing protocols such as [7], a data sink broadcasts a beacon to its direct neighbors, which will in turn broadcast it to their neighbors, and so on. Each node will compute its "levels" (measured by the number of hops from the sink). A level-(k+1) node may send a packet to the sink by forwarding it to a level-k node, or by forwarding it to its sibling (i.e., another level-(k+1) neighbor node).

The node that receives the packet will then repeat the same procedure until the packet reaches its destination. These routing protocols can be adapted to implement hint routing. Assume that A is in level-(k+1), and its next-hop neighbor N is in level k. If A has two level-k neighbors, A can forward p to N and forward h to the other level-k node with an attached constraint that it should not route h to N. Otherwise, A may probabilistically forward h to a level-(k+1) node with an attached constraint that it should not forward the packet to N. That node will then either forward h to a level-k node that is not N or probabilistically forward it to another level-(k+1) neighbor node. A "time-to-live" counter may be attached to the hint packet to ensure that it will either reach its destination or be dropped after a specified number of hops.

3.3 Diagnosis and Response

Based on the packets and hints received, the base station attempts to detect and to locate disruptive routers. When the base station detects a disruptive router, it may securely broadcast the information to other nodes, e.g., using [13]. Here, we present a diagnosis procedure to identify disruptive routers, and a response procedure to reconfigure the network to logically disconnect the disruptive routers.

The base station should never rely on single packets or single sources to draw a conclusion. First, packets may be lost because of a benign cause such as collisions and traffic congestion. Second, it is possible that a malicious node may attempt to "frame" a good node X by just sending a hint with X in the exclusion list. Third, a packet may be discarded by any nodes on the path. If the base station receives a hint sent by A (for testing N) but not the corresponding packet, it may be caused by any node on the path from N to the base station. As a result, the base station should depend on multiple tests/events and multiple sources to detect and to locate disruptive nodes.

The base station constructs a directed graph, called *accusation graph*, $G = (V, E)$, where V corresponds to the set of sensors and the sink in the network. There are three types of edges in G, namely accusation edges E_a, candidate edges E_c, and vindication edges E_v, such that $E = E_a \cup E_c \cup E_v$, and E_a, E_c, and E_v are disjoint. Informally, an edge $(P, Q) \in E_a$ means that a significant portion of the packets that P forwards to Q cannot reach the base station intact. An edge $(P, Q) \in E_c$ means that more hints from P are needed for the base station to perform diagnosis on Q. An edge $(P, Q) \in E_v$ means that most of the packets that P forwards to Q can reach the base station intact. The base station computes these edges based on the hint and data packets it received during the last period.

Let P be a sensor node in the network and Q be the next-hop router to which P forwards packets. Suppose the base station received n_2 distinct hints sent by P for testing Q (i.e., P forwarded the corresponding packets to Q), and received n_1 distinct packets that match those hints. Two thresholds, t_1 and t_2, define the edges of the accusation graph. The base station creates an *accusation edge* (P, Q) if the following conditions are met:

1. $n_2 > t_1$
2. $n_1/n_2 < t_2$

Condition 1 ensures that the number of hints is considered "significant" enough for disruptive node detection. Condition 2 concerns what the "acceptable" ratio on the number of hints and the corresponding packets pertaining to a node that can reach the sink is. If Condition 1 is not met, the edge (P, Q) is a *candidate edge*, representing that more hints from P are needed for the sink to diagnose Q. If Condition 1 is met and Condition 2 is not met, the edge (P, Q) is a *vindication edge*.

Notations: We denote the number of edges incident to a node N (i.e., the in-degree of N in G) by $in(N)$. We denote the number of accusation edges incident to N by $in_a(N)$. Moreover, $in_c(N)$ and $in_v(N)$ are similarly defined for candidate edges and vindication edges respectively. We denote the distance between a node N and the sink S by $dist(N, S)$. The set of nodes involved in forwarding data packets from a node N to the sink S, excluding N, is denoted by $F(N, S)$.

The sink maintains a profile for the ratio $n1/n2$ for edge (P, Q) to characterize the ratio of the packets P forwards Q that can reach the sink intact to those that P forwards Q during normal operation (i.e., when there is no disruptive node in the network). The statistics may be gathered, for example, during the initial phase of the deployment, when the probability of having a compromised node is small. The hint-based approach can be viewed as using random sampling without replacement to compute the mean of a finite population with two values—0 represents a packet that P sends to Q that fails to reach the sink intact, and 1 represents otherwise. When the population size (i.e., the number of data packets P sends to Q) is large compared to the sample size (i.e., the number of hints pertaining to the edge (P, Q) received by the sink), the ratio of the variance of the population and that of the sample is approximately equal to the sample size [3].

The value of the thresholds t_1 and t_2 are determined as follows. We first compute an estimate of the average of the population during normal operation based on the sample size (n), the sample average (μ_s), and the standard deviation of the sample (σ). The confidence interval for the average of the population can then be computed. For example, a 99.7%-confidence interval for the average of the population is $\mu_s \pm \frac{3\sigma}{\sqrt{n}}$, which is 3 standard errors from the sample average. In other words, one can compute a close estimate for the average of the population (μ) with high confidence level using a reasonably large n. In particular, we choose $\mu = \mu_s - \frac{3\sigma}{\sqrt{n}}$. By choosing t_2 to be 3 standard errors from μ, the probability that the sample mean n_1/n_2 is less than t_2 during normal operation is less than 0.15%. The standard error depends on the sample size, t_1. By choosing t_1 to be 100, we can set t_2 to $\mu - 0.3\sigma$.

Based on the accusation graph, the diagnosis procedure (Algorithm 1) computes a set of suspicious nodes V_s that contains the set of nodes that have one or more accusation edges incident to it. If V_s is non-empty, the diagnosis protocol will choose a node in V_s that is closest to the base station, say X. Note that a good node A may appear in V_s because there exists a disruptive node downstream (i.e., closer to the sink). Our protocol avoids the problem of classifying A as disruptive by finding a node in V_s that is closest to the base station. If there exists a candidate edge (M, N) such that the distance between N and the sink S is less than that between X and S, and there does not exist a vindication edge incident to N, then the protocol will need to wait for more hints to resolve the candidate edge before making a diagnosis. Otherwise, (1) if $in_a(X) > 1$, the diagnosis protocol will label X as a disruptive router; (2) if $in_a(X) = 1$, it will infer that one of the endpoints of the edge (Y, X) in E_a is a disruptive router— Either X is a disruptive router, or Y is a disruptive router that only sends hints to the sink (but not the corresponding data packets to X) to wrongly accuse X of being a disruptive node.

Algorithm 1. *(Disruptive Router Diagnosis)*

> *Let $G = (V, E)$ be the accusation graph*
> $\qquad V_s = \{N \in V \mid in_a(N) \geq 1\}$
> $\qquad\quad X$ *be a node in V_s such that $dist(X, S) \leq dist(N, S)$, $\forall N \in V_s$*
> **If** $\forall N \in F(X, S) \cdot in_v(N) > 0$
> $\qquad$ **If** $in_a(X) > 1$
> $\qquad\quad X$ *is a disruptive node*
> $\qquad$ **else** X *or Y is a disruptive router, where $(Y, X) \in E_a$*

Using the diagnosis result, the network reconfigures itself to logically remove disruptive routers from the network. For the case $in_a(X) > 1$, all the neighbors of X will stop using X to forward packets. For the case $in_a(X) = 1$, X and Y will cease their neighbor relationship. Although one cannot determine whether X or Y is a disruptive router, the connectivity of the disruptive router will be decreased after the reconfiguration. By repeated applications of this diagnosis and response procedure, the disruptive node will eventually be logically isolated from the rest of the network, if it keeps misbehaving.

3.4   Diagnosis Example

Figure 2 depicts an example accusation graph. In this scenario, node M is a disruptive router. Packets forwarded from A to B to M are affected by M. As a result, the two edges (A, B) and (B, M) are accusation edges. Suppose M does not send any hints for D, which is the next-hop router for M. The edge (M, D) is a candidate edge, because there are not enough hints corresponding to the edge. Assume that there are enough packets sent along the path $C \to D \to S$, and that the number of hints received by the sink for the edges (C, D) and (D, S) exceed t_1. The edges (C, D) and (D, S) are vindication edges. In Algorithm 1, the node set $V_s = \{B, M\}$. Because $dist(M, S) < dist(B, S)$, X is chosen to be M. Because $in_v(D) > 0$ and $in_a(M) = 1$, Algorithm 1 will report that either B or M is a disruptive router.[3] We note that it is possible to have a different

---

[3] Based on this diagnosis, the network may reconfigure itself so that B will not forward packets to M to prevent M from disrupting B's traffic.

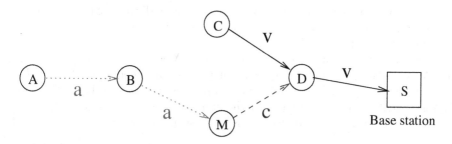

Fig. 2. Disruptive Router Diagnosis: An Example

scenario in which B is a disruptive router, and we have the same accusation graph. In that case, node B may generate hints for the edge (B, M) without sending the corresponding packets to S. Thus, the diagnosis protocol cannot pinpoint whether B or M is a disruptive node. However, if there is another accusation edge incident to M, one could conclude that M is a disruptive node.

4   Analysis

This section presents an analysis for the correctness of Algorithm 1 and analyzes the costs of the proposed scheme. We assume that a disruptive router drops or corrupts all of its transit packets, except those pertaining to routing control and those explicitly marked for diagnosis to avoid being diagnosed as non-functional or disruptive. Moreover, we assume that there is at most one disruptive router in the network. Section 6 discusses how these assumptions may be relaxed.

Lemma 1. *If there is a vindication edge from node P to node Q, then Q is not a disruptive router.*

Proof. If P is a disruptive router, Q cannot be a disruptive router, because there is at most one disruptive router in the network. If P is not a disruptive router, Q cannot be a disruptive router. Otherwise, we have $n_1/n_2 = 0 < t_2$, which contradicts that (P, Q) is a vindication edge. $\square$

Theorem 1. *If a router is diagnosed as a disruptive router by Algorithm 1, the router is a disruptive router. If two routers are diagnosed as potential disruptive routers by Algorithm 1, one of them is a disruptive router.*

Proof. Let X be a node that is closest to the sink S and has one or more accusation edges incident to it. Moreover, for all nodes N involved in forwarding packets from X to S, $in_v(N) > 0$.

We now show that an edge $(Y, X) \in E_a$ implies that X or Y is a disruptive node. If an accusation edge (Y, X) exists, we know that a node in $\{X, Y\} \cup F(X, S)$ is a disruptive router, based on the definitions of the thresholds t_1 and

t_2. Because all nodes in $F(X, S)$ have at least one vindication edge incident to them, Lemma 1 implies that $F(X, S)$ does not contain a disruptive router. Thus, X or Y is the disruptive router.

If the number of accusation edges incident to X is more than one, we argue that X is the disruptive node. Suppose there exists another edge $(Z, X) \in E_a$, where $Y \neq Z$. By repeating the above argument, we have either X or Z is the disruptive router. Together with the statement that either X or Y is the disruptive router, we conclude that X is the disruptive router.   □

In the following, we analyze the costs of the proposed scheme, and show that they are acceptable for sensor networks.

Network Bandwidth: The transmission overhead of sending hints depends on the hint generation probability δ, the average distance of sensors from the base station d_{avg}, and the length of the paths used for routing hints.

Consider a sensor node A that is d hops away from the base station S. Because A and the intermediate nodes that forward a data packet from A to S generate hints independently, the expected transmission overhead for sending hints with respect to this packet is

$$\sum_{k=2}^{d} \delta(1+p)k$$

where p is the average proportional increase in length for the hint route with respect to the shortest route. If the braided multipath scheme is used for routing hints, for example, simulation results [4] show that p is less than 0.4 for the range of node density and source-sink distance experimented.

Thus the proportional increase in network transmission cost is less than

$$\delta(1+p)(d_{avg}+1)/2$$

Computation: The main recurring computation cost for the sensor nodes is from hint generation. For every hint, the node needs to perform an encryption operation and a hash computation.

On a typical sensor platform such as TinyOS/Mica2, the cryptographic operations associated with a hint can be computed in a few milliseconds [8]. Also, sensor networks are typically bandwidth-constrained (ranging from 10's to 100's packets per second), and only a fraction of these packets trigger hint generation. Thus the computational overhead for sensors to generate hints will be acceptable.

Memory: For sensor nodes, the memory overhead involves storing two additional keys for securing the hints and keeping a small constant-sized buffer to add a random delay to the hints, as discussed in Section 3.1. The more memory-intensive tasks of buffering and matching hints with the corresponding packets are performed by the resource-rich base station.

5 Related Work

Existing techniques for detecting disruptive routers in a network are either not suitable for sensor networks or have limitations. In [10], the authors propose a protocol, called watchdog, for detecting routers that fail to forward packets in wireless networks. By exploiting the broadcast nature of wireless networks, a node may detect whether a neighbor fails to forward packets by monitoring the packets it receives and the packets it forwards. This technique has weaknesses. When node A forwards a packet to node B, the packet may not be received by B (e.g., because of packet collision or weak signal strength). Thus, the technique may fail in some cases. Our approach does not suffer from these problems because the detection and diagnosis are performed at the end point.

In [1,2], a technique based on conservation of flow is used for detecting routers that drop packets to cause DoS. In particular, the neighbors of a node maintain counters to keep track of the amount of traffic going into and going out of a node, and they exchange information by flooding to determine whether the node drops many packets. In [6], Huang and Lee presented an approach for detecting and identifying routing attacks in mobile ad hoc networks based on applying anomaly detection and signature-based detection in sequence. An attack not addressed by [1,2,6] involves a disruptive node modifying packets in transit. Because the amount of traffic flow is not changed by this attack, the disruptive node will not be detected by these techniques. On the other hand, our approach can detect unauthorized packet modification using hint packets, which contain cryptographic checksums of data packets.

In [12], a secure traceroute-based approach has been proposed to detect and locate the source of routing misbehavior. The scheme assumes that any node can establish a secret key with any other nodes in the network (e.g., using a public-key infrastructure or a PGP-style web-of-trust technique). This assumption may not hold for sensor networks because of the resource constraints. Furthermore, our approach is lightweight for sensors; most of the detection and diagnosis work is performed by the base station, which typically does not have the resource constraints of the sensor nodes.

Recently, McCune et al. [11] studied a DoS attack that prevents nodes from receiving a broadcast message in sensor networks. To detect that attack, McCune et al. presented a protocol, called secure implicit sampling, in which a subset of nodes is selected (deterministically), using a cryptographic technique, to send authenticated acknowledgements to the base station. Although both [11] and our work are on detecting DoS attacks caused by disruptive routers, they have a different, complementary focus, and use different techniques. In particular, this paper focuses on the convergecast problem, is based on a probabilistic approach, and exploits network redundancy to facilitate disruptive router diagnosis.

A related problem is on detecting "selfish" nodes that send their packets through a multihop wireless network, but refuse to forward packets for other nodes, thus saving their own resources. For detecting these misbehaving nodes, [9] proposes a solution based on the watchdog technique (see above).

6 Concluding Remarks

We present a new technique for detecting disruptive routers in sensor networks. A characteristic of the technique is that it exploits heterogeneity of sensor networks by shifting most of the detection workload from resource-constrained sensor nodes to resource-rich nodes such as the base station. The technique is also applicable for addressing unauthorized packet modification attacks, which cannot be handled by some prior work (e.g., [2]). In the following, we discuss some limitations of our results and future work items.

In Section 4, we assume that disruptive routers drop or corrupt all transit packets. We note that the diagnosis protocol and the associated analysis are applicable to disruptive routers that probabilistically drop or corrupt packets, assuming that the proportion of transit packets affected by a disruptive router is "significantly" different from that of the norm as defined by the threshold t_2 in Section 3.3. On the other hand, our technique may not be able to detect disruptive routers that do not affect enough packets to cause the threshold to be exceeded.

There are other disruptive router models that are not covered by our model. For example, a disruptive router may choose to affect packets with certain payload or addresses. We note that our work may be extended as follows: The sources may encrypt the packet payload, which can be decrypted by the sink only. Also, the sink may construct separate accusation graphs for different sources so that a disruptive router cannot evade detection by exhibiting acceptable aggregate behavior, and affecting packets from certain sources.

Another assumption we have in Section 4 is that there is at most one disruptive router in the network. The core idea of the diagnosis protocol may be adapted for the case in which there are multiple disruptive routers. Specifically, if there are at most k disruptive routers in the network, we may need $k+1$ accusation edges incident to a node to diagnose it as a disruptive router. Otherwise, a set of disruptive routers may cooperatively cause a good router to be misclassified as disruptive. Also, we need to use multiple paths for routing hints. If the disruptive routers can prevent hint packets from reaching the sink, diagnosis cannot be performed. A future work item is to investigate an efficient algorithm for constructing multiple paths (e.g., [4]) and extend the analysis in Section 4 for the multiple disruptive router case.

Acknowledgements

We thank Alfonso Valdes for helpful discussions, and the anonymous reviewers for their suggestions and comments.

References

1. K. A. Bradley, S. Cheung, N. Puketza, B. Mukherjee, and R. A. Olsson. Detecting disruptive routers: A distributed network monitoring approach. *IEEE Network*, Sept./Oct. 1998.
2. S. Cheung and K. N. Levitt. Protecting routing infrastructures from denial of service using cooperative intrusion detection. In *Proceedings of the New Security Paradigms Workshop*, Sept. 23-26, 1997.
3. J. E. Freund and R. E. Walpole. *Mathematical Statistics*. Prentice-Hall, 1987.
4. D. Ganesan, R. Govindan, S. Shenker, and D. Estrin. Highly-resilient, energy-efficient multipath routing in wireless sensor networks. *ACM Mobile Computing and Communication Review (MC2R)*, 1(2), 2002.
5. Q. Huang and Y. Zhang. Radial coordination for convergecast in wireless sensor networks. In *Proceedings of the IEEE EmNetS-I*, Nov. 2004.
6. Y. Huang and W. Lee. A cooperative intrusion detection system for ad hoc networks. In *Proceedings of the 1st ACM Workshop on Security of Ad Hoc and Sensor Networks (SASN '03)*, pages 135–147, Fairfax, Virginia, Oct. 2003.
7. C. Karlof, Y. Li, and J. Polastre. ARRIVE: Algorithm for robust routing in volatile environments. Technical Report UCB//CSD-03-1233, Department of EECS, University of California at Berkeley, Mar. 2003.
8. C. Karlof, N. Sastry, and D. Wagner. TinySec: A link layer security architecture for wireless sensor networks. In *Proceedings of the Second ACM Conference on Embedded Networked Sensor Systems (SenSys 2004)*, Nov. 2004.
9. R. Mahajan, M. Rodrig, D. Wetherall, and J. Zahorjan. Sustaining cooperation in multi-hop wireless networks. In *Proceedings of the Second Symposium on Networked Systems Design and Implementation (NSDI)*, Boston, MA, May 2–4, 2005.
10. S. Marti, T. Giuli, K. Lai, and M. Baker. Mitigating routing misbehavior in mobile ad hoc networks. In *Proceedings of the 6th Annual International Conference on Mobile Computing and Networking (MOBICOM 2000)*, pages 255–265, Boston, MA, Aug. 2000.
11. J. M. McCune, E. Shi, A. Perrig, and M. K. Reiter. Detection of denial-of-message attacks on sensor network broadcasts. In *Proceedings of the 2005 IEEE Symposium on Security and Privacy*, pages 64–78, Oakland, California, May 8–11, 2005.
12. V. N. Padmanabhan and D. R. Simon. Secure traceroute to detect faulty or malicious routing. In *ACM SIGCOMM Workshop on Hop Topic in Networks (HotNets-I)*, Princeton, NJ, Oct. 2002.
13. A. Perrig, R. Szewczyk, V. Wen, D. Culler, and J. Tygar. SPINS: Security protocols for sensor networks. In *Proceedings of the 7th Annual International Conference on Mobile Computing and Networks (MOBICOM 2001)*, pages 189–199, Rome, Italy, July 2001.
14. Y. Shang, M. P. Fromherz, Y. Zhang, and L. S. Crawford. Constraint-based routing for ad-hoc networks. In *Proceedings of International Conference on Information Technology: Research and Education (ITRE '03)*, pages 306–310, Newark, NJ, Aug. 10-13, 2003.
15. A. D. Wood and J. A. Stankovic. Denial of service in sensor networks. *IEEE Computer*, 35(10):54–62, Oct. 2002.

Energy-Efficient Data Dissemination in Sensor Networks Using Distributed Dynamic Tree Management

Kwang-il Hwang and Doo-seop Eom

Department of Electronics and Computer Engineering, Korea University
5-1ga, Anam-dong, Sungbuk-gu, Seoul, Korea
Tel.: +82-2-3290-3802; Fax.: +82-2-3290-3895
brightday@final.korea.ac.kr

Abstract. In this paper, an energy-efficient data disemination protocol designed for mobile sink applications in sensor networks, is proposed. The dissemination scheme exploits the Distributed Dynamic Tree (DDT), which is able to identify current location sinks locally and dynamically transform the tree shape according to sink movement. In addition, the Dynamic Shared Tree(DST), which is an extension of the DDT, is presented. The DST is able to accomocate multiple mobile sinks. The DST, based on the DDT, creates a two-tired network composed of a sensor data dissemination level and communication level between sinks. The simulation results demonstrate that the DST performs considerably energy-efficient data dissemination with relatively low delay, compared to other dissemination protocols.

1   Introduction

Wireless sensor networks are able to sense their environments, gather and process information and communicate their results either to one or more base stations (sink node) or each node. In such a large scale unattended network, energy is considered as one of the most expensive resources.

Recently, advances in wireless technology combined with the proliferation of sensor technology have led to an increase in the number and variety of applications based on sensor networks. One of the most important goals in sensor network applications is how to disseminate data from various sensors to sink nodes using minimum energy. There has been substantial literature discussing the solutions to this problem. The literature [2,5,6] considered mainly one or more stationary sink models. However, in the case of applications such as the battle field or rescuer activity, a data dissemination protocol that can cope with more dynamic sink environments is required.

Data dissemination protocols considering sink mobility may be classified into two categories: Sink oriented protocols and Source initiated protocols. SPIN [5] and Direct Diffusion [2] are representatives of classic sink oriented protocols. These protocols can efficiently aggregate data with low delay by constructing

T. Kunz and S.S. Ravi (Eds.): ADHOC-NOW 2006, LNCS 4104, pp. 32–45, 2006.

one or more aggregation paths on the basis of the sink. In-network processing [6] from sources on paths is also enabled. However, these protocols must propagate the sink location continuously through a sensor field in order to keep all sensor nodes updated in the direction of data forwarding.

In order to reduce flooding caused by sink mobility, SAFE [3], which uses geographically limited flooding, is proposed. However, in the case of highly mobile sinks, the SAFE cannot avoid increasing local flooding to retrieve the gate connecting itself to the tree.

In contrast to these sink oriented protocols, source initiated dissemination protocols, such as TTDD [1] and SEAD [4], use a method that mobile sinks access on the dissemination path constructed on the basis of each source. Each data source in TTDD proactively builds a grid structure, enabling mobile sinks to continuously receive data regarding the move, by only flooding queries within a local cell. SEAD creates a near-optimal dissemination tree by considering the distance and packet traffic rate among nodes. Each agent node continuously tracks sink movement. Evidently, these source initiated protocols perform energy-efficient data dissemination. However, the path per source makes in-network processing impossible. In addition, due to the sinks' access time, more delay is required to aggregate data. No existing approaches provide both a low delay and energy-efficient solution to mobile sink problems.

In this paper, an energy-efficient data dissemination protocol with low delay, designed for mobile sink environments, is proposed. In order for a dissemination tree to continuously pursue a dynamic sink, a Distributed Dynamic Tree (DDT) based on the *periodic local update* and *distributed local link reversal* algorithm, is used. In the proposed protocol, the shape of the tree is dynamically transformed according to master sink movement, and the tree is shared with other slave sinks. This is enabled by the Dynamic Shared Tree (DST), which is an extension of the DDT. The DST conserves a considerable amount of energy, despite maintaining robust connection from all sources to sinks, since tree maintenance of the DST is accomplished by distributed local exchanges.

The remainder of this paper is organized as follows. Section 2 presents the basic model and background idea in the proposed Distributed Dynamic Tree (DDT). Section 3 describes the detailed operation of the DDT. In section 4, the Dynamic Shared Tree (DST), an extended version of the DDT, is presented. In section 5, the energy efficiency of the DST is analyzed in terms of total communication cost compared with DD and TTDD. A comparative performance evaluation through simulation is presented in Section 6. Section 7 concludes this paper.

2   Overview of the Algorithm

In this section, the basic model of the target sensor network is presented. Based on the basic model, two basic ideas, *sink's periodic local update* and the *distributed local link reversal algorithm*, are introduced.

2.1 Basic Model

The application model that is mainly concentrated on is somewhat different from general sensor network applications which take one or more stationary sink nodes into account as a whole. The application model, in particular, focuses on the mobile sink application, where several sinks enter the sensor field directly and perform their roles based on source data dynamically refreshed from sensor fields. Such an application model is very useful for battle field or rescue activity.

The application model also makes the following basic assumptions:

- Homogeneous sensor nodes are densely deployed.
- Sensor nodes communicate with each other through short-range radios. Long distance data delivery is accomplished by forwarding data across multiple hops.
- Each sensor node is aware of its own location (for example through receiving GPS signals or through localization techniques such as [8]).
- Sensor nodes remain stationary at their initial location.
- Sink nodes possess much more abundant resources in memory, energy, and processing than that of general sensor nodes.

2.2 Sink's Periodic Local Update

In general, identifying the current location of a mobile sink is crucial in a sensor network with a mobile sink, since each source data should be propagated to the sink. However, it is widely accepted that global propagation of location information of a sink to an entire network can bring about tremendous traffic and energy wastage.

In proposed DDT and DST, however, the location update of a sink is achieved not globally but locally. This is derived from the fact that sink movement can be specified by a trajectory of nodes that receive sink update messages.

The sensor network model consists of a set N of sensor nodes. A^i is defined as a set of nodes receiving the sink's update message at time i. Thus, A^i and A^{i+1}, satisfying A^i, $A^{i+1} \subset N$, have each independent set $A^i = \{a_1^i, a_2^i, a_3^i, ..., a_k^i\}$ and $A^{i+1} = \{a_1^{i+1}, a_2^{i+1}, a_3^{i+1}, ..., a_k^{i+1}\}$ at time i and $i+1$, respectively. It is important to note that A^i depends on sink's location at time i and radio range of sink node. Therefore, the sink update interval can be calculated as follows:

$$T\,(Interval) = K\frac{D}{V_{s-\max}} - (T_p + \alpha) \quad, \text{ where } K \leq 1 \qquad (1)$$

K is a density factor. $V_{s-\max}$ and D denote the maximum sink node speed (m/s) and the maximum radio range, respectively. In addition, T_p and α represent the propagation time and additional delay caused by the MAC layer and processing, respectively. In assuming that T_p and α are negligible, $T = K\frac{D}{V_{s-\max}}$ can be obtained. It is important to note that interval (T) is the local update rate. In addition, the T value reveals that the sink node can move by maximum distance of its radio radius during the time from i to $i+1$. Therefore, under the assumption

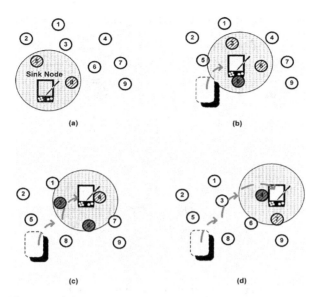

Fig. 1. Identification of the current sink location by the sink's periodic local update

that sensor nodes are densely deployed in the sensor field, the information that a common set exists between time i and $i+1$ can be obtained and that there are at least one or more common elements, i.e. $A^i \cap A^{i+1} \neq \phi$. Fig. 1 illustrates the change of common elements according to the local update of a sink. As the sink moves from (a) to (d) with local updates, the current location of the sink is specified by a trajectory connecting common elements from time $i-1$ to i, as presented in fig. 1. Therefore, a mobile sink's current location can be identified with minimum energy simply by a sink's local update in the network.

2.3   Distributed Local Link Reversal

From the periodic local update of a sink described in the previous subsection, the current location of the sink can be tracked. However, the following question still remains: "*How can the trajectory between common elements connect in a distributed fashion?*" In this subsection, an efficient solution to solve the problem is presented. First, it is assumed that a tree rooted at a temporary root node is already constructed, only the root node has a connection to a sink, and other nodes have only one parent individually, as presented in fig. 2 (a). As the sink moves with a periodic local update message, as presented in fig. 2 (b), node 8 identifies that the sink is close. However, node 5, the current root node, still does not realize this fact. Therefore, a timer relating to the update interval is employed.

$$Timeout = Interval + C \qquad (2)$$

Where C is a compensation value with respect to additional delay. By Timeout, node 5 comes to realize that sink moved further away and notifies this fact immediately by broadcasting locally. It is important to note that node 5, which

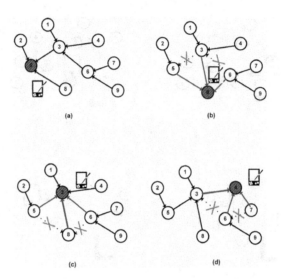

Fig. 2. Distributed local link reversal

was an element of the last sink update, is directly reachable node 8, the current root candidate, since node 8 is common element between the last update and current update, and nodes are densely deployed. Therefore, node 8 receives the Sink_Lost broadcast message of the old root, node 5. As soon as node 8 receives the message, it also forces a change in the parent of nodes in the local area to its id, node 8, by broadcasting. That is, the link of all nodes in the local area on the basis of the new root node is directed to the new root node as presented in fig.2 (b). In addition to a simple distributed local link reversal operation, all links are maintained loop free.

The periodic local update enables dynamic change of a temporary root of the tree according to the sink's current location, and local link reversal allows all parent directions in local area be directed to new root node.

In combining the two simple algorithms, the shape of a tree can be dynamically changed, according to sink movement, as shown in fig.2.

3   Operation of Distributed Dynamic Tree (DDT)

In this section, the operation of *Distributed Dynamic Tree* (DDT), which is based on the *periodic local update* and *distributed local link reversal algorithms*, is presented. The DDT consists of two of important phases: DDT construction and sink mobility management.

3.1   Distributed Dynamic Tree (DDT) Construction

As referred to in the previous section, sink nodes do not always reside in the sensor field, only when it is required. Therefore, all nodes lie in a dormant state with periodic listening if no sink exists in the field. As presented in fig. 3(a),

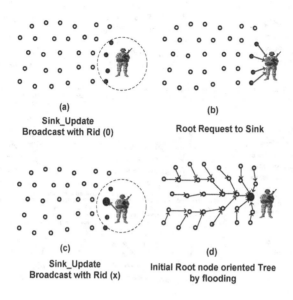

Fig. 3. DDT construction

the DDT operation starts with a sink entering the sensor field, and broadcasting Update_Request, which is a periodic local update message. Since the sink still has no Root node, the Rid (Root id) field in initial Update_Request is zero. In addition, the User's query is encapsulated in the Update_Request message. All Candidate nodes, hearing the sink's Update_Request with Rid (0), immediately transmit a Root_Request message to the sink, as shown in fig. 3(b). The sink only appoints one node as a Root node and then broadcasts the Update_Request with Rid (new Root node id), as shown in fig. 3(c). When the node receives the message, it starts to construct a reverse tree by flooding the entire network as in (d). Each node, receiving the tree construction message, registers the ID of the sender to its parent. Thereby, the reverse path to Root node at each node is established and a tree rooted at an initial Root node is constructed as shown in fig. 3 (f).

3.2   Sink Mobility Management of DDT

When constructing a tree, each path on the tree is used as a path conveying sensor information. If the sink is stationary at its initial location, no change in the tree occurs. However, in the application considered, a sink will move from place to place in the sensor field to perform operations according to dynamic source information. Fig.4 describes management operation of the DDT with respect to sink movement. As a sink moves away from the current Root node, new nodes come to hear the periodic Update_Request message of the sink as in (b), and the Current Root node cannot hear the message. Thereby, the Sink_Lost

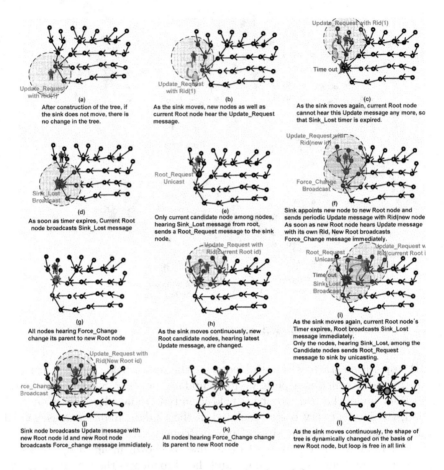

Fig. 4. Management of sink mobility

timer expires, as in (c). Therefore, the Root broadcasts a Sink_Lost message in its local area, as in (d). Some nodes of current root candidate nodes, hearing the latest local update message, can hear the Sink_Lost message of old Root node and send their Root_Request message to the sink immediately as in (e). On receiving the Root_Request message, the sink appoints the first arriving node as a new Root node and broadcasts the Update_Request with the new Root node id as in (f).

As soon as the new Root node receives the Update message with its own Rid, it immediately broadcasts the Force_Change message in its local area. All nodes receiving the Force_Change message change the parent to the new Root node. Therefore, the shape of initial tree is transformed as shown in fig. 4 (g). Like this, as the sink moves continuously, the shape of the tree is dynamically transformed as shown in fig. 4 (h) to (l).

4   Energy Efficient Data Dissemination to Multiple, Mobile Sinks

The DDT is intentionally designed to cope well with the mobile sink, so that the data dissemination cost from each source to mobile sink can be accomplished with the $O(n\sqrt{N})$ along the tree, where n is the number of source nodes and N is the number of all deployed sensor nodes. In addition, the cost can be more reduced by use of in-network processing when many sources exist. In this section, the Dynamic Shared Tree (DST), which is an extension of the DDT to multiple mobile sinks, is presented.

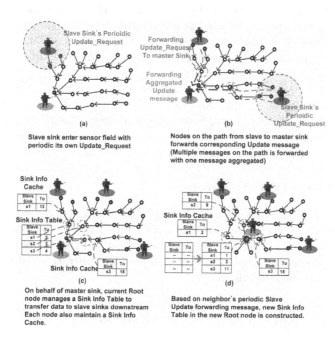

Fig. 5. Support of multiple, mobile sinks of DST

4.1   Extension of DDT to Dynamic Shared Tree (DST)

The DDT focuses on solving the mobility of a sink node. The DST is an extension of the DDT to support multiple mobile sinks. In the DST scheme, multiple mobile sinks share a DDT, which has already been constructed by a sink. In the presented DST, two types of sink nodes are defined: Master sink and Slave sink. The former is a sink, which is the first attended in the sensor filed and constructed the tree. The latter are slave sinks, which join the tree that have already been constructed from a master sink. Evidently, the master sink must be the only master sink in a sensor field.

Fig. 5 describes the DST operation for multiple mobile sinks. First, a slave sink enters the sensor field, broadcasting its Update_Request, as shown in fig. 5 (a).

The nodes receiving the message forward it to master sink node along the tree. In the process, the identical sink message obtained from different route is aggregated to a single message at their parent node and re-forwarded to the master sink node along the tree. This aggregated Update_Requset forwarding can save considerable energy by reducing the number of redundant transmissions. In addition, each node on the path from a slave sink to master maintains a Sink Info Cache, which stores a pair of the corresponding sink and its current child id. This means the node is on its way to the corresponding sink and is available through its child node. If a node lies on the path to multiple slave sinks, the Sink Info Cache of the node includes all information of the multiple sinks. In addition, on behalf of the master sink, the current Root node manages a Sink Info Table based on the neighbor's Update_Request, with respect to slave sinks. It is important to note that a new Root node, changed due to sink movement, can fill the new information in the Sink Info Table from its neighbors, as shown in fig. 5 (d).

A master sink aggregates information from each sensor and propagates the result to all slave sinks on the basis of Root's Sink Info Table and Sink Info Cache in the nodes on the route. Since each node remembers the latest down link information for each individual slave sink, the additional data transmission cost to all slaves is $O(m\sqrt{N})$, where m is the number of slave sinks.

4.2   Two-Tiered DST Architecture

As shown in fig. 6, the presented DST creates two-tired network which is composed of sensor data dissemination and communication level between sinks. In particular, the communication level maintains a logical star topology between master and slave sinks. This structure can facilitate operations of a group of military units on the battlefield.

5   Communication Overhead Analysis

In this section the energy efficiency of the DST is analyzed. For high-level comparative analysis, a specific metric is measured: *total communication cost* for data aggregation of the DST, DD and TTDD, respectively.

It is assumed that a square sensor field of area A in which N sensor nodes are uniformly distributed, so that on each side there are approximately $\sqrt{N}$ sensor nodes. There are m event sources and n sink nodes. For grid structure analysis, it is assumed that sensor fields are divided into cells; each cell has αsensor nodes. N_X is defined as the number of cells on the X-axis, and N_Y is the number of cells on the Y-axis.

Directed Diffusion: For event forwarding, Directed diffusion requires four steps: Interest flooding, Exploratory data forwarding, Reinforcement, and Data forwarding. Since Interest forwarding by sink node uses flooding, its cost, in the worst-case, is expressed as mN. The cost of exploratory data forwarding to setup multi-paths from source to sink can be approximately given by $\sqrt{N} \times \sqrt{N} = N$.

The cost of reinforcement to select single path among the multi-path by exploratory forwarding is $\sqrt{N}$. Eventually, cost data forwarding along the path is simply given by $n\sqrt{N}$. Accordingly, total communication cost C_{DD} for DD is given by

$$C_{DD} = (m+1)N + (n+1)\sqrt{N} \tag{3}$$

Therefore, in a sink mobile environment, total communication cost is represented as $O(mN)$, including all costs required to construct and maintain a dissemination path and data forwarding with respect to sink movement.

TTDD: TTDD exploits local flooding within the local cell of a grid, which sources builds proactively. Each source disseminates data along the nodes on the grid line to the sinks. The TTDD can be divided into three independent steps: Geographical forwarding for grid construction, Query forwarding by sinks, and Data forwarding from sources. Initially, only nodes on the grid line take part in the forwarding process during geographical forwarding. In addition, since the grid is independently constructed by each source, the cost of geographical forwarding is expressed as $n \times N_x \times N_Y \times \sqrt{\alpha}$.

Next, the query is flooded using a sink within a cell and then forwarded along the grid line. Therefore, the cost for query forwarding becomes $\alpha m + m(\sqrt{2N})$. Finally, data forwarding from each source to sink is expressed by $n\left(\sqrt{2N} + \frac{\sqrt{2\alpha}}{2}\right)$. This is because, in the worst-case, a sink will be found at the edge cell of the diagonal line of the source. Eventually, the total communication cost for TTDD is given by

$$C_{TTDD} = n\left(N^*\sqrt{\alpha} + \frac{\sqrt{2\alpha}}{2} + \sqrt{2N}\right) + m(\sqrt{2N} + \alpha) \quad , Where N^* = N_x \times N_y \tag{4}$$

Therefore, the total cost is $O(m\sqrt{N})$ or $O(n\sqrt{N})$ where $N^* << N$, however, TTDD' cost largely depends on the cell size. In addition, in the case of many sources, the cost rapidly increases.

DST: Similar to DD, the DST begins with flooding. However, the DST does not require additional flooding with regard to increasing sinks. Since the tree constructed by flooding is shared with other sinks, the tree construction cost is only $N + m\sqrt{N}$. In addition, the cost to maintain the tree is negligible, since the DST maintains the tree with some messages exchanged locally when the sink moves. Data is forwarded upstream along the tree and then forwarded to each slave sink along the tree, so the data forwarding cost is expressed by $(m+n)\sqrt{N}$. Eventually, the total communication cost for the DST is given by

$$C_{DST} = N + (2m+n)\sqrt{N} \tag{5}$$

However, in spite of sink's continuous movement, since flooding is required only at the initial tree construction phase, the actual communication cost becomes $O(n\sqrt{N})$, where $m_¡n$.

In summary, the DST is similar to sink oriented data aggregation approaches in the shape. However, additional flooding as well as sizeable communication overhead is not generated. Therefore, the energy efficiency for the DST is as good as the TTDD. However, the realistic TTDD' cost largely depends on the cell size as well as the number of sources. More realistic measures are presented by simulation in Section 6.

6   Performance Evaluation

In this section, the performance of the DST, based on DDT, is evaluated through simulations. Simulation metrics and methodology are described in Section 6.1. The main goal in simulating the DST is to evaluate how well it actually conserves energy, maintaining the robust DST connection and low delay in highly mobile environments. The parameters affecting the robustness of the DST are first studied. Then, the performance of the DST is compared to DD, TTDD, and SEAD.

6.1   Methodology

The DST is implemented as an independent routing agent module in ns-2.27. In the basic simulation setting, the same energy model is used, which is two-ray ground model and omni-directional antenna, as adopted in Directed Diffusion, and TTDD implementation in ns. A 802.11 DCF is used as the underlying MAC protocol. A sensor node's transmitting, receiving, and idling power consumption rate is set to 0.66W, 0.395W and0.035W, respectively. The network in the simulation consists of 400 sensor nodes randomly or uniformly distributed over a 1000m x 1000m area. Each simulation run lasts for 500 seconds. Each query packet is 36 bytes and each data packet is 64 bytes in length, in order to facilitate comparisons with other protocols.

6.2   Impact of Sink's Mobility

In this subsection, the impact of sink mobility on the performance of the DST is evaluated. In the simulation setting, 8 mobile sinks and 30 randomly chosen sources, in the sensor field, were used. Energy consumption and average end-to-end delay according to varying the maximum speed of a sink are measured from 0 to 30 m/s.

Figure 6 (a) presents the average dissipated energy as the sinks' speed is varied. In this Figure, the DST presents superior energy consumption over the other protocols. This is because the DST maintains the dissemination tree dynamically as the sinks migrate. In addition, the DST does not require additional flooding or location notification to access nodes or agents. In DD, the entire topology is changed so the new location of the mobile sink is propagated throughout the sensors, filed in order for all nodes to obtain the sink's location. The TTDD is designed for mobile sinks, but cannot avoid rebuilding a new multi-hop path between the sink and the grid to track the sink's location. Although SEAD based on the source initiated steiner tree shows smooth energy increase, SEAD has the overhead that each sink must recognize their specific location. This allows continuous access of nodes and changing of access nodes.

Since the DST only uses local interactions on the basis of the sink, for maintaining the dynamic tree, the increase of energy consumption in terms of the entire sensor network is moderate, as presented in (a) of Fig. 6. Figure 6 (b)

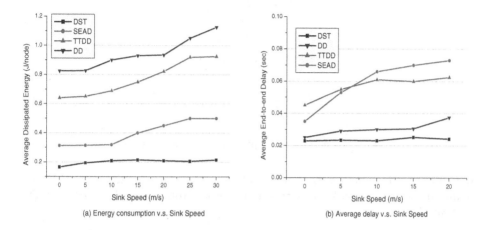

(a) Energy consumption v.s. Sink Speed   (b) Average delay v.s. Sink Speed

Fig. 6. Impact of sink's mobility

presents the average end-to-end delay as the sinks' speed is varied. This figure demonstrates that DD and DST maintain relatively lower delay than TTDD and SEAD. TTDD and SEAD are source initiated data dissemination protocols so they require finding a valid path from the source to each sink, each time a source generates an event. Finding the valid path adds extra delay to the protocols. Conversely, the DST and DD, which are sink-oriented approaches, do not require such additional delay, since all sensor nodes already know the path to each sink. Nevertheless, as a sink's speed increases, DD reveals a gradual increase in delay. This is because of the flooding effect according to sink mobility. However, the DST which dynamically maintains a dissemination tree oriented to sink, shows almost constant delay variation, as presented in (b) of fig. 6.

6.3 Impact of the Number of Sinks

In this subsection, the impact of the number of sinks on the performance of the DST, is evaluated. In this simulation, the sinks' speed is set at 10 m/s and energy consumption is measured as the number of sinks increases to 8.

Figure 7 presents the energy consumption as the number of sinks varies. This figure demonstrates that in case of a single sink TTDD and SEAD outperform the sink-oriented protocols, DST and DD. This is because the DST requires basic energy consumption to maintain the tree using the periodic Update message. However, as the number of sinks increases, in contrast to the other protocols, energy consumption in the DST only slightly increases. This is because the

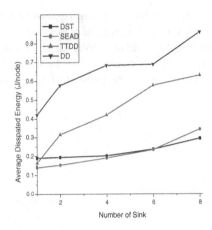

Fig. 7. Energy consumption v.s. number of sinks

dynamic tree is shared with other multiple sinks. As a result, there is little additional energy per sink, in contrast to the other protocols, such as DD, TTDD, and SEAD.

7   Conclusion

This paper mainly focuses on the mobile sink application where several sinks enter the sensor field directly and perform roles based on source data, dynamically refreshed from sensor fields. The proposed DDT is designed for coping well with such dynamic environments, by performing data dissemination with minimum energy using local update and distributed local link reversal. In addition, in order to efficiently support multiple, mobile sinks, the DDT was extended to the DST.

Moreover, the DST created a two-tired network. This two-tired model is very useful in a battle field or for rescue activity. Finally, the simulation results demonstrated that the DST performs excellent energy-efficient data dissemination with relatively low delay, compared to other dissemination protocols.

References

1. Haiyun Luo, Fan Ye, Jerry Cheng, Songwu Lu, Lixia Zhang, "TTDD: Two-tier Data Dissemination in Large-scale Wireless Sensor Networks," *ACM/Kluwer Mobile Networks and Applications, Special Issue on ACM MOBICOM*,2002.
2. C. Intanagonwiwat, R. Govindan, and D. Estrin "Directed diffusion for Wireless Sensor Networking," *IEEE/ACM Transaction on Networking*, Vol. 11, 2003.
3. Sooyeon Kim, Sang H. Son, John A. Stankovic, Shuoqi Li, Yanghee Choi, "SAFE: A Data Dissemination Protocol for Periodic Updates in Sensor Networks," In *Proceedings of the 23 rd International Conference on Distributed Computing Systems Workshops (ICDCSW'03)*,2003.

4. Hyung Seok Kim, Tarek F. Abdelzaher, Wook Hyun Kwon "Minimum-energy asynchronous dissemination to mobile sinks in wireless sensor networks," In *Proceeding of Embedded Networked Sensor Systems (SenSys03)*, Los Angeles, California, USA, 2003.
5. W. Heinzelman, J. Kulik, and H. Balakrishnan, "Adaptive Protocols for Information Dissemination in Wireless Sensor Networks," *ACM International Conference on Mobile Computing and Networking (MOBICOM'99)*, 1999.
6. B. Krishnamachari, D. Estrin, and S. Wicker, "The Impact of Data Dissemination in Wireless Sensor Networks," In *Proceedings of the 22nd International Conference on Distributed Computing Systems Workshops*, 2002.
7. C. Busch, S. Surapaneni, and S. Tirthapura. "Analysis of Link Reversal Routing Protocols for Mobile Ad Hoc Networks," *SPAA 2003*, pp. 210-219, San Diego, California, June 2003.
8. K. Langendoen and N. Reijers, "Distributed Localization in Wireless Sensor Networks," Computer Networks (Elsevier), special issue on Wireless Sensor Networks, August, 2003.

Virtual Coordinates with Backtracking for Void Traversal in Geographic Routing*

Ke Liu and Nael Abu-Ghazaleh

Computer Science Department
Binghamton University,
Binghamton, NY 13902
{kliu, nael}@cs.binghamton.edu

Abstract. Geographical routing protocols have several desirable features for use in ad hoc and sensor networks but are susceptible to voids and localization errors. Virtual coordinate systems which overlay a coordinate system on the nodes offer an alternative that is more resilient to localization errors. However, we show that it is vulnerable to different forms of the void problem where packets reach nodes with no viable next hop in the forwarding set. In addition, it is possible for nodes with the same coordinates to arise at different points in the network in the presence of voids. This paper identifies and analyzes these problems. It also compares several existing routing protocols based on Virtual Coordinate systems. Finally, we propose a routing algorithm that uses geographic routing in the greedy phase and virtual coordinates with backtracking to overcome voids and achieve high connectivity in the greedy phase with higher overall path quality and more resilience to localization errors. We demonstrate these properties using extensive simulation studies.

1 Introduction

Traditional ad hoc routing protocols (e.g., AODV [1]), are not a good fit for Wireless Sensor Networks (WSNs) which often require data dissemination patterns that are not efficiently mapped to the unicast connections assumed by ad hoc protocols. Further, nodes need to maintain routing state specific to destinations of active routes; this state may become invalid due to changes that are not near to the node. Finally, because of the need to maintain non-local state, traditional protocols require that nodes have globally unique identifiers. In contrast, WSNs favor routing protocols that support data-centric operation (e.g., global identifiers not necessary), localized interactions (e.g., maintaining only local state) and supporting arbitrary data-driven dissemination with in-network processing.

Geographic routing algorithms [2,3,5,6], provide attractive properties for WSNs. In these algorithms, nodes exchange location information with their neighbors. Packets addressed to a destination must provide its location. At every intermediate hop, the subset of the neighbors that are closer to the destination is called the forwarding set.

---

* This work is partially supported by NSF grant CNS-0454298 and US Army project W911SR-05-C-0014.

T. Kunz and S.S. Ravi (Eds.): ADHOC-NOW 2006, LNCS 4104, pp. 46–59, 2006.

Routing simply forwards a packet to one of the nodes in the forwarding set. This process is repeated greedily until the packet reaches the destination. Thus, interactions are localized to location exchange with direct neighbors and there is no need for global identifiers.

Geographical routing protocols suffer from significant problems under realistic operation. First, voids – intermediate nodes whose forwarding set (FS) relative to a destination is empty – can cause the greedy algorithm to fail[2,3,4,5]. Voids require a complex and inefficient complementary perimeter routing algorithm that is invoked when they are encountered[6]. Moreover, geographic routing has been shown to be sensitive to localization errors [7], especially in the perimeter routing phase [6,8]; such errors can cause routing anomalies ranging from suboptimal paths to failure to deliver packets. Making geographic routing practical is difficult [6].

Routing based on Virtual Coordinate Systems (VCS) has been recently proposed [13,14,15,16,18]. A VCS overlays virtual coordinates on the nodes in the network based on their distance (typically in number of hops) from fixed reference points; the coordinates are computed via an initialization phase. These coordinates serve in place of the geographic location for purposes of Greedy forwarding. Because it does not require precise location information, VCS is not sensitive to localization errors. Further, it is argued that VCS is not susceptible to conventional voids because the coordinates are based on connectivity and not physical distance [14]. On the negative side, VCS may be sensitive to collisions and or signal fading effects in the initialization phase. Furthermore, the initialization phase requires a flood from each reference point.

The first contribution of the paper is to identify three problems that arise in VCS routing. In practice, the three problems occur in common situations resulting in VCS failing to deliver packets. In the first problem, a set of neighbors are all of equal distance to the destination and greedy forwarding fails. In the second problem, we show that nodes with identical coordinates that are far from each other may arise in the presence of voids. In the third problem, a node is closer than any of its neighbors to a given destination, no matter how the distance is measured. We analyze the frequency of these problems using a number of random deployment scenarios of different densities.

The second contribution of the paper is to propose a hybrid routing protocol in Section 4. The protocol uses Greedy Forwarding (GF) [2,3] because of its superior path quality in the greedy mode. When voids are encountered, it switches to a VC based backup algorithm. The intuition is two fold: (1) since VC is based on connectivity it is effective in handling physical voids; and (2) Greedy VC is more efficient and less sensitive to localization errors than traditional void traversal algorithms such as planarization. However, since VC is susceptible to its own anomalies outlined above, routing may still fail; we use a simple backtracking algorithm where the packet is forwarded backwards towards an anchor to address such possibilities (other approaches such as localized flooding may be applied here).

We show experimentally that this approach attains high delivery ratio and tolerates localization errors better than geographically based protocols. Our experimental results also show that for all scenarios we used, the Hybrid geographic and virtual coordinate Greedy Routing (HGR) provides a 100% reachability, and much better path quality than Greedy Perimeter Stateless Routing (GPSR)[2,3]. The greedy nature of HGR makes its implementation practical and efficient. The experiments and results are presented in Section 5. Finally, Section 6 presents some concluding remarks.

2 Background and Related Work

Shortest Path (SP) routing is a commonly-used for sensor networks. In this protocol, data sinks flood the network with periodic beacons. As nodes receive the beacon, they set their next hop to the neighbor advertising the shortest distance to the sink. SP can provide the optimal path in terms of path length. However, it is a stateful, requiring routing state for each destination. Furthermore, it is vulnerable to mobility or other changes in the topology that might make the discovered shortest path unavailable.

In ad hoc and sensor networks, stateless routing, where the routing state is independent of the connection pattern, is desirable. Geographical routing protocols with the stateless property [2,3]; in the base mode, they use Greedy Forwarding (GF) which works as follows. Each node tracks only the location information of its neighbors. The *forwarding set* for a given destination is the subset of neighbors closer to the destination than the current node. GF proceeds by picking a node from this set, typically the closest one to the destination. If the forwarding set is empty, a *void* is encountered. Typically, a complementary algorithm is used to traverse the void. Face routing (or perimeter routing) is an approach based planar graph theory often used for void traversal. The general idea in face routing is to attempt to route around the void using a right hand rule that selects node around the perimeter of the void (details may be found in the original paper [2,3]). Face routing stops when a node closer to the destination than the void origin is encountered; at this stage, operation switches back to greedy forwarding.

Since GPS devices are costly, they may not be feasible for sensor networks; often, localization algorithms are employed that significantly increase the uncertainty in the location estimate (e.g., [21,23]). Both GF and face routing are susceptible to localization errors [7,8]. While some approaches to tolerate location errors have been suggested, in general, this remains a weakness of this class of protocols. Further, the paths constructed by face routing are typically not the best path available to cross the void. Thus, additional routing protocols have attempted to optimize the face routing phase of operation [5,4,19]. Kim et al [6] recognize effects that arise in practice during geographic routing and suggest a protocol which uses more control packets to planarize the network. The algorithm requires more resources and is stateful.

Routing based on a coordinate system was first proposed by Rao et al [10]. Their algorithm requires a large number of nodes to serve as virtual coordinate anchors (sufficient to form a bounding polygon around the remaining sensors). A large number of anchor nodes increases the overhead as well as the state maintained by each node. The location estimated by the virtual coordinates is used for geographic routing. Their approach is more accurately described as a localization mechanism for use in an otherwise geographically based algorithm. It is unclear if face routing will be effective with a coarse-grained location estimate.

Caruso et al recently proposed the Virtual Coordinate assignment protocol (VCap) [16]; several similar protocols were proposed by others [13,15,14,17,18]. In this approach, coordinates are constructed in an initialization phase relative to a number of reference points. Following this initialization phase, packets can be routed using the Greedy Forwarding principles, replacing node location with its coordinates. The paper uses 3 reference points to assign the virtual coordinates, constructing a 3-dimensional VCS. The authors identified the *VC Zone* (several nodes assigned to the same virtual coordinates) problem and provided bounds on its size. (The original research work on VC Zone can be found in [9,17] with more details.) The authors do offer a heuristic for situations where voids are encountered: they suggested a *local detour* to forward packet to some neighbor farther away to destination several times in hope of reaching some node with a different path to the destination; this approach may lead to longer paths as packets are misrouted. The evaluation in the paper shows that VCap performs worse than GPSR both in delivery ratio and path stretch.

Qing et al proposed a similar protocol to VCap with 4 reference nodes (4D) each located at a corner for a rectangle area [14]. The authors suggested a backtracking approach to packet delivery when facing any routing anomalies, which requires each hop in the forwarding path of each packet to be recorded. The authors did not analyze either why or when these anomalies happen. Although this backtracking approach converges, in the worst case, it will go through all the nodes in the network.

GEM [11] proposes routing based on a virtual polar coordinate space (VPCS). A tree-style overlay is then used for routing. Using the tree overlay results in poor path quality. Since it uses the VPCS to localize the network first, GEM tolerates only up to 10% localization error [11].

Using Manhattan-style distance (MD) in place of Euclidean distance (ED) was proposed by Rodrigo et al in BVR[18]. On a VCS with a very high number of reference nodes (typically 10 to 80), BVR suggested a different backtracking approach to send packets to the reference node closest to the destination when greedy forwarding fails. As we show in this paper, neither Manhattan distance or the one proposed in BVR[18] (we called semi-Manhattan distance) are a good measure of distance compared to a Euclidean distance. Having that many reference nodes require much more resources and may also hurt the performance of any practical sensor networks.

Papadimitriou and Ratajczak [12] conjecture that every planar 3-connected graph can be embedded on the plane so that greedy routing works. If this conjecture holds, then for such networks, a guaranteed greedy routing may exist.

3   Greedy Forwarding in VCS

Our approach is essentially Geographic Forwarding, with the use of greedy VC when voids are encountered. This approach is motivated by the shortcomings in the planarization procedure typically used for void traversal, which has high cost and complexity, is susceptible to localization errors, and results in suboptimal routes. Although the greedy phase of geographic routing works well under an up to 40% of the localization error [7], face routing may fail with some very small localization error [8].

In contrast, Virtual coordinate systems are attractive because they are more resilient than geographic routing to localization errors, and because it is thought that they reduce the effect of voids. We use as the basis of description the Virtual Coordinate System (VCS) for introduced by VCap [16]; however, the problems identified and the solutions generalize to other virtual coordinate systems. The use of virtual coordinates introduces a different set of problems, which we identify in this section.

3.1 Effect of Number of Anchors

The authors [16] argue that for a 2 dimensional geographical coordinate system (GeoCS), a 3-dimensional VCS is sufficient to accomplish effective Greedy Forwarding (GF). We show that in practice this does not occur and VCS is susceptible to routing problems resulting in suboptimal paths, packet misrouting, or routing failure. These problems do not necessarily coincide with geographic voids: for example, we show that GF based on VCS may fail in a network without geographic voids.

Figure 1(a) shows a VCS for a network where 25 nodes are deployed along the vertices of a grid. The radio range makes each internal node have 9 one-hop neighbors. For example, node 13 has neighbors 7, 8, 9, 12, 14, 17, 18 and 19. Perimeter node 1 has 3 neighbors: 2, 6 and 7. The numbers at the left of each node are IDs, and the triples in brackets under each node represent their coordinate values. It can be show that there is no geographical void in this network (the forwarding set is never empty). VCS anchors are chosen consistent with the requirement described by its designers [16].

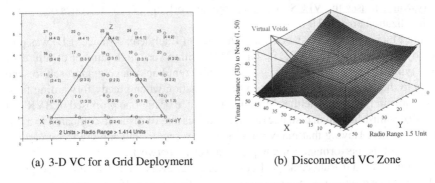

(a) 3-D VC for a Grid Deployment   (b) Disconnected VC Zone

Fig. 1. Routing Anomalies caused by VC Zone (3D VCS)

3.2 Expanded VC Zone Problem

Consider a packet P at node 11 destined to node 15. Table 1 shows the distances of each neighbors to the destination node 15. P is first forwarded to node 17 greedily. However, among the one-hop neighbors of 17 (node 11, 12, 13, 16, 18, 21, 22, 23), there are none with a shorter distance to node 15 measured in virtual coordinates – GF fails.

Nodes with the same virtual coordinate value are called a *VC Zone*. This problem occurs when VC Zones cross a contour line (the lines connecting of the VCS anchors).

Table 1. The Euclidean distances of node 11, 17 and their neighbors on VCS to node 15

node	6	7	11	12	13	16	17	18	21	22	23
distance	$\sqrt{14}$	$\sqrt{11}$	$2\sqrt{2}$	$\sqrt{5}$	2	$\sqrt{5}$	$\sqrt{3}$	$\sqrt{3}$	2	$\sqrt{5}$	$2\sqrt{2}$

Around these contour lines, the possibility of VC Zones that are larger than 2-hop across arises. The routing algorithm cannot deliver packets through the VC zone: we call this problem the *Expanded VC Zone* problem. One possible solution is to broadcast a path request within the VC zone.

3.3 Disconnected VC Zone Problem

The second problem occurs because it is possible for nodes with equal coordinates to occur in geographically disparate locations. Consider node 21 and 25; they have the same virtual coordinates but are not connected by any other nodes with the same virtual coordinates. We call this the *Disconnect VC Zone* problem. Node 22 and 24 are in the same situation. These nodes occur symmetrically around the contour line in a uniform deployment such as the grid. Note that the Expanded Zone problem is an instance of the Disconnected Zone problem where the two disconnected zones are neighbors.

   If a data packet produced by node 21 needs to be delivered to node 25, GF will obviously fail. Further, data packets produced by any one-hop neighbors of node 21 can not be routed to node 25 either. Broadcasting within a VC zone cannot solve this problem because the VC zone is not connected. Even if the Z anchor goes to infinity (is arbitrarily far), a disconnection remains whose size is larger than the limit argued in the original VCap paper (2.3 times the radio range) [16]. Note that the limit in the original paper is derived under infinite density assumptions. We visualize this problem in figure 1(b), where 2500 nodes are deployed in 50×50 grids, each per one. 3 reference nodes are located at grid $(1,1)$, $(50, 1)$, $(25, 50)$. For each point on the surface in figure 1(b), x and y values denote their physical location, z values denote the Euclidean distances on VCS of each node to the node located in grid $(1,50)$. The virtual distances to node$(1,50)$ of node $(50, 50)$ is 0, which leads to a virtual void caused by disconnected VC Zone. And there are still other virtual voids caused by expanded VC Zone.

   We argue instead that the contour lines connecting VC anchors should be a polygon containing all nodes of the network inside it. If this occurs, then nodes occur only on one side of the contour line making both problems above impossible. Thus, a 4-dimensional VCS is needed to guarantee the success of the greedy algorithm (with 4 corner nodes as anchors).

3.4 Effect of Distance Metric

Instead of using Euclidean distance to measure the distance between nodes in VCS, BVR [18] proposed a variant of Manhattan distance. We compare the effect of the different distance metrics using an example (Figure 2(a)). Node 1, 5, 12, 16 are 4 reference nodes. Suppose the data source is node 14 in this figure, while destination is node 3. All the distances of any type of the neighbors of node 14 is far away from the destination than itself, shown in table 2. We also visualize this problem in figure 2(b) with similar scenario as figure 1(b) except a physical void in the center. As we can see, the Manhattan-style distance does not help the distance measurement, while it may make the problem much worse. Euclidean distance performs better than the Manhattan-styled distance but still with some virtual voids (figure 3).

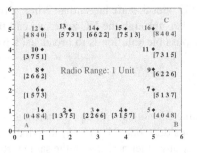  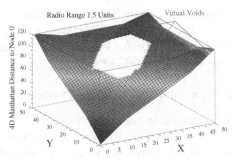

(a) 4D VCS with physical void   (b) Virtual Voids with Manhattan Distance

Fig. 2. Anomaly despite of Diversity of Distances (4D VCS)

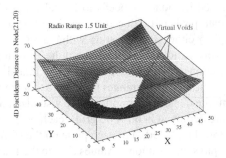

Fig. 3. Virtual Voids measured in Euclidean Distance (4D VCS)

Table 2. Different types of Distances to Node 3

Distance Diversity	Node 12	Node 13	Node 14	Node 15	Node 16
Euclidean (VCap, LCR)	$\sqrt{80}$	$\sqrt{68}$	$\sqrt{50}$	$\sqrt{68}$	$\sqrt{80}$
Manhattan	16	16	16	16	16
Semi-Manhattan (BVR) $\delta+$	8	8	8	8	8
Semi-Manhattan (BVR) $\delta-$	8	8	8	8	8

4   Hybrid Greedy Routing (HGR)

Our simulation result show that for any path GF constructs successfully, the path quality is almost identical to the optimal found by SP; moreover, this phase of the algorithm is tolerant to localization errors. Thus, our algorithm uses Geographic Forwarding as the base. HGR uses virtual coordinates for void avoidance because they naturally protect against geographical voids. However, in the presence of the problems outlined in the previous section, VC routing may fail. As a result, we use a simple backtracking technique in response to VC routing anomalies.

4.1   Void Avoidance Phase

To resolve the void problem, HGR uses virtual coordinates. The VCS of the network is initialized by the same procedure introduced by VCap [16]. The intuition of HGR is that if any node in the network with a VC value smaller than infinity, is reachable. Once a local maximum (void) is encountered, HGR switches to *hybrid-mode*. In the hybrid mode, it picks one of the coordinate axes and attempts to route using it towards the destination. If a point is reached where no neighbor closer to the destination is found, then backtracking is necessary.

We elect to carry out VC as well as backtracking in an axis by axis basis – alternative approaches are possible and will be a topic of future research. More specifically, if we reach a point along the current axis where no node leads closer to the destination on the same axis, the direction is reversed and we backtrack. In the current implementation, we backtrack until a void (relative to the same axis) is found while backtracking, or we reach coordinate 0. Alternative approaches for terminating backtracking are also possible. If either of the cases occur, we switch to the next axis and repeat the process. If all three axes are exhausted, the routing fails[1]. At any point in the algorithm if a node geographically closer to the destination than the node at which the hybrid mode is entered is found, the protocol switches back to the greedy mode (the void had been traversed).

The hybrid mode operates as follows. When a packet is faced by a void, it is labeled as hybrid-mode and the geographical coordinates of the current node, $GC_{initial}$, are entered into the packet header. The node also records the dimension index i (initially, $i = 1$) and the VC direction for dimension i into the header, where the VC direction is decided by the VC of the current node and the destination ($VC^i_{initial}$ and VC^i_{sink}) according to the function:

$$direction^i = -1 \; if \; VC^i_{sink} \leq VC^i_{entering} \; else \; 1$$

On receiving a *hybrid-mode* data packet, (ED stands for Euclidean distance)

1. **If** $ED_{entering,sink} > ED_{current,sink}$, to label *greedy-mode* to the data packet, switch back to using greedy forwarding, **else**:
2. **If** sink is in the neighbor list of current node, forward the data packet to sink and routing succeeds, **else**:
3. **If** $VC^i_{last_{hop}} = VC^i_{current}$, goto 4), **else**: Among all neighbors n with $VC^i_n = VC^i_{current} - direction^i$ (same VC^i as last hop), **if** $ED_{N,sink} = min(ED_{n,sink})$ and node N is not the last hop, then $direction^i = -direction^i$, forward to N; **otherwise**:
4. Among all neighbors n with $VC^i_n = VC^i_{current}$ (same VC^i as current node), **if** $ED_{N,sink} = min(ED_{n,sink})$ and $N \neq current$, forward to N; **otherwise**:
5. Among all neighbors n with $VC^i_{current} + direction^i$, forward the data packet to the Node N where $ED_{N,sink} = min(ED_{n,sink})$ for all n, **otherwise: if** no such neighbor exists, reverse the direction: $direction^i = -direction^i$, goto 2); **If** $direction^i$ has been reversed once in previous routing procedure (either current node, or other nodes, but a reverse in Step 3 does not count), increase the coordinate index i, if $i \leq max(dimension)$ goto 2); **otherwise**, label this data packet with *HGR fails*, **Quit**;

---

[1] A backup algorithm such as localized flooding may be used here, but in practice, we observe very few failures.

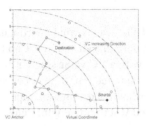

Fig. 4. HGR routing Demo

Note that if current node and destination have finite virtual coordinate values, they should be connected in the network. Figure 4 shows a sample path obtained by HGR.

5   Experiment

In this section, we present an experimental evaluation that illustrates the problems with existing geographical and VC protocols. The evaluation also characterizes the performance of the proposed HGR protocol. To allow scalability to very large networks, we use a custom simulator written for this study; the simulator abstracts away the details of the channel and the networking protocols which may affect performance such as the reach-abilities of routing protocols. Our results validate successfully with the NS-2 simulator.

We study both random and controlled deployment. Each point represents the average of 30 scenarios of 200 nodes that are deployed in a $1000 \times 1000m^2$ area; the number of scenarios was sufficient to tightly bound the confidence intervals (for random scenarios, each node's location was generated uniformly in the whole simulation area). We simulate the different densities by varying the radio transmission range. For every scenario, reachability is determined by testing whether a packet can be delivered between each pair of nodes in the network. Recall that the stateful SP is the optimal routing in terms of number of hops; for this reason it is used to derive the ideal performance in terms of path quality.

We also implemented the GPSR (with GG and RNG planarization algorithms)[3,6], Shortest Path (SP), Greedy Forwarding on VCS [16,14] and BVR on 4D VCS [18], and study their performance against HGR. To enable a fair comparison, we use the same number of anchors for all protocols (which, in fairness to BVR, is much lower than the recommended number).

5.1   Greedy Forwarding (GF)

Void Frequency Analysis. This first study shows the frequency of voids for both geographic and VC routing. Previously, it was demonstrated that voids occur in different situations for the two types of protocols. Figure 5(a) shows the ratio of node pairs facing void problem. In general, we can see that sparse networks suffer from void problem much more than dense ones. The frequency of voids is significantly higher in VCS systems than in Geographic Coordinate (GeoC) ones even when using 4 coordinate axes.

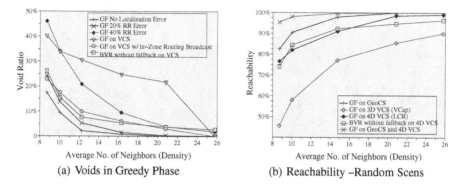

(a) Voids in Greedy Phase   (b) Reachability –Random Scens

Fig. 5. Greedy Forwarding Analysis

Thus, while the analysis in the original VCap [16] paper may apply under asymptotic conditions (infinite density), it does not hold under practical situations. It is clear that VCap on its own does not improve performance relative to pure GF even with 4 coordinates such as LCR without backtracking [14].

Localization errors affect the performance of GF on GeoCS significantly. Since the VCS does not need the location information to initialize, it does not suffer from this problem. However, the expanded VC zone plays an important effect even when network density is high. This can be seen in the graph where the zone broadcast is implemented (figure 5(a), curve labelled as "GF on VCS w/ in-Zone Routing Broadcast"); zone broadcast floods a packet in a VC zone, which eliminates the expanded zone problem but incurs some additional overhead. This curve suffers much fewer voids than just regular VC. The remaining voids in this curve are due to the disconnected VC Zone problem.

Analysis of Greedy Forwarding. Figure 5(b) shows the reachability of all pairs in the same random deployment scenario, using only GF as routing (no void traversal algorithm). GF based on a 3D VCS (as in VCap) shows the worst reachability. The 4D VCS (as in LCR) shows a much higher reachability than 3D VCS, but worse than GF on GeoCS. We also use a combination of the GeoCS and 4D VCS: GF on GeoCS first, when it fails, GF on 4D VCS is used. The result shows this combination works much better than any one independently. The reachability of it is higher than 95% even in a sparse network which leads much smaller cost of backtracking. Although our experiment results show a higher-dimensional VCS working better than a lower-dimensional one for GF routing, we found that this does not hold beyond 4 dimensions (graph not shown due to space limitation).

5.2 HGR Performance

Figure 6(a) shows the path quality obtained across all nodes under different densities. SP routing provides the optimal solution, which cannot be obtained by any stateless greedy solution in general. The performance of HGR is much better than GPSR with either GG or RNG planarization. When the densities of networks are higher, less voids

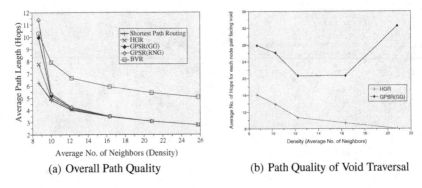

(a) Overall Path Quality   (b) Path Quality of Void Traversal

Fig. 6. Path Stretch

happen, leading to similar overall path qualities. The average path length of BVR is the highest. The reason may be that in BVR the backtracking path is much longer since it needs to forward packet back to one of the anchors.

Figure 6(b) shows the quality of paths facing void problems. As the density goes higher, loss of efficiency results due to the planarization algorithm (which forces the use of the nearest neighbor along the face). In contrast, HGR operates greedily even in the void traversal/hybrid mode. As a result, HGR performs well while the average performance of GPSR suffers. Figure 7(a) shows a sample path between 2 nodes in one of the 30 networks, generated by different routing protocols with radio range as 150m.

To study the behavior of the protocols under voids, we also create scenarios where 150 nodes are randomly deployed in a "C" region around the border of the area. In this case, a large portion of the paths are faced by voids. Figure 7(b) shows one path with radio range as 200m. The path length of HGR is longer than that of GPSR in Euclidean distance, but much shorter in number of hops. Figure 8 shows the average path length of routing protocols averaged across 20 randomly deployed "C" networks. Once the radio range is too small to cross the void, greedy forwarding faces voids. HGR performs much better than any of the GPSR flavors and BVR, roughly approximating

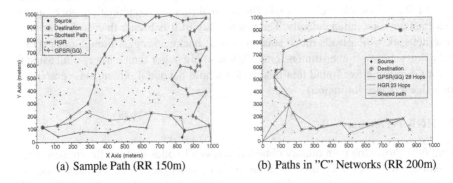

(a) Sample Path (RR 150m)   (b) Paths in "C" Networks (RR 200m)

Fig. 7. Sample HGR Path vs GPSR Path Anomalies

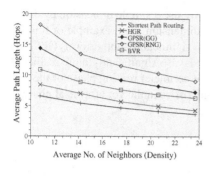

Fig. 8. Path Quality in "C" Networks

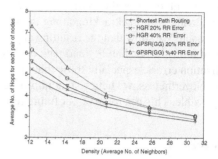

Fig. 9. Impact of Localization Errors

the optimal solution. BVR outperforms GPSR as well since virtual coordinate routing is effective in traversing physical voids. This scenario does not result in creating many VC voids.

We also study the impact of localization error on the different protocols. For routing, localization values uniformly distributed in a circle of radius $ratio \times range$ around the correct location are generated (resulting in average error of $ratio \times range$. GPSR may fail when the localization error is big, in either the greedy phase (causing an unnecessary switch to face routing), or in the face routing phase (causing routing failure). HGR is also susceptible to routing anomalies when it uses the geographic location in the greedy phase. We observed that the frequency of routing errors is much higher in GPSR compared to HGR. HGR uses VC once GF is faced with a void, allowing it to tolerate localization errors well in all but very sparse scenarios.

In order to study the effect on path quality, we planarized the graph based on symmetric connection of neighbors as used in [6]. The planarization is not affected by localization errors (which benefits only GPSR). HGR operates distributedly as before. Figure 9 shows the average path length in the 30 randomly deployed networks, with error of $ratio = 20\%$ radio range (effective location is 20% radio range away from the actual location) and 40% radio range under different densities. The reachability of HGR is still 100% and the path quality is much better than the centrally planarized GPSR.

6   Conclusion

In this paper, we first demonstrate problems that arise with Virtual Coordinate based routing and show that contrary to published conclusions, it is quite vulnerable to voids that arise during greedy operation. More specifically, we identify the expanded VC Zone problem, where nodes of equal VC coordinates span a multiple hop zone. We also identify the disconnected VC zone problem where nodes sharing the same VC value occur in geographically disparate locations. We show experimentally that these problems have a considerable effect on reachability provided by VCS algorithms.

The second contribution of the paper is to present a hybrid routing protocol that replaces the traditional face routing phase used for void traversal in geographic routing protocols with one based on virtual coordinates. We use a simple dimension by dimension hueristic with backracking along the same dimension as an example of this type of approach. We show that the resulting algorithm significantly outperforms GPSR and BVR in terms of reachability and path quality. It is also significantly more resilient to localization errors, especially those that affect the perimeter routing phase of the algorithm. Nevertheless, we believe that improved hueristics in the void traversal phase are still possible. This is a topic of our future research.

References

1. C. E. Perkins and E. M. Royer, *Ad hoc On-Demand Distance Vector Routing*, in Proceedings of the 2nd IEEE Workshop on Mobile Computing Systems and Applications, February 1999
2. P. Bose, P. Morin, I. Stojmenovic and J. Urrutia, *Routing with guaranteed delivery in ad hoc wireless networks*, DIAL M99, Auguest 1999
3. B. Karp and H.T. Kung, *GPSR: Greedy Perimeter Stateless Routing for Wireless Networks*, MobiCom 2000
4. Q. Fang, J. Gao and L. Guibas, *Locating and Bypassing Routing Holes in Sensor Networks*, INFOCOM 2004
5. Sophia Fotopoulou-Prigipa and A. Bruce McDonald, *GCRP: Geographic Virtual Circuit Routing Protocol for Ad Hoc Networks*, MASS 2004
6. Young-Jin Kim, Ramesh Govindan, Brad Karp and Scott Shenker, *Geographic Routing Made Practical*, the Second USENIX/ACM Symposium on Networked System Design and Implementation(NSDI'05), May 2005
7. T. He, C. Huang, B. Blum, J. A. Stankovic and T. Abdelzaher, *Range-Free Localization Schemes for Large Scale Sensor Networks*, MobiCom 2003
8. K. Seada, A. Helmy and R. Govindan, *On the Effect of Localization Errors on Geographic Face Routing in Sensor Networks*, IPSN 2004
9. Samir Khuller, Balaji Raghavachari and Azriel Rosenfeld, *Landmarks in graphs*, Journal of Discrete Appl. Math., volume 70, page 217–229, 1996
10. A. Rao, S. Ratnasamy, C. Papadimitriou, S. Shenker and Ion Stoica, *Geographic Routing without Location Information*, MobiCom 2003
11. J. Newsome and D. Song, *GEM: Graph EMbedding for Routing and Data-Centric Storage in Sensor Networks Without Geographic Information*, SenSys'03, Nov. 2003
12. C. H. Papadimitriou and D. Ratajczak, *On a Conjecture Related to Geometric Routing*, ALGOSENSORS 2004
13. T. Moscibroda, R. O'Dell, M. Wattenhofer, R. Wattenhofer, *Virtual Coordinates for Ad hoc and Sensor Networks*, ACM Joint Workshop on Foundations of Mobile Computing (DIALM-POMC), October 2004
14. Qing Cao and Tarek F. Abdelzaher, *A Scalable Logical Coordinates Framework for Routing in Wireless Sensor Networks*, RTSS 2004
15. D. M. Nicol, M. E. Goldsby and M. M. Johnson, *Simulation Analysis of Virtual Geographic Routing*, in Proceedings of the 2004 Winter Simulation Conference
16. A. Caruso, S. Chessa, S. De and A.o Urpi, *GPS Free Coordinate Assignment and Routing in Wireless Sensor Networks*, INFOCOM 2005
17. Mirjam Wattenhofer, Roger Wattenhofer and Peter Widmayer, *Geometric Routing Without Geometry.*, SIROCCO 2005
18. R. Fonseca, S. Ratnasamy, J. Zhao, C. T. Ee, D. Culler, S. Shenker and I. Stoica, *Beacon Vector Routing: Scalable Point-to-Point Routing in Wireless Sensornets*, NSDI'05, 2005

19. Q. Fang, J. Gao, L. J. Guibas, V. de Silva and L. Zhang, *GLIDER: Gradient Landmark-Based Distributed Routing for Sensor Networks*, INFOCOM 2005
20. R. Nagpal, H. Shrobe and J. Bachrach, *Organizing a global coordinate system from local information on an ad hoc sensor networks*, IPSN 2003
21. D. Niculescu and B. Nath, *Ad Hoc Positioning System (APS)*, GlobalCom 2001
22. J. Hightower and G. Borriella, *Location Systems for Ubiquitous Computing*, IEEE Computer, V34, 57-66, 2001
23. A. Haeberlen, E. Flannery, A. M. Ladd, A. Rudys, D. S. Wallach and L. E. Kavraki, *Practical Robust Localization over Large-Scale 802.11 Wireless Networks*, MobiCom 2002

A Distributed Protocol for the Bounded-Hops Converge-Cast in Ad-Hoc Networks[*]

Andrea E.F. Clementi[1], Miriam Di Ianni[1], Massimo Lauria[2],
Angelo Monti[2], Gianluca Rossi[1], and Riccardo Silvestri[2]

[1] Dipartimento di Matematica, Università degli Studi di Roma "Tor Vergata".
{clementi, diianni, rossig}@mat.uniroma2.it.
[2] Dipartimento di Informatica, Università degli Studi di Roma "La Sapienza".
{lauria, monti, silver}@di.uniroma1.it.

Abstract. Given a set S of points (*stations*) located in the d-dim. Euclidean space and a *root* $b \in S$, the h-HOPS CONVERGECAST problem asks to find for a minimal energy-cost range assignment which allows to perform the *converge-cast* primitive (i.e. node accumulation) towards b in at most h hops. For this problem no polynomial time algorithm is known even for $h = 2$.

The main goal of this work is the design of an efficient *distributed* heuristic (i.e. protocol) and the analysis (both theoretical and experimental) of its expected solution cost. In particular, we introduce an efficient parameterized randomized protocol for h-HOPS CONVERGECAST and we analyze it on *random instances* created by placing n points uniformly at random in a d-cube of side length L. We prove that for $h = 2$, its expected approximation ratio is bounded by some constant factor. Finally, for $h = 3, \ldots, 8$, we provide a wide experimental study showing that our protocol has very good performances when compared with previously introduced (centralized) heuristics.

1 Introduction

An *ad-hoc* (wireless) network consists of a set of radio stations connected by wireless links. In an ad hoc network, a transmission range is assigned to every station. The overall range assignment determines a transmission (directed) graph since one station s, with transmission r, can transmit to another station t if and only if t belongs to the *disk* centered in s and of radius r. The transmission range of a station depends, in turn, on the energy power supplied to the station. In particular, the power P_s required by a station s to correctly transmit data to another station t must satisfy the inequality

$$\frac{P_s}{d(s, t)^\alpha} > \gamma \qquad (1)$$

[*] Supported by the European Union under the Integrated Project IST-15964 AEOLUS ("Algorithmic Principles for Building Efficient Overlay Computers").

T. Kunz and S.S. Ravi (Eds.): ADHOC-NOW 2006, LNCS 4104, pp. 60–72, 2006.

where $d(s,t)$ is the Euclidean distance between s and t, $\alpha \geq 1$ is the *distance-power gradient*, and $\gamma \geq 1$ is the *transmission quality* parameter. The parameter α may vary from 1 to more than 6 depending on the environment conditions; in the ideal condition (empty space), $\alpha = 2$ (see [21]). Stations of an ad-hoc network cooperate in order to provide specific network connectivity properties by dynamically adapting their transmission ranges. A range assignment $r : S \to \mathcal{R}^+$ determines a directed *transmission graph* $G(S, E)$ where edge $(i, j) \in E$ if and only if $d(i, j) \leq r(i)$. A fundamental problem underlying any phase of a dynamic resource allocation algorithm in ad-hoc wireless networks is the following [10,16,7,15,19,23]: find a range assignment such that (1) the corresponding transmission graph G satisfies a given connectivity property Π, and (2) the overall energy power required to deploy the assignment (according to Inequality (1)) is minimized. The overall energy power (i.e. the *cost*) of a range assignment $r : S \to \mathcal{R}+$ is defined as

$$\mathsf{cost}(r) \;=\; \sum_{s \in S} r(s)^\alpha \tag{2}$$

In this work, we address the range assignment problem in which G is required to contain a tree directed towards a given *root* station b (called *root*), spanning S and of depth at most h. The relevance of this particular connectivity property is clear: feasible solutions for this problem, denoted as h-HOPS CONVERGECAST, allow minimal energy-cost *converge-cast* (i.e. node accumulation) operations towards b in at most h-hops.

It is easy to verify that the h-HOPS CONVERGECAST problem is a particular case of the well-known *Minimum h-hops Spanning Tree* problem (h-HOPS MST) defined as follows: given a graph $G(V, E)$ with nonnegative edge weights and a node $b \in V$, find a minimum-cost directed tree rooted at b, of depth at most h, and spanning G. In fact, h-HOPS CONVERGECAST corresponds to the restriction of h-HOPS MST in which nodes are the stations and there is an edge for any pair of stations i and j whose weight is $d(i, j)^\alpha$.

The main goal of this work is to design efficient *distributed* heuristics (i.e. protocols) for h-HOPS CONVERGECAST and then analyze (both theoretically and experimentally) their expected solution costs. We intend to investigate the problem, for any constant h, on *random* instances created by placing n points uniformly at random in a d-cube of side length L. Such instances will be simply called *random instances*.

Previous results. Almost all previous related works refer to the h-HOPS MST problem. It is known that it is MAX SNP-hard even when the edge weights of the input graphs form a *metric* and $h = 2$ [1].

The 2-DIM 2-HOPS MST problem can be easily reduced to the classic Facility Location Problem on the plane. Indeed, the distance of the root from vertex i can be seen as the cost of opening a facility at vertex i. It thus follows that all the (centralized) approximation algorithms for the latter problem apply to the 2-DIM 2-HOPS MST as well. In particular, the best result is the PTAS given by Arora et al in [3]. The algorithm works also in higher dimensions;

however, it is based on a complex dynamic programming technique that makes any (distributed) implementation very far to be practical. Several polynomial-time approximation algorithms for the METRIC 2-HOPS MST problem have been presented in the literature. The first constant factor approximation algorithm was given by Shmoys et al in [26], they presented a 3.16 approximation algorithm. After this, a series of constant factor approximation algorithms was published, see [5,18,13]. Currently, the best factor is 1.52 due to Mahdian et al [20].

The general h-HOPS MST problem was studied in [12,14,17] by providing exact but super-polynomial or $O(\log n)$-approximate solutions.

Another series of works have been devoted to evaluate and compare solutions for the d-DIM h-HOPS MST problem returned by some heuristics on *random* planar instances by performing computer experiments [8,11,12,25,27]. Almost all such works adopt *random instances*.

More recently, a tight analysis of the expected optimal cost for the d-DIM h-HOPS MST problem on random instances has been done in [9,6].

Given a rooted tree T the *cost* of T, denoted as $\mathsf{cost}(T)$, is the sum of the edge weights.

Theorem 1 ([9,6]). *Let h and d be fixed positive integers. Let S be a random instance of n points in a d-cube of side length L and let T be any tree of height h spanning S. Then, it holds that*

$$\mathsf{cost}(T) = \begin{cases} \Theta\left(L \cdot n^{\frac{1}{h}}\right) & \text{if } d = 1, \ \alpha = 1 \\ \Theta\left(L \cdot n^{1-\frac{1}{d}+\frac{d-1}{d^{h+1}-d}}\right) & \text{if } d \geq 2, \ \alpha = 1 \\ \Theta\left(L^2 \cdot n^{\frac{1}{h}}\right) & \text{if } d = 2, \ \alpha = 2 \end{cases} \quad \text{with high probability.}$$

Here and in the sequel the term *with high probability* (in short, *w.h.p.*) means that the event holds with probability at least $1 - e^{-c \cdot n}$, for some constant $c > 0$. So, according to our input model, claiming that a given bound holds w.h.p. is equivalent to claim that it holds *for almost all* inputs [2]. Theorem 1 shows that the optimal cost *quickly decreases* in h (even for small, constant values of h). Actually, an efficient, *centralized Divide and Conquer* heuristic h-PARTY (see Figure 2) is introduced in [9,6] that yields (w.h.p.) a constant approximation ratio.

1.1   Our Results

The asymptotically cost-optimal *divide and conquer* heuristic h-PARTY (see Figure 2) proposed in [9,6] requires global knowledge of the network and *centralized* decisions: at each recursive phase, a suitable grid partition of the d-cube Q is performed. In each element (*cell*) of the grid, the algorithm runs a *leader election* task in order to select a *base cell*. A cell base, elected in the recursive phase j, is connected to the root and becomes itself *root* for the $(h - j)$-HOPS CONVERGE-CAST problem on the sub-instance restricted to nodes inside the cell. And so on.

The key-issue of the optimality of this algorithm is the size of the cells (and, hence, the number of bases selected at each phase). It should be clear that, the above tasks are unfeasible (or extremely expensive) in our distributed model: once a (unique) base cell has been selected, all the other nodes of the cell must agree about that and must know its label and position.

We thus propose a distributed protocol that combines the grid partitioning and the leader election of h-PARTY with the greedy approach. Our protocol h-PROT "simulates" the "optimal" strategy of h-PARTY (based on grid partition and leader election), by using local *independent random choices*: every node (but the root) decides, *independently*, to be a leader (i.e. a base) with probability p. Then, we use a *greedy* approach in order to establish nodes-to-leader connections. The choice of parameter p depends on n (and h) and it is a key-ingredient in the quality of the returned solution.

We consider the (synchronous) ad-hoc network model in the following *distributed* fashion: at the starting time, every station knows only its label, its geographical position and the parameter L (i.e.a "good" bound on the diameter of region where nodes are located in). Furthermore, no (wireless or not) link exists before that time. All stations are in the quiescent status but the b: b starts the protocol by broadcasting a *start* message with 1-hop transmission of range L (as we will see, the cost of this root transmission is negligible w.r.t. the overall protocol and solution costs). We focus on the *global energy-cost* spent by the protocol and on the energy cost of the computed range assignment. So, we will not consider *interference* and *synchronization* problems [21]: we assume they will be eventually solved by using some of the techniques previously introduced in the literature [4,22,24].

Our main result is expressed by the following theorem.

Theorem 2. *Let S be a random set of n points in a 2-dim square Q of side length L and let* range^{Prot} *be the range assignment returned by 2-PROT(S, b). Then, for $h = 2$, the expected cost of* range^{Prot} *satisfies the following bounds*

$$\mathbf{E}\left(\text{cost}(\text{range}^{Prot})\right) = O\left(L^{\alpha} n^{\frac{2}{\alpha+2}}\right).$$

By comparing Theorem 1 with Theorem 2 we thus have that for $\alpha = 1$ and $\alpha = 2$, h-PROT achieves a constant expected approximation ratio.

We emphasize that the expected cost analysis of our protocol departs significantly from that of 2-PARTY, mainly because of two reasons. Firstly, our protocol makes no cell partition (it cannot!) and so it cannot guarantee that exactly one leader per cell will be selected. The analysis needs to be *amortized* and based on the expected *good* leader distribution on Q. On the other hand, we need to deal with *two* probabilistic distributions: the input one and the one determining our leader selection. The resulting random variables are not mutually independent. We also prove that the overall energy cost spent by the stations during the entire protocol is $\Theta(\text{cost}(\text{range}^{Prot}))$.

We believe that Theorem 2 can be extended to any *constant* h. This question is still open. However, we performed a large number of computer experiments that strongly support our conjecture. In particular, we have compared the performances of the h-PROT with those of the asymptotically-optimal h-PARTY and with those of a prim-based randomized heuristic, named RANDOMIZED h-PRIM, that gives good performance in practice (see [8]). The results of this experimental work are summarized in Figure 3.

Paper's Organization. In Section 2, we present our protocol. In Section 2.1 we prove that its expected approximation ratio is constant for $h = 2$. In Section 2.2 we provide an experimental analysis that compare the proposed protocol with the h-PARTY heuristic. Finally, in Section 3, we briefly address some open problems.

2   Distributed Protocol for the Ad-Hoc Model

We consider h-HOPS CONVERGECAST on the ad-hoc model and we propose a distributed randomized protocol h-PROT that, given a set S and a root b, constructs a feasible range assignment range^{Prot}. The protocol works in *phases* and assumes that, at the starting time, every node knows only its label, its geographical position and L. Each node (station) is equipped by an omni-directional antenna: it is able to change its transmission range. We will consider the case $d = 2$ (i.e. instances on the plane). The details of the protocol executed by station $s \in S$ are described in Figure 1.

Init phase (raws 3–10). The root station b sends a *start* message to all the other stations. All the other stations are waiting for this message.

Phase j (raws 13–28). If $j < h$, all the non-connected stations flip a bit x with $\mathbf{Pr}\,[x = 1] = f(n, h, j, \alpha)$. If $j = h$ or $x = 1$ each non-connected station s sends a *search* ("search:s") message (containing its label and its coordinates) at increasing range r (raw 23). Station s stops in sending messages as soon as it receives an *echo* message from the closest connected node v_s ("echo:v_s") at level $j-1$. Node s chooses v_s as its father by fixing range$^{Prot}(s)$ $= d(s, v_s)$. Finally s becomes connected and sets its level to j(raws 18–22). In the meantime, all the connected stations at level $j - 1$ are waiting for messages from non-connected stations. If a *search* message from a station v is received ("search:v") then, if s is the closest connected node to v, it sends an *echo* message to v containing its label and its coordinates (raws 25–28). Notice that, this last step can be easily performed by using the connection between the connected nodes.

The selection probability $\mathbf{Pr}\,[x = 1]$ is defined as

$$f(h, j, n, \alpha) = n^{-\lambda(h,j,\alpha)} \text{ where } \lambda(h, j, \alpha) = \frac{\sum_{i=1}^{h-j-1}(2/\alpha)^i}{\sum_{i=1}^{h-j}(2/\alpha)^i}.$$

```
01: h-Prot(S,b)
02: begin
03:    if (s = b) then begin
04:       send("start",L√2̄);
05:       connected := true;
06:       level := 0;
07:    end else begin
08:       wait("start");
09:       connected := false;
10:    end
11:
12:    for j = 1, ..., h do begin
13:       if (connected = false) then
14:          if ( j ≤ h − 2 ) then
15:             randomly choose a bit x with Pr[x = 1] = f(n,h,j,α);
16:          else x = 1;
17:       for r = 1, 2, ..., 2^ℓ, 2^⌈ log⌈ L √2̄ ⌉ ⌉ do
18:          if (connected = false and x = 1) then
19:             if (received("echo:v_s") = true) then begin
20:                range^Prot(s) := d(s,v_s);
21:                connected := true;
22:                level = j;
23:             end else send("search:s", r);
24:          else
25:             if (level = j − 1) then
26:                if (received("search:v") = true) then
27:                   if ( d(s,v) = min{ d( s̃,v ) : s̃ received "search:v" }) then
28:                      send("echo:s",d(s,v));
29: end
```

Fig. 1. The protocol h-PROT executed by station $s \in S$

This function has been inspired by the almost "optimal" centralized heuristic h-PARTY: The expected number of selected nodes during phase j equals the number of bases selected during the j-th recursive step of h-PARTY.

2.1   Probabilistic Analysis

Theorem 3 (Energy cost of the protocol). *Let $\widehat{r}(s)$ be the maximal transmission range used by $s \in S$ during any phase of the protocol. Then, it holds that*

$$\mathrm{cost}(\widehat{r}) = \sum_{s \in S}(\widehat{r}(s))^\alpha = \Theta(\mathrm{cost}(\mathrm{range}^{Prot}))$$

Proof. When a node s given in a Phase j is selected then its maximal range $\widehat{r}(s)$ will be no larger than twice its final range $\mathrm{range}^{Prot}(s)$. Furthermore, when a node $s \in S$ has been selected in Phase $j - 1$ (so it acts like a potential father), it will send *echo* and *father* messages, during Phase j, at ranges equal to those of the corresponding sons.[1] Let $\widehat{x}_s$ be the son of s at maximal distance in the final transmission graph. We have that

$$\sum_{s\in S}(\widehat{r}(s))^\alpha \leq 2\sum_{s\in S}(\max\{\text{range}^{Prot}(s), \text{range}^{Prot}(x_s)\})^\alpha \leq$$

$$\leq 4\sum_{s\in S}\text{range}^{Prot}(s)^\alpha = \Theta(\text{cost}(\text{range}^{Prot})). \qquad \square$$

Theorem 4 (Energy cost of the protocol solution). *Energy cost of the protocol. Let S be a random instance of n nodes selected from a square Q of edge size L and let $b \in S$. Then, for $h = 2$, the expected cost of range^{Prot} satisfies the following bounds*

$$\mathbf{E}\left(\text{cost}(\text{range}^{Prot})\right) = O\left(L^\alpha n^{\frac{2}{\alpha+2}}\right).$$

Proof. Without loss of generality, assume that the root b is the node of index n. We denote any node selected in the first phase as *base* and define, for any node i, D_i as the minimal distance between node i and the root and any base. We denote as p_b the probability that a node becomes a base. As for $h = 2$, this probability is equal to

$$f(n, 2, 0, \alpha) = n^{-\lambda(2,0,\alpha)} = n^{-\frac{\alpha}{\alpha+2}}.$$

It holds that,

$$\mathbf{E}\left(\text{cost}(\text{range}^{Prot})\right) \leq \sum_{i=1}^{n-1}[p_b(\sqrt{2}L)^\alpha + (1-p_b)\overline{\mathbf{E}}\left((D_i)^\alpha\right)]$$

$$\leq (n-1)(\sqrt{2}L)^\alpha p_b + \sum_{i=1}^{n-1}\overline{\mathbf{E}}\left((D_i)^\alpha\right). \qquad (3)$$

Where $\overline{\mathbf{E}}\left((D_i)^\alpha\right)$ denotes $\mathbf{E}\left((D_i)^\alpha\mid i \text{ is not a base }\right)$. In order to evaluate the value $\overline{\mathbf{E}}\left((D_i)^\alpha\right)$ we define

$$\Delta = \frac{L}{n^{\frac{1}{\alpha+2}}}$$

Then

$$\overline{\mathbf{E}}\left((D_i)^\alpha\right) \leq \overline{\mathbf{Pr}}\left[D_i < \Delta\right]\Delta^\alpha + \sum_{k=0}^{\infty}\overline{\mathbf{Pr}}\left[2^k\Delta \leq D_i < 2^{k+1}\Delta\right](2^{k+1}\Delta)^\alpha$$

$$\leq \Delta^\alpha\left[1 + 2^\alpha\sum_{k=0}^{\infty}\overline{\mathbf{Pr}}\left[2^k\Delta \leq D_i < 2^{k+1}\Delta\right]2^{k\alpha}\right]$$

---

[1] Observe that, in this case, we can assume that all such nodes (potential father) known the partial solution constructed so far. So, only the real father will send echo message to a son.

$$\leq \Delta^\alpha \left[1 + 2^\alpha \sum_{k=0}^{\infty} \left[\overline{\mathbf{Pr}} \left[D_i \geq 2^k \Delta \right] - \overline{\mathbf{Pr}} \left[D_i \geq 2^{k+1} \Delta \right] \right] 2^{k\alpha} \right]$$

$$\leq \Delta^\alpha \left[1 + 2^\alpha \sum_{k=0}^{\infty} \overline{\mathbf{Pr}} \left[D_i \geq 2^k \Delta \right] 2^{k\alpha} \right],$$

where $\overline{\mathbf{Pr}} \left[X \right]$ denotes $\mathbf{Pr} \left[X \mid i \text{ is not a base} \right]$. Observe that if $2^k \Delta \geq \sqrt{2}L$ then $\overline{\mathbf{Pr}} \left[D_i \geq 2^k \Delta \right] = 0$. So

$$\mathbf{E} \left((D_i)^\alpha \right) \leq \Delta^\alpha \left[1 + 2^\alpha \sum_{k=0}^{\lceil \log(\sqrt{2}L/\Delta) \rceil} \overline{\mathbf{Pr}} \left[D_i \geq 2^k \Delta \right] 2^{k\alpha} \right] \qquad (4)$$

We now need an upper bound for the probability that D_i is larger than a given parameter ρ, when $\rho \leq 2\sqrt{2}L$. It holds that,

$$\overline{\mathbf{Pr}} \left[D_i \geq \rho \right] = \left(1 - \frac{A_{i,\rho}}{L^2} \right) \sum_{j=0}^{n-2} \mathbf{Pr} \left[N_{i,\rho} = j \right] (1 - \mathbf{p_b})^j \qquad (5)$$

where $N_{i,\rho}$ is the number of nodes, different from i and the root, falling into the disk of radius ρ and centered at node i, while $A_{i,\rho}$ is the area of the intersection between the disk of radius ρ and centered at node i and the square Q. Then, it holds that

$$\mathbf{Pr} \left[N_{i,\rho} = j \right] = \binom{n-2}{j} \left(\frac{A_{i,\rho}}{L^2} \right)^j \left(1 - \frac{A_{i,\rho}}{L^2} \right)^{n-2-j} \qquad (6)$$

By combining (5) and (6), we get

$$\overline{\mathbf{Pr}} \left[D_i \geq \rho \right] = \left(1 - \frac{A_{i,\rho}}{L^2} \right) \sum_{j=0}^{n-2} \binom{n-2}{j} \left(\frac{A_{i,\rho}}{L^2} \right)^j \left(1 - \frac{A_{i,\rho}}{L^2} \right)^{n-2-j} (1 - \mathbf{p_b})^j$$

$$= \left(1 - \frac{A_{i,\rho}}{L^2} \right) \left(1 - \mathbf{p_b} \frac{A_{i,\rho}}{L^2} \right)^{n-2}$$

$$\leq \left(1 - \mathbf{p_b} \frac{\rho^2}{8L^2} \right)^{n-1}$$

where, in the last inequality, we use the fact that $A_{i,\rho} \geq \rho^2/8$ when $\rho \leq 2\sqrt{2}L$. By combining the last inequality with (4), we obtain

$$\mathbf{E} \left((D_i)^\alpha \right) \leq \Delta^\alpha \left[1 + 2^\alpha \sum_{k=0}^{\lceil \log(\sqrt{2}L/\Delta) \rceil} \left(1 - \mathbf{p_b} \frac{(2^k \Delta)^2}{8L^2} \right)^{n-1} 2^{k\alpha} \right].$$

By replacing the values of $\mathbf{p_b}$ and Δ, we get

$$\mathbf{E} \left((D_i)^\alpha \right) \leq \frac{L^\alpha}{n^{\frac{\alpha}{\alpha+2}}} \left[1 + 2^\alpha \sum_{k=0}^{\lceil \log(\sqrt{2}n^{\frac{1}{\alpha+2}}) \rceil} \left(1 - \frac{4^k}{8n} \right)^{n-1} 2^{k\alpha} \right]$$

$$\leq \frac{L^\alpha}{n^{\frac{\alpha}{\alpha+2}}} \left[1 + 2^\alpha \sum_{k=0}^{\lceil \log(\sqrt{2}n^{\frac{1}{\alpha+2}}) \rceil} e^{-\frac{4^k}{8n}(n-1)} 2^{k\alpha} \right]$$

$$\leq \frac{L^\alpha}{n^{\frac{\alpha}{\alpha+2}}} \left[1 + 2^\alpha \sum_{k=0}^{\lceil \log(\sqrt{2}n^{\frac{1}{\alpha+2}}) \rceil} e^{-\frac{4^k}{16}} 2^{k\alpha} \right].$$

We observe that there exists a constant $c = c(\alpha)$ (remind that α is a constant) such that

$$1 + 2^\alpha \sum_{k=0}^{\lceil \log(\sqrt{2}n^{\frac{1}{\alpha+2}}) \rceil} e^{-\frac{4^k}{16}} 2^{k\alpha} \leq c.$$

It thus follows that

$$\overline{\mathbf{E}}\left((D_i)^\alpha\right) \leq c \frac{L^\alpha}{n^{\frac{\alpha}{\alpha+2}}}.$$

By replacing this bound in (3), we get

$$\mathbf{E}\left(\text{cost}(\text{range}^{Prot})\right) \leq n(\sqrt{2}L)^\alpha p_b + \sum_{i=1}^{n} \overline{\mathbf{E}}\left((D_i)^\alpha\right)$$

$$\leq (\sqrt{2}L)^\alpha n^{\frac{2}{\alpha+2}} + cn \frac{L^\alpha}{n^{\frac{\alpha}{\alpha+2}}}$$

$$\leq cL^\alpha n^{\frac{2}{\alpha+2}}$$

$$= O\left(L^\alpha n^{\frac{2}{\alpha+2}}\right). \qquad \square$$

By comparing Theorem 4 with Theorem 1, we can state the following

Corollary 1. *The expected approximation factor yielded by* 2-PROT(S, b) *on random instances, for $h = 2$ and $\alpha = 1, 2$, is bounded by a constant.*

2.2   Experimental Evaluations

As mentioned in the Introduction, we believe that Theorem 3 holds for any *constant* $h \geq 1$. However, till now, we have not been able to extend our theoretical analysis to $h > 2$. The goal of this section is thus providing some experimental evidence of our conjecture by considering the cases $h = 2, \ldots, 8$ and $\alpha = 1$.

We compare the costs of the solutions generated by h-PROT with those generated by the centralized heuristics h-PARTY (see Figure 2 for a description) and with those of a prim-based randomized heuristic, named RANDOMIZED h-PRIM, that gives good performance in practice (see [8]). For practical reasons, we have implemented a variant of the protocol in which: (*i*) the selected node(s) of phase j may be connected to any already connected node (though not belonging to level $j - 1$); (*ii*) the probability according to which a node is selected as a base

procedure h-PARTY(h, V, p)
    if $h = 1$ **then** $T \leftarrow \{\{x, p\} | x \in V - \{p\}\}$;
    else begin
        $k \leftarrow \left\lfloor |V|^{\eta_\alpha(h)} \right\rfloor$; $T \leftarrow \emptyset$;
        Let l be the side length of the smallest square
            containing all points in V;
        Partition the square into a grid of square cells
            of side length $\frac{l}{\lfloor \sqrt{k} \rfloor}$;
        Let k' be the number of cells and let V_i be the
            points of V in the i-th cell, $1 \leq i \leq k'$;
        for $i \leftarrow 1$ **to** k' **do**
            if $|V_i| \geq 1$ **then begin**
                $a \leftarrow$ a random point in V_i;
                $T \leftarrow T \cup \{\{a, p\}\}$;
                if $|V_i| > 1$ **then**
                      $T \leftarrow T \cup h$-PARTY$(h - 1, V_i, a)$;
        end;
    end;
    output T

Fig. 2. The h-PARTY heuristic

is scaled by a factor f $(0.009 \leq f \leq 1)$ with respect to that adopted in the theoretical protocol.

All the experiments are carried out for several sizes n of the random instances (between 100 and 10,000) and for $h = 2, 3, \ldots, 8$. Node positions of any instance are chosen independently and uniformly at random in a square of side length 1. For each n and h, the number of instances for each n decreases from 100 to 50 as n grows. Moreover, for every instance five runs are executed (only for randomized algorithms with randomness). Then, we get the average value (on these runs) for a comparison among the algorithms.

As we can see from the Tables in Figure 3, h-PROT has performances equivalent to h-PARTY. This is not surprising since the selection probability in h-PROT is defined so as to *simulate* h-PARTY. So, it works very well for all $h \leq 5$.

Another important contribution of our experimental work is that of providing a "good" strategy for tuning the selection probability of h-PROT in order to overcome h-PARTY and reach RANDOMIZED h-PRIM's performances when h is large. The strategy relies on the following basic fact. RANDOMIZED h-PRIM works better when h is large since it generates much less *bases* (here, one may see a base as a node with a large transmission range) than h-PARTY: RANDOMIZED h-PRIM indeed chooses *just one* base per phase. The strategy is thus to reduce the selection probability f in order to get closer to RANDOMIZED h-PRIM's behavior as h grows. In Figure 3g, we can see the outperforming of our protocol with $f = 0.009$ and $h = 8$ with respect to both h-PARTY and RANDOMIZED h-PRIM and all implementations of h-PROT. Figure 3 indeed shows that, in order to improve the performance of h-PROT, f should decrease as h increases.

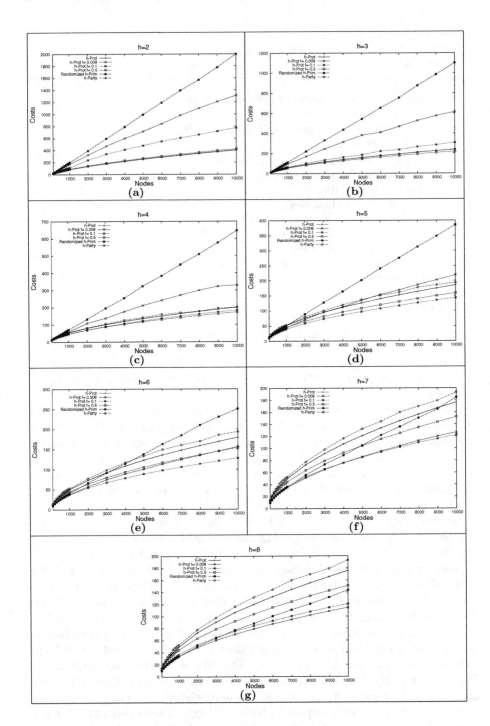

Fig. 3. Experimental results

3 Open Problems

The most important issue to be addressed is the probabilistic analysis of our protocol for $h \geq 3$. We believe that, for constant values of h, it is possible to prove that its expected approximation ratio is constant.

References

1. L. Alfandari and V.T. Paschos. Approximating minimum spanning tree of depth 2. *Intl. Trans. In Op. Res.* 6:607-622, 1999.
2. N. Alon and J. Spencer, *The probabilistic method.* Wiley, 1992.
3. S. Arora, P. Raghavan, and S. Rao. Approximation schemes for Euclidean k-medians and related problems. *Proc. 30-th ACM Symposium on Theory of Computing* 106-113, 1998.
4. B. Chlebus, L. Gasienec, A. Gibbons, A. Pelc, and W. Rytter. Deterministic broadcasting in unknown radio networks. *Proc. of 11th ACM SODA*, 2000.
5. F.A. Chudak. Improved approximation algorithms for uncapacitated facility location problem. *Proc. of the 6-th Conference on Integer Programming and Combinatorial Optimization* 1998.
6. A.E.F. Clementi, P. Penna, and R. Silvestri. On the Power Assignment Problem in Radio Networks *Mobile Networks and Applications (MONET)*, 9:125–140, 2004.
7. A.E.F. Clementi, G. Huiban, P. Penna, G. Rossi, and Y. Verhoeven, Some Recent Theoretical Advances and Open Questions on Energy Consumption in Static Ad-Hoc in Wireless Networks. *Proc. of the 3th Workshop ARACNE*, Carleton Scientific, 2002.
8. A.E.F. Clementi, M. Di Ianni, A. Monti, G. Rossi, and R. Silvestri, Experimental Analysis of Practically Efficient Algorithms for Bounded-Hop Accumulation in Ad-Hoc Wireless Networks, In *Proc. of the IEEE IPDPS-WMAN*, 2005.
9. A.E.F. Clementi, M. Di Ianni, M. Lauria, A. Monti, G. Rossi, and R. Silvestri, Divide and Conquer is almost optimal for the bounded-hop MST problem on random Euclidean instances, In *Proc. of SIROCCO*, 2005.
10. A. Ephremides, G.D. Nguyen, and J.E. Wieselthier. On the Construction of Energy-Efficient Broadcast and Multicast Trees in Wireless Networks. In *Proc. of the 19th INFOCOM*, 585–594, 2000.
11. L. Gouveia. Using the Miller-Tucker-Zemlin constraints to formulate a minimal spanning tree problem with hop constraints. *Computers and Operations Research* 22: 959-970, 1995.
12. L. Gouveia. Multicommodity flow models for spanning trees with hop constraints. *European Journal of Operational Research* 95:178-190, 1996.
13. S. Guha and S. Khuller. Greedy strikes back: Improved facility location algorithms. *Journal of Algorithms* 31:228-248, 1999.
14. L. Gouveia and C. Requejo. A new relaxation approach for the hop-constrain minimum spanning tree problem. *European Journal of Operational Research* 132:539-552, 2001.
15. A. Kesselman and D.R. Kowalski. Fast Distributed Algorithm for Convergecast in Ad Hoc Geometric Radio Networks. *Proc. of the 2nd International Conference on Wireless on Demand Network Systems and Service* 119-124, 2005.
16. L. M. Kirousis, E. Kranakis, D. Krizanc, and A. Pelc. Power Consumption in Packet Radio Networks. *Theoretical Computer Science*, 243:289–305, 2000.

17. G. Kortsarz and D. Peleg. Approximating the weight of shallow Steiner trees. *Discrete Applied Mathematics* 93:265-285, 1999.
18. M.R. Korupolu, C.G. Plaxton and R. Rajaraman. Analysis of a Local Search Heuristic for Facility Location Problems. *Proc. of the 9-th annual ACM-SIAM symposium on Discrete algorithms* 1-10, 1998.
19. E.L. Lloyd, R.Liu, M.V. Marathe, R. Ramanathan, S. S. Ravi. Algorithmic Aspects of Topology Control Problems for Ad Hoc Networks. *Mobile Networks and Applications* 10:19-34, 2005.
20. M. Mahdian, Y. Ye and J. Zhang, A 1.52-approximation algorithm for the uncapacitated facility location problem. *Proc. of APPROX 2002* LNCS 2462: 229-242, 2002.
21. K. Pahlavan and A. Levesque. *Wireless Information Networks*. Wiley-Interscience, 1995.
22. S. Ramanathan and E. Lloyd. Scheduling Broadcasts in multi-hop radio networks. *IEEE/ACM Trans. on Networking*, 1, 1993.
23. R. Ramanathan and R. Hain. Topology Control of Multihop Wireless Networks Using Transmit Power Adjustment. *Proc of IEEE INFOCOM* 404-413, 2000.
24. R. Ramaswami and K. Parhi, Distributed scheduling of broadcasts in radio networks. *Proc of IEEE INFOCOM*, 1989.
25. G. R. Raidl and B. A. Julstrom, Greedy Heuristics and an Evolutionary Algorithm for the Bounded-Diameter Minimum Spanning Tree Problem. *Proc. of the 2003 ACM symposium on Applied computing* 747-752, 2003.
26. D.B. Shmoys, E. Tardos and K. Aardal, Approximation algorithms for facility location problems. *Proc. of the 29-th Annual ACM Symposium on Theory of Computing (STOC)* 265-274, 1997.
27. S. Voss, The steiner tree problem with hop constraint, *Annals of Operations Research* 86:321-345, 1999.

Discrete Optimization Models for Cooperative Communication in Ad Hoc Networks

Carlos A.S. Oliveira

School of Industrial Engineering and Management,
Oklahoma State University,
322 Engineering North,
Stillwater, OK USA 74078
coliv@okstate.edu
http://www.okstate.edu/ceat/iem/iepeople/oliveira

Abstract. Ad hoc networks have recently been used as a communications medium on tasks that require high levels of communication and coordination. We study the problem of determining an optimal route for a group of wireless units, such that the total connection time among nodes in the resulting mobile ad hoc network is maximized, subject to a limit on the traveled distance. This problem, called the cooperative communication problem in mobile ad hoc networks (CCPM), is modeled using combinatorial optimization and mathematical programming techniques. Applications occur on the coordination of rescue groups, geographical exploration and recognition, unmanned air vehicles (UAVs), and mission coordination. The problem is shown to be NP-hard, and a dynamic programming algorithm is proposed for the problem. Mathematical programming models are presented and computational experiments performed.

Keywords: Optimization, cooperative control, integer programming, ad hoc networks, unit graphs.

1 Introduction

Mobile ad hoc networks have been developed in the last few years to provide improved communication infrastructure for wireless mobile devices. Such networks consist of loosely coupled units that communicate locally only to accessible neighbors. Routing of data among non accessible units in an ad hoc network is made possible through multi-hop retransmission [10]. Ad hoc systems have been used in situations where it is difficult or expensive to provide coverage using dedicated servers. Thus, it differs fundamentally from the communication techniques used in other mobile systems, such as cellular networks and other types of fixed coverage wireless systems.

An ad hoc network is based on wireless units that can act as clients as well as servers. Lacking a fixed network structure such as the one available for mobile phone systems, units participating in an ad hoc system must connect to neighbor

T. Kunz and S.S. Ravi (Eds.): ADHOC-NOW 2006, LNCS 4104, pp. 73–86, 2006.

wireless units in order to deliver information. The distribution and routing techniques used depend on dynamic protocols that discover information about the current network topology and use this information to deliver data [13]. Therefore, connectedness in this model of communication depends both on temporal and spacial components, determined by the current location of nodes.

Ad hoc networks have important applications, since they can be used to model real situations, where devices are not controlled by a central authority [16]. They are usually applied in very dynamic scenarios, where hosts may change positions, and communication is not guaranteed to reach a specific destination. Such applications range from commercial wireless access on urban locations to rescue forces and operations on military battlefield [2]. All these applications have in common the role of users in the dynamical state of the network.

We are concerned with the study of connectedness for an ad hoc network system from the point of view of a group of users that must cooperate to accomplish assigned tasks. With this objective we study a new problem, called the *cooperative communications problem*, which can be used to model the decisions of a set of wireless agents that need to maintain communication for as long as possible. In this paper we propose mathematical programming models and algorithms to solve this combinatorial problem.

1.1   Related Work

Group communication and synchronization is required in many situations where group members have specific tasks to accomplish. This type of problem occurs for example in the deployment of communications for rescue groups in military missions [17]. Similarly, task forces may use ad hoc services to maintain collaboration in distributed projects. In such applications, a group of users may have different but related targets, and coordination among participating agents may be crucial for the accomplishment of goals. Communication between group elements, being an important type of resource, must be maintained for as long as possible during the operation of the group.

Maximizing communication time is an important goal in any ad hoc network system, since one of the main objectives of an organization maintaining such systems is to keep the network connected for as long as possible. At the same time, users in any network can be viewed as having an initial position and final destination. Thus our results have general applicability in several types of ad hoc systems. Beyond the straightforward application of the study presented in this paper, one can apply these results to increase connectivity through the addition of members to an existing ad-hoc network. In this case, our results may be useful in defining the trajectory of these additional wireless units.

Much work has been done to create routing strategies for wireless devices on ad hoc networks [5,9,6]. Routing is particularly important, since traditional protocols are not able to handle dynamic issues related to mobility. Regarding this objective, [9] and [5] studied general issues related to the multi-hop protocols that can impact the efficient routing of packets in wireless systems.

Concerning optimization, much of the research has concentrated on structural properties of ad hoc networks [1,2,3,6,14]. In the study of dynamic properties of ad hoc network connectivity, a mathematical model of user mobility is important in order to make high level decisions based on actual usage patterns. Various mobility models have being proposed to account for changes in position of hosts in an ad hoc network [4,12]. Camp et al. [4] provided a survey of current mobility modes, with detailed comparison. An interesting study of the effects of mobility in ad hoc usage patterns was further provided by Rajan [12]. We discuss in this paper a special case of a mobility model encoded into a set of discrete steps, and formalized as a sequence of planar graphs. A previous discussion of a graph-based mobility model is presented by Tian et al. [15].

We present a mathematical formulation for the problem of maximizing the communication time for a group of clients with specified targets, given as a set of sources and destinations. The constraints of the problem require that the total distance traveled do not exceed some specified threshold. The paper is organized as follows. In Section 2, a formal definition of the cooperative communication problem on ad hoc networks (CCPM) is given. In Section 3 we show that the problem is NP-hard. An algorithm for the CCPM using a dynamic programming strategy is presented in Section 4. In Section 5, an integer programming model for the problem is introduced, along with some useful properties of the CCPM. This model will be important as a formalization tool as well as to provide upper bounds on the optimal solution of some instances of the CCPM. The results of computational experiments with the model are shown in Section 6. Finally, concluding remarks are given on Section 7.

2   The Cooperative Communication Problem

A model for the problem of maximizing connectivity for a set of wireless network units traveling from a set of sources to a set of destinations is proposed in this section. Initially we describe some useful concepts. Then, we discuss simplifying assumptions that will be made about the problem. A formal definition is provided at the end of this session.

2.1   Notation and Assumptions

Standard graph theory notation is used in this paper. The graphs employed represent location points, and therefore they are *planar*, that is, the set of nodes represent points in the plane. Let $d : V \times V \to \mathbb{R}$ be a function returning the *distance* between a pair of adjacent nodes. Then, a *unit graph* is a graph $G = (V, E)$ such that $d(v, w) \leq 1$, for $v, w \in V$. Unit graphs are used to model ad hoc networks in general. In this type of model, nodes represent wireless units and edges represent connectivity links between nodes. Assuming, without loss of generality, that the radius of transmission is equal to one unit of distance, a unit graph is a representation of the set of links between wireless units in the network. As additional notation, function $N : V \to 2^N$ returns the set of nodes adjacent to v, function $\hat{N} : V \to 2^N$ is defined as $\hat{N}(v) = N(v) \cup \{v\}$, and

$\delta : V \rightarrow N$ returns the degree of node v, i.e., $\delta(v) = |N(v)|$, for all $v \in V$. Also, let $\overline{n}$, $\overline{N}$, and $\overline{T}$ represent, respectively, the sets $\{1, \ldots, n\}$, $\{1, \ldots, |V(G)|\}$, and $\{1, \ldots, T\}$.

We make some simplifying assumptions regarding the type of communication problem that will be considered. The importance of these assumptions is that they will result in a subset of the general problem that is tractable using combinatorial optimization techniques. The main simplifying assumption used in this paper is that the area covered by the set of objects can be modeled as a planar graph $G = (V, E)$. Edges of the graph have the associated function $d : E \rightarrow \mathbb{R}_+$, which returns the distance between a pair of nodes in E.

A second assumption is that, in the graph G given as input, nodes represent the possible positions and edges exist only between nodes that can be reached by wireless units in one unit of time. This means in particular that all objects travel on a given edge at the same speed. As an example where this assumption is reasonable, consider the graph representing a road system, where objects are cars traveling at the maximum allowed speed. In this case, the distance covered in one unit of time is the same for all units traveling at a specific edge. A consequence of the assumption is that positions in the system change only in discrete units of time, each node in the graph representing a possible position. Moreover, at each node there is just a limited number of possible directions that can be taken. This is represented in the model graph by the fixed degree (number of neighbors) of each node.

The trajectory of wireless units is described in the following way. Let $U = \{u_1, \ldots, u_n\}$ be a set of mobile units, together with an origin s_i and a destination d_i for each unit $u_i \in U$. Each unit u_i must start its trajectory at node s_i of the graph and traverse the graph until node d_i is reached. Function $p_t : U \rightarrow V$ returns the position of wireless unit u_i at time t. The total time in the system is divided into T discrete time instants (represented by $\overline{T} = \{1, \ldots, T\}$). If at any particular instant $t \in \overline{T}$ the unit u_i is in node $v = p_t(u_i)$, then u_i can be at the next instant in node v or in any of the nodes adjacent to v. That is, $p_{t+1}(u_i) \in N(p_t(u_i)) \cup \{p_t(u_i)\} = \hat{N}(p_t(u_i))$.

Although these assumptions may seen to be very restrictive, notice that by increasing the number of nodes and edges in the graph one can simulate a continuous instance of the problem to any desired precision. For example, if the maximum error allowed in the position measurement is ϵ, one can create a graph with nodes separated by less a distance less than ϵ and achieve the required precision.

2.2   Problem Definition

As mobile units have a limited amount of available power [11], complete connectivity is difficult to achieve in most cases. Thus, we need to define a measure of connectivity in graphs resulting from the trajectories of the considered objects. Let $r : V \rightarrow \{0, 1\}$ be a function returning 1 if and only if the distance between nodes i and j is at most 1. The measure of connectivity used in this paper is given by the weighted sum of links established in the current network, i.e.,

$$\gamma_w(t) = \sum_{u,v \in U} w(p_t(u), p_t(v)) \, r(p_t(u), p_t(v)), \tag{1}$$

where $w : V \times V \to \mathbb{R}$ is a cost function returning the weight (or relative importance) of the link between a pair of nodes $a, b \in V$.

Given a maximum completion time $T \in \mathbb{N}$, the objective of the CCPM problem is to maximize connectivity defined in equation (1), for all periods of time:

$$\max \sum_{t=1}^{T} \gamma_w(t) = \sum_{t=1}^{T} \sum_{u,v \in U} w(p_t(u), p_t(v)) \, r(p_t(u), p_t(v)).$$

We also require that the total distance traveled by unit u_i be at most D_i, due to physical constraints (such as, for example, total amount of available fuel). Clearly we must have $D_i \geq d^*(s_i, d_i)$ for all $i \in \overline{n}$, where $d^*(s_i, d_i)$ represents the cost of the shortest path from s_i to d_i.

3   Computational Complexity of the CCPM

The CCPM problem requires the determination of multiple sequences of nodes among the pairs of source-destinations while maximizing connectivity of the ad hoc network over the time intervals $[1 \ldots, T]$. In this section we investigate the computational complexity of this problem. Initially, we show that the CCPM is hard to solve exactly.

Theorem 1. *The CCPM is NP-hard.*

Proof. Our proof consists of a reduction from 3SAT (which is well known to be NP-hard [8]) to the CCPM. An instance of 3SAT is a logical formula representing the conjunction of m clauses over n variables, where each clause is the disjunction of exactly three literals (a literal is either a variable or its negation).

Given an instance of 3SAT, we design an instance of the CCPM such that there is a truth assignment satisfying all clauses of the 3SAT instance if and only if there is a solution for the CCPM instance with a specific objective value. The construction of the CCPM instance is as follows. Let U be the set of mobile units. Each unit will either represent one of the n variables or be a separate unit u_0 that is used to define if a clause is satisfied or not. Therefore, we have $n + 1$ wireless units in the resulting CCPM problem.

The graph G of the CCPM instance is formed by requiring that each unit describe a trajectory from its source to destination with $m + 1$ stages and in $T = 3(m + 1)$ units of time (we can assume that $D_i = T$ for all $i \in \overline{n}$, such that it does not become a restricting factor). We describe in the following paragraphs the steps defined by the graph and their interpretation in the final solution. Initially, the trajectory of unit $u_i \in U$ will define if variable x_i is true or false. This is done by forcing the i-th unit to decide between two alternative paths in the first stage of the process, which is shown in Figure 1.

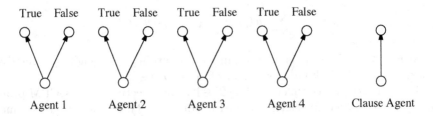

Fig. 1. First stage of the CCPM instance of Theorem 1

After the truth assignment is defined in this way, there will be one stage for each of the m clauses in the 3SAT instance. The true/false nodes from the initial stage lead to independent paths linking sources to destinations. Each of the next stages is formed by $6n + 8$ nodes, where $6n$ are used by the source–destination paths and represent the occurrence of literals in the clause, and 8 nodes are used to define if the corresponding clause is satisfied or not. The part of the graph representing the literals in stage i is composed of three arrays of $2n$ nodes each, one for each possible literal (the three rows are necessary to ensure that all wireless units will advance exactly one edge per time unit). Nodes in the first row are connected to nodes corresponding to the same literal in the third row of the previous stage and to some nodes in the second row of the current stage. The second row serves to connect the literals that *do not appear* in the current clause. Literals *appearing* in the current clause are linked from nodes in the first row to one of the three nodes in the separate gadget described bellow. Finally, the last row is connected to nodes representing literals in the second row, and leads to the next stage.

The part of the i-th stage representing the clause is a gadget with 8 nodes, $z_1, \ldots, z_8$. Node z_1 links to node z_8 in the previous stage of the graph and to z_2, z_3, and z_4 in the current stage. Nodes z_5, z_6, and z_7 link to the exactly three literals appearing in the clause (in the first and third rows described above). Nodes z_2, z_3, and z_4 link to the last node of the gadget (z_8). Finally, node z_8 links to the node z_1 in the next stage $i + 1$ (for $i < T$).

As seen in Figure 2, at the i-th stage, the clause unit (u_0) can make one of three choices: meeting the unit that is at the first, second, or third positions in clause i. It is important to stress that all nodes, except the three pairs of nodes (z_2, z_5), (z_3, z_6) and (z_4, z_7) in each stage, are separated by a distance greater than 1. This will constrain the number of wireless connections to at most one at each stage. If the clause is satisfied by the truth assignment defined in the first stage, then the connection will corresponding one of the literals satisfied in that clause. Clearly, the only way of improving the objective cost is to have a connection made at the clause unit for exactly one of the three pairs of nodes described above.

Finally, the last stage of the construction simply has $n + 1$ destinations. The destinations corresponding to variables receive edges from the nodes representing literals for that variable in the third row of the previous stage. The destination for the special clause unit u_0 receives an edge from node z_8 in the previous stage.

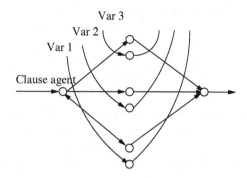

Fig. 2. Subgraph representing the satisfiability of the i-th clause

The construction above makes clear the fact that an instance of 3SAT is satisfiable if and only if the corresponding CCPM can be solved to a maximum objective function of m (equal to the number of clauses). Therefore, 3SAT reduces to CCPM, and the CCPM is NP-hard as well.

The above result shows that the CCPM cannot be solved efficiently in general. However, even removing the maximum distance requirements will not make the problem much easier, as shown in the following corollary:

Corollary 1. *The CCPM is NP-hard even when the distance threshold D_i is infinity for each wireless unit $u_i \in U$.*

Proof. In this case, there is no restriction in the distance that must be travelled by units, so we simply want to maximize the cost in the given time. The same transformation is still valid in this case, so the result is immediate.

3.1   Computing a Feasible Solution

It is interesting to investigate the conditions for the existence of a feasible solution for the CCPM. Looking at a solution for the problem as a sequence of decisions composed of T time steps, we see that decisions at each step must be feasible, and therefore we can decompose the solution as follows.

Definition 1. *A configuration c is a set of positions for wireless units $u_1, \ldots, u_n$ at a specific time step t, and is given by $p_t(u_1), \ldots, p_t(u_n)$. The set of all feasible configurations at time t is denoted by C_t.*

The following proposition describes an important condition that must be satisfied by any configuration. Let $d^* : V \times V \to \mathbb{N}$ be the minimum distance (in number of edges) between two nodes in G.

Proposition 1. *A configuration is feasible at time t if and only if $d^*(s_i, p_t(u_i)) \le t$, $d^*(p_t(u_i), d_i) \le T - t$, and $d(s_i, p_t(u_i)) + d(p_t(u_i), d_i) \le D_i$.*

Proof. Each wireless unit u_i must have moved from position s_i to the current position $p_t(u_i)$ during the time steps 1 to t. The minimum time required to do this is given by $d^*(s_i, p_t(u_i))$, and therefore a configuration with distance greater than this cannot be feasible. Similarly, unit u_i can reach destination d_i in at least $d^*(p_t(u_i), d_i)$ steps. Thus, $d^*(p_t(u_i), d_i) \leq T - t$. Finally, the total distance travelled is restricted to be at most D_i.

It is also possible to consider the above result for problems where the maximum distance traveled is not a restriction, but only the number of steps. In this case, we only need to consider the minimum distances between s_i and $p_t(u_i)$, and between $p_t(u_i)$ and d_i. This can be stated as follows.

Proposition 2. *If the instance of the CCPM has $D_i = \infty$, then a configuration is feasible at time t if and only if $d^*(s_i, p_t(u_i)) \leq t$, and $d^*(p_t(u_i), d_i) \leq T - t$.*

The proposition above leads to the question of recognizing if a configuration is feasible or not. It turns out that this can be done in polynomial time as shown bellow.

Theorem 2. *Determining if a configuration is feasible can be done in $\mathcal{O}(n|V|^3)$ time.*

Proof. For each position of the wireless unit u_i we need to check if its distance to s_i is at most t and the distance to d_i is at most $T - t$. This can be done in time $\mathcal{O}(|V|^3)$ using Floyd's algorithm for shortest path. The same algorithm can be used to check the maximum distance traveled. This step must be repeated n times, hence the upper bound.

4   A Dynamic Programming Algorithm

The structure of the CCPM is composed of a set of decisions that must be made at each time step, leading to the construction of a complete solution for the problem. This structure makes the problem suitable to be solved using dynamic programming techniques, which divide the problem into sub-stages and compute optimal decisions at each stage.

In a dynamic programming formulation, two concepts are necessary: the stage at which the problem is being solved, and the set of states in each stage of the system. These two decisions will lead to a representation of the solutions of the problem using a recursive formulation.

For the CCPM, a possible interpretation of the stage of the problem is given by the current time period t. Given a configuration c for time t (i.e., $c \in C_t$), let neigh(c) represent all feasible configurations that can occur in the next time step $t + 1$. The decision that must be made at each stage is the choice of the next configuration $c' \in$ neigh(c).

A property of the CCPM that is useful in creating a dynamic programming algorithm is the following.

Lemma 1. *At each time step t of the CCPM problem, the partial cost of the optimal solution starting from configuration c until the final configuration at* *time step T is independent of the particular trajectory followed by wireless units during the time interval $[1, t]$.*

Proof. Given an intermediate configuration c, the next configuration must be in $\text{neigh}(c)$. This new configuration depends only on c, as well as all other configurations generated in the interval $[t + 1, T]$.

For example, at time t we have only to worry about the cost of the feasible configuration $c \in C_{t-1}$ and from there we must compute the best partial solutions starting at c. Using this strategy, one can compute the optimum cost $f_t(c)$ for each of the at most $\binom{N}{n}$ feasible configurations $c \in C_t$ at time step t, where $N = |V|$ and $n = |U|$.

$$f_t(c) = \max_{c' \in \text{neigh}(c)} \{\text{cost}(c) + f_{t+1}(c')\}, \tag{2}$$

where $\text{cost}(c) = \gamma_w(t)$ for the current time t. Therefore, we can now immediately compute $f_T(c) = \text{cost}(c)$ if c is the (unique) feasible final configuration, and $f_T(c) = 0$ otherwise.

For a very simple bound on the number of steps necessary to compute the recurrence above, note that given a configuration $c \in C_t$ defined by $p_t(u_1), \ldots, p_t(u_n)$, the number of adjacent configurations can be at most

$$\prod_{1=1}^{n} (\delta(p_t(u_i)) + 1) \geq (\delta(G) + 1)^n \geq 2^n,$$

where $\delta(G)$ is the minimum degree of a node $v \in V(G)$, and the second inequality follows from the fact that G is connected. Therefore, the worst case results in an exponential number of possibilities, even for one step.

However, the dynamic programming technique can still yield good computational results for small values of n. To see this, note that we just need to compute the values of recurrence (2) for pairs of configurations occurring at times t and $t + 1$. Once the optimum value is found for each of the feasible configurations at the current time step, based on this analysis, the dynamic program can go to step $t - 1$ and repeat the procedure. The general algorithm is described in Figure 3. We can prove the following result about the algorithm.

Theorem 3. *Algorithm DYNAMICPROG computes an optimal solution for the CCPM in time $\mathcal{O}(T(\binom{N}{n}^2 + \binom{N}{n}n^2))$.*

Proof. Lines 4–10 of Algorithm DYNAMICPROG compute the best moves starting at time $t = T$. To do this, it needs to consider all pairs of feasible configurations $c \in C_t$, $c' \in C_{t+1}$, such that $c' \in \text{neigh}(c)$. There are at most $\binom{N}{n}$ feasible configurations in C_t (with $t \in \overline{T}$). Thus, the maximum number of evaluations of (2) is $\mathcal{O}(\binom{N}{n}^2)$. The function $\text{cost}(c)$ can be computed in $\mathcal{O}(n^2)$ time, for each of the $\mathcal{O}(\binom{N}{n})$ configurations. Finally, considering the loop of line 3 the complexity of the whole algorithm becomes $\mathcal{O}(T(\binom{N}{n}^2 + \binom{N}{n}n^2))$.

1   **Algorithm** DYNAMICPROG
2      $c_T \leftarrow$ configuration $p_T(u_i) = d_i$, for $i \in \overline{n}$
3   **for** $t \leftarrow T - 1$ downto 1 **do**
4   **for all** $c \in C_t$, compute $cost(c)$.
5   **for all** configurations $c \in C_t$ and $c' \in C_{t+1}$ **do**
6   **if** $c' \in$ neigh(c) **then**
7   Compute the cost of moving from c to c' using (2)
8   **end**
9   **end for all** $c \in C_t$, store the value of the best move starting at c.
10   **end**
11   Return best move starting from the unique configuration $c = C_1$.

Fig. 3. Dynamic programming algorithm for CCPM

The analysis above shows that the algorithm can be useful to solve instances of the CCPM with a small number of wireless units. We have the following corollary:

Corollary 2. *Algorithm* DYNAMICPROG *is a polynomial time algorithm for* $n = 2$ *and* $n = 3$, *with time complexity* $\mathcal{O}(TN^4)$ *and* $\mathcal{O}(TN^6)$, *respectively.*

Proof. For the case $n = 2$, the time complexity $\mathcal{O}(T(\binom{N}{n}^2 + \binom{N}{n}n^2))$ becomes $\mathcal{O}(TN^4)$, since $\binom{N}{2} = \mathcal{O}(N^2)$. Similarly, $\binom{N}{3} = \mathcal{O}(N^3)$, and for the case $n = 3$ the time complexity becomes $\mathcal{O}(TN^6)$.

5   Mathematical Programming Model

Mathematical programming techniques will be used in this section to study the CCPM problem. Such a formulation will be useful as a way to find exact and/or approximate solutions using standard integer programming solvers. Initially, let us define a set of variables describing the problem. Let $x_{ijt} \in \{0, 1\}$ be a binary variable, defined as $x_{ijt} = 1$ if wireless unit i is in node j at time t, and $x_{ijt} = 0$ otherwise, for $i \in \overline{n}$, $j \in \overline{N}$ and $t \in \overline{T}$. Additionally, consider the variable $y_{ijt} \in \{0, 1\}$ defined as $y_{ijt} = 1$ if nodes i and j are connected at time t (i.e., $d(i, j) \leq R$ and there are wireless units at nodes i and j), and $y_{ijt} = 0$ otherwise, for $i \in \overline{n}$, $j \in \overline{N}$, and $t \in \overline{T}$. Thus, the connectivity of graph G at time t can be represented as

$$\gamma(y, t) = \sum_{i=1}^{N} \sum_{j=1}^{N} w(i, j) y_{ijt},$$

where w is a weight function defined on the edges, as explained in the Section 2. Then, the CCPM can be described as follows:

$$\max \sum_{t=1}^{T} \gamma(y, t) \quad \text{subject to } y \in Y,$$

where Y is a polyhedron describing the possible solutions of the CCPM, and defined by the constraints presented in the following.

The first set of constraints define the binary value of the variable y_{ijt}. This value must be equal to 1 if and only if both nodes v_i and v_j are the current position of some pair of wireless units and these units can communicate, i.e., the distance between them is less than the radius of transmission R:

$$y_{ijt} \leq \sum_{u \in \overline{n}} x_{uit}, \qquad \text{for } i,j \in \hat{N}, \, t \in \overline{T}, \, d(i,j) \leq R \tag{3}$$

$$y_{ijt} \leq \sum_{u \in \overline{n}} x_{ujt} \qquad \text{for } i,j \in \hat{N}, \, t \in \overline{T}, \, d(i,j) \leq R. \tag{4}$$

The next set of restrictions account for the simple rules describing the movement of objects in the system. We assume that each wireless unit must travel to an adjacent node or continue at the same position. The constraint determines that, for each node $i \in \overline{n}$, at least one of the adjacent nodes in the previous graph configuration must be selected:

$$x_{uit} \leq \sum_{j \in N(i)} x_{ujt} \qquad \text{for } u \in \overline{n}, \, i \in \hat{N}, \text{ and } t \in \overline{T}. \tag{5}$$

Moreover, each wireless unit must be in exactly one position:

$$\sum_{i=1}^{N} x_{uit} = 1 \qquad \text{for } u \in \overline{n} \text{ and } t \in \overline{T}. \tag{6}$$

The following constraints give initial and final conditions that must be true for all units in the group

$$x_{ui1} = 1, \qquad \text{for } u \in \overline{n}, \text{ and } i = s_u, \tag{7}$$

$$x_{uiT} = 1, \qquad \text{for } u \in \overline{n}, \text{ and } i = d_u. \tag{8}$$

These constraints simply enforce each unit u to start in node s_u at time step 1, and finish in node d_j in the last step T.

The next set of constraints are related to the limit on distance traveled by objects in the system. To keep track of this information we need to use another binary variable r_{uijt}, for $u \in \overline{n}$, $i,j \in \hat{N}$, and $t \in \overline{T}$. This variable has value $r_{uijt} = 1$ if unit u is going from node v_i to node v_j at time t. These variables are used in the following set of constraints.

$$r_{uijt} \geq (x_{ui(t-1)} + x_{ujt}) - 1, \tag{9}$$

for $u \in \overline{n}$, $i,j \in \hat{N}$, with $i \neq j$, and $t \in \{2 \dots T\}$. This constraint requires that variable r_{uijt} be equal to one only when both nodes i and j (with $i \neq j$) are subsequently visited. Then, we can bound the traveled distance using the constraints

$$\sum_{i,j \in \hat{N}} \sum_{t \in \overline{T}} d(i,j) r_{uijt} \leq D_u \qquad \text{for } u \in \overline{n}. \tag{10}$$

Finally, the last set of constraints determine the type of values allowed for the variables in the problem

$$x_{uit} \in \{0,1\}, \quad y_{ijt} \in \{0,1\}, \quad r_{uijt} \in \{0,1\}, \tag{11}$$

for $u \in \bar{n}$, $i, j \in \hat{N}$, and $t \in \bar{T}$. This results in the following integer programming (IP) formulation for the CCPM, which we refer to as formulation Π:

$$\max \quad \sum_{t=1}^{T} \gamma(y,t)$$

subject to   constraints (3) through (6) and (7) through (11).

Next, we formally state the validity of the formulation given above:

Theorem 4. *The IP formulation Π is a correct formulation for the CCPM.*

Proof. The discussion of each of the constraints (3) through (11) shows why they are necessary to describe the CCPM problem. Conversely, they are sufficient, because if a solution to the IP model is given, we can readily create a solution for the CCPM. Thus, formulation Π correctly models the CCPM problem.

6   Computational Results

Computational experiments were performed with the proposed IP model to determine its effectiveness. In our experiments, a set of instances were generated using parameters based on real values. The instances tested bellow were generated for areas ranging from 10 to 20 miles. The generator is based on previous work done for the minimum size backbone problem [2,3].

The software used for IP modeling was the Mosel XPress package, which has an integrated modeler and LP solver. Integer programs can be solved by calling an embedded branch-and-bound routine, which also applies standard valid inequalities, such as Gomory cuts (see [7]). The experiments were run on a PC equipped with a 2.8GHz processor and 504MB of main memory.

The IP model was generated for instances with 20 to 120 nodes and each group with 5 units. The sources and destinations were chosen randomly from available nodes. The integer programming model was solved, yielding the results summarized in Table 1. The last three instances were stopped after 2 hours of computation, but the integrality gap reported was less than 5%.

To make a comparison possible, we created a simple local search algorithm for the problem. It starts from an initial solution which is equal to the union of shortest paths between sources and destinations. This solution is improved using a 2-opt method for increasing connectivity: if two units cannot communicate, we try to move then to a closer position, provided that the resulting configuration remains feasible. The algorithm was coded in C language, and compiled with the gcc compiler. It was executed in the same instances, and the resulting values reported in the last columns of Table 1.

Table 1. Results of computational experiments

instance	num. of nodes	IP opt	time (s)	LS opt.	time (s)
1	20	517	12	517	1
2	30	721	71	721	4
3	40	769	59	758	5
4	60	811	129	806	12
5	70	876	178	875	16
6	80	974	291	969	19
7	90	992	482	990	23
8	100	1184.7	7200	1153	34
9	110	1523.4	7200	1432	39
10	120	1589.6	7200	1496	46

From the results in Table 1 we see that the integer model, processed through standard methods such as branch-and-bound, is useful in solving instances of moderate size. We could not, however, achieve optimality for some larger instances. This suggests that other techniques must be applied to improve the performance in solving the CCPM. One possible method that we intend to explore in the future is to generate cuts based on polyhedral properties of the problem.

7   Concluding Remarks

In this paper, we introduced the CCPM, a problem occurring on mobile ad hoc networks on cooperative environments. The CCPM has been modeled in the context of discrete network optimization, and the assumptions made allowed the problem to be more easily tractable with tools from combinatorial optimization. Based on this model, an IP formulation for the problem was proposed, as well as a heuristic algorithm.

An characteristic of mobile ad hoc networks that has not been addressed here is the non symmetry that characterizes most existing networks. Asymmetry can occur as a result of physical limitations as, for example, different equipment used, interference, etc., or as a deliberate result of power control decisions. It would be interesting to develop models that consider these characteristics using asymmetric links.

References

1. K. M. Alzoubi, P.-J. Wan, and O. Frieder. Distributed heuristics for connected dominating set in wireless ad hoc networks. *IEEE ComSoc/KICS Journal on Communication Networks*, 4(1):22–29, 2002.
2. S. Butenko, X. Cheng, D.-Z. Du, and P. M. Pardalos. On the construction of virtual backbone for ad hoc wireless network. In S. Butenko, R. Murphey, and P. M. Pardalos, editors, *Cooperative Control: Models, Applications and Algorithms*, pages 43–54. Kluwer Academic Publishers, 2002.

3. Sergiy Butenko, Xiuzhen Cheng, Carlos A.S. Oliveira, and Panos M. Pardalos. A new algorithm for connected dominating sets on ad hoc networks. In S. Butenko, R. Murphey, and P. Pardalos, editors, *Recent Developments in Cooperative Control and Optimization*, pages 61–73. Kluwer Academic Publishers, 2003.

4. T. Camp, J. Boleng, and V. Davies. A survey of mobility models for ad hoc network research. *Wireless Communications and Mobile Computing (WCMC): Special issue on Mobile Ad Hoc Networking: Research, Trends and Applications*, 2(5):483–502, 2002.

5. Ching-Chuan Chiang and Mario Gerla. Routing and multicast in multihop, mobile wireless networks. In *Proceedings of ICUPC '97*, 1997.

6. B. Das and V. Bharghavan. Routing in ad-hoc networks using minimum connected dominating sets. In *International Conference on Communications*, pages 376–380, 1997.

7. Dash Optimization Inc. *Xpress-Optimizer Reference Manual*, 2003.

8. M. R. Garey and D. S. Johnson. *Computers and Intractability - A Guide to the Theory of NP-Completeness*. W. H. Freeman, San Francisco CA, 1979.

9. David B. Johnson. Routing in ad hoc networks of mobile hosts. In *Proceedings of the IEEE Workshop on Mobile Computing Systems and Applications*, 1994.

10. D. K. Kim. A new mobile environment: Mobile ad hoc networks (manet). *IEEE Vehic. Tech. Soc. News*, pages 29–35, 2003.

11. Carlos A.S. Oliveira and Panos M. Pardalos. A distributed optimization algorithm for power control in wireless ad hoc networks. In *Proc. of the 6th Workshop on Advances in Parallel and Distributed Computational Models (WAPDCM'04)*, Santa Fe, New Mexico, 2004.

12. Hridesh Rajan. Effects of applying mobility localization to source routing algorithms in mobile ad hoc networks. In *proceedings of IEEE Symposium on Computers and Communication (ISCC)*, 2003.

13. E. M. Royer and C.K. Toh. A review of current routing protocols for adhoc mobile wireless networks. *IEEE Personal Communications*, 6(2):46–55, 1999.

14. I. Stojmenovic, M. Seddigh, and J. Zunic. Dominating sets and neighbor elimination based broadcasting algorithms in wireless networks. In *Proc. IEEE Hawaii Int. Conf. on System Sciences*, 2001.

15. J. Tian, J. Hhner, C. Becker, I. Stepanov, and K. Rothermel. Graph-based mobility model for mobile ad hoc network simulation. In *Proceedings of 35th Annual Simulation Symposium*, San Diego, California, 2002.

16. P.-J. Wan, K. Alzoubi, and O. Frieder. Distributed construction of connected dominating set in wireless ad hoc networks. In *IEEE INFOCOM'02*, 2002.

17. Lidong Zhou and Zygmunt J. Haas. Securing ad hoc networks. *IEEE Networks*, 13(6):24–30, 1999.

ROAR: A Multi-rate Opportunistic AODV Routing Protocol for Wireless Ad-Hoc Networks

Kwan-Wu Chin and Darryn Lowe

Wireless Technologies Laboratory
University of Wollongong
Northfields Avenue, NSW, Australia
{kwanwu, darrynl}@uow.edu.au

Abstract. In this paper, we outline a simple approach, called ROAR, that enables the Ad-Hoc On-Demand Distance Vector (AODV) routing protocol to strengthen its routes by recruiting neighbors of nodes on the least cost path as support nodes during the route construction process, and working closely with the medium access control (MAC) to employ an opportunistic forwarding scheme that takes advantage of the node diversity at each hop. We have implemented ROAR in the *ns-2* simulator over the IEEE 802.11a physical layer. From our simulation studies conducted using various network topologies and realistic radio propagation model, we find that ROAR increases AODV's packet delivery ratio and end-to-end throughput several orders of magnitude, in particular for hop count based routes. Therefore, ROAR provides a simple add-on that allows routing protocols to reap the benefits of diversity without relying on physical layer approaches.

1 Introduction

Wireless Ad-hoc network routing protocols must react quickly given challenges such as mobility and unreliable wireless links. In the recent past, test-bed experiences such as [2] and [3] have demonstrated the inadequacy of using hop count as a route metric. This is because routes are usually constructed using unreliable links that frequently experience deep fades since the hop count metric does not reflect link quality. As a result, routing protocols spend a significant amount of time reconstructing routes whenever one of the links on a route fades, or whenever a shorter but unreliable link becomes available.

A promising concept that can be used to combat the vagaries of the wireless channel is node or macro diversity. A common exploit in all diversity techniques is the uncorrelated or independent communication paths provided by nodes in the vicinity of the receiver. Therefore, when the receiver's link goes into a fade, it can rely on its neighbors to receive packets on its behalf. Thus far, the majority of the works that exploit macro diversity lie at the physical layer (see [10] for more details). This paper departs from these existing works in that we show how the network layer can take advantage of diversity at minimal cost using existing hardware. As we will see later, our solution is capable of achieving significant performance gains that otherwise would only be possible using multiple antennae or maximal ratio combining techniques at the physical layer.

T. Kunz and S.S. Ravi (Eds.): ADHOC-NOW 2006, LNCS 4104, pp. 87–100, 2006.

An important factor we need to take into consideration when forwarding packets is the underlying data rate adaptation algorithm. For example, the IEEE 802.11b [6] standard offers data rates ranging from 1 to 11 Mbps, and even higher for the IEEE 802.11a [7], which provides eight data rates to choose from, ranging from 6 to 54 Mbps. Therefore, a routing protocol must be cognizant of the MAC's multi-rate capability and also the behavior of the link adaptation algorithm being employed. For example, a receiver at d meters away may be able to support a packet error rate (PER) of 1% at 6 Mbps but quickly degrades to 10% PER if 54 Mbps is used. Thus, depending on the underlying link-adaptation algorithm's behavior, a routing protocol will observe different link quality. This is particularly important during the route construction process since routes are constructed using the base data rate whereas data are transmitted using much higher data rates.

Our approach called multi **R**ate **O**pportunistic **A**d-hoc on-demand distance vector (AODV) **R**outing protocol or ROAR provides a practical method to exploit node diversity and works closely with the underlying link-layer adaptation algorithm. ROAR extends the well-known AODV [13] routing protocol with the ability to set up support nodes passively along a given path. By exploiting the independent reception probability and link quality of each support node, ROAR is able to increase the packet delivery ratio and throughput of a given route. A key feature in ROAR is that we relax the constraint that packets must be forwarded to the next hop node identified in the routing table. Instead, any support node that belongs to the same relay zone can become the next-hop node. In other words, packets are forwarded between relay zones rather than hop-by-hop. Lastly, since support nodes at a given hop have the same routing information, packets are forwarded along the least cost path to the destination; a feature that avoids unnecessarily increasing the network's contention level.

We have implemented ROAR in the *ns-2* simulator and conducted extensive simulation studies over different network topologies. Our results show that ROAR offers a marked increase in packet delivery ratio and throughput. This is mainly due to the improved packet error rate (PER) at each hop, thus allowing higher data rates to be supported and less transient link failures that cause AODV to unnecessarily tear down routes. In our experiments, we observe several orders of magnitude increase in packet delivery ratio and throughput. Further, these performance gains are for routes constructed using the hop count metric. ROAR therefore provides a case to continue using hop count as a route metric in ad-hoc networks despite previous works documenting its negative effect on performance. Interested readers are referred to [4] for 18 reasons why ad-hoc networks should use long hop routes.

This paper has the following structure. We start by describing ROAR's motivation in Section 2. After that, in Section 3 we present ROAR and its integration with the underlying MAC and link-adaptation algorithm. We then present our simulation studies and results in Section 4. After that we outline relevant works in Section 5 before concluding in Section 6.

2 Motivation

Before presenting ROAR, we first quantify the benefits of node diversity; see Section 4 for simulation parameters. In the experiments to follow we use a simple point-to-point topology where initially there is only a sender and receiver nodes. The nodes then move apart by 10m after each simulation run. The traffic is constant bit rate (CBR) at 8 Mbps. After obtaining a baseline performance of a two nodes topology, we place additional support nodes in the vicinity of the receiver.

Figure 1 shows the benefits of having an additional node where the sender is able to sustain a given data rate for much longer distance. This is particularly significant for the base data rate, 6 Mbps, which see a 50 meter increase in transmission range before packet delivery ratio starts to decline.

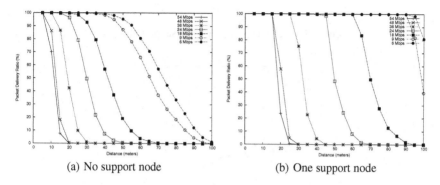

(a) No support node   (b) One support node

Fig. 1. The above figures show the benefits of having the receiver supported by an additional node

To determine the usefulness of diversity, an important question is the number of supports nodes required to yield a reasonable amount of performance gains. Figure 2 shows the benefit of adding up to six support nodes. It is clear that a limited number of nodes is sufficient to improve performance, however, as more nodes are added performance gains tend to become less.

Another investigation is to determine the benefits of node diversity on link-adaptation algorithms. Figure 3 shows the performance of two link-adaptation algorithms, AARF [9] and RBAR [5]. As expected, having more support nodes mean a given data rate is supported over a larger transmission range. Also, we see that AARF has a higher throughput than RBAR when the receiver is less than 50m since AARF does not incur the RTS/CTS exchange overhead. However, after 50m, RBAR becomes slightly superior due to data rate feedback from the receiver. Note, we restrict our comparison to these two algorithms since in [9] it has been shown that AARF has comparable performance to RBAR; minus the RTS/CTS overheads in addition to being deployable on current hardwares.

To summarize, we see that node diversity effectively strengthens a wireless link, as demonstrated by the increased packet delivery ratio and throughput for a given distance. We can therefore conclude that diversity will have benefits to any link adaptation algorithms. Apart from that, we see that the distance before a data rate's packet error rate worsen and the optimum data rate to use will depend on the diversity at the next hop. In the next section, we present a new protocol called ROAR that helps AODV reap the aforementioned performance gains.

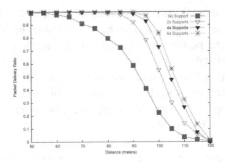

Fig. 2. The benefits of having more support nodes

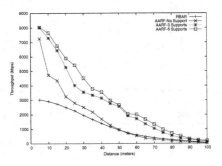

Fig. 3. AARF versus RBAR

3   ROAR: Protocol Description

3.1   AODV Background

AODV [13] is a reactive routing protocol. To construct a route, a route request (RREQ) is flooded across a MANET. Nodes with a route entry to the required destination send a route reply (RREP) message. To prevent unnecessary flooding of RREQ packets, nodes record the sequence number in each RREQ when they create a route entry to the RREQ originator. If a RREQ message with a lower or equal sequence number is received, it is discarded. Further, each node maintains a precursor list that, by recording the nodes using a given route, allows the determination of what neighboring nodes should receive a route error (RERR) message if the link to a given destination fails. In ROAR, we require each node record the number of duplicate RREQ messages transmitted by neighboring nodes, their IP addresses and link quality. We also augment both the RREQ and RREP messages to include the IP address of the previous and next hop node recorded in the routing table. For example in Figure 5, node-B would include node-C's IP address in the RREP before forwarding it to node-A.

We like to point out that ROAR can also be applied to other on-demand routing protocols, since it relies only on the route request/reply process; a function performed by all on-demand based routing protocols.

3.2   Relay Zone

A key concept in ROAR is *relay zones*. We like to point out that the term "zone" as applied here is different to conventional definition, which usually refers to a cluster of nodes managed by a leader and is used mainly to promote routing scalability. In ROAR, a relay "zone" is defined as follows.

Definition 1. *A zone is the set of nodes within the transmission range of one or more nodes that are on the least cost path to a given destination.*

That is to say, nodes within the transmission range of a next-hop node for a given destination are classified into the same zone. Figure 4 shows two zones where nodes in zone-B are within node-B's reception area and similarly, nodes within node-A's reception range belong to zone-A. Further, we see that node-3 belongs to two zones. It is worth pointing out that nodes A and B, referred to as *core* nodes, lie on the least cost path to a given destination and they are chosen during the route request and reply process. As will become clear later, no nodes play an active role in maintaining a zone. In fact, zone membership is defined as nodes that are listening to a multicast address constructed using the core node's IP addresses and a prefix indicating whether its upstream or downstream, denoted as $MCAST_U$ and $MCAST_D$ respectively (see Section 3.2 for details).

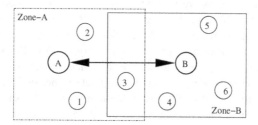

Fig. 4. Nodes within each zone play a support role to the node on the least cost path; nodes A and B

In the zone creation process we need to choose support nodes that will progress packets toward their intended destination quickly. This is because a support node may not have a good link to the next zone. In Figure 5, node-2 belongs to relay zones managed by node-B and node-A respectively. If node-A uses a high data rate when transmitting packets to node-B, it is likely that node-2 will receive a majority of the transmitted packets. However, notice that node-2 is much farther away to node-C (the next hop node) compared to nodes 3 to 6. Therefore, although node-2 can support a high data rate, it does not necessarily mean packets forwarded by it will make good progress toward their destination nor indicate it is in a good position to exploit node diversity in the next relay zone.

Relay Zone Construction. Relay zones are created once the node on the least cost path or core node is selected, which happens after receiving a RREP message. As the RREP

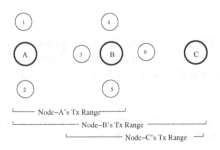

Fig. 5. Example used to construct relay zones

traverses from one node to the next, nodes overhearing the RREP message confirm their zone membership as follows.

Assume node-C has a route entry to the required target node and generates a RREP message to node-B. From Figure 5 we see nodes 3 to 6 are within node-C's transmission range, thus they will receive node-C's RREP message to node-B. Nodes 3 to 6 check to see whether they have created a route entry to the source due to an earlier RREQ message. If they have, the next step is determining whether they should join node-C or node-B's zone. The criteria for doing so must ensure node 3 or 6 has a good link to the next relay zone, thus ensuring packets are only forwarded on high-rate links.

In the discussion to follow we will define node i's next hop core node to a given destination or source as N_{i+1} or N_{i-1} respectively. We first consider setting up zones that ensure good progress toward the destination. To meet this requirement, support nodes must be located near or have good signal-to-noise ratio (SNR) to the core node; the node that sent the RREQ message previously. Once a neighboring node overhears the RREP message transmitted by the core node, it determines its SNR to the core node, and becomes a support node if its SNR is higher than a given threshold.

For packets going from the destination to the source, we used a slightly different procedure. Unlike before, we need to have support nodes close to node-B or N_{b-1} (node-A in this example). This is easily obtained since each RREQ contains the IP address of node N_{b-1}, therefore this information is already available when the RREQ came past. As stated above, nodes join zone-B if their SNR to node-B is higher than a predefined threshold or they can hear node N_{b-1}.

In the above cases, support nodes do not forward RREP messages. This ensures only core nodes forward the RREP message back to the source node. Each support node also updates their route entry to either node N_{i-1} or N_{i+1}'s IP address as the next hop to either the source or destination. This means support nodes have the same route entry for the given source and destination as the core node of the zone they belong to.

Route Maintenance. Route maintenance is trivial since we rely on AODV's RERR messages or route timeout. Alternatively, we could also rely on HELLO messages to indicate that a node is no longer a neighbor. In all cases, the routing layer removes the corresponding route information for a given neighbor and destinations using that neighbor as the next-hop node. Finally, for each destination, the routing layer informs the MAC to unsubscribe from the corresponding multicast address, thus removing itself from the relay zone. Note that, instead of unsubscribing from the relay zone, a support nodes within the relay may become the core node, hence enable quick local recovery from node or link failure. We leave this as future work.

Ethernet Multicast Address. As described earlier, each node in a given zone, including the core node, subscribes to an Ethernet multicast address. To construct a multicast address, we simply prepend the prefix $MCAST_U$ or $MCAST_D$, both 16-bits in length, to the hash of the core node and destination IP addresses. Notice that we have two sets of multicast address which we use to indicate support nodes that forward packets toward the source or destination node. This is done to ensure that only those nodes that contribute to the progress of packets join the zone. The prefix $MCAST_U$ and $MCAST_D$ denote Ethernet multicast address prefixes. The reason we need to hash the destination's IP address is to ensure that support nodes that are participating in other communication sessions do not intercept transmitted packets of other sessions since they may be not be within transmission range of a packet's next relay zone.

3.3 The MAC Layer

In the previous section we have shown how relay zones are created and also defined what it means to belong to a given zone. In this section, we show how the IEEE 802.11 MAC makes use of relay zones.

Before describing our modifications, we first highlight some key design criteria.

1. *Minimal modifications to the IEEE 802.11b standard.* This criterion ensures that ROAR is practical and is deployable using current hardware. As we will see later, ROAR's only modification consists of introducing a slight random delay in the clear channel assessment process to avoid acknowledgment message collision.
2. *Minimal probing.* We want to avoid having to obtain data rate estimates from nodes in the next relay zone, since this incurs non-negligible delay and signaling overheads, in addition to requiring changes to IEEE 802.11's RTS and CTS exchange.

Data Packet Transmission. The forwarding process at the network layer proceeds as normal where the routing daemon obtains the next-hop or core node's IP address from the routing table which is then passed to the MAC to construct an Ethernet multicast address. In the descriptions to follow we assume all control messages are transmitted at the base rate. Further, we will use Figure 6 to illustrate the steps involved.

The packet transmission proceeds as usual. A node senses the channel, and if the channel is free, it transmits the packet using the next relay zone's multicast address. The data rate used will be dependent on the packet transmission history to the given neighbor. In this example, since there are three nodes listening to the multicast, we need to resolve which one of them will send an acknowledgment message to the sender. To resolve this issue, each node performs the clear channel assessment operation for a small random period. At the end of the period, if the channel is free, it transmits the acknowledgment message back to the sender. The packet is then forwarded to the routing layer. Otherwise, if another transmission is heard, it discards the data packet.

Link Adaptation Algorithm. The multi-rate capability of current wireless systems has spurred researchers to develop link adaptation algorithms. For example, OAR [15]

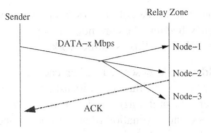

Fig. 6. Data transmission

determines the best data rate to use by probing the channel using a request-to-send (RTS) message and having the receiver return its estimated data rate in the returning clear-to-send (CTS) message. Interestingly, in [1] and also in [9] the authors commented that, based on their test-bed experience, estimating data rate using instantaneous SNR is unreliable, thus giving more credence to opportunistic forwarding schemes.

Due our aforementioned criteria and the above reason, we have chosen to use AARF [9], thus avoid the use of the RTS/CTS exchange. AARF's key advantages is its simplicity and zero signaling overheads, which is particularly important when dealing with a large number of support nodes. Further, it does not require any modifications to the RTS/CTS handshake nor changes to packet formats. It has also been shown to have comparable performance to RBAR [5]; a scheme that relies on the exchange of RTS and CTS messages in order to obtain a suitable data rate estimate before transmission. Lastly, it can be deployed immediately without any modifications to the MAC.

AARF works as follows [9]:

– After M successful transmissions, the sender increases its data rate to the next level. Initially, M is set to 10.
– Define the first packet transmitted using the new data rate as the probe packet. If the probe packet fails at the new data rate, reduce data rate immediately and double the value of M. Note, M has a maximum value of 50.

AARF has an important characteristic. That is, the ability to settle on an M value suited for a given path loss. This optimization is particular important in the context of routing in wireless ad-hoc networks because increasing the link rate aimlessly may result in a higher packet loss, thus causing the routing protocol to incorrectly conclude that the link to a neighbor is unstable.

4   Experiments and Results

This section presents our experiments and results. We equip *ns-2* (v2.28)'s [12] with IEEE 802.11a [7], thus giving us eight data rates; 6 to 54 Mbps. To ensure a realistic channel model, we used the Ricean propagation model modifications of [15] and fixed the velocity value to 2.5. Unless indicated otherwise, all our experiments are performed with $K = 0$. Further, a new seed is used to initialize the random number generator for each simulation run. Lastly, we extended AODV with ETX [3]. Note, for simulation involving ETX, we allow each node to average up to 600 ETX values before starting the route construction process.

4.1   Linear Topology

We start our experimentation with a linear topology, see Figure 7, to first measure the performance gains in a simple setting before moving onto more general topologies in Section 4.2. In this experiment, node-A sends a CBR stream at 10 KB/s to node-E. We increase the distance between nodes by five meters after 50 runs and record the average packet delivery ratio (PDR) and throughput. In addition, depending on the experiment, we add one to three support nodes in the vicinity of nodes B, C and D. In our simulation, AODV's local repair feature is disabled.

Fig. 7. Linear topology. Additional support nodes are added near the vicinity of nodes B, C and D.

Figure 8(a) shows the PDR when AODV uses hop count as a route metric with three support nodes. In this experiment we have omitted ETX since the clustering of nodes meant that there are no alternative reliable paths that can be constructed using ETX. An important aspect of ROAR is improving the link when it is unreliable. For example, see Figure 8(b), as nodes become more distant from each other, i.e., PER gradually becomes higher, ROAR is able to increase a route's PDR significantly. This is important for hop based routes since they tend to select unreliable next hop node, therefore, by using ROAR the benefits hop count are maintained; with the negative effects due to wireless fading minimized.

Figure 9 shows the average throughput for the linear topology. Due to the lower PER, AARF has a better chance of using a higher data rate. Ensuring a link's PER is low is important. This is because AARF reduces the current data rate whenever the MAC has given up retransmitting a data packet (7-retries) or when the probe packet is lost. Another benefit is that AARF is able to reach steady state sooner, recalling that the IEEE 802.11a physical layer supports eight data rates and AARF only increases the data rate after 10 consecutive successful transmissions. In other words, since there is less packet loss, AARF is able to quickly settle on a data rate commensurate with link condition.

4.2   Node Density

In this experiment we investigate the effect of node density. We use a grid size of 200x50 meters. We fix the position of the sender and receiver coordinate $< 10, 10 >$ and $< 190, 90 >$ respectively. We then increase the number of nodes from 10 to 60. In this experiment, we set the Ricean propagation model's K value to three. Also, in each simulation, the sender generates a CBR traffic with a maximum of 2000 packets. We record the PDR of 50 simulation runs, and plot the median PDR for each network density. The reason we plot the median instead of the mean is due to the variability caused by wireless errors and varying path loss from random node positions.

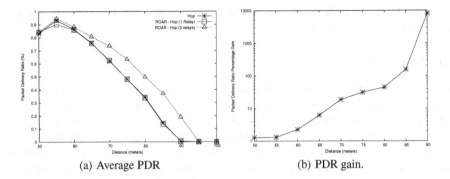

| (a) Average PDR | (b) PDR gain. |

Fig. 8. Figure (a) shows the PDR for the linear topology, and (b) is the corresponding PDR gain for ROAR with three support nodes over a scenario with no support nodes

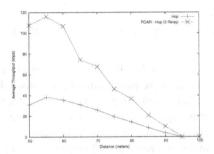

Fig. 9. Average throughput for linear topology

Figure 10 shows the performance of AODV with and without ROAR for two route metrics. For both metrics, their PDR increased up to 25% compared to AODV without ROAR. As network density rise, most of the links on a route will be within transmission range and in close proximity. Further, they have neighbors close by which ROAR can exploit. All these mean that wireless losses become less due to more reliable links on a given route, thus leading to a gradual performance improvement as node density rises. Similar to PDR, ROAR has a significant impact on throughput, as illustrated in Figure 11. Observe that as node density rises, the performance of AODV using ETX becomes lower due to increasing path length. However, in both ETX and hop count cases, ROAR sees up to 50% increase in throughput over basic AODV.

4.3   Recruitment Threshold

In this experiment we adjust the data rate thresholds used to recruit support nodes. Threshold refers to the minimum data rate a neighbor must support to qualify as a support node. For example, a threshold of 50 means only those neighbors that can support a data rate of 54 Mbps, whereas a threshold of 10 means those neighbors that can support a data rate of at least 10 Mbps, i.e., 12 to 54 Mbps. In this experiment we place

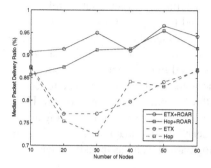
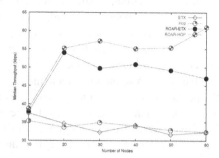

Fig. 10. Effect of node density on packet delivery ratio

Fig. 11. Effect of node density on throughput

50 nodes randomly in a grid size of 200x50 meters, and calculate the average packet delivery ratio for 50 simulation runs.

Figure 12 shows the effect of using different thresholds in two different network densities; 20 and 40 nodes. We see that for both network densities, having a threshold at 40, i.e., those neighbors capable of supporting 48 and 54 Mbps, is sufficient to obtain a high PDR. At the highest threshold of 10 we see that PDR drops dramatically. This is due to increased contention resulting from duplicate packets since support nodes are out of each other's transmission range, thereby unable to sense that the other has transmitted an acknowledgment for a given data packet. As a result, packets are dropped, followed by the MAC informing AODV that the link to a neighbor is broken. This then causes the removal of packets buffered on the said link and re-initiation of the route construction process.

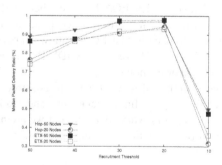

Fig. 12. The effect of support node recruitment threshold on PDR

5   Related Work

Researchers are beginning to investigate new packet forwarding paradigms that exploit node diversity to combat the vagaries of the wireless channel. Recently Biswas et al. [1] presented a protocol where the sender transmits a batch of packets without first having to probe the channel condition nor the need to specifically target a set of relays that will forward the packets onward. Packets that failed to reach the receiver are retransmitted by relays in priority order set by the sender. Further, these relays are required to track which packets have been received by the destination to minimize redundant transmissions. In a different work, Yu et al. [18] proposed that a neighboring node of the receiver send an acknowledgment message to the sender if the receiver failed to receive the packet. ROAR has similar objectives to both of the aforementioned works. However, ROAR requires less state exchanges and coordination between forwarding nodes. Moreover, we consider the dynamics of the underlying link adaptation algorithm. Finally, packets are forwarded using the shortest hop path, thus it has the advantage of not adding to the contention level.

ROAR is also closely related to two other works. In [17], the authors use two hops neighborhood information to select relay nodes with good link quality. On the other hand, [8] relies on a multicast routing protocol to obtain candidate relays' addresses. In both cases, the relay addresses are included in RTS messages to determine the relay with the best channel condition. The authors of [8] also suggested using other information such as queue length and geographical distance of neighboring nodes. The list of addresses embedded in the RTS message are then used by the relays to determine which node has the highest priority to transmit a CTS message. We improve upon these works as follow. ROAR eavesdrop on AODV's RREQ and RREP messages, thus relays are recruited passively and they are identified using a Ethernet multicast address. Secondly, a sender does not need to know relays' IP addresses since ROAR uses an opportunistic forwarding paradigm that does not require it to determine the best relay or support node to use. Finally, ROAR considers the effects of link adaptation.

The zone creation aspect of ROAR resembles AODV-BR [11]. AODV-BR establishes backup nodes at each hop where upon a route failure, they are called upon to provide an alternative path to the next node on the shortest hop path. AODV-BR also relies on RREP messages to set up backup nodes. ROAR on the other hand establishes so called backup nodes for a different reason, that is to increase node diversity which the MAC can then exploit. Another difference is that ROAR allows support nodes in different relay zones to communicate directly. However, in AODV-BR, a backup node's next hop node is always a node on the shortest hop path.

There have been quite a number of works related to constructing multiple paths in MANETs to increase reliability [16], limit packet loss and load balancing [14]. Although the end goals are similar, our works defer in that we are not concern with end-to-end disjoint paths nor require a multicast routing protocol.

6 Conclusion

ROAR incorporates AODV with two interesting concepts: node diversity and opportunistic forwarding. These concepts enable ROAR to exploit the vagaries of the wireless channel and to do so with minimal signaling overheads. Furthermore, these concepts are realized with minimal modifications to AODV and the IEEE 802.11 MAC.

Through extensive simulation studies, we evaluated ROAR in various network topologies and found that ROAR provides orders of magnitude performance gain at minimal cost. Apart from that, we show that ROAR strengthens routes constructed using hop count. Finally, ROAR provides a simple network layer solution that offers some of the performance gains that have been touted by cooperative diversity techniques at the physical layer.

Acknowledgment

This project is partially funded by ARC Discovery Project DP0559769.

References

1. S. Biswas and R. Morris. ExOR: Opportunistic multi-hop routing for wireless networks. In *ACM SIGCOMM'2005*, Pennsylvania, USA, Sept. 2005.
2. K.-W. Chin, J. Judge, A. Williams, and R. Kermode. Implementation experience with MANET routing protocols. *ACM/SIGCOMM Computer Communications Review*, 32(5), Nov. 2002.
3. D. S. J. DeCouto, D. Aguayo, J. Bicket, and R. Morris. A high throughput path metric for multi-hop wireless routing. In *MOBICOM'03*, San Diego, USA, Sept. 2003.
4. M. Haenggi and D. Puccinelli. Routing in ad-hoc networks: A case for long hops. *IEEE Communicaitons Magazine*, 44(10):93–101, Oct. 2005.
5. G. Holland, N. Vaidya, and P. Bahl. A rate-adaptive MAC protocol for multi-hop wireless networks. In *ACM MOBICOM'2001*, Rome, Italy, Oct. 2001.
6. IEEE. Wireless LAN medium access control (MAC) and physical layer (PHY) specifications. IEEE Standard, Draft 802.11, 1997.
7. IEEE. Part 11a: Wireless LAN medium access control (MAC) and physical layer (PHY) specifications: High speed physical layer in the 5 GHz band, standard specification. IEEE Std 802.11a-1999, 1999.
8. S. Jain and S. R. Das. Exploiting path diversity in the link-layer in wireless ad-hoc networks. In *Proceedings of IEEE WoWMoM*, Taormina, Italy, June 2005.
9. M. Lacage, M. Manshaei, and T. Turletti. IEEE 802.11 rate adaptation: A practical approach. In *ACM MSWiM'2004*, Venice, Italy, Oct. 2004.
10. J. Laneman, D. Tse, and G. Wornell. Cooperative diversity in wireless networks: Efficient protocols and outage behaviour. *IEEE Transactions on Information Theory*, 50(12):3062–3080, Dec. 2004.
11. S.-J. Lee and M. Gerla. AODV-BR: backup routing in ad-hoc networks. In *IEEE Wireless Communications and Networking Conference (WCNC'00)*, Chicago, IL, Sept. 2000.
12. S. McCanne and S. Floyd. ns network simulator-2. http://www.isi.edu/nsname/ns/.
13. C. E. Perkins, E. M. Royer, and S. R. Das. Ad hoc on-demand distance vector (AODV) routing. draft-ietf-manet-aodv-06.txt, July 2000.
14. M. Perlman, Z. J. Haas, P. Sholander, and S. Tabrizi. On the impact of alternate path routing for load balancing in mobile ad-hoc networks. In *IEEE/ACM MobiHOC'2000*, Chicago,IL, Aug. 2000.
15. B. Sadeghi, V. Kanodia, A. Sabharwal, and E. Knightly. Opportunistic media access for multirate ad-hoc networks. In *ACM MOBICOM*, Atlanta, Georgia, USA, Sept. 2002.

16. A. Tsirigos and Z. Haas. Analysis of multipath routing - part ii: Mitigation of the effects of frequently changing network topologies. *IEEE Transactions on Wireless Communicaitons*, 3(2):138–146, Mar. 2004.
17. J. Wang, H. Zhai, W. Liu, and Y. Fang. Reliable and efficient packet forwarding by utilizing path diversity in wireless ad-hoc networks. In *Proceedings IEEE Milcom'04*, 2004.
18. C. Yu, K. G. Shin, and L. Song. Link-layer salvaging for making routing progress in mobile ad-hoc networks. In *ACM MobiHOC*, Urbana-Champaign, Illinois, USA, May 2005.

Energy-Efficient Face Routing on the Virtual Spanner*

Héctor Tejeda[1], Edgar Chávez[1], Juan A. Sanchez[2], and Pedro M. Ruiz[2]

[1] Escuela de Ciencias Físico-Matemáticas, Universidad Michoacana, México
{elchavez, htejeda}@fismat.umich.mx
[2] Facultad de Informática, University of Murcia, Spain
{jlaguna, pedrom}@dif.um.es

Abstract. Geographic routing protocols are one of the most common routing schemes for sensor networks. These protocols consist of two different modes of operation: greedy routing to forward data to the destination using neighbors which are closer to the destination than current node and face routing to avoid voids in the network. Face routing requires the graph to be planar, which usually means that some crossing links of the original network cannot be considered when routing in face mode. In this paper we introduce a new localized scheme to build a virtual spanner which is planar by construction and is guaranteed to be connected if the underlying network is connected as well. Unlike previous works, by performing face routing over this spanner we can reduce energy consumption in face mode because the elimination of any of the original links in the network is not required. Thus, the most energy-efficient paths can be selected when the protocol enters face mode. The virtual spanner is easy-to-build and uses only local information, making it scalable to large-scale networks. Routing is always performed in real nodes; virtual nodes are used only as routing anchors when the agent is in face mode. In addition, our simulation results show that the proposed scheme outperforms the best energy-efficient geographic routing protocol for different network densities and energy models.

1 Introduction and Related Literature

Mobile ad hoc networks (often referred to as MANETs) as well as wireless sensor networks have become one of the hotest topics in communications. One of the main reasons is their wide applicability to many scenarios such as disaster relief, battlefield environments, etc. These networks consist of wireless nodes that communicate with each other in the absence of a fixed infrastructure. When a node needs to send a message to another host which is outside of its radio range, it uses other intermediate hosts as relay nodes. Those intermediate nodes are dynamically selected by the routing protocol according to some metric.

Among all routing protocols for these networks, geographic routing [BMSU01] has emerged recently as a very efficient way to provide guaranteed delivery routes

---

* Partially supported by CONACyT and the Spanish MEC by means of the Ramon y Cajal work program and the SAVIA (CIT-410000-2005-1) project.

T. Kunz and S.S. Ravi (Eds.): ADHOC-NOW 2006, LNCS 4104, pp. 101–113, 2006.

without flooding the whole network with control messages. A node is only required to be able to know its position, the one of its neighbors (e.g. using periodic becons) and the position of the destination. When a node wants to send a message to some destination, each intermediate node selects locally its *best* neighbor to forward the message towards that destination. The best node depends on the particular routing metrics. For instance, in networks with battery-operated devices, energy consumption is one of the most important. This operation of selecting the best neighbor is called *greedy mode*.

There has been a number of energy-efficient geographic routing solutions reported in the literature. Stojmenovic and Lin [SL01] proposed localized power-aware algorithms based on the consideration of different neighbor selection schemes in greedy mode. For instance, one of the proposed schemes was selecting the nearest neighbor to current node among those providing advance towards destinations. More recently, those schemes were enhanced in [KNS04] by considering a cost over progress neighbor selection. The cost is determined by the energy to reach a candidate neighbor while the progress is the reduction in distance towards the destination by selecting that neighbor. The neibghbor with lowest ratio is selected at each step. The best solution among all alternatives reviewed is iPowerProgress, which uses the same idea but iteratively enhances the selection by trying to find another neighbor so that reaching the selected next hop through that neighbor reduces the overall energy required.

However, regardless of the greedy function used to select neighbors, all of them end up using face routing [BMSU01] to escape from local minima (e.g. a void zone). Face routing sends packets along the faces of a planar graph representing the underlying network graph following the right hand rule, and changing faces when the line between the node where face started and the destination is crossed. Correct operation of face routing requires the underlying graph to be planar. Thus, some of the original (crossing) edges are not considered while in face mode because they are eliminated by planarization tests.

There are several methods to extract a planar subgraph from a given Unit Disk Graph (UDG), which models the entire network. A UDG is a graph in which an edge $[u, v]$ exists only if $dist(u, v) \leq r$ being r the radio range. The *Relative Neighborhood graph*, RNG [Tou80] is obtained by applying the RNG test to every edge of the UDG: an edge $[u, v]$ is retained in $RNG(G)$ if there is no vertex z such that $\max\{d_G(u, z), d_G(v, z)\} < d_G(u, v)$. That is, if there is no vertex in the intersection of their disks. The *Gabriel graph*, GG [GS69], applies a slightly different test to every edge of the graph. It retains an edge $[u, v]$ in $GG(G)$ if there is no node in the disk with diameter $\overline{uv}$. Finally, the *Morelia test* [BCG+04] manages to preserve some long edges by using a stronger condition for the removal of edges. An edge $[u, v]$ is not included in $MG(G)$ if there is a couple of points $[x, y]$ so that one of them (or both) is in the disk with diameter $\overline{uv}$ and $[x, y]$ crosses $[u, v]$. Given a UDG G we have $RNG(G) \subseteq GG(G) \subseteq MG(G)$.

Energy-efficient optimizations for geographic routing reported to date, have just focused on the greedy part of the protocol. In this paper, we propose the creation of a planar virtual spanner of the original graph using a tesselation planar by construction. Face routing will then be performed on the virtual graph. Once the next hop virtual neighbor is selected according to face routing, real nodes will route the message greedily towards the representative (e.g. centroid) of the selected neighboring tessel. Given that real nodes will route using all available links (no links are eliminated), it is possible to find an energy-efficient path towards that representative which enhances the performance in face mode of the protocol. In our case, we try both iterative PowerProgress [KNS04] and Dijkstra's shortest path (applied to the local neighborhood based on energy cost) to reach the representative of the next tessel using energy-efficient links. We shall see in our simulation results how the use of this virtual spanner manages to reduce energy consumption for all network desnsities tested.

The remainder of the paper is organized as follows: Section 2 presents our network model and the problem formulation. Section 3 illustrates how the virtual spanner is built. We explain how to route based on the virtual spanner in section 4. Finally we present some simulation results in section 5 and give some conclusions and future work in section 6.

2   Network and Energy Model

This section introduces the notation and the model we use throughout the paper. We consider routing algorithms on Euclidean graphs. As usual, a graph G is defined as a pair $G := (V, E)$ where V denotes the set of vertices and $E \subseteq V^2$ denotes the set of edges. The number of nodes is denoted by $n := |V|$.

In our evaluations we will use the general energy model proposed by Rodoplu and Meng [RM98] in which the power consumption between two nodes at distance d is $u(d) = d^\alpha + c$ for some constants $\alpha (2 \leq \alpha \leq 4)$ and c. We also consider a more realistic model for some devices which also accounts for the energy of reception. $u(d) = d^\alpha + c + c_r$, where c_r is usually $(r^\alpha + c)/3$ being r is the maximum transmision range.

In this paper we consider the standard UDG model for ad-hoc networks, but we assume that nodes can asjust their transmission range (r) to reach only a subset of their neighbors. Thus, given two nodes $v_1, v_2 \in V$, the edge $[v_1, v_2] \in E \Leftrightarrow ||[v_1, v_2]|| \leq r$, being $||[x, y]||$ the euclidean distance between vertices x and y.

As in previous geographic routing works in the literature, we assume that nodes know their positions and those of their neighbors. It is also assumed that sources of data packets know the position of the destination.

3   The Virtual Spanner

We explain in this section, how each node locally builds its partial view of the virtual spanner. For clarity, we first explain the overall idea and its topology.

We divide the plane in regions with a regular tessellation, which is a tessellation (or planar subdivision) made up of congruent regular polygons. The idea is that an entire region may be represented by a single virtual point, the centroid of the regular polygon. If we link the centers of the polygons we observe a peculiar behavior: the centers define a dual tessellation that is also planar. The dual of a triangle tessellation is a hexagonal tessellation, while a square tessellation is auto dual.

Only three regular polygons tessellate the Euclidean plane: triangles, squares or hexagons, from elementary geometry. They are depicted in figure 2.

The virtual node for a polygon is chosen as the centroid of the polygon. Two virtual nodes will share an edge if in their respective cells two real nodes are mutually reachable. Thus, we need to choose a suitable polygon size to build the virtual graph, so that we achieve a good trade-off between the simplicity to build the virtual graph (guaranteeing that is planar), and the number of cells to be checked for possible links to preserve connectivity. We have analyzed three options as we show in figure 1.

a) The transmission radius does not cover all the cell.
b) Any two points in the cell are within radio range.
c) A node in one cell can reach any other node in a neighboring cell.

Case a) complicates the design because it may require multihop routing within a cell. In case c) there may be a very big number of cells in which to look for neighbors. We decided to use case b) because it is the configuration which avoids multihop within a cell in which the number of cells to look for virtual neighbors is reduced.

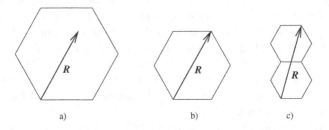

Fig. 1. Variation of polygon size with transmission radius fixed

If the graph is dense enough, there will be at least one node in each cell. Thus, each virtual node will be connected to all neighboring virtual nodes. The resulting virtual graph is exactly the dual of the graph, which is planar. In real situations we cannot guarantee that every cell will have a node. Thus, to preserve connectivity we must find all possible virtual neighbors. They may be in cells which are not contiguous to the current one. In figure 2 we show for each different tesselation (triangular, square and hexagonal) the possible cells that may contain real nodes which are neighbors of nodes in the current cell t. The cell t can reach more cells when using a triangular configuration (24 cells). With a square configuration 20 cells are candidates and when using hexagonal cells only 18 cells are candidates. Please note that adding virtual links to all those cells having real nodes which are neighbors of nodes in current cell may result in

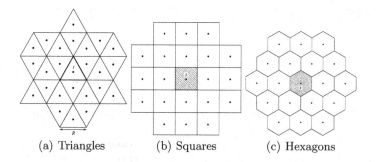

(a) Triangles   (b) Squares   (c) Hexagons

Fig. 2. Regular tessellations in the plane and cell centroids. Additionally, the cells shown are those reachable from t using a radius R equal to the diameter of the cell.

a non-plannar virtual spanner. Unlike other traditional planarity tests, crossing links can be detected and avoided using a static table containing all the possible configurations.

The grid with triangles, squares or hexagons is located arbitrarily in the plane. Each cell is identified by a coordinate pair as we show in Figure 3. Note that real nodes only need to know the type of tesselation and the communication radius at deployment time. Based on that, and given their current position they can easily compute the coordinates of their centroid. In addition, only with local information about the position of its neighbors they can compute their local view of the virtual graph (virtual edges). This has no additional overhead because position of real neighbors is already required for geographic routing to operate.

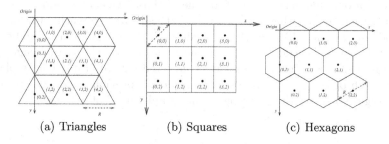

(a) Triangles   (b) Squares   (c) Hexagons

Fig. 3. Fixing the origin, the virtual coordinates are computed with elementary calculations

We show below some elementary calculations for the node to compute the coordinates of the virtual node for each real node and the tesselation to which it belongs. The node only needs the transmission radius R and its position (x, y).

Once a node computes its tessel and its centroid, it individually and locally computes its view of the virtual graph. The process consists of two different stages:

Type of tilling	Position of centroid	Tessel
Hexagonal	$y_c \leftarrow 3R(y + 1/3)/4$ **if** $y \bmod 2 = 0$ **then** $x_c \leftarrow \sqrt{3}R(x + 1/2)/2$ **else** $x_c \leftarrow \sqrt{3}Rx/2$	$y \leftarrow$ **truncate** $(3y_n/4/R)$ **if** $y \bmod 2 = 0$ **then** $x \leftarrow$ **truncate** $(2x_n/\sqrt{3}/R)$ **else** $x \leftarrow$ **truncate** $((2x_n + \sqrt{3}R/2)/\sqrt{3}/R)$
Triangular	$x_c \leftarrow Rx/2$ **if** $(x + y) \bmod 2 = 0$ **then** $y_c \leftarrow \sqrt{3}R(y + 2/3)/2$ **else** $y_c \leftarrow \sqrt{3}R(y + 1/3)/2$	$x \leftarrow$ **truncate** $(x_n/2/R)$ $y \leftarrow$ **truncate** $(2y_n/\sqrt{3}/R)$
Square	$x_c \leftarrow (x + 1/2)R/\sqrt{2}$ $y_c \leftarrow (y + 1/2)R/\sqrt{2}$	$x \leftarrow$ **truncate** $(\sqrt{2}x_n/R)$ $y \leftarrow$ **truncate** $(\sqrt{2}y_n/R)$

Table 1. Formulas for a node to compute position of its centroid and its tessel

1. Test surrounding cells that are neighbors by their side.
2. Test all other cells that are reachable from current cell but are not neighbors by their side.

The first stage is easier than the second one because it always produces a planar graph. There are no edge crossings, as it is depicted in figure 4. In the first stage, a virtual edge is added between centroids of two cells adjacent by the side if there are two mutually reachable real nodes, one in each of those cells. Unfortunately, the virtual graph produced after the first stage may not be connected (see Fig. 4). Thus, we need to apply the second stage to obtain a connected graph without crossings of virtual edges.

For the second part, we start testing if we can add a virtual edge to the centroid of those cells (see figure 2) which are second degree neighbors (side neighbors of our side neighbors). If for one of those, we cannot add the virtual edge (i.e. there is no other real node in that cell directly reachable from any real node in current cell) then we try again with those cells being neighbors by side of this particular cell we couldn't find nodes to add the virtual edge. This condition guarantees that the resulting connected virtual graph will be planar, because each of those edges is tested not to cross previous edges before being added. Figure 5 shows the resulting virtual graph after both stages.

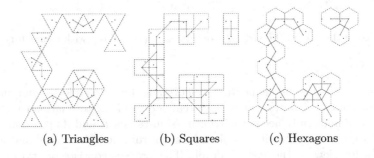

(a) Triangles   (b) Squares   (c) Hexagons

Fig. 4. First connectivity test. Natural neighbors.

As an example, we give the concrete algorithms used in each stage to add edges to the virtual spanner with an hexagonal tilling. The algorithms for squares and triangles are similar and are not included in here due to space limitations.

The algorithm for the first connectivity test using hexagons is given in algorithm 1.

Algorithm 1. Algorithm for the first stage with hexagons

```
1: procedure ReviewHexagonsStage1(I, t)              ▷ I are the neighbor cells
2:     k ← 0                        ▷ by their side to t, they are enumerated from 0 to 5
3:     while k < 6 do
4:         if isThereEdge(t, I_k) then        ▷ real edge among nodes of those tessels?
5:             addEdge(t, I_k)
6:         end if
7:         k ← k + 1
8:     end while
9: end procedure
```

Algorithm 2. Algorithm for the second stage with hexagons

```
1: procedure ReviewHexagonsStage2(I, E, t)           ▷ E are the rest of the cells
2:     k ← 0                          ▷ reachable by t, enumerated from 0 to 11
3:     while k < 6 do                            ▷ Review for the odd cells from E
4:         a ← 2k
5:         b0 ← !isThereEdge(t, I_k)
6:         b1 ← !isThereEdge(I_k, E_a)
7:         b2 ← isThereEdge(t, E_a)
8:         if b0 AND b1 AND b2 then
9:             addEdge(t, E_a)
10:        end if
11:        k ← k + 1
12:    end while
13:
14:    k ← 0
15:    while k < 6 do                          ▷ Review for the even cells from E
16:        a ← (k + 1) mod 6
17:        b ← (2k + 1) mod 6
18:        b0 ← isThereEdge(t, I_k)
19:        b1 ← isThereEdge(I_k, E_b)
20:        b2 ← isThereEdge(t, I_a)
21:        b3 ← isThereEdge(I_a, E_b)
22:        b4 ← isThereEdge(t, E_b)
23:        if !(b0 AND b1) AND !(b2 AND b3) AND b4 then
24:            addEdge(t, E_b)
25:        end if
26:        k ← k + 1
27:    end while
28: end procedure
```

For the second stage with an hexagonal tilling, all reachable cells which are not side neighbors of the current cell (t) are tested. The test needs to take into account existing virtual links which have been added before, to avoid creating a non-planar virtual spanner. The detailed algorithm is given in algorithm 2.

As we stated, the goal of this virtual graph is enhancing the energy consumption of face routing. Thus, we will explain in the next section how real nodes perform energy-efficient face routing over the virtual graph.

4   Energy Efficient Face Routing on the Virtual Spanner

When the geographic routing protocol finds a void zone, we perform face routing over the virtual spanner. In this section we explain how real nodes route those messages in face mode to reduce energy consumption.

When a real node enters into face mode, it locally builds its view of the virtual spanner. Then, it applies face routing algorithm considering its centroid as the current node, and all other centroids connected through virtual edges as neighbors. The application of face routing gives a new virtual neighbor, representing the tessel to which the message needs to be sent in face mode. Now, real nodes are used to create a path towards that tessel. In addition, given that all edges in the original graph are preserved, we can try to find the path which minimizes energy consumption. In particular, we have considered two different alternatives to compute that path:

- Use greedy routing with an energy metric to go towards the closest node in the selected tessel.
- Compute Dijkstra's shortest weighted path algorithm considering energy as cost using local information.

A brief description of the proposed algorithm is presented below. At each step of the algorithm the node currently trying to send the packet to the next neighbor in face mode performs the following operations:

1. Based solely on its coordinates, the node finds out his cell and corresponding virtual node according to expressions given before.
2. Using the information from neighbors (obtained by any geographic routing protocol using periodic beacons), the node finds out which virtual edges exist according to the procedures explained in the previous section. As we explained before, a virtual edge can only exist to a virtual node if there is a real neighboring node in the corresponding cell.
3. In *face mode* the current real node routing in face mode will use the virtual graph to select (according to face routing) the proper virtual edge to follow. Once it is selected, it uniquely defines the next cell that needs to be reached using *real nodes*. The node then computes the shortest path tree in energy consumption towards the nearest real node in the next cell. If real nodes of the selected cell are not directly reachable by current real node, it computes the the shortest energy consumption path towards the most distant real node

in its own cell in the direction of the next cell. So when the message reaches that node it can continue computing the shortest energy consumption path to reach some node from the next cell. To route along the lowest energy path, we use either iterative power progress or Dijkstra's shortest path tree.

4. Once a real node in the destination cell (the next cell in the path) received the packet it will forward the packet by repeating the process.
   Inside a cell the packets can be forwarded greedily because all nodes in a cell are mutually reachable.

Figure 6 shows an example of the application of the algorithm above. The source node and the target are labeled as 22 and 15 respectively. With greedy routing the packet can only travel until node 2. At that point the algorithm switches to face mode using the left hand rule. Therefore, the packet is sent to the nearest node in the virtual cell $(1, 0)$ using iterative power progress or finding the

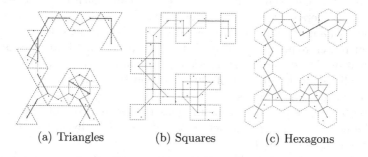

(a) Triangles   (b) Squares   (c) Hexagons

Fig. 5. Second test. Reachable neighbors.

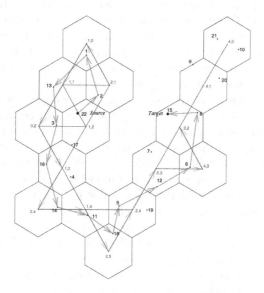

Fig. 6. Face routing using virtual hexagonal graph

local shortest weighted path tree. The face routing in the virtual nodes demands reaching cell $(1,1)$, hence a real node in that cell must be selected. The rest of the example follows similarly. The sequence of virtual nodes, and the sequence of real nodes are depicted in figure 6.

5   Performance Evaluation

We used *connected* random unit disk graphs for our simulations. We considered scenarios consisting of 1000 nodes. The network density is defined as the average number of neighbors per node, which was varied between 4 and 18 with increments of 2. For each density the nodes were placed randomly in the simulation area and 100 different graphs were generated. The size of the simulation area was adapted to preserve the density of the network. Finally, for every graph, we selected 100 different (source,destination) pairs. Thus, each point in the graph represents the average over 10000 routing tasks.

The performance metric we have used to compare our algorithm is the total amount of power required to successfully route a message from the source to the destination. We define *power dilation* as the ratio of the power requirement of the specific algorithm to that of Dijkstra's shortest weighted path algorithm applied over the whole network.

We present the results of our simulations for different densities and energy models based on the general model $u(d) = d^\alpha + c + c_r$, by just adjusting the values of constants α, c and c_r.

Figure 7 shows the power dilation of the proposed schemes as the density of the network increases. As we can see in all three graphs the lower density, the better the performance of our virtual spanner compared with the traditional Gabriel Graph (GG)[GS69] and the more recent Morelia Graph (MG)[BCG+04] planar tests. Neither of the above graphs can guarantee energy optimality of the preserved links. Morelia Graph will preserve longest links to diminish hop distance (perhaps along with short links, but without certainty), while Gabriel Graph cannot guarantee neither long nor short links.

For the GG the density of the underlying unit disk graph will drive the preserved links, for the sparse case most of the links will be preserved. For the dense case only very short links are preserved which are suboptimal when $c > 0$ because a large number of relays are required. For $c = 0$ all the schemas will be nearly the same because short links will be present. Since MG cannot guarantee short edges the energy consumption in this model will be higher for all densities.

The virtual graph, by considering all possible links when finding the shortest path route to the selected node in face mode, is particularly better for low density networks because that is the case when energy can be reduced by using multiple short links rather than a single (eventually long) link. For higher densities, the spanner is also better, because it can use energy efficient links, without using too short edges. However, the overall improvement in dense networks is lower because when density increases, the amount of routing performed in face mode is very low.

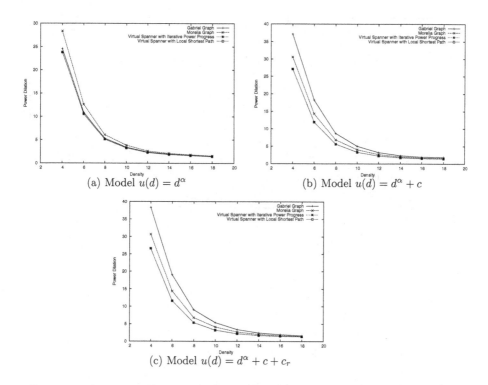

(a) Model $u(d) = d^\alpha$ (b) Model $u(d) = d^\alpha + c$

(c) Model $u(d) = d^\alpha + c + c_r$

Fig. 7. Comparing the efficiency of the Virtual Spanner vs Gabriel Test and Morelia Test when $r = 250$, $\alpha = 2$ and $c = 10000$. Lower is better.

If we compare Fig. 7(a), Fig. 7(b) and Fig. 7(c) we see that the more detailed the energy model, the better the improvement achieved by the virtual spanner both with greedy routing or Dijkstra's algorithm. The reason is that when $c = 0$ and $c_r = 0$ the shorter the edges, the lower the energy consumption. Thus, GG performs particularly well for higher densities because only very short edges are preserved. In fact, in that case our protocols only manage to improve over GG a little bit in low density networks, where links preserved by GG may be longer and face routing may select a link whose cost is higher than the use of two or more links to reach the same node. However, when $c > 0$, GG is suboptimal at increasing density because the reduction of d does not compensates the additional number of transmissions that are needed to reach the destination (each of them having a cost of at least c units). We can see that for energy models in which $c > 0$ our proposed scheme is able to offer up to a 40% energy reduction compared to the same protocol which currently uses face routing based on GG and is reported in the literature as the best solution to date. MG is in disadvantage in the simplest model but as the model becomes mode accurate it outperforms GG because the longer edges preserved achieves a better trade-off in terms of energy.

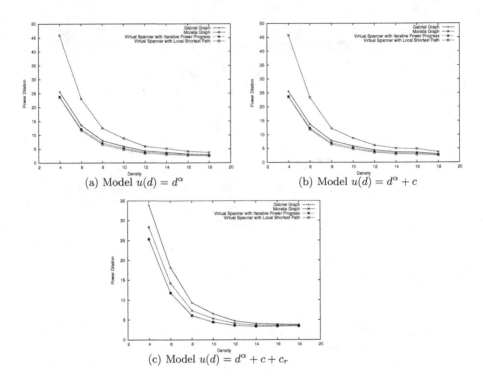

(a) Model $u(d) = d^\alpha$

(b) Model $u(d) = d^\alpha + c$

(c) Model $u(d) = d^\alpha + c + c_r$

Fig. 8. Comparing the efficiency of the Virtual Spanner vs Gabriel Test and Morelia Test when $r = 250$, $\alpha = 4$ and $c = 10000$. Lower is better.

To analyze the impact of the radio attenuation factor (α) we conducted experiments in which $\alpha = 4$. As we see in figure 8, again a lower density produces a higher difference across approaches for the reasons given above. However, the difference in performance is reduced compared to the previous case. The reason is that with $\alpha = 4$, again reducing distance becomes more effective (i.e. relative importance of the constants c and c_r compared to d^α is lower). However, as we see in Fig. 8(a), 8(b) and 8(c), our proposed scheme still outperforms the GG and MG for all densities.

6   Conclusions

We have presented the first localized scheme to reduce energy consumption in face mode for geographic routing protocols. Our strategy is to retain all the links in the wireless network, giving the protocol the ability to select energy-efficient paths even in face mode. Links are retained by using a regular tesselation of the plane, and defining a virtual planar spanner superimposed to the underlying network graph. A virtual link is defined between two cells if there is at least a node in each of them so that those nodes are within radio range one of another. The virtual spanner is used to perform face routing, selecting virtual nodes according to the traditional face routing algorithm. Nevertheless the routing is performed in the underlying network all the time, virtual nodes are used only as routing anchors. Energy efficiency is achieved by selecting energy efficient paths between cells connected by the selected virtual edge.

This scheme does not require any additional control overhead compared to existing geographic routing protocols and all computations are performed based solely on local information. The best performance in all the experiments is achieved with the Local Shortest Path in our virtual spanner. This schema outperform iterative power progress with face routing in both Gabriel Graph and Morelia Graph. In particular, for low densities our experiments report that energy consumption can be reduced up to 40%.

References

[BCG+04] P. Boone, E. Chavez, L. Gleitzky, E. Kranakis, J. Opartny, G. Salazar, and J. Urrutia. Morelia test: Improving the efficiency of the gabriel test and face routing in ad-hoc networks. *Lecture Notes in Computer Science*, 3104:23–24, January 2004.

[BMSU01] P. Bose, P. Morin, I. Stojmenovic, and J. Urrutia. Routing with guaranteed delivery in ad-hoc wireless networks. *ACM/Kluwer Wireless Networks*, 7(6):609–616, 2001.

[GS69] K. Gabriel and R. Sokal. A new statistical approach to geographic variation analysis. *Systematic Zoology*, 18:259–278, 1969.

[KNS04] J. Kuruvila, A Nayak, and I. Stojmenovic. Progress based localized power and cost aware routing algorithms for ad hoc and sensor wireless networks. *Lecture Notes in Computer Science*, 3158:294–299, January 2004.

[RM98] V. Rodoplu and T. Meng. Minimum energy mobile wireless networks. In *ICC'98: Proceedings of the 1998 IEEE International Conference on Communications*, volume 3, pages 1633–1639. ACM Press, June 1998.

[SL01] I. Stojmenovic and X. Lin. Power aware localized routing in ad hoc networks. *IEEE Trans. Paralled Distributed Systems*, 12(10):1023–1032, 2001.

[Tou80] G. Toussaint. The relative neighbourhood graph of a finite planar set. *Pattern Recognition*, 12(4):261–268, 1980.

Energy Efficient Multipath Routing Using Network Coding in Wireless Sensor Networks

Li Shan-Shan, Zhu Pei-Dong, Liao Xiang-Ke,
Cheng Wei-Fang, and Peng Shao-Liang

School of Computer, National University of Defense Technology,
ChangSha, 410073, China
{shanshanli, pdzhu, xkliao, wfangch, slpeng}@nudt.edu.cn

Abstract. Multipath is a sought-after mechanism to achieve reliability along the error-prone channel in wireless sensor networks. However, this technique is not energy efficient since sensor networks are subject to strict resource constraints. In this paper, we propose an energy aware method to combine multipath routing with practical network coding technique. Through this method, we can guarantee the same reliability while reduce much energy consumption by decreasing the number of paths needed to delivery data. This method only needs little metadata overhead and some small-scale linear operations. Simulations under different circumstances verify the theoretical results. The paper also discusses other advantages of network coding in multi-receiver case to illustrate our future work.

1 Introduction

Wireless sensor networks are a distributed, self-organization solution to provide sensing and computing in various environments where conventional networks are impractical [1][2][3]. Such distributed sensor networks have been widely used for applications ranging from environment monitoring to military surveillance and disaster rescue. However, reliability in wireless sensor networks is an intractable issue due to the limited power and computing capability of little sensors. In many applications such as chemical leak diagnosis, the leaking information is critical and requires a highly reliable delivery. Therefore, some measures should be taken to guarantee the desired reliability. Here desired reliability is defined as the probability with which the source wants the packet to be delivered to the sink. Recent research often adds redundancy to data transmission such as multipath or retransmission to help provide the desired reliability [5-12].

However, the energy consumption is one of the major constraints in sensor networks. These reliability mechanisms are not energy efficient since they introduce much redundancy that shortens the network lifetime. Network coding [13-19] can solve this problem effectively. Network coding is a novel mechanism proposed to improve network throughput, transport reliability, energy consumption especially in broadcast environment. Network coding has many potential applications in wireless sensor networks. In this paper, we mainly consider its effectiveness in reliable data

T. Kunz and S.S. Ravi (Eds.): ADHOC-NOW 2006, LNCS 4104, pp. 114–127, 2006.

delivery. We can see in the following that through this coding technique, network would be less sensitive to the sudden failure of nodes, thus enhance the reliability of data delivery especially for sensor networks such an error prone environment. We always guarantee reliability through redundant data transmission. With network coding, we can guarantee the same reliability while reduce redundancy as much as possible. In section 3 we will give the theory of Practical Network Coding in detail.

In this paper, we combine the multipath routing with the network coding to obtain the desired reliability in an energy-efficient way. In order to provide reliable data delivery while reduce the redundancy, a probability method is used to calculate the minimum number of paths to transmit the encoded data segment. It's well known that communication dominates the energy consumption in wireless sensor networks. Therefore, we should reduce the amount of transmission as much as possible. Network coding can achieve this by reduce number of paths needed, which means that it reduce the number of data copy needed to be transmitted. As long as the number of packets that sink receives is equal to or large than a lowest bound h, the original messages can be reconstructed at the sink in a desired probability r.

The rest of this paper is organized as follows. In Section 2, we briefly describe some related work. A practical network coding method is presented in Section 3. Section 4 gives an overview of multipath routing. In Section 5, we show the combination of these two schemes using probability method and focus on how to compute the path number. Several simulations described in Section 6 verify the obtained theoretical results. Section 7 discusses other applications of network coding in sensor networks. The paper ends with conclusion and directions for the future.

2   Related Work

The network coding theory started with Ahlswede *et al.*'s seminal paper [13], where they show that the multicast capacity, which is defined as the maximum rate that a sender can communicate common information to a set of receivers, can be achieved via network coding, but cannot be achieved by routing in general. Li et al. [14] show that linear network coding is sufficient to achieve the multicast capacity. Koetter and M´edard [15] provide an algebraic framework for linear network coding and prove the existence of linear time-invariant codes achieving the multicast capacity. An overview of network coding and its possible Internet applications is given in [18]. However, most of these previous works on network coding are largely based on theoretical calculations that assume a detailed knowledge of the network topology, and a centralized knowledge point for computing the distribution scheme. However, little effort has been made to study the practical aspects of the implementing network coding on a real distributed setting. In this regard [16] proposes a practical network coding system for streaming content. Based on the theory in [16], [19] provides a practical end-system cooperative architecture that uses network coding to enable efficient large-scale content distribution. To the best of our knowledge, little work applies the practical network coding to the wireless sensor networks.

Much of the existing work related to reliability in sensor networks deals with multipath and retransmission techniques. In this paper, we choose multipath technique. The same principle may work on retransmission technique. Since network coding can combine with most of multipath techniques, we haven't studied any specific multipath routing protocol and give a brief overview of current multipath technique. In [5], multiple paths from source to sink are used in diffusion routing framework [12] to quickly recover from path failure. The multiple paths provided by such protocols could be used for sending the multiple copies of each packet. [6] proposes a multipath-multipacket forwarding protocol for delivering data packets at required reliability based on data priority using a probabilistic flooding scheme. [7] proposes a lightweight end-to-end protocol called ReInForM to deliver packets at desired reliability. Similarly for mobile ad-hoc networks, different multipath extensions to well known routing algorithms have been proposed [8][9].

3 Practical Network Coding

In this section, we explain concisely the theory of Practical Network Coding [16]. Suppose that a node has a data collection that has been divided into several groups. Each group contains h data segments with equal length; Each segment is tagged with a group identifier and a h-bit *coefficient vector* $\vec{e}_i$. For segment Bi ($i = 1...h$), its coefficient vector $\vec{e}_i$ is defined as: the value of all the elements in the vector is 0 except that the ith one is 1. So all vectors in each group form an $h \times h$ identity matrix. Figure 1 gives the transmitting packet format:

Group Identifier	Coefficient Vector	Data Segment

Fig. 1. The packet format when using network coding

By default, all data segments described in the following part belong to the same group. When the node wants to transmit information, it first produces h random numbers $r_1, r_2 \cdots, r_h$ and codes the h data packets by equation (1) and (2):

$$E = r_1 B_1 + r_2 B_2 + \cdots + r_h B_h \tag{1}$$

$$\vec{\alpha} = r_1 \vec{e}_1 + r_2 \vec{e}_2 + \cdots + r_h \vec{e}_h \tag{2}$$

Then new data segment E and coefficient vector $\vec{\alpha}$ are produced which forms a new data packet. We can see an example in figure 2, when node S wants to send packet to node A and node B, it produces two groups of the new segments and coefficient vectors and transmit them to A and B respectively. Furthermore, Not only the source node but also the intermediate node can encode the data packets when it has received multiple data packets. For any receiver t, if it has received h (or more than h) pieces of data packet that belongs to the same group, suppose they are E_1, E_2, $\cdots$, E_h, then t judges whether the h coefficient vectors α_1, α_2, $\cdots$, α_h are linear independent. If

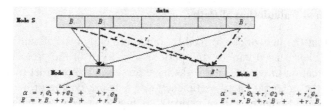

Fig. 2. An example for data encoding

$$
\begin{bmatrix} E_1 \\ E_2 \\ \vdots \\ E_h \end{bmatrix} = \begin{pmatrix} r_1^1 & \cdots & r_h^1 \\ \vdots & \ddots & \vdots \\ r_1^h & \cdots & r_h^h \end{pmatrix} \begin{bmatrix} B_1 \\ B_2 \\ \vdots \\ B_h \end{bmatrix} \Rightarrow \begin{bmatrix} B_1 \\ B_2 \\ \vdots \\ B_h \end{bmatrix} = \begin{pmatrix} r_1^1 & \cdots & r_h^1 \\ \vdots & \ddots & \vdots \\ r_1^h & \cdots & r_h^h \end{pmatrix}^{-1} \begin{bmatrix} E_1 \\ E_2 \\ \vdots \\ E_h \end{bmatrix}
$$

Fig. 3. Original data recovery

yes, the original data segments can be recovered through the matrix G that is made up of α_1, α_2, $\cdots$, α_h, as we can see in figure 3.

Therefore, the h original data pieces can be decoded by solving a set of linear equations after collecting any h data segments, assuming that the coefficient vectors are linearly independent. This will be true with high probability if the random numbers lie in a finite field of sufficient size Fq [20]. For example, if $q = 2^8$, the success ratio of network coding is 99.6%, marginally affected by h. The relation of the probability of linear independency as a function of finite field size can be found in Table 1. It's implied that the coefficient vectors does not need much storage space, which is well suitable for the energy constraint characteristic of little sensors.

4 Integration of Network Coding with Multipath Routing

In this section, we present how to predict the number of paths needed that will successfully deliver a message among the multiple paths in the desired reliability r using network coding. According to the probability estimate in table 1, a 2^8 Fq size is chosen for coefficient vector.

Table 1. Probability of linear independency as a function of finite field size (q)

q	Probability	q	Probability	q	Probability
2^1	0.288788	2^5	0.967773	2^9	0.998043
2^2	0.688538	2^6	0.984131	2^{10}	0.999022
2^3	0.859406	2^7	0.992126	2^{11}	0.999511
2^4	0.933595	2^8	0.996078	2^{12}	0.999756

4.1 Theoretical Calculation of Paths

We assume that the network is formed by random deployment of sensors in a field. Sensors send their sensed data to sink periodically. Most of the protocols for multipath communication in wireless networks have proposed to construct routes [5]. Here we only care the potential energy efficiency that network coding brings to multipath data delivery because even without network coding, the process of path construction cannot be avoided, thus we assume this process is well done in advance, resulting in multiple paths that are to be used. If source has fewer neighbors than the number of paths needed, each neighbor will relay more than one packet to simulate multiple paths. The shortest path from source to sink is k hops. For ease of exposition, each hop has the same error rate e. The reliability desired by the application is r.

We first consider the case without network coding. Using above values, the source could compute the number of paths N_s required for delivering each packet at desired reliability to the sink. We suppose that in our studies several paths exist in a thin band between the source and the sink such that the average path length is almost equal to the shortest path length. Thus, to compute the number of paths we can use some formula such as (3) [7]. That's mean we need at least $\lceil N_s \rceil$ paths to transmit each packets (maybe some paths have more than k hops and various error rate e).

$$N_s = \frac{\log(1-r)}{\log(1-(1-e)^k)} \tag{3}$$

When using network coding, data segments can be first grouped and encoded, then transmitted through multiple paths. If only sink receives more than h packets (to form $h*h$ reversible matrix), the h data segments could be decoded and recovered. There are two parameters to be set. The first is how many data segments are proper to form a group? If h is too small, we are unable to take advantage of the potential capability of network coding. But if h is too large, it will take up much storage space that is impractical for little sensors in that a data packet is not more than two hundred bytes in common. Different applications may have different optimal value. Suppose that h is set, the second parameter is how many paths are needed to guarantee at least h successfully delivering paths in the reliability desired r? We could use equation (4):

$$C_m^h(1-(1-e)^k)^{m-h}(1-e)^{kh} + C_m^{h+1}(1-(1-e)^k)^{m-h-1}(1-e)^{k(h+1)} + \cdots + C_m^m(1-e)^{km} = r \tag{4}$$

m is the number of paths needed to transmit h packets. It's not always to find m paths with k hops simultaneity, so we could use some paths to transmit more than one packet. From probability aspect, it's obvious to see that m is less than N_s*h, at most equal. That is to say each packet only needs $\frac{m}{h}$ paths instead of N_s to obtain the desired reliability. It's well known that communication dominates the energy consumption in wireless sensor networks. Thus, in order to make optimal use of energy, communication should be minimized as much as possible. Network coding technique only needs a few linear operations and several bytes storage to reduce much energy consumption, increasing the lifetime of sensor networks.

However, it's not convenient to solve equation (4), especially for the little sensor. The following part will present an approximate way to simplify computation but keeping the veracity.

4.2 Approximate Measure

Suppose we need m paths in total to transmit the h packets. Each path has some rate $(1-e)^k$ that corresponds to the probability of successfully delivering a message to the destination. So the probability of n ($1 \le n \le m$) successfully delivering paths out of m paths corresponds to the Bernoulli distribution. Then, the expectation for the total number of successful paths is given by equation (5):

$$E(n) = m(1-e)^k \tag{5}$$

We use normal distribution to approximate the Bernoulli distribution. Here we need at least h successfully delivering paths to reconstruct the original packets in the desired reliability r, so we want to decide the value of m to guarantee $P(n \ge h) \ge r$ holds. In order to approximate by normal distribution $N(\mu,\sigma)$, the mean μ and the standard deviation σ are needed. In our case, μ will be given by the expectation for n, i.e. by the sum of the probabilities of successful delivery along each path, and thus we set:

$$\mu = E(n) = m(1-e)^k \tag{6}$$

$$\sigma^2 = m(1-e)^k (1-(1-e)^k) \tag{7}$$

Obviously, each combination of the multipath degree m and different probabilities $(1-e)^k$ will yield a different normal distribution. To overcome this problem, we transform to the standard normal distribution $N(0,1)$. The random variable

$$n^* = \frac{n-\mu}{\sigma} \tag{8}$$

is $N(0,1)$-distributed. Now, consider a given bound r for the desired probability of being able to reconstruct the original message at the destination after being sent along m paths. For the standard normal distribution, the values of the bound x_r for any given r such that the probability $P(n^* \ge x_r) \ge r$ holds are known. We can see from table 2 some value-pairs. Note that these values are independent from the number of paths m used to send data. Then we transform the argument from equation (8) to get

$$\frac{n-\mu}{\sigma} \ge x_r \Rightarrow n \ge x_r\sigma + \mu \tag{9}$$

$$h = x_r\sigma + \mu = x_r\sqrt{m(1-e)^k (1-(1-e)^k)} + m(1-e)^k \tag{10}$$

paths m. Then we transform the argument from equation (8) to get the formula (9) and obtain the inequation $P(n \geq x_r\sigma + \mu) \geq r$. Now we can therefore use $x_r\sigma + \mu$ as the expectation of the distribution. From equation (10) we get the proper value of m.

4.3 Adaptive Adjustment

Hereto, we have found a way to compute reduced number of paths to transmit data. Actually, when h doesn't approach to infinity, using normal distribution will compute a larger m than the ideal value in equation (4). To approach the ideal value, the source sensor should treat m with some heuristic processing. However, it is even inaccurate after this initialization. In order to minimize the energy consumption, we adopt a dynamic adaptive adjustment to deal with it. Therefore, the sink will check the attained reliability periodically and adjust the path number based on it. How to do adjustment is dependent on the application, some multipath algorithms [6, 7, 10] achieve the optimal value by adjusting the forwarding probability, here we use equation (11) until the desired reliability is met:

$$m_{i+1} = \frac{r * m_i}{r_i^r} \tag{11}$$

5 Performance Evaluation

We have performed several simulations in order to verify the theoretical results. Our goal is to get a better understanding on how well our solution works when using network coding under different situation compared with what is computed by equation (3). The simulations are done on a uniform topology consisting of 200 nodes spread in a square area of 100m*100m. The channel error rate is normally distributed across the field with the mean and deviation varying for different simulation. In all simulations, the source is located in lower left zone of the field and the sink is adjusted for different simulation. The normal packet size is 100 bytes. For the sake of simplicity, MAC protocol can assign a unique channel for every node to prevent possible collisions. Here we assume that in a given sensor network there exists a multipath constructing algorithm that returns several paths from a source sensor to the sink [5] and a multipath routing algorithm that can transmit data packets concurrently.

When transmitting each packet, the source sensor produces m_0 groups of random number to encode h packets and transmit them Once the sink have received more than h packets and the coefficient vectors of which are linear independent, these h original packets can be recovered. We use 2^8 Fq size for coefficient vector. As is explained in the section 5.3, normal distribution will compute a larger m than the ideal value. Figure 4 plots their difference under different r and e. The error means the difference of the number of path between the normal distribution and the ideal case per packet under h = 4.Thus, we use a heuristic method to modify m in simulation to counteract

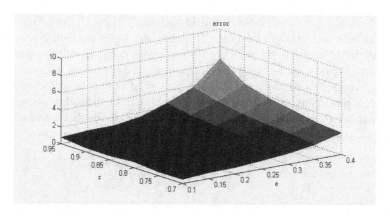

Fig. 4. Difference between the approximate measure and the ideal case under different e and r

this difference in equation (12). Any unpredictable change or inaccuracy incurred by the measurement adopted in this paper will be left to the adaptive adjustment by equation (11).

$$m = \left\lfloor \frac{m}{h} \right\rfloor * h \tag{12}$$

Figure 5 presents the number of paths needed per packet under different desired reliability. Energy consumption of data transmission is decreased dramatically since the decrease of number of paths means decrease of data packets to be transmitted. The black line that represents the transmission without network coding is drawn through equation (3). The red line is the average value after entering the stage through dynamic adjustment when using network coding. With the increase of the desired reliability, the number of paths needed for one packet is increased. However, through network coding, we need not guarantee the delivery of each packet. Instead, only part of

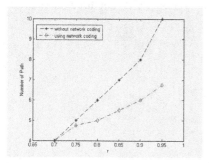

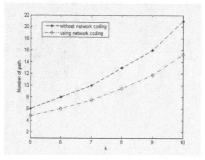

Fig. 5. Number of path needed per packet under different r, here e=0.2,k=6,h=4

Fig. 6. Number of path needed per packet under different k, here e=0.2,r=0.9,h=4

encoded packets is necessary to rebuild the original data packets. Besides, these encoded packets are independent with each other. Therefore, the higher the reliability is desired, the greater the difference of the number of paths needed between with and without network coding. Figure 6 shows the number of paths needed per packet when the distance between source and sink is different. With the increase of number of hop between source and sink, the probability of successfully deliver packet along the path is decreased, and then the number of paths needed for one packet is increased. Similarly, network coding can save much energy consumption by reducing the amount of data transmission.

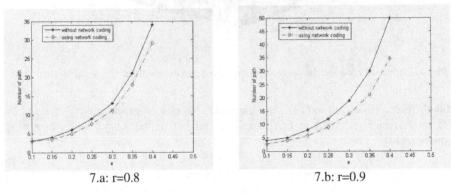

7.a: r=0.8   7.b: r=0.9

Fig. 7. Number of paths needed per packet under different e, here h=4,k=6

Figure 7 gives the comparison under different channel error rate when r is 0.8 and 0.9. We can see that when the channel condition is bad, network coding shows its predominance again, because the worse the channel condition, the more path needed to guarantee the reliability. All these experiments validate the energy efficiency that network coding brings. It's concluded that network coding can save much energy consumed by communication when the network scale is large, channel condition is bad, desired reliability is high.

Figure 8 plots the same comparison when a data group contains different number of packets. The more packets one group contains, the more paths can be saved for each packet. Actually, this can be easily proved in theory. Here we do not discuss it in detail.

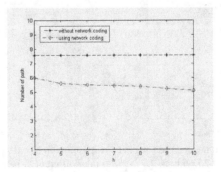

Fig. 8. Number of paths needed per packet under different h, r=0.9,e=0.2,k=6

Finally, we will discuss the additional energy consumption that network coding brings. Since the matrix and vector operation are just linear, their energy consumption is little and neglectable compared with data transmission. Here we only care the energy consumption for metadata transmission (coefficient vectors). In order to see whether this additional energy consumption may counteract the energy saving that network coding brings, we make some comparisons on their amount. Suppose each group contains 4 packets, that is to say the size of a packet will be 104 bytes when using network coding, 4 additional bytes should be transmitted together with all the encoded data packets. For ease of exposition, the comparisons are made in terms of number of bytes (for the reliable transmission of each packet) that network coding can save and metadata overhead.

Figure 9 and figure 10 make the comparisons under different desired reliability and channel error rate respectively. The data transmission overhead incurred by the metadata is very little compared with the energy saving that network coding brings. This can be easily explained since the space that metadata needed is very little compared with the size of a packet.

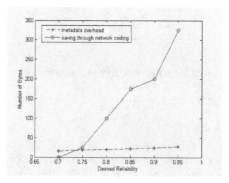

Fig. 9. Number of bytes saved by network coding vs. consumed by metadata per packet under different r,e=0.2,k=6, h=4

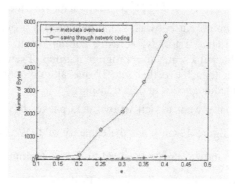

Fig. 10. Number of bytes saved by network coding vs. consumed by metadata per packet under different e,r=0.9,k=6, h=4

But there is an exception in figure 10 when r is 70%. The reason for that is when the desired reliability is low, not many paths are needed, in these condition network coding cannot take effect. Therefore, network coding may not be used when the channel condition is good and the desired reliability is low.

Figure 11 shows the same comparison under different h. From the figure we can see that even if metadata need more memory space with the increase of h, the energy saving is still largely more than the metadata overhead. To sum up, the overhead of metadata needed by encoding and decoding may be neglectable compared with the energy saving that network coding brings.

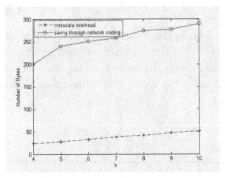

Fig. 11. Number of bytes saved by network coding vs. consumed by metadata per packet under different h, r=0.9,e=0.2,k=6

6 Multi-receiver Case

Now we confirm that network coding can minimize energy consumption in achieving desired reliability from source to one sink. Actually, this advantage may not exhibit its distinguished predominance. Let's return to the original intention of network coding that for a directed graph (V, E), a sender $s \in V$ can communicate common information to a set of receivers $T \subseteq V$ at a rate achieving the broadcast capacity h (the value of the minimum cut between s and any $t \in T$) provided one allows network coding [13,16]. That is to say, network coding would be more effective in **multi-sink** case or when sink wants to reprogram a group of sensors. Conversely, it is generally not possible to achieve this rate if one allows only routing or copying messages at the intermediate nodes in data transport.

Figure 12 shows that s want to deliver two data packet b_1 and b_2 to two sink t_1 and t_2. Each edge in figure 12 only has the capacity of 1. We can see with network coding in node v_6, b_1 and b_2 can be received by t_1 and t_2 simultaneously. Obviously this cannot be achieved without coding in v_6. In this case, network coding improve network throughput.

Suppose in such a situation, two intermediate nodes A and B may both receive two same packets α and β (due to multicast to multiple receivers). Maybe in the downstream delivery of A and B, β is lost due to lossy channel and only α is delivered. While using network coding, A and B can encode α and β respectively again into α_1, β_1 and α_2, β_2. Evidently, the reliability is enhanced since losing any two of these four packets still have chance to recover the original packets. In this case, network coding improves reliability further than in one sink case due to the encoding in intermediate nodes.

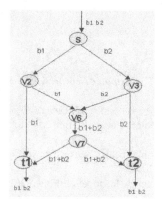

Fig. 12. Throughput improvement

Fig. 13. Energy consumption reduction

Let's see another simple example in figure 13 when sink want to reprogram or "re-task" a group of sensors. Sensor A and sensor B may want to exchange packet information to get complete data block through sensor C. Supposed one transmission consumes one unit energy, exchange of a and b need 4 unit energy in common. While using network coding, a and b can be encoded to a+b in C and then broadcast to A and B, thus only need 3 unit energy. In this case, network coding improves energy consumptions.

Therefore, network coding is much more effective in multi-receiver situation, mainly due to the encoding operations in intermediate nodes. In our future work, we will mainly study how to integrate network coding in multi-receiver case perfectly.

7 Conclusion

This paper describes the combination of multipath routing with network coding technique for energy-efficient reliable data delivery. This technique can guarantee the same reliability while consume the least energy. Actually, it uses a little computational cost to get energy saving in return and just needs several bytes of metadata (coefficient vectors). This feature is important in sensor networks since it can increase network lifetime.

Moreover, due to the large deployment of sensors, data collection in sensor networks may be highly redundant. As part of our future work, we will also study how to utilize the data correlation of dense deployment sensors and apply the practical network coding on clustering techniques. That is to find a way to integrate distributed source coding and practical network coding. We believe this combination will bring on a good tradeoff between data compression efficiency and network robustness.

Acknowledgement

This work is supported by National High-Tech Research and Development Program (Grant No. 2005AA121570, 2002AA1Z2101 and 2003AA1Z2060) and Grant Project 51415040205KG0151.

References

[1] F. Akyildiz et al., "Wireless Sensor Networks: a survey, " *Computer Networks*, Vol. 38, pp. 393-422, March 2002.

[2] H. Karl and A. Willig, "A short survey of wireless sensor networks," *Technical Report TKN-03-018*, Telecommunication Networks Group, technical University Berlin, October 2003.

[3] Dazhi Chen and Pramod K. Varshney, " QoS Support in Wireless Sensor Networks:A survey",Internatioal Conference on Wireless Networks, 2004.

[4] R. Ahlswede, N. Cai, S.-Y.R. Li and R.W. Yeung, "Network Information Flow," IEEE Trans. On Info. Theory, vol. 46, pp. 1204-1216, 2000.

[5] D.Ganesan, R.Govindan, S.Shenker, and D.Estrin, "Highly-resilient, energy-efficient multipath routing in wireless sensor networks", In Mobile Computing and Communications Review (MC2R) Vol 1., No.2,2002.

[6] S.Bhatnagar, B.Deb, and B.Nath, "Service differentiation in sensor networks", In Wireless Personal Multimedia Communication, 2001.

[7] B. Deb, S. Bhatnagar and B. Nath, "ReInForm: Reliable Information Forwarding using Multiple Paths in Sensor Networks", Proc. of IEEE LCN, 2003.

[8] M. Marina and S. Das, "On-Demand Multipath DistanceVector Routing in Ad Hoc Networks", In Proc. of IEEE ICNP, 2001.

[9] M. Pearlman, Z. Haas, P. Sholander and S. Tabrizi, "The Impact of Alternate Path Routing for Load Balancing in Mobile Ad-Hoc Networks", In Proc. of the ACM MobiHoc, 2000.

[10] B.Deb, S.Bhatnagar, and B.Nath, "Information Assurance in Sensor Networks", WSNA, 2003.

[11] Fred Stann, John Heidemann, "RMST: Reliable Data Transport in Sensor Networks", In 1st IEEE International Workshop on Sensor Net Protocols and Applications, 2003.

[12] C.Intanagonwiwat, R.Govindan, and D.Estrin, "Directed diffusion:a scalable and robust communication paradigm for sensor networks", In MOBICOM,pages 56-57,2000.

[13] R. Ahlswede, N. Cai, S.-Y.R. Li and R.W. Yeung, "Network Information Flow," IEEE Trans. On Info. Theory, vol. 46, pp. 1204-1216, 2000.

[14] S.-Y. R. Li, R.W. Yeung, and N. Cai, "Linear network coding," IEEE Trans. Info. Theory, vol. 49, no. 2, pp. 371–381, Feb. 2003.

[15] Ralf Koetter, Muriel Medard. "An algebraic Approach to Network Coding", IEEE/ACM Trans. On Networking, Vol. 11(5), pp.782-795 Oct. 2003.

[16] Philip A. Chou, Yunnan Wu, Kamal Jain, "Practical network coding", In: 41st Annual Allerton Conference on Communication Control and Computing, Oct. 2003.

[17] J. Widmer and J. Boudec, "Network Coding for Efficient Communication in Extreme Networks", in Proc. ACM SIGCOMM WorkShop WTDN'05, Philadelphia, PN, Aug. 2005.
[18] Philip Chou, Yunnan Wu, and Kamal Jain, "Network coding for the Internet," in *IEEE Communication Theory Workshop*, Capri, 2004, IEEE.
[19] C. Gkantsidis, P.R. Rodriguez, "Network coding for large scale content distribution", INFOCOM 2005.
[20] M. Medard, S. Acedanski, S. Deb and R. Koetter, "How good is Random Linear Coding Based Distributed Networked Storage?", in *Proc. NETCOD'05*, Italy, Apr. 2005.

Formal Verification and Simulation for Performance Analysis for Probabilistic Broadcast Protocols

Ansgar Fehnker and Peng Gao

National ICT Australia* and University of New South Wales

Abstract. This paper describes formal probabilistic models of flooding and gossiping protocols, and explores the influence of different modeling choices and assumptions on the results of performance analysis. We use PRISM, a model checker for probabilistic systems, for the formal analysis of protocols and small network topologies, and use in addition Monte-Carlo simulation, implemented in MATLAB, to establish if the results and effects found during formal analysis extend to larger networks. This combination of approaches has several advantages. The formal model has well defined synchronisation primitives with clear semantics for modeling synchronous and asynchronous communication between nodes. Model checking of the probabilistic model determines exact probabilities and performance bounds, even if the model is non-deterministic; results that cannot be obtained by simulation. However, Monte-Carlo simulation can then be used in addition to study effects that only emerge in larger networks, such as phase transition.

1 Introduction

Wireless sensor networks is an emerging field that has received a lot of attention in recent years. Characteristic features of wireless networks are that each they gather information about the network and the environment in a distributed fashion, and that nodes use multi-hop communication on an unreliable medium. Each node has its own processor, radio, antenna, and clock. The processors operate under tight energy restrictions because nodes are either battery operated or rely on abient energy sources. This combination of characteristics calls for simple and robust distributed algorithms, that require few computing cycles, and thus little processing power.

Many protocols for wireless sensor networks employ some form of the very simple flooding protocol to acquire or distribute information. Flooding a message means that each node that receives a message, propagates it to all its neighbours by broadcast. This introduces an unnecessary redundancy, because a node may receive a message multiple times. Gossiping protocols introduce a random element to reduce this redundancy. Gossiping means that each node decides with a

---

* National ICT Australia is funded through the Australian Government's *Backing Australia's Ability* initiative, in part through the Australian Research Council.

T. Kunz and S.S. Ravi (Eds.): ADHOC-NOW 2006, LNCS 4104, pp. 128–141, 2006.

certain probability to forward a message on or not. This reduces the probability for node to receive a message multiple times, thus the redundancy and cost.

The main tool in model based development to evaluate the performance of a protocol is simulation. Common simulators in the wireless domain are ns-2, Opnet, and Glomosim, but it is not uncommon to build a customised simulator, for example, in Java or Matlab. Flooding and gossiping protocols have been studied before, for example in [1,2,3,4]. It has been observed in [5] that different simulators can produce vastly different results, even for a simple protocol such as flooding. The reason is that simulators employ different models for the MAC and physical layers. The results of the simulators say as much about the protocol, as about the particular the lower level implementation of the simulator.

Protocols are a traditional subject for analysis in formal methods, and wireless protocols are no exception. The model checker PRISM, for example, has been used to analyse the randomised backoff procedure in the 802.11 wireless protocol [6]. Flooding and gossiping protocols were examined in [3] and [4]. The latter combine simulation studies and manual analytic evaluation. The formal analysis in [3] deals with the correctness of the protocol, while simulation is used to assess the performance. The analysis in [4] is manual for a general model of possible topologies, however, for a specific set of assumptions on the synchronisation.

This paper uses model checking for performance analysis of flooding and gossiping protocols. Model checking has been used before for performance analysis, for example, [7] describes for example how PRISM can be used to evaluate a strategy for dynamic power management, and [8] describes a tool based on model checking timed automata that evaluates schedulers for embedded systems.

The main motivation of this paper is the observation that different papers on wireless protocols such as gossiping use different assumptions, which makes comparison of results difficult. For example, [3] and [5] both evaluate the flooding protocol, but sending and receiving is perfectly synchronised in [3], while [5] assumes a random waiting period in-between sending and receiving. In this paper we examine what effect this and similar choices have on the outcome of the analysis. This paper explores in particular the effect of collisions, unreliable channels versus probabilistic broadcast, and the influence of timing. This analysis can explain why the performance result of gossiping for a perfectly synchronised network without collision, are similar to the result on a network with randomised delay and collision, although the actual behaviour is vastly different.

The next section of this paper gives an introduction to flooding and gossiping protocols. Section 3 and 4 introduce the model checker PRISM and Monte-Carlo simulation respectively. The models and the results will be presented in Section 5, and followed by a discussion and summary in Section 6.

2   Flooding and Gossiping

Gossiping is a simple protocol that uses probabilistic broadcast to send or request information in a wireless network. The gossiping protocol can be informally summarised as follows:

– The source node broadcasts the message or request to all its neighbours. It then proceeds to sleep.
– Nodes that receive a message chooses with probability psend to forward the received message, and with probability 1 − psend to ignore the message. In either case the node proceeds to sleep.

Gossiping is equal to flooding, if psend is equal to 1.

An interesting property of gossiping protocols is that for sufficiently large values of the sending probability, the reliability is barely reduced, while the cost of propagating the message is. If the sending probability psend is reduced further, the reliability drops suddenly. This phenomenon is called *phase transition* and the value at which this effect occurs is the *phase transition threshold*.

Numerous modifications of the gossiping protocol have been proposed to improve reliability and efficiency. Halpern et al, for example, propose that nodes within k (k is a constant) hop distance from the source node broadcast received messages with probability 1, reducing the chance that the message dies out completely in the first few steps. This and other modification are presented in Halpern et al. in [1]. Other protocols modifying or building on gossiping protocols can be found in [3,5,9,2].

This paper uses the gossiping protocol described above as baseline. That description, however, is incomplete. It does not specify what happens in case of a collision, i.e. in the case that a node receives two messages at the same time. It does also not specify if the protocol has to deal with an unreliable medium. And it does not mention if the nodes in the network are synchronised.

For our basic model we assume perfect synchronisation and no collision. We will extend it to a model with collision, to determine its effect on the performance results. Another modification is that we introduce lossy channels to model a unreliable medium. Next, we introduce an asynchronous model in which each node can take transitions at a non-determined pace, which covers any possible clock drift or jitter. This is a very conservative assumption, hence we compare in addition different probabilistic models to capture limited drift and jitter.

An important assumption in our models is that the network topology is static. We also assume for simplicity that the source node initiates the protocol with probability psend, i.e. it uses probabilistic broadcast like the other nodes. Note, that the probability of a node to receive the message if the source node sends with probability psend, is psend times the probability if the source node would send with probability 1.

3   The Prism Model Checker

For formal modeling and analysis of the flooding and gossiping protocols we use the probabilistic model checker PRISM, developed at the University of Birmingham [10]. PRISM supports three types of probabilistic models: *Markov decision processes* (MDP), *discrete-time Markov chains* (DTMC), and *continuous-time*

Markov chains (CTMC) [10]. We use the discrete time modeling frameworks of Markov decision processes or discrete-time Markov chains for our models.

A MDP consists of a set of states S, an initial state s_0, and a probabilistic transition relation $Steps \subseteq S \times Dist(S)$, where $Dist(S)$ is the set of distributions over S. The successor state of a state s is determined by first choosing non-deterministically a step $(s, \mu) \in Steps$, and then choosing a successor state probabilistically according to distribution μ. States may be labeled with *atomic propositions*. DTMCs can be viewed as a restriction of MDPs, where all non-deterministic choice has been replaced by probabilistic choice.

The PRISM input language is a state based language for modeling systems as a composition of modules that act on shared variables and synchronise on common actions. Specifications for PRISM models of MDPs and DTMCs are defined in a probabilistic extension of CTL. A probabilistic specification might be that the probability for a certain event to happen is smaller than a certain threshold. PRISM also allows to compute the probability for a given specification to become true for DTMCs, and the minimal and maximal probability for MDPs. The maximal and minimal probability for an MDP refer to the worst and best case resolution of the non-deterministic choices. PRISM can for example compute the (maximal or minimal) probability for a certain node in the network to receive the message. The underlying model checking algorithm of PRISM uses a MTBDD package to store and manipulate transition matrices and distributions efficiently.

4   Monte-Carlo Simulation

Monte-Carlo simulation is a common statistical sampling scheme that solves the problem by generating suitable random (or pseudo-random) numbers and observing what fraction of these random runs obey given properties. In a stochastic process that either has a very complex time evolution or is too big in scale, Monte-Carlo is one of the few feasible and consistent ways to obtain approximate results.

In our particular protocol evaluation scenario Monte-Carlo is used to approximate the DTMC models described in PRISM, but on a larger and more realistic scale, beyond the capabilities of PRISM. The precision of the results and the computation time are both satisfactory as we are sampling the protocol behaviour thousands of times. The randomness of the sampling is guaranteed by the pseudo-random numbers generated in MATLAB. Probabilistic choice is simulated by comparing the random number output of the MATLAB function **rand** with a predetermined probability threshold specified as a parameter of the protocol.

5   Models and Results

5.1   Gossiping Without Collision

The baseline model for our experiments is a PRISM model for gossiping without collision. We furthermore assume that all nodes are perfectly synchronised. We

```
module node4
active4: bool init true;
send4:   bool init false;

[tick]   active4 & !send4 & (send1|send3|send5|send7)
               ->    psend :(active4'=true)  & (send4'=true)
             + (1-psend):(active4'=false) & (send4'=false);
[tick]   active4 & !send4 & !(send1|send3|send5|send7)
               -> (active4'=true)  & (send4'=false);
[tick]   active4 &  send4
               -> (active4'=false) & (send4'=false);
[tick] !active4 -> (active4'=false) & (send4'=false);
endmodule
```

Fig. 1. PRISM model for the gossiping protocol without collision and sending probability psend

choose for the network topology a 3 by 3 square grid. Each node can receive packets from its immediate neighbours. The source node is in one of the corners of the grid. The nodes are numbered 0 to 8, increasing with hop distance form the source. Fig. 1 depicts the model for the central node 4.

The module for node 4 has two state variables active4 and send4. The node is active and listing when active4 is true, and ready to send if send4 is true. The variables are readable by the other nodes. The nodes synchronise on label tick, i.e. they all update their state vector at the same time.

The first transition in Fig. 1 fires when the node is active and not sending, and when one of its neighbours is ready to send. In this case the node will choose with probability psend to remain active and to be ready to send in the successor state. Alternatively, it chooses with probability 1-psend to become inactive and to not send. This transition models the probabilistic choice to either propagate a received message, or to become inactive.

The second transition models that a node remains active and does not send a message, if none of its neighbours is currently ready to send. The remaining transitions model that a node becomes inactive once it broadcasts a message, or remains inactive, if it is inactive. The modules for the other nodes differ only in the number and names of the neighbours. The composition has no non-deterministic transition; the composition hence falls in the class of DTMCs. The composed model has just 65 states and 140 transitions.

The results for the PRISM model are depicted in Fig. 2(a). The nodes one hop from the source receive the message with probability psend. This probability declines with an increasing hop distance. The central node 4 has a higher probability of receiving the message than node 3 and 5 with the same hop distance. The pairs 1 and 2, 3 and 5, and node 6 and 7, have exactly the same receiving probability, due to the symmetry of the grid. Note, that the probabilities that were computed for the PRISM model are exact probabilities, rather than averages over a big number of experiments.

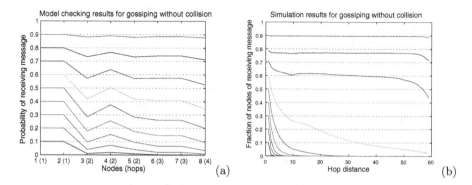

Fig. 2. (a) Probability of receiving the message for each node in the PRISM model. Sending probability ranges from 0.1 to 1 in increments of 0.1. (b) Results of the Monte-Carlo Simulation for a 20 by 50 network in MATLAB.

For the Monte-Carlo simulation we model the protocol in MATLAB to explore the behaviour of gossiping protocol on a medium size 20×50 gird. The MATLAB simulation model makes the same communication assumptions as the PRISM model, in particular we assume that there are no collisions. The gossiping protocol is initiated by the node on the middle of the narrow edge of the grid. For a reasonable precision, we computed the average for 3000 runs.

Similar to the PRISM model, the simulation model has 2 state variables, $active(i)$ and $send(i)$, for each node i. The node is active and listening to the medium if $active(i)$ is true, and ready to send if $send(i)$ is true. All active and sending state of all nodes updated sequentially in the same iteration. This loop through the sending and active vector leads to a model in which sending and receiving is perfectly synchronised.

The results for the MATLAB model are presented in Fig. 2(b). As for the PRISM model, we find that the fraction of nodes with hop distance 1 receiving the message is approximately equal to the sending probability psend. A phase transition can be observed between the sending probability 0.6 and 0.7. If the sending probability is 0.7, the fraction of nodes receiving the message remains mostly above 0.6, while if the sending probability is 0.6 most part of the receiving probability curve remains below 0.3. Also note that there is an interesting 'bump' phenomenon at a hop distance of about 10 and sending probability 0.6 to 0.8. This is the point where the propagation of the message reaches the boundary of the 20×50 grid. Neither the phase transition nor the 'bump' effect can be observed in PRISM model, as the grid size is too small for those effects to occur.

5.2   Gossiping with Collision

An important characteristic of the wireless domain is that transmissions are prone to collisions with messages from other nodes. A simple modification of the baseline PRISM model takes collisions into account (Fig. 3). The model uses two integer variables active4 and send4, instead of boolean variables, to denote

```
module node4
active4:[0..1] init 1;
send4:   [0..1] init 0;

[tick] active4=1 & send4=0 & send1+send3+ send5 +send7 = 1
                   ->      psend : (active4'=1)&(send4'=1)
                    + (1-psend):(active4'=0)&(send4'=0);
[tick] active4=1 & send4=0 & send1+ send3+ send5 +send7 !=1
                   -> (active4'=1) & (send4'=0);
[tick] active4=1 & send4=1 -> (send4'=0)& (active4'=0);
[tick] active4=0  -> (send4'=0)& (active4'=0);
endmodule
```

Fig. 3. PRISM model for the gossiping protocol with collision

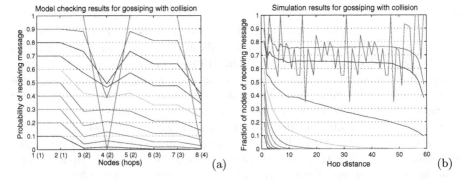

(a)   (b)

Fig. 4. Results for gossiping with collision. (a) Model checking results. (b) Monte-Carlo simulation.

if a node is active or sending. A node receives a message if it is active and not sending, and if the number of sending neighbours is equal to 1. It will not receive a message if there is no or more than one sending neighbor. The resulting model has 72 states and 146 transitions.

The results for the PRISM model in Fig. 4 (a) show that the receiving probability is significantly smaller for nodes 4, 6, 7 and 8, compared to the result for gossiping without collision. For nodes 4 and 8 we observe even a sudden decline of reliability when the sending probability exceeds 0.7, due to an increased chance for collision. The probability that those nodes receive a message is even equal to zero for a sending probability of 1, due to collisions.

Monte-Carlo simulation confirms these results. Compared to gossiping in the ideal no-collision scenario, a smaller fraction of nodes receives the message. The fraction even jumps erratically in a band from 0.4 to 1 for sending probability 1 (i.e. flooding) due to the vast amount of collisions. The phase transition is still visible, now between sending probability 0.8 and 0.7, rather than between 0.7 and 0.6 as in the ideal model.

```
module scheduler
turn: bool init true;

[tick]  !turn ->(turn' = !turn);
[tock]  turn -> (turn'= false);
endmodule

module chan01 //lossy channel between node 0 and 1
buff01: [0..1] init 0;

[tock]  send0=1 | send1=1 -> precv:(buff01'=1)+(1-precv):(buff01'=0);
[tock]  send0!=1 & send1!=1 -> (buff01'=0);
endmodule
```

Fig. 5. PRISM model for scheduler and lossy channel

5.3   Probabilistic Broadcast and Lossy Channels

Another important characteristic of wireless networks is the unreliable medium.
Each message has a positive chance to get lost due to external disturbances. The
wired equivalent is a lossy channel in which the message is send with probability
one, but received with probability `precv`. Although there exist no fixed channels
between nodes in the wireless domain, the link between nodes can be suitably
modeled as a lossy channel.

The PRISM model for flooding on lossy channels includes for each channel a
module with a one place buffer; variable `buff01` in Fig. 5. The buffer will take
the value 1 with probability `precv`, if a node sends via this channel. The model
includes a scheduler to ensure that the nodes and the channels are updated alter-
natingly. Transitions labeled `tick` update the state of the nodes and transitions
labeled `tock` update the state of the channels. The resulting PRISM has 12856
states and 76732 transitions.

The results for the PRISM model are depicted in Fig. 6. The probability of
receiving the message increases compared to gossiping with collision for all nodes
in the network. The lossy channel cancels some of the effects of collisions. If two
nodes send a message at the same time, one of the two messages might get lost,
such that the other can be received uncorrupted. The probability that node 4
receives a message for `precv=0.5` is 0.44 on lossy channels, compared to about
0.30 for gossiping with collision and `psend=0.5`, and 0.375 for gossiping without
collision.

On first sight this is a surprising result. The neighbours of a node that sends
with probability p on a reliable channel receive the message with probability
p. Sending with probability 1 on a lossy channels with probability p, yields the
same probability that the neighbours receive the message. In [2] it is conjectured
that probabilistic broadcast is equivalent to removing links probabilistically from
the network. The latter is equivalent to a lossy channel, if we assume that only
one message is send. The results, however, show that this is not true for a

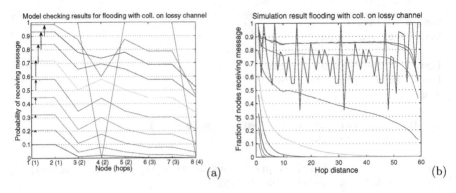

Fig. 6. (a) Probability of receiving the message for each node in the PRISM model of flooding on lossy channels with collision. Probability `precv` ranges from 0.1 to 1 in increments of 0.1. The arrows on the left hand side denote the increase of the receiving probability in nodes 1 and 2 with respect to probability `precv` of the lossy channel. (b) Results of the Monte-Carlo simulation.

multi-hop network. In general, probabilistic broadcast and lossy channels (and probabilistically removing links) show different behaviour.

A reason for this difference is that in a probabilistic broadcast scheme the probability that two nodes receive a message is correlated. If node 0 decides to send a message with probability 0.5 to node 1 and 2, then either both nodes will receive a message or none of the nodes. The probability for the message to die out in this step is 50%. If node 0 sends a message to node 1 and 2 on lossy channels with receiving probability 0.5, then there is only a 25% chance that the message dies out, i.e that none of the nodes receives the message. If one node does not receive the message directly, there is still a positive chance that it receives it via another route. This explains that node 1 and 2 receive the message with a probability of about 0.58 rather than 0.5.

The result of the PRISM model is confirmed by the Monte-Carlo simulation. Fig. 6(b) also shows phase transition as expected, but similar to gossiping without collision between receiving probability 0.6 and 0.8. Note also, that the result for receiving probability `precv`=1, is the same as for gossiping with collision and `psend`=1, as both are equal to flooding.

5.4   Non-deterministic Execution Order

All models presented thus far have been synchronous in the sense that all nodes (and all channels) update their state synchronously. Nodes that receive a message at the same time, will respond to it at the same time. In a wireless network such synchronicity is typically absent. Nodes operate independently and have each their own clock that might drift or jitter.

Asynchronous models are traditionally used in Formal Methods to model concurrent systems that operate independently and synchronise on shared actions and variables. Transitions in the different components may interleave in any

```
module node4
active4:[0..1] init 1;
rcv4:   [0..2] init 0;

[msg1] active4=1 & rcv4= 0 ->   psend: (active4'=1)&(rcv4'=1)
                              + (1-psend):(active4'=0)&(rcv4'=0);
(...)
[msg1] active4=1 & rcv4!=0 -> (active4'=1) & (rcv4'=2);
(...)
[]      active4=1 & rcv4= 2 -> (active4'=1) & (rcv4'=0);
[msg1] active4=0           -> (active4'=0) & (rcv4'=0);
(...)
[msg4] active4=1 & rcv4= 1 -> (active4'=0) & (rcv4'=0);
endmodule
```

Fig. 7. PRISM model with non-deterministic execution order

order, unless they synchronise explicitly. The model checking problem is then to check a specification for any possible interleaving and execution order.

The corresponding PRISM model for gossiping with a non-deterministic execution order is a MDP. All previous models were DTMCs. MDPs combine non-determinism and probabilistic choice, and the model checking problem becomes to compute the maximal and minimal probability for a specification to become true. The specification in our case is that a given node receives the message. The maximal and minimal probability correspond to the best and worst case execution order. Model checking thus gives firm upper and lower bounds on these probabilities. Monte-Carlo simulation cannot be used to produce similar results. Any simulation has to choose particular execution orders, and the Monte-Carlo simulation can produce at best averages, but no conservative bounds.

Fig. 7 depicts part of the PRISM model for the central node 4. We assume that a node receives the message as soon as its neighbour sends it. However, receiving a message and responding to it may be interleaved with any number of transitions of other nodes. The first transition of module node4 models reception of a message from node1. Both nodes synchronise on label msg1. There are similar transitions to model reception of messages from the other neighbouring nodes, which were omitted from Fig. 7. The integer variable rcv4 records if no message, one message, or more than one message has been received, with rcv4=2 if a collision occurred. If a node detects a collision, i.e. active4 =1 & rcv4=2, it may return to the initial state. Since this is an internal action, it has no synchronisation label. If a node did receive exactly one message it can broadcast. Node 4 synchronises on label msg4 with the neighbouring nodes to realise a broadcast.

The execution order has little impact on the performance of gossiping without collision. Synchronous execution realises the best result for this model. We get a different picture for the gossiping with collisions. The results in Fig. 8(a) show that the difference between upper and lower bound increases as the sending

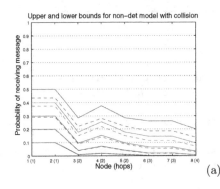
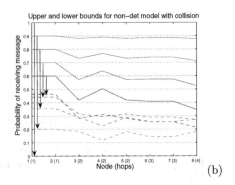

Fig. 8. (a) Upper bound (solid line) and lower bound (dashed) on the probability of receiving the message. Sending probability ranges from 0.1 to 0.5 in increments of 0.1. (b) Upper and lower bounds for sending probability from 0.6 to 1. The arrows indicate which upper bound belongs to which lower bound.

probability `psend` increases from 0.1 to 0.5. The upper bound is equal to the probability of the synchronous model without collision. The lower bound does not correspond to a particular execution order. It is reached for each node by a different execution order. In contrast, the result of Monte-Carlo simulation, such as depicted in Fig. 4(b), shows only some possible behaviour. Monte-Carlo simulation gives neither the best, nor the worst case, and cannot give any firm lower bounds.

The results in Fig. 8(b) for sending probabilities between 0.6 and 1 are even more telling. The lower bound of gossiping with `psend` of 0.6, for example, is higher than the lower bound for `psend`=0.8. Sending with 0.6 thus gives a higher guaranteed performance, while it is at the same time more efficient. Its performance is also in a narrower band, making it more predictable under different execution orders. Different orders may be the consequence of clock drift and jitter, different battery levels and thus clock speeds, or the placement of the nodes.

5.5   Unreliable Timing

Assuming any execution order is a very conservative assumption. Although clocks may drift and jitter, they all proceed positive rate in about the same range. In this section we will consider different timing models for gossiping that contain different types of random delay. This random delay can also be viewed as part of a randomised protocol that introduces a random waiting period to reduce collisions. A variant of the flooding protocol with a randomised delay was for example used in [5].

A simple way to model unreliable timing is to postpone sending the message with a certain probability `pdelay`. If sending has been postponed, it will be send with probability 1-`pdelay` in the next step, or be postponed another time with probability `pdelay`, etcetera. This is very simple memoryless model of unreliable

```
[tick] active4=1 & send4=0 & send1+send3+send5+send7 =1
               ->(1-psend):              (active4'=0)&(send4'=0)
             +    psend*(1-pdelay):  (active4'=1)&(send4'=1)
             +    psend*   pdelay:  (active4'=2)&(send4'=0);
[tick] active4=2 ->(1-pdelay):  (active4'=1)&(send4'=1)
             +    pdelay:  (active4'=2)&(send4'=0);
```

Fig. 9. Transition modeling sending and delay of the PRISM for gossiping model with collisions and simple memoryless delay

timing, and the delay is governed by a geometric distribution with an expected waiting time of $1/(1\text{-pdelay})$. Fig. 9 shows the two transitions that model sending and delay. The rest of the model is the same as the model for gossiping with collision, shown in Fig. 3.

Fig. 10 shows the results of PRISM for a sending probability psend=0.8, and a delay probability pdelay ranging from 0.1 to 0.9. We chose a sending probability of 0.8 for this comparison because it is above the threshold for phase transition. The results show that the receiving probability increases in nodes 4, 6, 7 and 8, as the probability of delay increases. This, because those nodes are prone to collisions, and more delay reduces the chance of collision. In contrast, there is very slight decrease in the receiving probabilities of node 3 and 5 in the corner of the grid. They have only two neighbours, each with a different hop distance to source, such that collisions are impossible in a synchronous setting. Probabilistic delay now introduces a slight chance for collision. But despite the decrease, the receiving probability is for all nodes close to the upper bound which was obtained for the model with non-deterministic execution order. The receiving probability of node 8 was 0.41 for the synchronous model with collision, and it increased to 0.69. This is an increase of 68%, just 3.4% under theoretical upper bound, obtained with the non-deterministic model.

We considered also two other models for unreliable timing. One is similar to the previous one, except that a node waits for a fixed amount of time after receiving the message. After this waiting period it decides probabilistically to delay or to send. If the waiting time is for example 10, there is a chance of 1-pdelay that the node sends at 10, and pdelay chance sending will be postponed. Since we assume no fixed time scale, this would be equivalent with a model in which some nodes send at 1, others at 1.1, 1.2 etcetera. We found that this additional waiting time lead only to a slight increase of the performance. We obtained similar results for a model that did send uniformly distributed in an interval, up to 10 time units. Apparently, all three approaches are effective in reducing the impact of collisions.

For the simple model of randomised delay we find for the 20 by 50 grid that increasing the delay increases the performance on average. The fraction of nodes that receives the message in a synchronous setting with collision was on average 22% lower than the fraction for the synchronous protocol without collision. Increasing the probability of delay to 0.9 reduces this gap to 1.2%. The performance for the protocols with a delay probability of 0.8 and 0.9 overlap, due to sampling

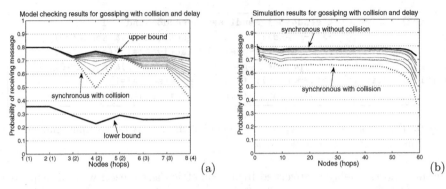

Fig. 10. (a) Result for gossiping with probabilistic delay. The probability of delay ranges from 0.1 to 0.9 in increments of 0.1. Bold lines are the upper and lower bound for non-deterministic execution order, the dotted line is the performance of the synchronous model. (b) Results of the Monte-Carlo simulation for a 20 by 50 network.

error. By its nature, results of Monte-Carlo simulations are only approximations. The model checking results, in contrast, show that the performance is strictly ordered. It increases in nodes that are prone to collision monotonically with an increasing delay, and decreases monotonically in nodes on the edge of the grid that have little chance for collision.

6   Conclusion

This paper uses a combination of model checking for probabilistic automata and the more traditional method of Monte-Carlo simulations. The formal probabilistic automaton model helps to make certain assumptions in the model explicit that are often hidden in simulation models. Synchronisation between nodes, for example, has to be explicitly defined in a formal PRISM model. It cannot happen that one inadvertently implements a perfectly synchronised model. Monte-Carlo simulation, however, can give a fast feedback and show if the results obtained for a small network extend to networks of a more realistic size.

We considered different models that make different assumptions, their influence on outcome of performance analysis. Collisions can have a big impact on the performance, especially in protocols with a high sending probability. Probabilistic delay mitigates the effect to an extend that it almost vanishes for high delay probabilities. However, even if the performance of an ideal synchronised network without collision is within the range of accuracy of the Monte-Carlo simulation from the performance of a network with collision and simple probabilistic delay with probability 0.9, the networks show very different behaviour otherwise. Sending a message takes one time unit in the ideal model, while the expected duration between receiving and sending in the other is 10 time units. Similarly, gossiping protocols, which employ probabilistic broadcast, show a very different behaviour from flooding protocols on lossy channels.

Using the model checker PRISM for the analysis has the advantage that it delivers exact results. In addition, the model checker is able to compute firm upper and lower bounds. Construction of DTMC or MDP took PRISM at most 45 seconds for the model with lossy channels, on a Pentium M 1.8GHz with 512 MB RAM. Checking this model took at most 4 seconds. Constructing the models with probabilistic drift took at most 5 seconds and checking them took at most 10 seconds. All other models were constructed in less than a second each, and checking them took also less than a second as well, in more than 50% of the cases even less than 0.1 seconds. The size of the model ranged from 65 states and 140 states for the ideal model to 12856 transitions and 76732 transitions for flooding on lossy channels. Constructing the latter model for a 4 by 4 grid exceeded the capabilities of PRISM. To be able to compare results we used the 3 by 3 grid for all PRISM models.

The model checker PRISM has the option to export the DTMC as a sparse matrix. This sparse matrix however conceals the structure of the system, and we chose to implement the simulator separately. A straightforward translation of the PRISM model to MATLAB is possible, however, this yields a fairly inefficient simulator. An obvious drawback of having a custom build simulator and a formal model is that we had to develop and maintain two version of the model. Future work will be to provide an interface for modeling wireless networks, such that the formal model and the simulation model are build from the same source. This interface should give access to as well model checking as Monte-Carlo simulation.

References

1. Li, L., Halpern, J., Haas, Z.: Gossip-based ad hoc routing. In: INFOCOM. (2002)
2. Sasson, Y., Cavin, D., Schiper, A.: Probabilistic broadcast for flooding in wireless mobile ad hoc networks. In: WCNC 2003. (2003)
3. Cardell-Oliver, R.: Why Flooding is Unreliable (Extended Version). Technical Report UWA-CSSE-04-001, CSSE, University of Western Australia (2001)
4. Viswanath, K., Obraczka, K.: Modeling the performance of flooding in multihop ad hoc networks. Computer Communications Journal (2005)
5. Sasson, Y., Cavin, D., Schiper, A.: On the accuracy of manet simulators. In: Principles of Mobile Computing (POMC 2002). (2002) 38–43
6. Kwiatkowska, M., Norman, G., Sproston, J.: Probabilistic model checking of the IEEE 802.11 wireless local area network protocol. In Hermanns, H., Segala, R., eds.: PAPM/PROBMIV'02. Volume 2399 of LNCS., Springer (2002) 169–187
7. Kwiatkowska, M., Norman, G., Parker, D.: Probabilistic model checking and power-aware computing. In: PMCCS'05. (2005)
8. Amnell, T., Fersman, E., Mokrushin, L., Pettersson, P., Yi, W.: Times: A tool for schedulability analysis and code generation of real-time systems. In: FORMATS. (2003) 60–72
9. Kumar, R., Paul, A., Ramachandran, U.: Fountain broadcast for wireless networks. Technical report, CERCS, Georgia Tech (2005)
10. Hinton, A., Kwiatkowska, M., Norman, G., Parker, D.: PRISM: A tool for automatic verification of probabilistic systems. In: TACAS'06. (2006)

Policy-Based Route Optimization for Network Mobility of Next Generation Wireless Networks

Moon-Sang Jeong, Yeong-Hun Cho, and Jong-Tae Park

School of Electrical Engineering and Computer Science, Kyungpook National University
1370, Sankyuk-Dong, Buk-Gu, Daegu, 702-701, Korea
{msjeong, yhcho, jtpark}@ee.knu.ac.kr

Abstract. Recently, due to the recent advance of ubiquitous wireless networking technology, there has been a great deal of research efforts for seem-less support of the movement of mobile networks. The IETF NEMO WG's approach to network mobility has the so called 'pinball routing problem,' and the most of research attempts to solve this problem has limitations in the efficiency of both signaling and data communication. In this paper, we have presented policy-based mobility management architecture for next generation ubiquitous wireless networks, which can provide faster signaling and data transmission. In the proposed architecture, either care-of-address or home address of the root mobile router can be used as a binding address, which is decided by the policy on the route optimization. We present the architecture, model of the policy, and management procedures. Finally, the comparative analysis is provided by simulation to sow the efficiency of the proposed approach.

1 Introduction

Recently, there has been a great deal of research on network mobility management that can support the movement of a mobile network consisting of several mobile nodes. The emerging next generation mobile networks such as B3G/4G personal wireless network, ubiquitous sensor networks, telematics, and vehicular networks require a new paradigm for the management of network mobility. The Internet Engineering Task Force (IETF) Network Mobility (NEMO) working group proposed a basic support protocol (BSP) which defines a methodology for supporting network mobility by using bi-directional tunneling between the home agent (HA) and the mobile router (MR) [1], [2], [3], [4]. It extends binding mechanism of Mobile IPv6 [5] and the data transmission of a mobile network can be achieved by using the MR which is the egress interface of a mobile network.

The BSP scheme does not require an additional signaling in case of a root-MR's handover, so that the signaling complexity of the BSP scheme can be very small. If a depth of mobile network increases, however, both the data transmission time and the packet overhead can be drastically increased by multiple encapsulations which are caused by the so-called pinball routing problem; that is, all traffic to or from the MNN

T. Kunz and S.S. Ravi (Eds.): ADHOC-NOW 2006, LNCS 4104, pp. 142–155, 2006.

in the nested mobile network should pass through the home agents (HAs) of all preceding mobile routers [6], [7].

In order to solve this problem, various methods have been proposed [8], [9], [10], [11], [12], [13]. The most of these methods use direct tunneling between the root-MR and the HA of a nested MR which is a subordinate MR to the root-MR. The direct tunneling mechanism, however, requires an additional handover procedure, such that the binding address of the nested mobile network should be updated when the mobile network moves along with the root-MR. This is because the care-of-address (CoA) of the root-MR, which is used for the binding procedure, is changed even though the nested MRs does not change their point of attachments. In a vehicular environment, this root-MR handover usually generates mass signaling for all nested mobile networks, and network services might thereby be disconnected until the handover procedure is finished.

Additionally, the root-MR experiences a very heavy load because it must maintain full paths for all nested MRs [9], [12], [13], [14], [15]. The root-MR suffers from a large processing load, and can be a communication bottleneck. Furthermore, the direct message exchange between the mobile network nodes under the same root-MR, which is called intra-domain communication, is not supported. If an MR or MNN which has a mobile-HA, that is an MR playing the roles of HA, moves into a foreign network, the node also suffers from the pinball routing problem because the packet, to or from the nodes, must pass through many HAs, such as its own HA, the HA of the mobile-HA, and the HA of the visited MR. Thus, most of the previous route optimization methods have serious shortcomings such as the signaling overhead, additional pinball routing problem for mobile-HA, and concentrated traffic and load in the root-MR [16].

In order to solve aforementioned problems, we had previously presented a hierarchical mobile network binding (HMNB) scheme which uses a home address (HoA) of root-MR as a binding address of tunneling [16]. The HMNB scheme can provide better performance in signaling complexity and service discontinuity time than those of other route optimization techniques which are using the CoA of root-MR for tunneling. However, the HMNB scheme should pass through a two-hop path to exchange the data between the correspondent node (CN) and MNN, while other route optimization methods using the CoA of root-MR pass through one-hop path, resulting in faster data transmission.

In this paper, we propose an efficient route optimization technique by extending the HMNB scheme, which can provide an optimal performance in both signaling complexity and data transmission time. More specifically, we have employed a policy-based approach for the management of network mobility. The idea of the proposed approach is that either CoA or HoA of the root-MR can be used as a binding address, and the selection of the appropriate binding address is determined by the policy on the route optimization. This extension scheme overcomes the two-hop path problem of the HMNB scheme, so that the performance can be improved with respect to both signaling delay and data transmission time.

Policy-based technology has been extensively used for the inter-domain IP routing such as BGP [17], the management of distributed systems [18], QoS and security management in the IP networks [19], [20], as well as applications management [21]. Zhung, et. al [20] presented a policy-based QoS management architecture for the integrated UMTS and WLAN domains. Cho et. el [22] applyed the policy-based technique to the management of a multimedia conferencing service [22]. There have been research attempt for applying the policy to the mobile IP handover decision making among heterogeneous wireless networks [23]. MosquitoNet [24] addressed the Mobile IPv4 policy issues, applying policies on choosing the most desirable packet delivery path from multiple network interfaces, which is based on the characteristics of traffic flows. However, there have not been any attempts, to our knowledge, to employ the policy-based approach to route optimization for the mobility of a group of mobile network nodes.

We present a detailed architecture, model of the policy, and management procedures. According to the pattern of handover, each MR chooses the HoA of root-MR and CoA of root-MR dynamically by policy of route optimization. We have also evaluated by simulation the performance of the proposed route optimization scheme, comparing with IETF's BSP and other route optimization techniques. The simulation results show that the proposed scheme demonstrates high performance in both signaling and data transmission time.

The rest of the paper is organized as follows. In Section 2, we introduce the basic concept of network mobility, and describe other related works. In Section 3, we propose a policy-based route optimization scheme for efficient support of network mobility. In Section 4, we comparatively evaluate the performance by simulation. Finally, the conclusion follows in Section 5.

2   Background

2.1   Basic Concept of IETF NEMO Basic Support Protocol

Network mobility is defined as the capability to support mobility of whole network which is transparent to the nodes inside the mobile network [1], [2], [4], [25]. The IETF NEMO working group has defined a basic protocol operation to support the network mobility of a mobile network based on Mobile IPv6. A mobile network consists of one or more mobile routers (MRs) and several mobile network nodes (MNNs). An MR is one of mobile node which plays the role of router within a mobile network. MNNs are the nodes which are connected to the MR, and they can be fixed to the MR (LFN: local fixed node) or can be mobile nodes (LMN: local mobile node). If a local mobile node moves into a foreign network, the node is called visited mobile node (VMN) to the foreign mobile network. The MR communicates with the external network by accessing the access router (AR) which is the edge router of external network.

The goal of the NEMO basic support protocol is to provide the handover procedures for the whole mobile network movement. The data between the MNN and correspondent node (CN) is transmitted over bi-directional tunnel between the HA and the MR of the mobile network which the MNN belongs to. IPSec is used for secure signaling between the MR and HA [4].

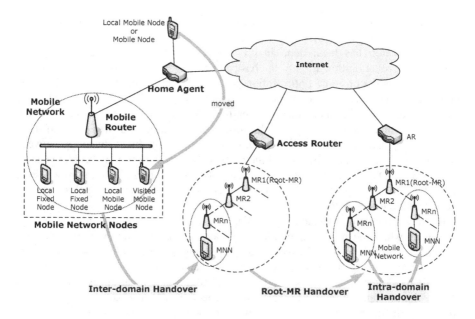

Fig. 1. Mobile network handover scenarios of the IETF NEMO WG

Fig. 1 shows the components of the mobile network and its handover scenarios of the IETF NEMO WG. In Fig. 1, mobile network handovers can be divided into several cases; Inter-domain, intra-domain and root-MR handovers. NEMO basic support protocol only considers the inter-domain handover procedure. When a mobile network moves into a foreign network, a bi-directional tunnel is established between the HA and the MR of the mobile network. The MR of the mobile network is only involved in the acquisition of CoA for the handover of the mobile network, and all the MNNs in the mobile network can keep their home network addresses. The MR operates both as a router for the data transmission of MNNs and as a mobile node itself. If a mobile node is attached to an MR on a visited foreign network, signaling and data transmission can be achieved by using the Mobile IPv6 protocol. For the construction of a bi-directional tunnel, the basic support protocol extends the binding message of Mobile IPv6. The extended binding update (BU) message contains a network prefix instead of a home address, and an egress interface address of the MR for CoA. By using these extensions, network mobility can be supported without changing the addresses of MNNs in a mobile network.

2.2 Related Work

The NEMO BSP defines the minimal procedures to support network mobility, and it excludes issues in route optimization, multi-homing, and so on. These issues are being investigated as a NEMO extended support, but the work is not yet completed. In BSP, the tunnel of a nested mobile network is constructed through all preceding mobile network tunnels, and all traffic of the nested mobile network passes through the HAs of all preceding mobile networks. This causes a serious problem that is called the 'pinball routing problem' [6], [7].

Much research has been done in order to solve the pinball routing problem, including the route optimization of the basic support protocol [7] to [26]. The most of the previous works on route optimization have used bi-directional tunneling between the HA of the nested mobile network and the root-MR [9], [10], [11], [16], [21]. They use two bi-directional tunnels, one between the HA of the nested mobile network and the root-MR, and the other between the root-MR and the nested MR. We call these methods as a direct tunneling mechanism (DTM). By using direct tunneling between the HA of nested mobile network and the root-MR, the pinball routing problem can be solved. In the DTM, the HAs of every nested MRs are tunneled with the root-MR of them. Therefore, the packets can be transmitted to the CN just passing through only one HA of the nested MR.

The direct tunneling mechanism (DTM), however, requires an additional handover procedure, when the mobile network moves along with the root-MR. In this case, when the CoA of the root-MR is changed, the binding addresses of all the nested MRs should be also changed accordingly. In a vehicular environment where this kind of root-MR handover often occurs, the amount of additional signaling overhead due to MRs's binding updates could be very large. The root-MR also suffers from a large processing load, becoming a communication bottleneck, and the non-negligible service discontinuity may occur, because the signaling delay of the nested MR becomes larger as the depth of the MR increases. Furthermore, the direct message exchange between the mobile network nodes under the same root-MR, which is called intra-domain communication, is not supported.

Table 1. Characteristics of various network mobility support protocols

	BSP	DTM	Moble IPv6	HMNB extension
Signaling Amount	Small	Large	Large	Small
Signaling Latency	Large	Medium	Small	Small
Signaling Complexity	Small	Large	Small	Small
Handover Delay	Large	Medium	Large	Small
Transmission Delay	Large	Small	Small	Medium
Intra-domain Data Transmission	N/A	Possible	Possible	Well-supported
Scalability	Large	Medium	Large	Medium
Micro-mobility	N/A	Possible	N/A	Support
Additional Routing Repository	Not Necessary	Root-MR only	Not Necessary	All MRs
Number of Tunnels	Number of Nesting Level	Number of Nesting Level (Intra-domain)	N/A	One or Two DRH
Processing Load of MR	Small	Large at Root-MR	Small	Medium

As stated previously, in order to solve these problems, we had presented a hierarchical mobile network binding (HMNB) scheme which uses a home address (HoA) of root-MR as a binding address of tunneling. The HMNB scheme can provide better performance in signaling complexity and service discontinuity time than those of other route optimization techniques using the CoA of root-MR for tunneling. However, the HMNB scheme should pass through a two-hop path to exchange the data between the correspondent node (CN) and MNN, which might result in slower data transmission time that that of the direct tunneling mechanism.

In this paper, we have proposed an extension of HMNB scheme, which can use either CoA or HoA of the root-MR as a binding address. This extension scheme overcomes the two-hop path problem of the HMNB scheme, so that the performance can be improved with respect to both signaling delay and data transmission time. The selection of care-of-address or home address of the root mobile router as a binding address is decided by the policy on the route optimization. Table 1 compares the characteristics of the representative network mobility protocols, which include the IETF BSP, Mobile IPv6, the direct tunneling approach (DTM), and HMNB extension. The NEMO BSP and HMNB extension have the least amount of signaling overhead, as they do not need to perform binding procedures when root-MR handover occurs. Therefore, they have the advantages in terms of both signaling complexity and scalability. However, the BSP has shortcomings in terms of the signaling latency, handover delays, and data transmission delays due to the dog-leg problem.

3   Policy-Based Network Mobility Management

In this section, we present policy-based network mobility management for seem-less handover over large-scale mobile networking environment. We present the policy-based route optimization architecture and the policy model.

3.1   Policy-Based Route Optimization Architecture

Fig. 2 shows a configuration of a policy-based HMNB extension scheme for route optimization. In the HMNB extension scheme, it uses the home address (HoA) of the root-MR as a binding address to support route optimization of the nested mobile network. The root-MR binds its CoA to its HoA and advertises the HoA to the nested mobile networks. The nested MR binds its network prefix to the HoA of the root-MR; thus, the data from the correspondent node (CN) to the MNN passes through the HA of the MNN and the HA of the root-MR, as shown in dotted line in Fig. 2. By using this mechanism, the nested mobile networks can be independent of the root-MR handover because the binding address of the nested mobile networks, which is the HoA of the root-MR, does not change.

To support the binding and routing of the HMNB extension scheme, a router advertisement message extension is required to discover the root-MR and to advertise its information. Fig.3 shows the packet head structure of root-MR option which is delivered with the extended RA message in the HMNB extension scheme. When the mobile network which includes nested mobile networks does handover within

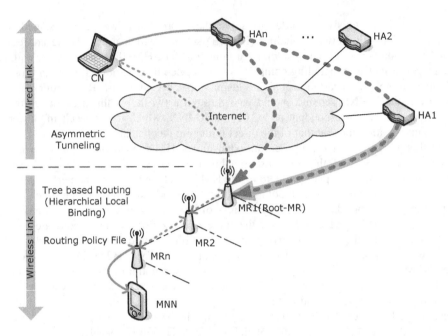

Fig. 2. Policy-based route optimization architecture of HMNB extension scheme

intra-domain, the highest level MR can only acquire the CoA, and the other subordinate MRs are not able to update their changed addresses. In this case, by using local update sequence, an MR can inform its subordinate MRs of the change of its address. Subordinate MRs compare the new incoming Local Update Sequency (LUSeq) of RA message and root-MR option with the existing ones. If the new LUSeq is different form the existing LUSeq, subordinated MRs performs the local binding update procedure, and can deliver their route information to their superior MR.

In the proposed HMNB extension scheme, both HoA and CoA of the root-MR can be delivered together by RA message extension. Each MR can choose one of the HoA or the CoA of root-MR in order to perform a binding update. There is some problem in performance if a HoA or a CoA of root-MR is only considered as a factor for mobility type. This is because the binding update overhead may drastically increase as the pattern of HoA and CoA selection is alternate. For example, let us suppose that a MR first moves into a visited network using HoA as a binding address, and next moves into another network using CoA, and this pattern is repeated. If the HoA is used constantly, when a root-MR handover occurs, there is not necessary to perform a binding update because the binding address is not changed. In the aforementioned movement scenario, however, binding address must be changed whenever a handover takes place. Therefore, the binding update must be performed, and a signaling overhead increases drastically. In order to avoid this catastrophic phenomenon, the selection of HoA or CoA should be decided by the mobility pattern, such as type of handover, the number of handover times, and the nested level of depth. Each MR requires to choose the HoA or the CoA of root-MR, dynamically by policy of route optimization.

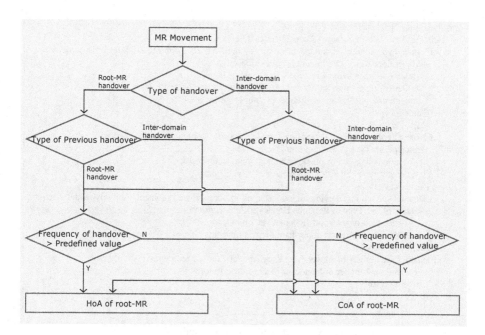

Fig. 3. Policy decision procedure

If the root-MR handover occurs frequently, the HoA of root-MR can be chosen as binding address of the nested MR. In this case, although the nested-MR moves, a signaling for binding update of each nested MR is not necessary. Therefore, the signaling delay can be obviously reduced. If the inter-domain handover occurs frequently, the CoA of root-MR instead of HoA is chosen as the binding address of the nested MR. In this case, as the CoA of root-MR is used as binding address of nested MR, the HMNB extension scheme can provide the shortest path, i.e., one-hop path as shown in Fig. 2, without changing the architecture. In other words, as the CoA of root-MR is used as binding address of nested MR, the HMNB extension scheme can provide the shortest path. The route between the MR and its HA is shorter than that of the BSP with the same signaling complexity because the HMNB extension scheme has at most two intermediate HAs, regardless of the nested level of the mobile network. On the other hand, the outbound packet from the MNN can be directly tunneled to the CN by using an asymmetric tunneling mechanism, without passing through any HAs.

Fig. 3 shows the policy decision procedure and Fig. 4 shows the XML schema of the route optimization policy. According to the policy of a mobile network, the nested MR can choose the route optimization methods selectively. Moreover, according to dynamic alteration of mobility pattern and network environment, an appropriate routing policy in the route optimization file can be selected, and the nested MR can appropriately choose HoA or CoA of the root-MR by delivering the binding update message to the HA of nested MR. Each mobile route has a Routing Policy File (RPF) which includes the route optimization policy as shown in Fig. 5, and the RPF is specified by Policy Information Base (PIB) which is defined by IETF.

```
<?xml version="1.0" encoding="UTF-8"?>
   <Schema xmlns="http://www.w3.org/2001/XMLSchema-instance">
      <ElementType name="RouteOptimization" model="closed" content="eltOnly" order="many">
         <description>Route Optimization Policy</description>
            <element type="handover-type" minOccurs="1"/>
            <element type="previous-handover-type" minOccurs="0"/>
            <element type="handover-frequency" minOccurs="0"/>
      </ElementType>
      ... skip ...
      <ElementType name="handover-type" model="closed" content="eltOnly" order="seq">
         <description>Type of Handover</description>
               <element type="rootMR-handover" minOccurs="0"/>
            <element type="interdomain-handover" minOccurs="0"/>
      </ElementType>
      <ElementType name="previous-handover-type" model="closed" content="eltOnly" order="seq">
         <description>Type of Previous Handover</description>
               <element type="rootMR-handover" minOccurs="0"/>
            <element type="interdomain-handover" minOccurs="0"/>
      </ElementType>
      <ElementType name="handover-frequency" model="closed" content="textOnly">
         <description>Number of Handover Occur</description>
      </ElementType>
      ... skip ...
      </ Schema >
```

Fig. 4. The XML schema of the route optimization policy

```
frwkBasePibClasses
            OBJECT IDENTIFIER ::={ frameworkPib 1 }
frwkRouteOptimizationPolicy OBJECT-TYPE
   SYNTAX         SEQUENCE OF FrwkBindingAddressEntry
   PIB-ACCESS    notify
   STATUS        current
   DESCRIPTION
         " "
   ::= { frwkBasePibClasses 1 }
frwkBindingAddressEntry OBJECT-TYPE
   SYNTAX        frwkBindingAddressEntry
   STATUS        current
   DESCRIPTION
         " "
   PIB-INDEX { frwkHandoverTimes }
   UNIQUENESS { frwkHandoverType }
   ::= { frwkRouteOptimizationPolicy 1 }
FrwkBindingAddressEntry ::= SEQUENCE {
      frwkHandoverType        OCTET STRING,
      frwkHandoverTimes         InstanceId }
```

Fig. 5. PIB-based routing policy file of extended HNMB scheme

In the HMNB extension scheme, a hierarchical local binding scheme is employed between the HA of the nested mobile network and the root-MR in order to reduce the signaling overhead of the nested mobile network. The root-MR can only de-capsulate the packets from the CN. The HMNB scheme can provide better route than the BSP, and can reduce an additional handover and signaling complexity in a vehicular network environment.

3.2 DRH and Intra-domain Handover of HMNB Extension

In order to support asymmetric tunneling and intra-domain routing, the extended HMNB scheme defines the destination routing header (DRH) by extending the IPv6 routing header option. The destination routing header in the encapsulated header has only the destination address of the original packet as a routing option, and it is used to check whether the packet reaches the destination or not. If the packet does not reach the destination, the MR changes the address of the encapsulated packet to the next-hop address in the local binding cache entries. If there is no local binding cache entry to the destination, the MR changes the address to the default routing entry which is the address of the parent MR. The root-MR, which has received the packet without reaching the destination, changes the destination address of an encapsulated header as the destination address in the destination routing header, and the packet is forwarded to the CN by using a normal IP routing mechanism. By using this mechanism, intra-domain communication can be efficiently supported without passing through any HAs. An intra-domain handover occurs when a mobile network moves within the root-MR domain. In this case, the MR does not need to update the binding to its HA and it only performs a local binding procedure. If an MR which plays the role of the HA receives a binding message, the MR forwards this message to its own HA after updating its local binding cache information. By using this mechanism, the packet from the CN is transmitted to the HA of the foreign root-MR without passing through additional intermediate HAs including its own HA. Therefore, the traffic to the MR is always passed through only two HAs.

3.3 The De-registration Procedure of Extended HNMB Scheme

When a handover of mobile network occurs, MRs in previous domain still has local binding information of the MR which is already moved. Therefore, data from the previous domain cannot be delivered to the moved MR. The deregistration procedure of MR's local binding cache (LBC) is different whether intra-domain handover or inter-domain handover is occurred. In the case of intra-domain handover, the LBU is delivered to a crossover MR or a root-MR, then the crossover MR or root-MR sends the deregistration message to its subordinate MR using the old path. The MR which receives the deregistration message, sends the message to its old path to the moved MR and updates the LBC recursively. Then, all MRs in the previous domain can update the local binding information.

In the case of inter-domain handover, the MR send the deregistration message to its previous root-MR. The previous root-MR delivers the de-registration message recursively to its subordinate MRs using its old path information. If an MR which plays the role of the HA receives a binding message, the MR forwards this message to its own HA after updating its local binding cache information. By using this mechanism, the packet from the CN is transmitted to the HA of the foreign root-MR without passing through additional intermediate HAs including its own HA. Therefore, the traffic to the MR is always passed through only two HAs.

4 Simulation

We evaluated the performance of the proposed HMNB scheme by using discrete event simulation with the topology as shown in Fig. 6. All HAs are assumed to have the same wired link with a 10ms delay, and mobile nodes also have the same wireless link with a 10ms delay. The routing advertisement interval of an MR is assumed to be 5ms, and each MR performs movement detection within 15ms. MRs and HAs are built by implementing the following protocols: BSP, DTM method, and the proposed extended HMNB schemes.

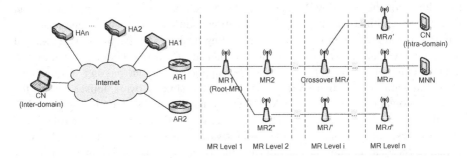

Fig. 6. Test configuration of mobile networks for simulation

Fig. 7 (a) and (b) show the performance results of the inter-domain and intra-domain data transmission times for the TCP traffic, as the depth of the destination MR increases. In the inter-domain communication environment, the data transmission delays of DTM and the extended HMNB schemes were smaller than those of the BSP. The intra-domain communication times of the extended HMNB schemes show that the delays become smaller than other schemes, if the crossover MR is located below the root-MR.

Fig. 8 shows the data transmission delays and the service discontinuity times for the CBR (100 KBps) traffic and the TCP data transmission under the various handover scenarios. Fig. 8(a) shows the performance analysis for CBR traffic. In the case of the CBR handover scenario, the delays of the extended HMNB scheme are larger than those of the TLMR schemes because it uses two intermediate HAs. The service discontinuity time of extended HMNB scheme is, however, the smallest in the root-MR handover scenario. In the case of the root-MR handover or whole mobile networks movement from AR1 to AR2, both extended HMNB and BSP schemes can accomplish the handover procedure with only accessing the root-MR, so that the discontinuity times of these schemes are the smallest. In the case of the BSP scheme, the nested depth of the MR is deeper, so that the service discontinuity time becomes larger than other handover scenarios. For the cases of intra-domain handover (for example, "MR5 moving to the subordinate to MR4") and sub-tree handover scenarios (for example, " MR4 and sub-tree moving to the subordinate to MR4"), the extended HMNB scheme shows the best performance in data transmission, since it can provide fast handover by using local binding update. Thus, the DTM and the BSP may not be suitable for a vehicular environment where the root-MR moves frequently.

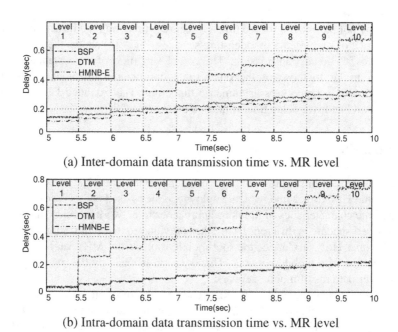

(a) Inter-domain data transmission time vs. MR level

(b) Intra-domain data transmission time vs. MR level

Fig. 7. Performance analysis in accordance with the increase of the depth of the MR

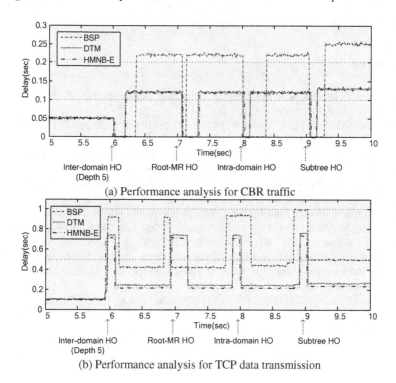

(a) Performance analysis for CBR traffic

(b) Performance analysis for TCP data transmission

Fig. 8. Data transmission time and the service discontinuity time for various handover scenarios

They may not be suitable for the time-critical real-time multimedia services such as wireless VoIP or multimedia broadcasting services. Fig. 8(b) shows the performance analysis for TCP data transmission. The round trip time of the extended HMNB scheme is similar to those of the other route optimization methods, because the extended HMNB scheme uses asymmetric tunneling. The extended HMNB scheme also has small handover delay time similar to the case of CBR.

5 Conclusion

In this paper, we have presented an efficient route optimization technique by extending the HMNB scheme, which employs a policy-based path selection. We have compared the characteristics of previous approaches to network mobility management. The proposed scheme is different from other previous route optimization approaches, in which either HoA or CoA of the root mobile router (root-MR) can be used as the binding address of a mobile network. This decision is made by the routing policy on the handover patterns of mobile network.

The policy-based extended HNMB scheme can overcome the two-hop path problem of the HMNB scheme, so that the performance can be improved with respect to both signaling delay and data transmission time. Finally, we have performed extensive simulation to compare the proposed HMNB scheme with the other network mobility support protocols. Future work may include the detailed analysis on the security issues related to the proposed scheme.

References

[1] T. Ernst, "Network Mobility Support Goals and Requirements," IETF, Internet draft <draft-ietf-nemo-requirements-05>, 2005.

[2] T. Ernst, and H. Y. Lach, "Network Mobility Support Terminology," IETF, Internet draft <draft-ietf-nemo-terminology-04>, 2005.

[3] P. Thubert, R. Wakikawa, and V. Devarapalli, "NEMO Home Network Models," IETF, Internet draft <draft-ietf-nemo-home-network-models-05>. 2005.

[4] V. Devarapalli, R. Wakikawa, A. Petrescu, and P. Thubert, "Network Mobility (NEMO) Basic Support Protocol," RFC 3963, January 2005.

[5] D. Johnson, C. Perkins, and J. Arkko, "Mobility Support in IPv6," RFC 3775, June 2004.

[6] E. Perera, V. Sivaraman, and A. Seneviratne, "Survey on Network Mobility Support," Mobile Computing and Communications Review, vol. 8, no. 2, pp. 7-19, April 2004.

[7] J. D. Wells, "A Network Mobility Survey and Comparison with a Mobile IP Multiple Home Address Extension," M.S.thesis, Virginia Polytechnic Institute and State Univ., Virginia, USA, 2003.

[8] P. Thubert, and M. Molteni, C. Ng, H. Ohnishi, and E. Paik, "Taxonomy of Route Optimization models in the Nemo Context," IETF, Internet draft <draft-thubert-nemo-ro-taxonomy-02>, 2004.

[9] H. S. Kang, K. C. Kim, S. Y. Han, K. J. Lee, and J. S. Park, "Route Optimization for Mobile Network by Using Bi-directional Between Home Agent and Top Level Mobile Router," IETF, Internet draft <draft-hkang-nemo-ro-tlmr-00>, 2003.

[10] H. Ohnishi, K. Sakitani, and Y. Takagi, "HMIP based Route Optimization method in a mobile network," IETF, Internet draft <draft-ohnishi-nemo-ro-hmip-00>, 2003.

[11] P. Thubert and M. Molteni, "IPv6 Reverse Routing Header and its application to Mobile Networks," IETF, Internet draft <draft-thubert-nemo-reverse-routing-header-05>, 2004.

[12] H. S. Cho, E. K. Paik, and Y. H. Choi, "R-BU: Recursive Binding Update for Route Optimization in Nested Mobile Networks," IEEE Vehicular Technology Conference 2003, Orlando, USA, October 2003.

[13] J. K. Na, S. H. Cho, C. K. Kim, S. J. Lee, H. J. Kang, and C. H. Koo, "Route Optimization Scheme based on Path Control Header," IETF, Internet draft <draft-na-nemo-path-control-header-00>, 2004.

[14] Z. J. Gu, D. M. Yang, and C. H. Kim, "Mobile IPv6 Extensions to Support Nested Mobile Networks," in the 18th Int. Conf. Advanced Information Networking and Applications, Fukuoka, Japan, March 2004.

[15] M. H. Park, C. M. Park, S. H. Kim, S. B. Hong, and J. S. Choi, "A Novel Routing Protocol for Personal Area Network Mobility (PANEMO) Environment," in Int. Conf. Advanced Communication. Technology, Phoenix Park, Korea, 2004.

[16] M. S. Jeong, and J. T. Park, "Hierarchical Mobile Network Routing: Route Optimization and Micro-mobility Support for a NEMO," in Int. Conf. on Embedded and Ubiquitous Computing 2004, Lecture Notes in Computer Science 3207/2004, pp. 571-580, August 2004.

[17] Y. Rekhter and T. Li, "A border gateway protocol 4 (BGP-4)," IETF RFC 1771, Mar. 1995.

[18] E. C. Lupu and M. Sloman, "Conflicts in policy-based distributed systems management," IEEE Transactions on Software Engineering, Vol. 25, Issue 5, Nov/Dec, 1999.

[19] E. Law and A. Saxena, "Scalable Design of a Policy-Based Management System and its Performance," IEEE Communications Magazine, Vol. 41, Issue 6, June 2003.

[20] W. Zhuang, Y. Gan, K. Loh, and K. Chua, "Policy-Based QoS Management Architecture in an Integrated UMTS and WLAN Environment," IEEE Communications Magazine, Vol. 41, Issue 11, November 2003.

[21] M. Katchabaw, H. Lutfiyya, and M. Bauer, "Administrative Policies to Regulate Quality of Service Management in Distributed Multimedia Applications," IEEE/IFP 6th International Conference on Management of Multimedia Networks and Services (MMNS), 2003.

[22] Y. H. Cho, M. S. Jeong, J. W. Nah, W. H. Lee, and J. T. Park, "A Policy-based Distributed Management Architecture for Large-Scale Enterprise Conferencing Service Using SIP," IEEE Journal on Selected Areas in Communications(JSAC), Vol. 23, No. 10, Oct. 2005.

[23] S. Aust, D. Proetel, N. A. Fikouras, C. Pampu, and C. Gorg, "Policy based Mobile IP handoff decision (POLIMAND) using generic link layer information," Proceeding of the 5th IEEE International Conference on Mobile and Wireless Communication Networks (MWCN 2003), October 2003.

[24] X. Zhao, et. al, "Flexible Network Support for Mobile Hosts," Mobile Networks and Applications, Kluwer Academic Publishers, Vol. 6, 2001

[25] H. Y. Lach, C. Janneteau, and A. Petrescu, "Network Mobility in Beyond-3G Systems," IEEE Commun. Mag., vol.. 41, pp. 54-57, July 2003.

[26] C. Ng, E, Paik, T. Ernst, and M. Bagnulo, "Analysis of Multihoming in Network Mobility Support," IETF, Internet draft <draft-ietf-nemo-multihoming-issues-04>. 2005.

Geographic Pattern Routing for MANETOR in IVC*

Jiang Hao[1], Jian-Jin Li[2], Kun Mean Hou[2], and Chen Lijia[1]

[1] Wuhan University, School of Electronic Information, Wuhan, Hubei, 430079, P.R. China
jianghaow@263.net
[2] LIMOS UMR 6158 CNRS, Campus des Cézeaux, BP 10125,
63173 AUBIERE CEDEX, France

Abstract. This paper proposes a novel routing protocol dedicated to IVC. This routing protocol, named as geographic pattern routing (GPR). The protocol does not need to search route frequently, so sufficient bandwidth is reserved for application packets, and QoS is easily guaranteed. It has ingredients to support geographic pattern search and packet forwarding assisted by geographic pattern, and it considers whether sufficient nodes are on the path represented by geographic pattern to allow packets forwarding to the destination. Besides vehicle location, it does not need other information. The simulation shows that GPR is better than other routing protocols.

1 Introduction

Vehicles are part of people's life in modern society, into which more and more high-tech devices are integrated. With the development of business, people spent more time on vehicles. The IVC(Inter-Vehicle Communication) may be used to communicate the road information between vehicles, reduce the risk of accidences, improve road traffic and access Internet.

Mobile Ad Hoc Networks (MANET) equipped with high-bandwidth, and short range wireless access medium are well suitable for vehicles. In recent years, the IVC based on MANET has attracted the interests of many automobile manufactures and researchers and was investigated in various research projects: FleeNet [1], CarTALK [2]. MANET in IVC was denoted as MANETOR (MANET on road).

However, MANETOR will behave in fundamentally different ways from the models that previous MANET research. Mobility constraints and high speeds create unique characteristics in MANETOR [3]. These characteristics have important implications for design decisions in these networks. In fact, MANETOR differs from typical MANET models in four key ways, which are characterized by rapid topology changes, with frequent fragmentation, a short path lifetime, and limit capacity [4].

---

* This work is supported by NSFC (No.60502028).

T. Kunz and S.S. Ravi (Eds.): ADHOC-NOW 2006, LNCS 4104, pp. 156–169, 2006.

The IVC routing protocol must efficiently handle the features of MANETOR. Current MANET routing protocols fail to fully address these specific needs. Proactive protocols will be impacted fiercely by the extremely rapid change topology. Reactive protocols spend too much time and bandwidth to discover a route before sending a message. Location-based routing protocols [5, 10, 11] do not take into account the constraints on vehicle movements and the build obstacle of wireless signal. The geographic forwarding will fail in case of a topology hole [6, 10]. Specific adaptations of location-based routing for IVCs propose the routing of packets along the road network [6]. Due to the constraints on vehicle movement imposed by the roadway, greedy location-based forwarding algorithms may result in suboptimal routes, but also it needs addition support from GIS(Geography Information System). GPCR [12] can be applied to a city scenario without assuming that nodes have access to a static street map. The detecting crossroad methods of GPCR can not guarantee to find crossroad in any situation, and GPCR does not consider whether there is a sufficient number of nodes on the street to allow packet forwarding to the destination.

The purpose of this study was to design a novel routing protocol dedicated to IVC. This routing protocol, named as geographic pattern routing (GPR), could be suitable for MANETOR. The geographic pattern contributed to route packets is stable and has a long lifetime. The protocol does not need to search route frequently, so sufficient bandwidth is reserved for application packets, and QoS is easily guaranteed. It has ingredients to support geographic pattern search and packet forwarding assisted by geographic pattern, and it considers whether sufficient nodes are on the path represented by geographic pattern to allow packets forwarding to the destination. Besides vehicle location, it does not need other information.

This paper assumes that all vehicles knew their own locations from location service, and the source could obtain the approximate location of the destination from location service. From locations, all vehicles infer their speed and movement direction.

2   Overview of GRP

The key advantage of GPR is that the protocol could detect crossroads or corners, characterized as geographic element in this paper, and acquire the geographic pattern from source to destination in MANETOR through route request. The geographic pattern consists of an ordered sequence of geographic elements of networks and the approximate location of destination, similar to IP loose source routing information. Route request not only obtains geographic pattern from source to destination, but also discover intermediate nodes density along path. A geographic element is the approximate position of a crossroad or corner. If the road from source to destination is straight, geographic elements are empty. A constraint flooding is adopted for disseminate result of geographic element as soon as possible. To avoid unawareness of geographic element in routing and invalid transmission in crossroad, a novel beacon method is proposed in this paper.

To send packets from a source to a destination, the route request sets the geographic pattern in packets' header, and the source writes the geographic pattern into the header of data packet. The geographic pattern is like itinerary instructions given to every relay node, which tells the node how to select next hop. The packet is sent by intermediate nodes in the direction of the next geographic element in the list until it reaches a node locates in crossroad represented by the next geographic element and forwards to the node, at which point the next geographic element becomes the following in the list. After passed the last geographic element, packets are forwarded to intermediate nodes in the direction to the location of the destination.

GRP uses beacon to exchange neighbors' states, reserve the geographic element parameters in a zone. In GRP, new fields are inserted into beacon message, the interval of beacon also varies in different region.

GPR can suit any kind of road, even circuitous streets. The acquirement of geographic pattern does not consume too much networks bandwidth, which guarantees the QoS of the application. The geographic pattern requires much less states than number of nodes, thus GPR can be deployed in large networks. The geographic pattern is stable, and data sources with GPR do not need to discover routes frequently to deal with topology change.

3   Beacon in General

Beaconing algorithm provides all nodes with their neighbors' parameters. Periodically, each node transmits a MAC broadcast packet, containing its own identifier, speed, location and movement direction.

The Euclidean position (x, y) is encoded by single precision floating-point quantities, for x and y coordinate values. To avoid synchronization of neighbors' beacons, the protocol jitters each beacon's transmission by 25% of the interval B between beacons, such that the mean inter-beacon transmission interval is B, uniformly distributed in $[0.75B, 1.25B]$. A node maintains a neighbor entry for each neighbor. Each neighbor entry contains following attributes:

Table 1. Neighbor entry

Item	Description	property
ID	Identifier	general
L_n	Location	general
V_n	Velocity	general
D_n	Movement direction	general
T_n	Received MAC broadcast packet time	general
P_c	Position of crossroad	optional
F_c	Flag bit for being in crossroad	optional

In general, a node receives a MAC broadcast packet containing speed, location, and movement direction of its neighbor. It decodes these parameters, and updates the entry of this neighbor. With location, speed and movement direction, the node could estimate current location of neighbors using L_n, V_n, D_n, T_n at current time, these states are enough to detect geographic element. Pc, Fc are used for reservation geographic element parameters and announce the node in crossroad that could relay packets.

3.1 Detecting Geographic Elements

GPR detects the crossroad according to the movement direction, shown in figure 1:

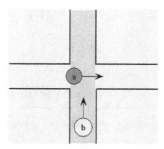

Fig. 1. Detect crossroad based on movement direction

Since buildings obstruct wireless signal, only when the node a broadcasts a beacon message in crossroad, the node b receives this beacon message. According to node b's movement direction and movement direction of node a contained in node a's beacon message, node b discovers that node a moves impossibly along the road that node b moves along. Then node b can infer that there is a crossroad which node a locates in. This paper adopts a heuristics function to decide whether node a moves along the road same with node b.

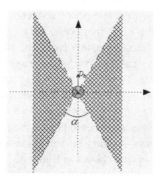

Fig. 2. Illustrate the function for detecting crossroad

The function is illustrated in figure 2. If the movement direction of node a is in the shadow area, node a moves along different road with node b. The detail expression is

that if $\left| D_a - D_b \right| > \dfrac{\alpha}{2}$ or $\left| D_a - (D_b + \dfrac{\pi}{2}) \right| > \dfrac{\alpha}{2}$, then node b infer that node a

moves along other road and there is a crossroad. As nodes on the same road with node b move in direction little different from node b, α is used to avoid defective judgment of crossroad. The centre of crossroad is intersection point of two lines, node *a* and node *b* are separately on these two lines, two lines propagate in movement directions of node *a* and node *b*. In this paper, we use the position of crossroad center as location of crossroad. Like figure 3.

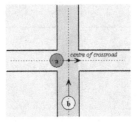

Fig. 3. Center of crossroad

Due to only seldom nodes broadcast beacon message in crossroad, not all nodes near crossroad could discover crossroads. So a method for distribution and reservation geographic elements representing the crossroad is necessary.

3.2 Constraint Flooding in Cross Zone

We define a zone --- centered the cross, radius is transmission range *R*, called cross zone. To disseminate the geographic element parameter, we propose a novel method, called constraint flooding in the zone, that broadcast a "geographic element message" when a node in a cross zone detects the new crossroad or received a "geographic element message" describing this new crossroad. "geographic element message" contains the position of crossroad center. To reduce broadcast packets, nodes outside the cross zone just record the new crossroad in their geographic element list, while they received the geographic element message, then nodes locates in the region with double R radius could know the crossroad. This method could extend the region that nodes are aware of crossroad at the price of few broadcasting packet. It is useful in route request and data packets forwarding. Two regions are shown in figure 4.

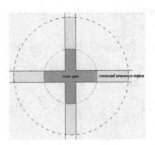

Fig. 4. Cross zone and crossroad awareness region

3.3   Reservation Crossroad in Cross Zone

To make nodes in crossroad know the crossroad and avoid redundant constraint flooding, GPR need a method to reserve crossroad information in cross zone. We define a field of beacon message to reserve crossroad parameters --- P_c, shown in table 1. P_c contains the position of crossroad center, the node sending beacon locates in the cross zone of crossroad. If the node locates in two or more cross zones simultaneously, P_c has these crossroads' center.

As radius of crossroad awareness region is double transmission range, nodes know how far it is from crossroad and which neighbor is the best next hop, missing crossroad or getting across the crossroad can be avoided.

Nodes in crossroad play a key role for packets relaying. Geographic elements work as landmarks, packet relaying depend on nodes in crossroad. The problem of packet replying has two folders. One is to announce that a node is in crossroad, while the node just comes in crossroad. Another is that make nodes in crossroad now neighbors along different roads.

3.4   Announce Node in Crossroad

The node need to announce that there is an intermediate node in crossroad, once it is in crossroad. GPR uses Euclidean distance to detect whether the node is in crossroad, and send additional beacon message to broadcast "I am in crossroad".

After detect crossroad successfully, geographic element message will be disseminated and reserve in cross zone, so nodes have acquire the geographic element parameters—position of crossroad center.

While distance between node and crossroad center, $|L_n - p_c| < \beta$, β is radius of crossroad, the node sends additional beacon message containing position of the crossroad center, and set F_c bit of beacon message to 1. Nodes received the beacon know that an intermediate node is in crossroad and store relevant information into node in crossroad list, shown in table 2.

Table 2. Node in crossroad list entry

Item	Description
ID	Identifier of node in crossroad
P_c	Position of crossroad center
T_e	expire time of node in crossroad

In this entry, $T_e = \dfrac{2\beta}{V_n} + current\ time$, T_e is estimated approximately.

3.5 Double Beacon Frequency to Promote Probability of Valid Transmission

Relay nodes in crossroad need neighbor states along different roads connecting to crossroad. Since relay nodes in crossroad pass crossroads in few seconds, GRP doubles beacon frequency when nodes are in cross zone, interval $B = \dfrac{B}{2}$. This way could provide more choice neighbors to forward packets without increasing beacon message size.

4 Route Request

Route request is the process of acquiring the geographic pattern from source to destination. The approximate location of destination can be obtained from location service. The main part of geographic pattern is an ordered sequence of geographic elements. In GPR, the geographic element g is represented by the position of crossroad center.

For the IVC application, at least two types of information are needed: road traffic status and accident alarm. In general, the road traffic status is a forward request sending by a node and the accident alarm is a backward request. Note that the two requests are operated by the same manner except that the sending direction is different. In this paper, we only describe the forward route request. To consume less bandwidth and transmit less route request packets, the route request is forwarded greedy.

The header of route request consists of $< L_S, L_D, C_c, G, density_{min} >$. G is the sequence of geographic elements, $G = [g_1, g_2, g_3, \cdots , g_m]$. L_S is the location of source. L_D is the approximate location of destination. C_c is the circuitous counter, which is used to control discover range of route request, its function likes TTL of IP. G is the sequence of geographic elements. $density_{min}$ is the minimal nodes density of roads.

Since obstacles (e.g., buildings) block wireless signals, route request packets should be routed along roads. Crossroads represented by geographic elements are the only places where actual routing decision is taken. Therefore packets should always be forwarded to a node in a geographic element rather than being forwarded across it, as described in paper [12]. Due to reservation crossroad information in cross zone by beacon, intermediate nodes can be aware of the crossroad when they are far two hops from the crossroad. According to section 3.4, nodes in cross zone can choose a node as a relay from node in crossroad list. It is very important for packet transmission. If node in crossroad list is empty, route request is sent to neighbor that is closer to crossroad. If there are valid nodes in node in crossroad list, the node with the largest T_e. To a certain extent, this way can solve the problem due to the inaccuracy of T_e. While forwarding route reply packets and data packets in similar situation, nodes will do like this.

There are topologies in which the only route to a destination required a packet move temporarily farther in geometric distance from the destination.

We also consider whether there are enough intermediate nodes on road as relay. So after nodes in crossroad received route requests, they forward route request to variant nodes on different road except node which are on the same road with packet's last hop. We illustrate forwarding of node in crossroad in figure 5.

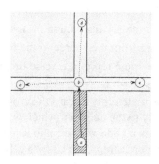

Fig. 5. Illustrate forwarding of node in crossroad

After node b received route request from node a, node b record coordinate of crossroad center into g_i of route request and forward route request duplicates to node c, node d and node e respectively, node b never send route request to nodes that are on the shadow road--- the same road with node a.

To avoid endless route request, intermediate node compares distance from itself to destination---$d(t, D)$ with distance from neighbor to destination---$d(n, D)$. If $d(t, D) < d(n, D)$, the circuitous counter C_c is incremented by 1. When the C_c is larger than a constant, the route request will not be sent to the neighbor.

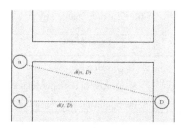

Fig. 6. Controlling circuitous transmitting of route request

Vehicles do not distribute uniformly along road. Packets may be dropped on road, as no choice neighbor. To find a reliable route to destination, intermediate nodes record nodes density along road. We present a method to implement it. When source writes number of neighbors in its range into $density_{min}$ of route request, at each hop intermediate node compares number of its neighbors $density_n$ with $density_{min}$. If $density_n < density_{min}$, intermediate node replaces $density_{min}$ with $density_n$. $density_{min}$ will be brought back source by route reply for choosing reliable route.

5 Route Reply

While destination receives a route request, it will store the sequence of geographic elements G into its cache and returns a route reply to source. Hence, the route reply contains G obtained from the route request, d_{min} and L_S, L_S is the location of source. $G = [g_1, g_2, g_3, \cdots, g_m]$ is the itinerary instruction from source to destination, in which g_1 represents the first crossroad from the source to the destination, and g_m represents the last one. It is also used for route reply in reverse order. The header of route reply consists of $< density_{min}, L_S, G >$.

First, the destination sends route reply toward the crossroad represented by g_m. The route reply is forwarded greedy to the neighbor which was the closest to g_m.

When the route reply attains a node whose radio range covers the place of g_m, the node looks for the node in the cross represented by g_m. If a neighbor satisfying above constraints exists, it will be selected as the next hop of route reply. Otherwise the neighbor closer to the place of g_m will be chosen as next hop of route reply.

After crossing g_m, the route reply is forwarded to g_{m-1}. The operation is similar as above.

If there are no geographic element, or g_1 is crossed, the route reply is directly forwarded greedy to the source.

6 Data Sending

When the source receives a route reply, it will restore destination ID, the geographic pattern and the minimal nodes density of roads d_{min} into its route cache entry, Table 3 is detail.

Table 3. Route cache entry

Item	Description
DID	Destination ID
G	Sequence of geographic elements from S to D
$density_{min}$	the minimal nodes density of roads

As multiple path, the source maybe receive two or more route reply with different route and $density_{min}$. Some routes are long with high reliability, others are short, but are broken easily. How to select the optimal route is not discussed in this paper.

The source chooses the neighbor node located near the place of g_1 as next hop node for transmitting data packets toward g_1. When data packets attains a node whose radio range covers the place of g_1, the node looks for the next hop in the cross represented by g_1. If a neighbor satisfying with above constraints exists, it is selected as the next hop of data packets. Otherwise the neighbor closer to the place of g_1 will be chosen as next hop of data packets.

After data packets crossing g_1, g_2 is seen as the next temporary destination. The operation is similar as above until data packets cross the g_m. If there are no geographic element, or g_m is cross, data packets are directly forwarded greedy to the destination.

7 Simulation

To evaluate the achievement of our goals for GRP, two scenarios are simulated: highway scenario and city scenario. To compare the performance of GRP with prior related work in wireless routing, we also did simulation with GPSR, which have been shown better performance than several other ad-hoc routing protocols.

7.1 Simulation Environment

We simulated GPR in ns-2. GPR and GPSR are evaluated by two metrics: packet delivery ratio and delay.

Packet delivery ratio is the ratio of the data packets delivered to the destinations to those generated by the CBR sources.

In this paper, we collect delay from sources to destinations while the passed distances between sources and destinations are in different ranges. The passed distance means the distance that packets actually passed, it was not the line distance from source to destination.

Based on the classical vehicular traffic theory, the velocity was followed normal distribution [7]. The density of nodes on 1km road followed the Linear Model described in [8]. If the speed was high then the density of nodes was low.

In the case of highway scenario, the wireless nodes are distributed randomly along a straight road and the number of nodes is 160. The highway is 4 km long and 20m wide and has 4 lanes. Nodes move along these lanes toward two opposite directions.

In the case of city scenario, the nodes move in the area, illustrated by the figure 7.

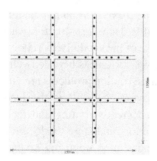

Fig. 7. City scenario

In the city scenario, there are totally 400 nodes. The city is crossed by four main roads, thus there are four crossroads. In this case, nodes may go through crossroads directly or turn around at crossroads.

We simulate average velocities 30km/h, 40km/h, 50km/h in the city scenario, 70km/h, 90km/h, 110km/h in the highway scenario. The wireless transmission range was 250m.

In both scenario, we simulate 20 CBR traffic flows, originated by 20 sending nodes. Each CBR flow is sent at 2 Kbps with 256-byte packets. Sources and destinations are selected randomly. In our simulation, the beaconing interval was 3 seconds same as GPSR in [5].

7.2 Packet Delivery Ratio

The following two figures show how many application packets GPR delivers successfully for various velocity in two scenarios ,while results from for GPSR are included for comparison.

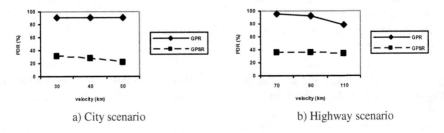

a) City scenario   b) Highway scenario

Fig. 8. Packet delivery ratio for two scenario with various velocity

The results of the packet delivery ratio suggested that the packet delivery ratio of GPR is much better in both scenarios, and difference in average velocities has a little impact on the packet delivery ratio especially in city scenario. The simulation result proves that the geographic pattern could overcome the frequently change topology and short path lifetime problems of MANETOR, and transmit the packets successfully. But the packet delivery ratio of GPR decreases slightly, when the velocity is 110km/h in highway scenario. The decrease of packet delivery ratio may be caused mainly by error location estimation and the larger distances between vehicles. A recent research paper has suggested that location based routing protocols are vulnerable to location errors [9].

Why does GPSR perform worse? GPSR is the landmark of routing algorithms combining two concepts called greedy routing and face routing. GPSR adopts beaconing to exchange nodes location, and store the location in itself, the interval is seconds. In city scenario, some nodes move opposing, so the relative speed is large. When GPSR is in greedy mode, according to locations which is gotten from beaconing, it selects the neighbor that is the closest to destination as next hop. Due to non real-time location estimation and large relative speed, GPSR is easy to send data to neighbor that is out of transmission range practically. Beside above reason, GPSR often misses neighbors on other roads, and is unaware of crossroads, due to buildings obstacle radio signal, so packets are send to error direction, as illustrated in figure 9.

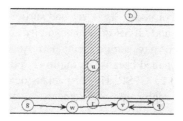

Fig. 9. Problem of GPSR

In this figure, node *S* wants to send packets to node *D*, with GPSR. Building blocks beacon message from node *u* to node *w*, even node *r* is in crossroad. node *w* does not know crossroad and neighbors in crossroad, it still sends packets greedy to node *v*. node *v* switches to perimeter mode and forwards packets to node *q*, node *q* discovers that node *v* is closer to destination, it sends back packets to node *v* and switches to greedy mode. Packets will jump between node *v* and node *q*, until be dropped or discovering new neighbors. The modification of GPSR, such as GOAFR+, can not yet solve the problem.

7.3 Delay

Delay was an important parameter of QoS. Below figures showed the delays at different passed distances for GPR and GPSR, in which the x axis was passed distance range. Because the city scenario had a larger area and the roadway was not straight, its passed distance range is also larger. And avg(v) indicates the average velocity.

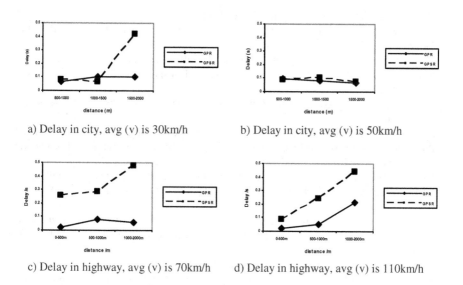

a) Delay in city, avg (v) is 30km/h   b) Delay in city, avg (v) is 50km/h

c) Delay in highway, avg (v) is 70km/h   d) Delay in highway, avg (v) is 110km/h

Fig. 10. Delay vs distance for two scenario with various velocity

Due to minimized neighbor nodes discovery and sufficient bandwidth reserved, the delay performance of GPR and GPSR are better in city and highway scenario. In the city scenario, GPR and GPSR have similar delay performance, but in highway GPR is better than GPSR. This happened because of the lower complexity of GPR, GPR did not planarize the topology like GPSR. Higher packets delivery ratio and lower delay could guarantee the QoS of application requirement.

8 Conclusion

We presented a new position-based routing approach, GPR, which is able to deal with the challenges of IVC. Our approach does not require external information such as a static street map to avoid the problems that existing position-based approaches face in city environment. The geographic pattern contributed to route packets is stable and has a long lifetime. The approach does not need to search route frequently, so sufficient bandwidth is reserved for application packets. It has ingredients to assist geographic pattern search and packet forwarding, and it considers whether sufficient nodes are on the path represented by geographic pattern to allow packets forwarding to the destination.

The simulation shows that GPR performs much better than other position-based routing, such as GPSR. It does so by maintaining its routing overhead low. It is suitable to IVC.

References

1. FleetNet Documents, www.et2.tu-harburg.de/fleetnet
2. CarTALK, http://www.cartalk2000.net, european Project contract-number IST-2000-28185.
3. Jeremy J. Blum, Azim Eskandarian, and Lance J. Hoffman. Challenges of Inter-vehicle Ad Hoc Networks. IEEE Transactions on Intelligent Transportation Systems, Vol.5, No.4, December 2004.
4. Jiang Hao, Kun-mean Hou et al. Capacity and packets delivery analysis of MANET on road. ICWN 2005.
5. B. Karp and H.T. Kung. GPSR: Greedy perimeter stateless routing for wireless networks. In Proceedings of 6th Annual Internation Conference on Mobile computing and Networking (MobiCom 2000), Boston, MA, USA, pages 243{254, Feb 2000.
6. J. Tian, L. Han, K. Rothermel, and C. Cseh. Spatially aware packet routing for mobile ad hoc inter-vehicle radio networks. The Intelligent Transportation Systems Conf., Shanghai, China, Oct. 2003.
7. Rudack M., Meincke M. Lott M. On the Dynamics of Ad-Hoc Networks for Inter-Vehicle Communications (IVC). The 2002 International Conference on Wireless Networks ICWN 2002, Las Vegas, USA, June 2002
8. Denos C. Gazis. Traffic Theory. Kluwer publisher, 2002.
9. Yongjin Kim, Jae-Joon Lee, and Ahmed Helmy. Modeling and analyzing the impact of location inconsistencies on geographic routing in wireless networks. SIGMOBILE Mob. Comput. Commun. Rev., 8(1):48-60, 2004.

10. J.Yves L.Blazevic and S.Ciordano. A location-based routing method for mobile ad hoc networks. IEEE Transaction on Mobile Computing, 3(4):243-254, Feb 2005.
11. F. Kuhn, R. Wattenhofer, Y. Zhang, A. Zollinger. Geometric Ad-Hoc Routing: Of Theory and Practice. In Proc. Of the 22nd ACM Symposium on the Principles of Distributed Computing(PODC), July 2003
12. C. Lochert, M. Mauve, H. Füßler, H. Hartenstein. Geographic routing in city scenarios. ACM SIGMOBILE Mobile Computing and Communications Review, Volume 9, Issue 1, January 2005.

Evaluation of the Energy Consumption in MANET

Géraud Allard[1], Pascale Minet[1], Dang-Quan Nguyen[1], and Nirisha Shrestha[2]

[1] INRIA Rocquencourt, Domaine de Voluceau, 78153 Le Chesnay Cedex, France
{geraud.allard, pascale.minet, dang-quan.nguyen}@inria.fr
[2] Macquarie University, North Ryde, NSW 2109, Australia
nirisha@ics.mq.edu.au

Abstract. In mobile ad hoc wireless networks, energy consumption is an important issue as most mobile nodes operate on limited battery resources. Existing models for evaluating the energy consumption in a mobile ad hoc network have shown that the various components of energy related costs include transmission power as well as the reception power.

In this paper, we extend the model for calculating the energy spent at a node due to a flow in the network. We include the transmission and reception costs if the node belongs to a flow, and reception costs if it is near a flow. This model gives the energy costs of nodes in ideal conditions where interferences are absent. It is then extended to evaluate the interference effect on energy consumption in more realistic conditions. The collisions due to concurrent flows are also measured. We then show how the extra energy spent due to collisions can be calculated by predicting the collisions in the nodes of the network. This prediction is shown to be capable of accurate calculation of the extra energy consumption.

1 Introduction

A Mobile Ad hoc Network (MANET) is a decentralized infrastructure-less network where wireless nodes move arbitrarily. The nodes communicate directly with nodes in transmission range, and take part in multi-hop communication with others. Such networks are gaining more and more popularity because of their ease of deployment. However, due to the shared wireless medium and its dynamic nature, these networks face various problems like unpredictable topology, increased interference and congestion, and limitation of resources like bandwidth and energy.

The nodes in this type of networks are usually power constrained since they depend on limited battery resources while wireless communications consume lot of energy. Wireless also means that there is a shared environment, and some energy is consumed due to neighborhood transmissions: nodes are spending their batteries not only by sending their own packets, but also by just overhearing packets from other nodes. As ad hoc network is based on multi-hop communication, energy is also spent by forwarding packets for others. The uniqueness of this decentralized network requires different energy management strategies.

T. Kunz and S.S. Ravi (Eds.): ADHOC-NOW 2006, LNCS 4104, pp. 170–183, 2006.

Design of ad hoc networks while considering the energy issues has spurred a great deal of research interest recently.

Feeney *et al.* [2, 1] have presented an energy consumption model in terms of the costs of both sending and receiving traffic. Those results have become a basis for comparing routing protocols considering the overhead due to routing as well as the data traffic. In this paper, we extend this model by calculating the energy consumed at each node due to flows in the network. We consider not only the transmission and reception costs of the nodes belonging to a flow, but also the costs due to interference of the flows in these nodes and other nodes in the neighborhood of the flows. By simulation, we show the accuracy of the model, then use this model to evaluate the effect of interference and collisions on the energy consumption. Such a model can be used to analyze the performance of various protocols, or can be used by routing protocols to assign paths for existing flows in an energy-efficient way.

The remainder of this paper is organized as follows. In section 2, we outline various energy management techniques and the energy measurement model presented by Feeney. In section 3, we present our model for calculating the energy costs at various nodes. Section 4 describes the simulation methodology, and section 5 presents the results and their discussion. A method for predicting collision and measuring the energy consumption due to collisions is discussed in section 6. Finally, we conclude in section 7.

2   State of the Art and Motivation

2.1   Energy-Aware Routing Protocols

The mobile nodes in MANET usually need to be untethered, and are thus powered by batteries which provide limited energy. In the absence of a central control entity like a base station, each node has to participate in distributed protocol(s) for building routes, causing them to spend more energy. In the past few years, many energy-aware protocols have been proposed for wireless MANET and sensor networks. They are aiming at solving different goals as described below.

The first set of protocols try to minimize broadcast traffic, and the energy thus spent. BIP [3] and LMST [4] are based on minimum-energy broadcasting using Minimum Spanning Tree (MST) development. Similarly, [5] uses Shortest-Path Tree development. There has also been some work done on the formation of a virtual backbone, like Connected Dominating Sets and other forms of clustering algorithms for routing purposes, e.g. [6].

The second set of protocols are based on topology control and minimum energy routing problems. The nodes can alter their transmit power level to maintain the connectivity of the network topology, while increasing the network capacity, and reducing the interference and energy consumption. Here, Signal to Interference Ratio (SIR) threshold, transmit power and received signal strength are generally used to find the minimum transmission power required between any two nodes in the network. Such protocols usually prefer several short hops to one long hop, as these have been found to reduce interference and contention in the network. COMPOW [7] is a protocol for finding the common lowest transmit power for all the nodes in the network, while maintaining network connectivity. Topology control algorithms based on the geometric properties of the network are presented in [8, 9]. The protocol in [10] tries to limit the degree of the nodes by forming MST based on local connectivity. PARO [11] uses a cost function based on transmit power level, and if intermediate nodes can forward with less total power, they offer to become the re-directors of packets. Similarly, [12] uses hello messages to distribute transmission power, and takes into account the minimum power required to connect to a neighbor, while considering the costs of reception of a packet at the neighboring nodes.

Next set consists of routing for maximum network lifetime. In papers like [13, 14, 15, 16], routing is done by using an appropriate metric for optimization of power consumption of the network interface per packet, while taking into account the battery reserves at the nodes.

Few other protocols make use of the power save mode available in the network interface to reduce energy consumption. Since there is no central entity to control the sleep and wake up periods, some distributed control mechanism is required for such protocols. Example of such protocols are PAMAS [17], SPAN [18], and GAF [19].

2.2   Measurement of Energy Consumption

Since energy is a scarce and non-renewable resource in wireless ad hoc networks, energy-efficient protocol design is a key concern. The design and performance analysis of such protocols require proper modeling for the measurement of energy consumption. Feeney *et al.* [2, 1] presented some results of measuring energy consumptions of various network interfaces. Four possible energy consumption states are identified: transmit, receive, idle and sleep. The first two states are when the node is transmitting and receiving packets respectively, the idle state is when the node is waiting for any packet transfers, and the sleep state is a very low power state where the node can neither receive nor transmit.

The cost associated with each packet at a node is represented as the total of incremental cost m proportional to the packet *size* and a fixed cost b associated with channel acquisition:

$$Cost \; = \; m \times size + b \qquad (1)$$

Developing on this energy model, we present a simple model to calculate the energy spent at each node due to the flows present in the network. Using the power consumption values for transmit and receive state as measured in Feeney's results, we show through simulations in ns-2 that in an ideal network without any interference and collisions, our model gives exact measurement of energy consumption in nodes due to a flow in the network. Then this model is used to compare the energy consumption in non-ideal simulation settings to evaluate the effect of interference and collision on energy. In fact, a similar model can be used to measure other parameters of the network like bandwidth [20] in order to carry out performance evaluation of routing protocols.

3   Model and Measurement

The energy spent at each node due to a flow can be calculated in a simple way according to our model. Figure 1 shows the model that is used. Depending on whether the node N belongs to a flow or not and where in the flow the node M affecting it is situated, the total energy expenditure at a node N due to another node M in the network can be calculated as follows:

$$E_{N/M} = \mathbb{1}_{n>0}(\mathbb{1}_{M=N}E_{T_{ack}} + \mathbb{1}_{M \neq N}E_{R_{ack}})$$
$$+\mathbb{1}_{m>0}(\mathbb{1}_{M=N}E_{T_{pck}} + \mathbb{1}_{M \neq N}E_{R_{pck}}) \tag{2}$$

where

$E_{N/M}$ = energy spent at node N due to node M,

$E_{T_{ack}}$ = energy spent for transmission of one acknowledgment (ACK) packet,

$E_{T_{pck}}$ = energy spent for transmission of one data packet,

$E_{R_{ack}}$ = energy spent for reception of one ACK packet,

$E_{R_{pck}}$ = energy spent for reception of one data packet

$$\mathbb{1}_p = \begin{cases} 1 & \text{if } p \text{ is true,} \\ 0 & \text{otherwise.} \end{cases}$$

This model simplifies the packet exchanges by including data and the ACK packets only. If other packets are also included, like RTS/CTS (Request to Send/Clear to Send) and ARP (Address Resolution Protocol) packets, the related costs for these packets can be simply added to extend this model.

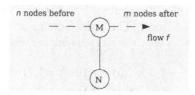

Fig. 1. Effect on node N due to node M in flow f

4   Simulation

Using the model presented in the previous section, a number of simulations were done in ns-2 to experiment with energy consumption in wireless ad hoc networks, where the flows affect/interfere with each other. The main aim was to measure energy at the nodes in the flow and see how the energy consumption is affected by interference from other flows, so that effect of collisions can be observed. The simulation settings and results will be discussed next.

4.1 Simulation Settings

Nodes have the default transmission range of 250m and a Carrier Sense range of 500m. The RTS/CTS packets are turned off in the Medium Access Control (MAC) layer. Various Constant Bit-Rate (CBR) flows at a bit rate of 250kbps are considered for the simulation time of 600 seconds each. The size of the data packets is set at 1.5 kilobytes.

4.2 Calculation of Energy for Transmission and Reception of a Single Packet

For Data Packets

Packet length = 1500 bytes, bit rate = 250 kbps (20.8 packets/sec),
Total packet size = size of (preamble + PLCP header + MAC header + IP header + data) = $(144 + 48 + 28 \times 8 + 20 \times 8 + 1500 \times 8)$ bits (default values, as used in ns-2).
The preamble and PLCP header are transmitted at 1Mbps while the rest is sent at 11 Mbps. Thus, we have 144 + 48 bits sent at 1 Mbps, with a transmission time for single packet = 0.19 ms.
With 8×1548 bits sent at 11 Mbps, the transmission time for a single packet is 1.128 ms.
Hence, the total transmission time for a single packet = $1.128 + 0.19 = 1.318$ ms.

For ACK Packets

Packet length = 14 bytes, bit rate = 250 kbps.
Total packet size = size of (preamble + PLCP header + ACK) = $(144 + 48 + 14 \times 8)$ bits.
So, the transmission time for a single acknowledgement is 0.304 ms.

Calculation of Energy Spent

For the simulation, the typical transmission and reception cost for Lucent Silver card as specified in [1,2] is used. The transmission power used is 1.3 W, and reception power is 0.9 W. Thus, the various energy cost components for each packet are:

$$E_{T_{pck}} = 1.3 \times 1.318 \times 10^{-3} = 1.713 \ mJ$$
$$E_{R_{pck}} = 0.9 \times 1.318 \times 10^{-3} = 1.186 \ mJ$$
$$E_{T_{ack}} = 1.3 \times 304 \times 10^{-6} = 0.395 \ mJ$$
$$E_{R_{ack}} = 0.9 \times 304 \times 10^{-6} = 0.274 \ mJ$$

Thus, using the energy calculation equation (2), the energy for the flow as shown below is calculated for each node in the flow:

$$0 \longrightarrow 1 \longrightarrow 2 \longrightarrow 3 \longrightarrow 4 \longrightarrow 5 \longrightarrow 6$$

For node 0, $E_{0/0} = E_{T_{pck}} = 1.713$ mJ, $E_{0/1} = E_{R_{ack}} + E_{R_{pck}} = 1.46$ mJ and $E_{0/2} = E_{R_{ack}} + E_{R_{pck}} = 1.46$ mJ. So the energy spent at node 0 is the sum of these values and is given by $E_0 = 4.633$ mJ. Similarly, energy is calculated for other nodes.

The calculated values are compared with the energy expenditure of the nodes per second (first, the results above have to be multiplied by 20.8 for the number of packets being generated by the flow per second). The difference in the two energy levels gives the excess energy lost due to packet collisions during the simulation.

Other more complex scenarios are then tested. The first one consists of three flows, each flow consisting of 7 nodes (see Figure 2), and the second one consists of four flows in a network with 200 nodes created randomly in a square network of 1800 meters length.

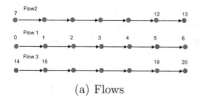

(a) Flows

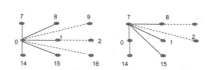

(b) Neighbors and interfering nodes (solid and dashed lines respectively) for nodes 0 and 7

Fig. 2. Simulation network with 3 flows

5   Discussion and Analysis of Results

5.1   Single Flow

In the simple case of a single flow, the objective is to find the correctness of the formula calculated above (2). Since there is no interference involved, it is expected that the simulated and calculated values should match. It is found that the calculated and simulated energy levels are exactly the same, proving the correctness of the formula.

5.2   Three Flows

For the case with 21 nodes and three flows, two cases are considered, one where all the flows are in the same direction (Figure 2) and one where flow 1 is in the opposite direction (note that, although the flow is still from node 0 to node 6, the position of these nodes are interchanged so that flow 1 is now from right to left). The result of the simulation is shown in Figure 3.

In the results, there is a larger difference in the calculated and simulated energy consumption in the source nodes (up to about 83mJ/sec), showing that these nodes have to transmit and receive more, most probably due to collisions. This can be explained, for example, for flow 1, nodes 0 and 3 can transmit at the same time, since they are not in the carrier sense region of each other, but node 3 acts as a hidden node to node 0, causing collisions at node 1. This is

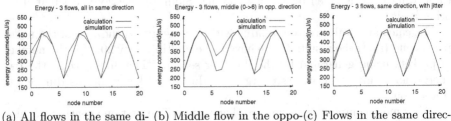

(a) All flows in the same di- (b) Middle flow in the oppo-(c) Flows in the same direc-
rection   site direction   tion with jitter

Fig. 3. Energy consumption rate of nodes in the network with 3 flows

verified by figure 4, which shows the number of packets destined/forwarded to
the nodes that were lost due to collisions and also the number of retransmissions
by the nodes. It is seen that for the first flow, some of the packets forwarded
to node 1 from node 0 are lost due to collisions, requiring node 0 to retransmit
them. Since transmission requires more energy than reception, node 0 shows
more energy consumption due to retransmission than node 1 due to reception of
the retransmitted packets. Similar behavior is seen in nodes of other flows, with
more energy loss at nodes that retransmit and some energy loss in the colliding
nodes due to reception of the retransmitted packets. Please see section 6 for
more discussion of these results on collision.

The results show that source nodes experience about 31% increase in energy
consumption due to collisions. It is also noticed that some nodes actually consume less energy than that calculated. This is due to the fact that, when doing
the calculation, we assume that each flow is continuous, generating/forwarding
a new packet for each packet interval at each node. However, because of various factors like backoff due to the medium being busy, packets being lost and
needing retransmissions due to collisions, and delay before an ACK is received
due to the previous reasons, the flow is not actually continuous. The final energy
measured, which is the average energy spent by nodes each second, can be lower
if a node is not able to transmit some packets during the simulation time due to
long delays.

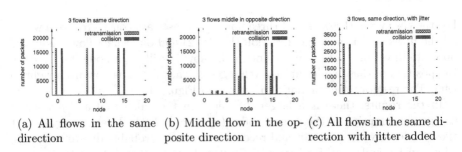

(a) All flows in the same (b) Middle flow in the op- (c) All flows in the same di-
direction   posite direction   rection with jitter added

Fig. 4. Number of packets that cause extra energy consumption (due to collisions and
retransmissions) of nodes in the network with 3 flows

When the flow in the middle is in the opposite direction, the difference in energy is greater, up to 42%. In the middle flow, the effect of collisions and retransmissions is maximum for the end nodes (node 4, 5, 6), whereas, for the other two flows, it is maximum for the source node and other nodes at the beginning of the flow. It shows that the source nodes in flows 2 and 3 highly effect nodes 4, 5, 6. As mentioned above, node 3 still causes collisions in node 1 in flow 1, but due to collisions in the flows 3 and 4, the source nodes transmit/retransmit more packets causing more energy loss at these three nodes. This shows that inter-flow collisions can cause a significant amount of energy consumption.

When evaluating flows at the same rate and when all the flows are started at the same time, collisions are more likely to take place than when there is some delay between the flows. When a jitter is added, collisions decrease by a large amount. In the simulation (see Figure 3(c)), some jitter is added between the flows so that the second flow starts just after the first flow finishes the transmission of one packet (assuming no collision), so that inter-flow collisions are greatly reduced. In this case, the increase in energy is now reduced to about 19%, but overall, the energy consumption is more balanced, and introduces less collisions than in no jitter cases. The number of retransmissions are thus greatly reduced, as can be seen in figure 4.

5.3   Four Flows in a Random Network

For the case of 200 nodes and 4 flows, 4 cases are considered - SIR is cumulative with no jitter between the flows, SIR is cumulative with jitter present between the flows, SIR is non-cumulative with no jitter between the flows, and SIR is non-cumulative with jitter present between the flows. By default, ns-2 uses a non-cumulative SIR model, where the noise from the longest signal is taken to calculate the SIR for each reception. This model does not accurately represents the real environment. In our model, a cumulative SIR model is added, where the noise taken is the sum of reception power of all signals that can be heard at the node receiving a packet, and reception is successful only if the new SIR is greater than the carrier sense threshold.

The results of the simulations for the energy consumption rate of the nodes belonging to the flows, among 200 randomly placed nodes, are shown in Figure 5 for the case with jitter and cumulative SIR model. For each node, the calculated (left) and simulated (right) energy values are presented. More detailed results and discussions on the experiments can be found in [21].

For this network and flow settings, the results are similar to that of the three flows cases: the source nodes show the most energy loss. In the default case, with no jitter and SIR non-cumulative, collisions may increase the energy consumption by up to 45.4%. The introduction of jitter into the model reduces the number of collisions, decreasing the energy consumption by up to 26.5%, and about 6% in average.

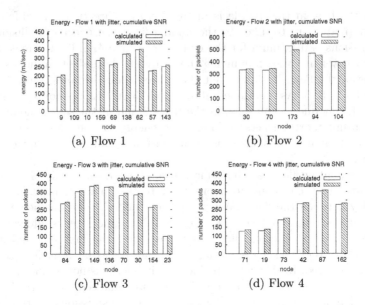

(a) Flow 1 (b) Flow 2

(c) Flow 3 (d) Flow 4

Fig. 5. Energy consumption rate of the nodes of the 4 flows, with jitter and with cumulative SIR Model

Both the cases with jitter and no jitter were also compared to the case where a cumulative SIR is used, but the energy consumption while using cumulative and non-cumulative SIR model did not show significant differences.

Thus, it is seen that though collisions between different flows can increase the energy consumption significantly, introduction of jitter can decrease such collision probabilities. It would be interesting to extend the experiments to include more flows in very dense networks, where the carrier sense diminishes at the center of the network due to extensive interference and collisions [22, 23].

6   Collision Awareness

It can be deduced from the results presented here, that collisions can affect the energy consumption of nodes in the network. The energy measurement model presented in this paper takes into account all the nodes that are in the reception area of the flow. So, not all collisions may result in more energy consumption. The main effect of collision will be at the nodes in the flow, where the packets that is forwarded to them collides, so that it has to be resent. So only these collisions are measured. As predicted, the nodes at which such collisions occur and the nodes which have to retransmit as the result, show more decrease in energy. Since transmission requires more energy than reception, nodes retransmitting show more energy consumption due to retransmission than the colliding nodes which consume extra energy due to reception of the retransmitted packets.

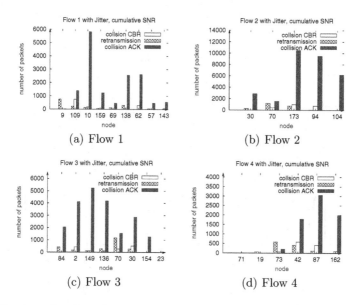

(a) Flow 1   (b) Flow 2

(c) Flow 3   (d) Flow 4

Fig. 6. Number of packets that cause extra energy consumption (due to collisions and retransmissions) at the nodes of the four flows, with jitter and cumulative SIR Model, obtained by ns-2 simulations

Due to the carrier sensing mechanism and backoff procedure used when a node wants to send a packet but the channel is not idle, the probability that two transmitter nodes collide can be neglected when these nodes are neighbors. However, this is no longer the case when they do not hear each other. Therefore, in this paper, we are interested in the case of collisions on a link due to hidden nodes to that link, as we assume that no RTS/CTS is used. Let us consider a link (T, R) where T is the transmitter and R is the receiver. A packet being transmitted from node T to node R encounters a collision if and only if during its transmission, a hidden node also transmits. We recall that a hidden node to the link (T, R) is any node H that can be heard from node R and cannot be heard from node T. In the following, we assume that the number of hidden nodes to any link, denoted n_{hidden}, is available.

Thus, if the duration of transmission of a packet on the link (T, R) is t_{trans} and the amount of time the channel is sensed idle (no noise is detected) at node T is t_{idle}, where $t_{idle} > t_{trans}$, then the probability of collision on that packet is $p = \frac{t_{trans}}{t_{idle}}$, given that there is at least one hidden node.

Let t_{busy} be the amount of time the channel is busy at node T due to node T's own traffic and noises detected. The total amount of time node T spends due to retransmissions of a packet caused by i successive collisions is given by $\Theta_i = i(t_{trans} + EIFS) + \sum_{j=1}^{i} CW_j$, where $EIFS$ is the Extended Interframe Space following reception of an erroneous frame and CW_j is the average size of the contention window at j^{th} stage according to the IEEE 802.11b standard,

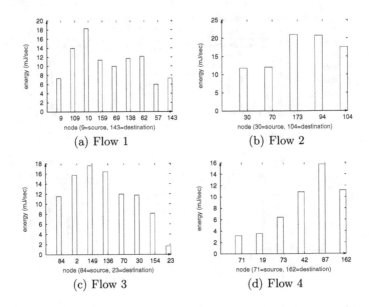

Fig. 7. Amount of extra energy consumed (due to collisions and retransmissions) at the nodes of the four flows, obtained by computation (see figures 5, 6 for comparison).

where CW_j is given by $CW_j = CW_{min} * 2^j - 1$. Therefore, the percentage of time the channel is sensed idle after i successive collisions on the same packet is $t_i = 1 - \frac{t_{busy} + \Theta_i}{t_{arrival}}$, where $t_{arrival}$ is the inter-arrival time of packets on link (T, R).

We can now compute the amount of energy consumption at a node N due to collisions on the link (T, R), denoted $E^N_{(T,R)}$, using the collision model presented in [20], under the assumption that a packet collides at most once with a given hidden node:

$$E^N_{(T,R)} = \frac{\sum_{i=1}^{n_{hidden}} i(E^N_T + E^N_R)\delta_i}{1 + \sum_{i=1}^{n_{hidden}} \delta_i}$$

where

$$\delta_i = \frac{t^i_{trans}}{\prod_{j=0}^{i-1} t_j}$$

is the probability of encountering i successive collisions on a packet, given that at least one collision is encountered,

$$E^N_T = \begin{cases} E_{T_{pck}} & \text{if } N \equiv T \\ E_{R_{pk}} & \text{if } N \text{ can receive packets from } T \\ 0 & \text{otherwise,} \end{cases}$$

and

$$E_R^N = \begin{cases} E_{T_{ack}} & \text{if } N \equiv R \\ E_{R_{ack}} & \text{if } N \text{ can receive packets from } R \\ 0 & \text{otherwise.} \end{cases}$$

Let $\mathcal{L}$ be the set of links that interfere with node N; in other words, node N is in the reception area of either the transmitter or the receiver of each link in $\mathcal{L}$. The amount of energy consumed at node N due to packet collisions of $\mathcal{L}$ is $E_{\mathcal{L}}^N = \sum_{l \in \mathcal{L}} E_l^N$.

Figures 6 and 7 compare the results of packet collisions obtained by ns-2 simulations and by computation using the above formula. We use the same parameters of packet size, bit rates and node configuration for the simulations and computation as in section 4. We notice that the results obtained by computation match those obtained by ns-2 simulations. This shows that the collision model presented here can predict collisions in the nodes belonging to a flow and thus the energy spent because of the collisions. It should be noted again that retransmission requires more energy than reception causing retransmitting nodes to consume more energy than those receiving the packet again due to collision.

7 Conclusion

This paper mainly presents a new model for calculating the energy consumption at a node due to various flows in the network. Such a measurement can be used for calculating costs in energy aware routing, for performance evaluation of routing protocols and even in flows requiring Quality of Service (QoS) where, besides other resource guarantees for the flow being accepted, it is also required that the selected route has enough energy to support the new flow.

We evaluated the accuracy of the model, and found it to give exact measurements in ideal network conditions with no interference or collisions. The model was then used to calculate the energy consumption of more complex flows, and the comparison of calculated and simulated values allowed for an evaluation of how the flows interfere with each other. Through simulations, it was shown that the flows may interfere each other, causing the energy consumption to increase up to about 46% due to the collisions that occur. It was also noticed that introducing some jitter between the flows reduce the number of collisions, decreasing the energy consumption change to about 27%. The default SIR model in ns-2 was also changed to include a cumulative model where the total interference is the sum of all the noises in the background. It was however found to bring insignificant changes in the energy consumption as compared to the default model.

Finally, since we concluded that collisions are the main cause of discrepancies between our simulated results and calculated values, we tried to measure this extra energy consumed by predicting collisions in the nodes of the network, and calculating the energy spent due to these collisions. The prediction was also found to be accurate, with the calculated values matching the values simulated. This result further strengthens the correctness of the model for energy measurement.

The models presented here can be used to define the problem of minimum energy consumption and we are looking at it as a possible future extension.

Acknowledgment

Part of this work was done while Géraud Allard, Philippe Jacquet and Dang-Quan Nguyen were visiting Bernard Mans at Macquarie University in December 2004. The authors wish to thank Philippe and Bernard for their early comments and advices.

References

1. L. M. Feeney, *"Energy Efficient Communication in Ad Hoc Wireless Networks,"* Computer and Network Architectures Laboratory, Swedish Institute of Computer Science, 2003
2. L. M. Feeney, M. Nilsson, *"Investigating the energy consumption of a wireless network interface in an ad hoc networking environment,"* in Proceedings of IEEE Infocom, April 2001.
3. J.E. Wieselthier, G.D. Nguyen and A. Ephremides, *"On the construction and energy-efficient broadcast and multicast trees in wireless networks,"* in Proceedings of IEEE INFOCOM, Tel-Aviv, Israel, 2000.
4. N. Li, J. Hou, and L. Sha, *"Design and Analysis of an MST-Based Topology Control Algorithm,"* in Proc. of INFOCOM 2003.
5. P.J. Wan, G. Calinescu, X.Y. Li, and 0. Frieder, *"Minimum-Energy Broadcast Routing in Static Ad Hoc Wireless Networks,"* in Proceedings of 20th Ann. Joint Conf. IEEE Computer and Comm. Soc., INFOCOM,, Anchorage, Alaska, April 2001.
6. J. Wu and F. Dai, *"A Distributed Formation of a Virtual Backbone in Ad Hoc Networks using Adjustable Transmission Ranges,"* in Proceedings of ICDCS, 2004.
7. S. Narayanaswamy, V. Kawadia, R. S. Sreenivas, and P. R. Kumar, *"Power control in ad-hoc networks: Theory, architecture, algorithm and implementation of the COMPOW protocol,"* in European Wireless Conference, 2002.
8. R. Wattenhofer, L. Li, V. Bahl, and Y. Wang, *"Distributed topology control for power efficient operation in multihop wireless ad hoc networks,"* in Proceedings of IEEE INFOCOM. April 2001
9. W. Song , Y. Wang and X. Li, *"Localized algorithms for energy efficient topology in wireless ad hoc networks,"* in Proceedings of the 5th ACM international symposium on Mobile ad hoc networking and computing, Tokyo, Japan , May 2004.
10. R. Ramanathan and R. Rosales-Hain, *"Topology Control of multihop wireless networks using transmit power adjustment,"* in Proceedings of IEEE INFOCOM, Tel-Aviv, Israel, 2000.
11. J. Gomez, A. T. Campbell, M. Naghshineh and C. Bisdikian, *"Conserving Transmission Power in Wireless Ad Hoc Networks,"* in Proceedings of the Ninth International Conference on Network Protocols (ICNP'01), p.24, November 2001
12. B. H. Liu, Y. Gao, C. T. Chou and S. Jha,*"An Energy Efficient Select Optimal Neighbor Protocol for Wireless Ad Hoc Networks,"* Technical Report, UNSW-CSE-TR-0431, Network Research Laboratory, University of New South Wales, Sydney, Australia, October 2004

13. S. Singh, M. Woo and C. Raghavendra, *"Power-aware routing in mobile ad hoc networks,"* in Proceedings of 4th Annual ACM/IEEE Int. Conf. Mobile Computing Networking (MobiCom), Oct. 1998.
14. J.-H. Chang and L. Tassiulas, *"Energy conserving routing in wireless ad-hoc networks,"* in Proceedings of IEEE Infocom, vol. 1, pp. 22 31, 2000.
15. C.-K. Toh, Maximum battery life routing to support ubiquitous mobile computing in wireless ad hoc networks, IEEE Communications Magazine, pp. 138-147, June 2001.
16. T. M. Trung and S-.L. Kim, *"ERA: Energy-Saving Routing Algorithms for Ad Hoc Networks,"* in Proceedings of Australian Telecommunications, Networks and Applications Conference, Melbourne, Australia, 2003
17. S. Singh and C. S. Raghavendra, *"Power efficient MAC protocol for multihop radio networks,"* in The Ninth IEEE International Symposium on Personal, Indoor and Mobile Radio Communications, pp. 153-157, 1998.
18. B. Chen, K. Jamieson, H. Balakrishnan, and R. Morris, *"Span: An energy-efficient coordination algorithm for topology maintenance in ad hoc wireless networks,"* in Proceedings of MOBICOM, July 2001
19. Y. Xu, J. Heidemann and D. Estrin, *"Geography-informed energy conservation for ad hoc routing,"* in Proceedings of 7th Annual International Conference on Mobile Computing and Networking, pp. 7084, July 2001.
20. D. Nguyen and P. Minet, *"Evaluation of the Bandwidth Needed at the MAC 802.11b Level,"* Research Report, INRIA-RR-5553, INRIA, France, April 2005.
21. G. Allard, P. Minet, D. Nguyen and N. Shrestha, *"Evaluation of the Energy Consumption in MANET,"* Research Report, INRIA-RR-5851, INRIA, France, February 2006.
22. P. Jacquet, *"Space-time Information Propagation in Mobile ad hoc Wireless Networks,"* 2004 IEEE Information Theory Workshop, Texas, October 2004
23. P. Jacquet, *"Geometry of Information Propagation in Massively Dense ad hoc Networks,"* in Proceedings of MobiHoc 2004, Tokyo, Japan, May 2004

Link Availability at Any Time in MANET

Jianxin Wang, Xianman Zhu, and Jianer Chen

School of Information Science and Engineering, Central South University,
Changsha 410083, Hunan, China
jxwang@csu.edu.cn, hnsfman@163.com, chen@cs.tamu.edu

Abstract. Path broken, which is caused by node mobility, leads to the rerouting operations and the degradation of network performance. How to select a reliable path becomes a critical issue in designing routing protocol. The metric for routing selection in terms of path availability which is based on link availability can be used to select reliable paths. Based on the random walk-based mobility model, we propose a prediction method for calculating the link availability at any time ($LA(t)$) on the basis of the duration T during which two nodes keep their movement status unchanged for the first time. Meanwhile, a general method to calculate the average LET ($\bar{\tau}$) is developed as well as the general expression of $LA(t)$ is given. Simulation results show that the results of our method are more approximate to the practical results as compared with those of other methods.

Keyword: link availability, mobile ad hoc networks, prediction.

1 Introduction

An ad hoc network is a dynamically reconfigurable wireless network with no fixed infrastructure, each of whose host acts as a router and moves in an arbitrary manner. For an active connection, the end host as well as the intermediate nodes can be mobile. Therefore routes are inclined to frequent broken, which lead to frequently rerouting to find another link to repair or a new available path to recover communication. These rerouting operations consume the limited resources such as power and bandwidth etc in mobile ad hoc networks and are likely to cause congestion and extra delay, which affect the QoS of some critical application such as voice and video and degrade network performance. However, the proposed routing protocols based on the shortest path and minimum hops cannot prevent the performance degradation resulting from link failure [1,2,3,4]. So it is very necessary to find a new routing metric to select the stable and optimum paths to minimize disruptions caused by the changing topology. The most stable path is the path with maximum route expiration time (RET) during which all links along the path keep available, that is to say, it holds the longest bottleneck link expiration time (LET), where bottleneck link is the link with minimum expiration time among all links along the whole path. The stability of a path strongly depends on the stability of the constituting links, because the break of any link will lead to the break of the whole

T. Kunz and S.S. Ravi (Eds.): ADHOC-NOW 2006, LNCS 4104, pp. 184–196, 2006.

path. Therefore, link stability is expected to be obtained before the determination of path stability.

In typical mobile networks, similar to human activities, nodes exhibit some degree of regularity in the mobility pattern. They are likely to maintain their direction and speed for some period of time before changing them. By exploiting a mobile user's non-random traveling pattern, we can predict the future state of a network topology and thus provide a transparent network access during the period of topology changes. So how to predict link stability by utilizing node's current mobility information becomes a critical problem and meaningful groundwork in ad hoc routing protocol design.

At present, there are three strategies proposed for link prediction:

(1) Link Expiration Time (LET) Prediction [5]

Assume that two nodes i and j are within the transmission range r of each other initially. Let (x_i, y_i) be the coordinate of mobile host i and (x_j, y_j) be that of mobile host j, v_i and v_j be the speeds, θ_i and θ_j $(0 \cdot \theta_i, \theta_j \cdot 2\pi)$ be the moving directions of nodes i and j, respectively. Then each of them moves at a constant speed and direction respectively. The link expiration time namely LET during which they can keep connected continuously can be predicted by:

$$LET = \frac{\sqrt{(a^2 + c^2)r^2 - (ad - bc)^2} - (ab + cd)}{a^2 + c^2}$$

Where $a = v_i \cos\theta_i - v_j \cos\theta_j$; $b = x_i - x_j$; $c = v_i \sin\theta_i - v_j \sin\theta_j$; $d = y_i - y_j$. The lifetime of the whole path referred as RET (Routing Expiration Time) is the minimum of the set of LETs along the route, namely, RET=Min(LET$_i$), where LET$_i$ represents LET of the *ith* link in the path. The prior condition for this prediction method is that the speed and direction of node should keep constant.

(2) Node Position Prediction [6]

Predict the future positions of two nodes according to their current mobility information (including position, speed and direction). It can be judged whether they are connected or not at time t according to their positions. The latest two renewed historical records should be stored when lacking of direction information. Essentially, it is the same as LET prediction method.

(3) Link Availability Prediction

The link availability model is the hot spot of link prediction research. Link availability is defined as the probability that two mobile nodes keep connected from time t_0 to time $t_0 + T$ $(T > 0)$ given that they can communicate directly at time t_0. Making use of the knowledge on probability theory and geometry, we can obtain link availability between two nodes on the basis of the assumption that nodes move obey some certain mobility models.

The prediction of link expiration time and node position are both on the basis of the assumption that node's velocity (speed and direction) should keep unchanged, which is unpractical in realistic ad hoc networks where nodes change their motion status regularly. Therefore, it is more practical to establish link availability model. This paper focuses on the research of probability that link stay active over any period of time t, which can be used as a new metric for routing protocols.

The remainder of this paper is organized as follows. Section 2 introduced the related work. Section 3 characterized the random ad hoc mobility model and gave the expression of link availability as a function over some parameters related to t, epoch and speed. The simulation results with comparison to the methods proposed in [10,11], and detailed analysis are reported in Section 4. Finally, we summarize our contributions and provide suggestions for further research in Section 5.

2   Related Work

The conception of link availability and path availability was firstly proposed and applied to routing designing by A. Bruce McDonald in[7,8]. The link availability was developed in two distinct cases: (1) Node activation, namely, node m becomes active at time t_0, and it is assumed to be at a random location within range of node n. (2) Link activation, namely, node m moves within range of node n at time t_0 by reaching the boundary defined by R_{eq}, and it is assumed to be located at a random point around the boundary.

A Prediction-based Link Availability $L(T_p)$ [10] with respect to the probability that the link will last to t_0+T_p was estimated by considering possible changes in nodes' movements occurring between t_0 and t_0+T_p, where T_p is a continuous time period that a currently available link will last from time t_0 by assuming that both nodes of the link will keep their current movements (i.e., speed and direction) unchanged. More precisely, $L(T_p)=L_1(T_p)+L_2(T_p)$, namely, the link availability consists of $L_1(T_p)$ with respect to 'unaffected T_p' with the above assumption and $L_2(T_p)$ with respect to 'affected T_p' with movement being changed. Although possible changes in the nodes' movements were considered, there were still some unreasonable assumptions and approximation which support the derivation of $L_2(T_p)$:

(1) The continuous period refers to LET that the link can stay active after two nodes' firstly change on movement status (i.e., position, speed and direction) is equal to the original LET depending on the original movement status which is different from the current movement status generally.

(2) Assume that two node will keep connected potentially only when they move close toward each other, but ignore the fact that the link will possibly stay active even if the two nodes move apart from each other if they change their movement status before the 'real' separation.

(3) Only two changes on movement status are considered during the prediction time T_p. Actually, the changes is more than two times under many circumstance according to the simulation results.

(4) A correction (refer as ε) to the link availability obtained by simulation experiment is introduced, which strongly depends on the simulation environment and parameters. Before computing the link availability, simulation experiment should be fulfilled first, moreover it should change if the parameters change.

An Enhanced Prediction-Based Link Availability was proposed in [11]. The T_p is divided into k link epochs during which both nodes associated with this link will not change their velocities. And k is Poisson with a mean rate equal to the reciprocal of the average link epoch. For each k, let link probability be the product of the conditional probability that the link keeps continuously available during the

$ith(0 \leq i \leq k)$ link epoch. Accordingly, the link availability is the sum of all link proba-bilities with different values of k ($k \geq 0$). Similar to their former method, estimation refer to average LET which comes from simulation experiment is introduced and makes the link availability lacking of practicability.

Different from the model developed in [7,8], a new prediction model with few restriction to node's movement was adopted in [12]. Given that nodes keep movement status unchanged, the link availability can be predicted by $P_a(r,\psi,t)=1-F_a(r,\psi,t)$, and it was extend to derive the availability to multiple-hop path. Here $F_a(r,\psi,t)$ represents the probability that the link available time smaller than t, which is equivalent to the probability that the relatively speed larger than V_{min} , where V_{min} is the minimum required velocity to cover the distance, over which one node moves out of the radio transmission range of the other, exactly after time t.

As mentioned in part one, the fatal weakness of the model mentioned above such as link expiration time prediction, node position prediction and link availability proposed in [12] is that they do not consider the possible changes of nodes' movement status. And the methods developed in [10,11] are simulation-dependent, that is to say, we need obtain the correction to link availability via simulation experiment before employing it, which disobeys the original intention and is unpractical.

The purpose of this paper is to resolve the issues mentioned above by proposing a link availability model over any time for mobile ad hoc networks with a random walk mobility model.

3   Link Availability Prediction

3.1 Network Model and Mobility Model

Due to dynamic topology, limited memory space and limited power in ad hoc networks, it is impossible to predict the future motion status of nodes according to a series of historical data because there is not enough space to store the data. Moreover, the stored data is likely to be stale. Consequently, we can only predict nodes' behavior if nodes move according to a certain regulation rather than thorou-ghly in random. In the paper, we adopt a random walk-based mobility model [15] which is slightly different from the one in [7,8,9,10,11]. In this model, the transmission range of nodes in ad hoc networks is exactly a circle with radius r. Any two nodes are connected if they locate within the transmission range of each other. Meanwhile, the communication between them is bidirectional. Nodes can move within a two dimensional space without boundary. Based on the random ad hoc mobility model, each node's movement consists of a sequence of random length intervals called mobility epochs during which a node moves in a constant direction at a constant speed. The velocity (i.e. speed and direction) of each node varies randomly from epoch to epoch. Consequently, during epoch i, node n moves in a straight line at a speed of V_n^i and an angle of θ_n^i. The parameters in this model are defined as follows:

(1) The epoch lengths are identically, independently distributed (i.i.d.) exponentially with mean $1/\lambda_n$.
(2) The direction and speed of the mobile node during each epoch remain constant only for the duration of the epoch, whose distributions are arbitrary.
(3) Speed, direction, and epoch length are uncorrelated.
(4) Mobility is uncorrelated among the nodes of a network, and links fail independently.
(5) According to the practical situation, the node can evaluate the interval $[t_0, t_0+T_n]$ during which it keep its velocity (speed and direction) unchanged.

3.2 Link Availability

Consider two node m and n with various epochs during a certain duration t, it is likely that node m keeps its velocity unchanged while node n changed its state and entered into its next epoch. For convenience, the link epoch is introduced as definition 1[11].

Definition 1. Link epoch is a time period during which both two nodes m and n associated with this link will not change their velocities.
The link epoch is identically, independently distributed (i.i.d.) exponentially with mean $1/\lambda'$ in terms of the characterization of exponential distribution, whose CDF (Cumulative Density Function) is given by $E(x)=P\{\text{link epoch} \le x\}=1-e^{-\lambda'x}$, Where $\lambda'=\lambda_m+\lambda_n$ and it is simplified to $\lambda'=2\lambda$ according to the assumption $\lambda_m=\lambda_n$, and consequently

$$E(x)=1-e^{-2\lambda x}.\qquad(1)$$

And its PDF (probability density function) is:

$$f(x)=2\lambda e^{-2\lambda x}.\qquad(2)$$

The duration T_m and T_n which evaluated by node m and node n respectively are likely to be different because the tasks of the two nodes are probably different. For convenience, we define $T=Min(T_m, T_n)$.

Definition 2. Link Availability at any time t_0+t (i.e. $LA(t)$) is the probability that two nodes keep connected at time t_0+t assuming that they are connected at time t_0, i.e.

$LA(t)=P\{two\ nodes\ keep\ connected\ during\ [t_0,\ t_0+t]|they\ are\ connected\ at\ t_0\}$

Assume that the duration during which both the two nodes keep their velocities unchanged for the first time is T (the first link epoch), and the expiration time of the link associated to these two nodes which predicted at time t_0 is LET, and the one predicted at time t_0+T is LET_2 (i.e. τ). The next link epoch after time t_0+T is φ (the second link epoch). Based on the assumption proposed above, $LA(t)$ can be derived depending on the relations between t, LET and T in various cases. Fig.1. shows the model for predicting $LA(t)$ in the case of $T<t<LET$, and other cases with different t can be obtained by changing the location of t in the horizontal axis according to the dotted line. We discuss the computation of $LA(t)$ under different conditions:

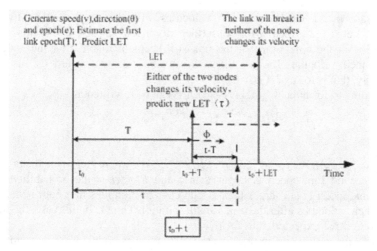

Fig. 1. The model for link availability prediction

1. *T≥LET*. It implies that two nodes keep their velocities constant before the expiration time *LET* when the link is expected to break up. Whereupon, depending on the relation between *t* and *LET* there are two distinct cases:

(1) *t≤LET*, i.e. *t≤LET≤T*. This case meets the assumption for the prediction of Link Expiration Time and meanwhile the expected expiration time, *LET*, is longer than *t*. So the link will stay available at time t_0+t. It is the same to say that *LA(t)=1*.

(2) *t>LET*. As opposed to case (1), the link has broken at time t_0+LET before the time t_0+t when we predict *LA(t)*, so *LA(t)=0*.

2. *T<LET*. It implies that the link will keep available probably because at least one of the nodes has changed its velocity before time t_0+LET when the link is expected to experience the possible break for the first time if neither of them has changed its velocity. Similarly, depending on the relation between *t* and *T* there are two distinct cases:

(1) *t≤T*. Similar to the case 1(1), the link will keep active at time t_0+t and thus *LA(t)=1*.

(2) *t>T*. As showed in Fig.1., the link is active at time t_0+T absolutely. Therefore, we only need to consider whether the link keeps available or not during the duration $[t_0+T, t_0+t]$. And the cases in which the nodes change their velocities for two times or more than two times should be considered further. Specifically, there are also two distinct cases:

(a) *t-T≤τ*

In the case, *LA(t)=P{φ≥t-T}*LA₁(t)+P{φ<t-T}*LA₂(t)*, where *P{φ≥t-T}* is the probability that the second link epoch is not less than *t-T* and *P{φ<t-T}* is the probability that the second link epoch is less than *t-T*. The case *φ≥t-T* infers that both nodes do not change their velocities before time t_0+t and the expected expiration time *τ* is later than time t_0+t because of *t-T≤τ*. So the link will not break during the duration *t-T* in the case *φ≥t-T*, that is to say *LA₁(t)=1*.

The case $\varphi < t\text{-}T$ indicates nodes' velocities will change more than two times possibly. Let L_{avg}, which will be further discussed in 3.3, represents the link availability on the condition that the link still keeps available after the nodes have changed their velocities for two times. So in this case, we regard L_{avg} as the link availability, that is to say $LA_2(t)=L_{avg}$.

According to formula (1), in this case $LA(t)$ can be rewritten as:

$$LA(t)=e^{-2\lambda(t-T)}+(1-e^{-2\lambda(t-T)})*L_{avg} .$$ (3)

(b) $t\text{-}T>\tau$

In this case, $LA(t)=P\{\varphi>\tau\}*LA_1(t)+P\{\varphi\leq\tau\}*LA_2(t)$, where $P\{\varphi>\tau\}$ is the probability that the second link epoch is larger than τ and $P\{\varphi\leq\tau\}$ is the probability that the second link epoch is not larger than τ. The case $\varphi>\tau$ implies that both nodes do not change their velocities after the time Link Expiration Time τ. In this case, the link will really break up as expected, thus $LA_1(t)=0$.

Just like the case $\varphi < t\text{-}T$ in (a), the case $\varphi\leq\tau$ indicates nodes' velocities will change more than two times possibly. So in this case, we regard L_{avg}, which is mentioned above, as the link availability, that is to say $LA_2(t)=L_{avg}$.

According to formula (1), in this case $LA(t)$ can be rewritten as:

$$LA(t)=(1-e^{-2\lambda\tau})*L_{avg} .$$ (4)

Table 1 shows how to calculate $LA(t)$ under different conditions.

Table 1. The expression of $LA(t)$ in different cases

Condition			LA(t)
T≥LET	T≤LET		1
	t>LET		0
T<LET	t>T	t≤T	1
		t-T≤τ	$e^{-2\lambda(t-T)}+(1-e^{-2\lambda(t-T)})*L_{avg}$
		t-T>τ	$LA(t)=(1-e^{-2\lambda\tau})*L_{avg}$

3.3 L_{avg}

Further discussion for the computation of L_{avg}, which represents the link availability on the condition that the link still keeps available after the nodes have changed their velocities for two times, is given as follows.

Let T_c be the continuous time interval from the original time t_0 and $L(\varphi)$ be the link availability. In the case that the link still keeps available after the nodes have changed their velocities for two times, T_c can be estimated as [10] .

$$T_c = T + \varphi + (t - T - \varphi)e^{-2\lambda(t-T-\varphi)}$$

and $L(\varphi)$ also can be calculated by [10].

$$L(\varphi) = \frac{T_c}{t} = \frac{T + \varphi + (t - T - \varphi)e^{-2\lambda(t-T-\varphi)}}{t} .$$ (5)

In the formulation above, T is the first link epoch, φ is the second link epoch.

Seeing that φ is a random variable, the average $L(\varphi)$ over φ can be used to estimate L_{avg}:

$$L_{avg} = \int_0^{t-T} L(\varphi)f(\varphi)d\varphi \cdot \tag{6}$$

According to (2) and (5), (6) can be changed to:

$$L_{avg} = \frac{1 + 2\lambda T + e^{-2\lambda(t-T)}(-1 - 2\lambda t + 2\lambda^2(t-T)^2)}{2\lambda t} \cdot \tag{7}$$

It is impossible to calculate τ after the second change on velocities because the locations, speeds and directions of the two nodes are all unknown at this moment. However, it is reasonable for us to replace τ with $\overline{\tau}$, which is the average LET in the network.

3.4 Average Let $\overline{\tau}$

Consider two nodes m and n which are located within each other's transmission range with the same radius r, assume v_1 and v_2 are the speeds of each other, and θ be the angle between v_1 and v_2. Let node m move with respect to node n. Then according to the Cosine Theorem, the relative speed of v_1 and v_2 is $V = \sqrt{v_1^2 + v_2^2 - 2v_1v_2\cos\theta}$. Node m moves at the constant relative speed V firstly and after the time LET it moves out of the transmission range of node n. The distance it traveled during this period is d. Hence, $LET=d/V$.

Apparently, the initial location of node n is uniformly distributed within the transmission range of node m, and the phase angle of the relative velocity, θ, is uniformly distributed over $[0,2\pi]$ and d is uniformly distributed over $[0,2r]$. Therefore, LET is a random variable function over d and V and its expectation $E(LET)$ is the average of LET, i.e. $\overline{\tau} = E(LET) = E(\frac{d}{V})$. The random variables d and V are independent of each other for v_1, v_2, θ and d are independent of one another, thus:

$$E(LET)=E(d)*E(1/V)$$

Due to the fact that d is uniformly distributed over $[0,2r]$, we can get $E(d)=(0+2r)/2=r$. It is clear that $E(1/V)= \int_0^{2\pi}\int_{-\infty}^{+\infty}\int_{-\infty}^{+\infty} g(v_1,v_2,\theta)f(v_1,v_2,\theta)dv_1dv_2d\theta$, so the average LET, $\overline{\tau}$, can be calculated by:

$$\overline{\tau} = r\int_0^{2\pi}\int_{-\infty}^{+\infty}\int_{-\infty}^{+\infty} g(v_1,v_2,\theta)f(v_1,v_2,\theta)dv_1dv_2d\theta \cdot \tag{8}$$

Where $g(v_1,v_2,\theta)=1/V$, $f(v_1,v_2,\theta)$ is the joint PDF of random variables of v_1,v_2 and θ which are independent of one another and so $f(v_1,v_2,\theta)=f(v_1)*f(v_2)*f(\theta)$, where $f(v_1)$, $f(v_2)$ and $f(\theta)$ are the PDF of v_1,v_2 and θ respectively.

Assume that the nodes' speed v is uniformly distributed over $[a,b]$, and the phase angle of the relative velocity, θ, is uniformly distributed over $[0,2\pi]$, then we can have $f(v)=1/(b-a)$, $f(\theta)=1/2\pi$. Consequently, $\bar{\tau}$ is given by:

$$\bar{\tau} = r\int_0^{2\pi}\int_a^b\int_a^b \frac{1}{\sqrt{v_1^2+v_2^2-2v_1v_2\cos\theta}} * \frac{1}{b-a} * \frac{1}{b-a} * \frac{1}{2\pi} dv_1 dv_2 d\theta$$

$$= \frac{r}{2\pi(b-a)^2} \int_0^{2\pi}\int_a^b\int_a^b \frac{1}{\sqrt{v_1^2+v_2^2-2v_1v_2\cos\theta}} dv_1 dv_2 d\theta$$

4 Simulation and Analysis

In order to illustrate the effectiveness of the link availability $LA(t)$ developed in this paper, the comparisons among the results calculated by $LA(t)$, Min_LT_p (see formula (9), which is developed in [10]), En_LT_p(see formula (10), which is proposed in [11]) and the practical results obtained by simulation experiment are given in this part. Formula (9) and (10) are the simplified forms of the original formulas with the corrections being set to 0 because they are strongly dependent on the practical simulation results.

$$Min\_LTp = \frac{1-e^{-2\lambda T_p}}{2\lambda T_p} + \frac{\lambda T_p e^{-2\lambda T_p}}{2} \qquad (9)$$

$$En\_LT_p = L(T_p) \approx e^{-\lambda T_p e^{-\lambda \tau}} \qquad (10)$$

The methods denoted by (9) and (10) are limited for the link availability prediction at time T_p (LET), so the comparisons will be taken among Min_LT_p, En_LT_p and $LA(T_p)$.

A discrete-events simulation is developed which generates two mobile nodes with the same random ad hoc mobility parameters according to the random walk-based mobility model. Given a set of parameters such as transmission radius r, average length of epoch, speed and direction as well as the first link epoch, T, let two nodes which located within the range of each other initially move according to the same mobility parameters. Then the statistical real duration that the link consisting of these two nodes keeps active can be obtained. We can calculate the practical link availability L_{sim} with these data. We calculate L_{sim} in different scenario, such as different radiuses, speeds and epochs.

Fig. 2. gives the simulation results L_{sim}, the analytical results of $LA(T_p)$ and other methods with the speed obeys the uniform distribution over $[a,b]$. In Fig.2, each plot represents a special scenarios and demonstrates the relation between *link availability* and *epoch* (measured by second), where r is the transmission radius of mobile nodes, v is the average speed and T_p is the link expiration time of the link related to these two nodes.

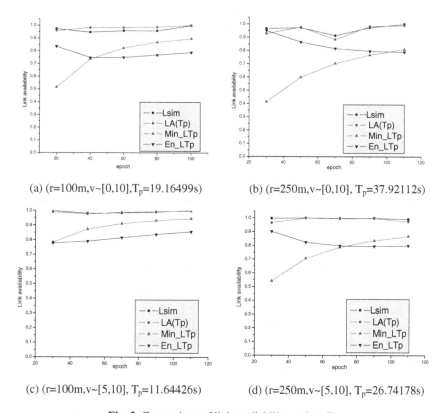

(a) (r=100m,v~[0,10],T_p=19.16499s)   (b) (r=250m,v~[0,10], T_p=37.92112s)

(c) (r=100m,v~[5,10], T_p=11.64426s)   (d) (r=250m,v~[5,10], T_p=26.74178s)

Fig. 2. Comparison of link availability at time T_p

It can be seen clearly that the result of $LA(T_p)$ is much more approximate to the simulation result L_{sim} as compared to that of Min_LT_p and En_LT_p. The gap between the result of En_LT_p and L_{sim} becomes larger and larger as the epoch increases, however, the results of $LA(T_p)$ and L_{sim} can nearly fit together well all through. Though in some certain scenarios (such as Fig.2(b)), the result of En_LT_p is slightly superior to that of $LA(T_p)$ while epoch is small, the difference between the result of $LA(T_p)$ and L_{sim} is insignificant. It also can be observed that in different scenarios with different parameters the results of $LA(T_p)$ outperforms those of Min_LT_p proposed in [10], especially in the case of small epoch (means that the node change their movement status frequently), which infers that $LA(T_p)$ is much more practical in the environment of high mobility and more flexible to adapt to the quick change of the network topology. Moreover, the trend of change between Min_LT_p and L_{sim} *along* with the increase of epoch is thoroughly inconsistent, which is very harmful for selecting stable path. It's very difficult to distinguish that whether a link is stable or not if the changing trend of link availability can not reflect that of the simulation results.

In order to validate the performance of $LA(t)$ at any time t, we give the comparisons between the results calculated by $LA(t)$ and the simulation result L_{sim} on the case of *epoch=30s* (see Fig.3.) In Fig.3, each plot represents a special scenario and demonstrates the relation between *link availability* and the predicting time t (measured by second).

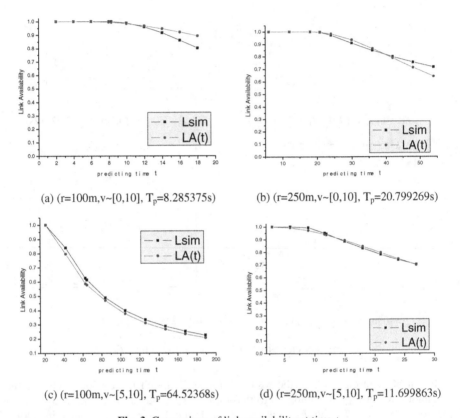

(a) (r=100m,v~[0,10], T_p=8.285375s)   (b) (r=250m,v~[0,10], T_p=20.799269s)

(c) (r=100m,v~[5,10], T_p=64.52368s)   (d) (r=250m,v~[5,10], T_p=11.699863s)

Fig. 3. Comparison of link availability at time t

It can be seen from Fig.3 that the result of *LA(t)* approximates to the simulation result L_{sim} extremely, which implies that the prediction accuracy of *LA(t)* is comparatively high. As t increases, the link availability drops off under the condition with the same value of transmission radius, speed, length of epoch and T_p.

5 Conclusion

Based on the random walk-based mobility model, we propose a new link availability method for predicting nodes' movement status at any time in mobile ad hoc networks in this paper. Compared to other methods, it is independent of practical results which given by simulation experiment and it can provide the prediction of movement status not only at time *LET* but also at any time t. The simulation results demonstrate that it is superior to other link availability prediction methods under different simulation conditions.

Link availability can be expanded into path availability which can be regarded as groundwork for further analysis of ad hoc network performance, as well as a guide to ad hoc network routing protocol design, e.g. in the optimization of route selection and route discovery algorithms. The routing protocol based on the metric of path availability is adaptable to the high dynamic topology and holds better scalability and expansibility. It can predict the breakage of links and select a stable path, thus reducing the overhead and delay which is caused by rerouting operations, improving the throughput and meeting such QoS requirement. As a result, using link availability can improve the whole performance of mobile ad hoc networks.

In future work, we plan to extend our algorithm in multiple directions. Firstly, we intend to study other mobility models. Secondly, applying our link availability to the analysis of path availability in ad-hoc networks is another future research topic. For mobile ad-hoc networks, constraints such as battery power and available bandwidth are not addressed in this paper. We will extend our current research to further analyze these issues.

Acknowledgments. This work is supported by the National Natural Science Foundation of China Grant No.90304010 and Program for New Century Excellent Talents in University (NCET-05-0683).

References

1. Jianxin Wang, Shuguang Deng, Songqiao Chen and Jianer Chen: A route recovery method based on anycast policy in mobile ad hoc networks. Journal on Communications, 24(10) (2003) 172–176
2. Jianxin Wang, Shuguang Deng, Songqiao Chen and Jianer Chen: A new routing based on mobility prediction in mobile ad hoc wireless networks, High Technology Letters (2002) 10–15
3. R.Ramanathan and J.Redi: A brief overview of Ad hoc networks challenges and directions. IEEE communications Magazine, 50th anniversary commemorative issue, May 2002.
4. Abolhasan M, Wysocki T, Dutkiewicz E, Abolhasan M: A review of routing protocols for mobile ad hoc networks. Ad Hoc Networks 2 (2004) 1–22
5. William Su, SungJu Lee, Mario Gerla: Mobility prediction and routing in Ad hoc wireless networks International Journal of Network Management 3 (2001) 3–30
6. Shah H, Nahrstedt K: Predictive location–based QoS routing in mobile A d hoc Networks. In Proceedings of IEEE International Conference on Communications (ICC 2002), New York, (2002)
7. A.B. McDonald and Taieb Znati.: Link availability models for mobile ad hoc networks, Technical Report TR99-07, University of Pittsburgh, Department of Computer Science (1999)
8. Bruce McDonald A, Taieb Znati: A path availability model for wireless Ad hoc networks In Proceedings of IEEE Wireless Communications and Networking Conference 1999 (WCNC'99) , New Orleans, LA (1999) 21–24
9. A.B.McDonald and T.Znati: A mobility based framework for adaptive clustering in wireless ad-hoc networks. IEEE Journal on Selected Areas in Communications, Special Issue on Ad-Hoc Networks, in press (1999)

10. Shengming Jiang, He D J, Rao J Q.: A prediction-based link availability estimation for mobile Ad hoc networks, IEEE INFOCOM (2001)
11. Shengming Jiang: An enhanced prediction-based link availability estimation for MANETs, IEEE Transactions on Communications, vol. 52, no. 2 (2004) 183–186
12. Yu D., Li H.: Path Availability in ad hoc networks, 10th ICT'2003, Tahiti, French Polynesia (2003)
13. W. Su and M. Gerla: IPv6 Flow Handoff in Ad–Hoc Wireless Networks Using Mobility Prediction, Proceedings of IEEE GLOBECOM' 99, Rio de Janeiro, Brazil (1999) 271–275
14. W.Su, S.–J,Lee and Gerla: Mobility Prediction in Wireless networks, Proc.of IEEE MILCOM2000 (2000)
15. T. Camp, J. Boleng, and V. Davies: A survey of mobility models for ad hoc network, research In Wireless Communications and Mobile Computing (WCMC): Special issue on Mobile Ad Hoc Networking: Research, Trends and Applications (2002)

Reputation Based Localized Access Control for Mobile Ad-Hoc Networks

Sangheethaa Sukumaran[1] and Elijah Blessing[2]

[1] Coimbatore Institute of Technology, Coimbatore- 641014, Tamilnadu, India
sangheethaa@gmail.com
[2] Karunya Institute of Technology and Sciences, Coimbatore-641114, Tamilnadu, India
elijahblessing@yahoo.com

Abstract. The absence of a router or a base station urges the need for better access control mechanism in the ad-hoc networks. This paper presents a localized approach for access control which implements ticket certification services through reputation mechanism. Reputation refers to, the opinion of one node as seen by other nodes. Ticket certification ensures that, only well behaving nodes (which forward and route packets of other nodes) can get the tickets and only they can access the network. The tickets are obtained from any node in the locality with high reputation instead of getting it from a centralized authority or from some k neighbors in the neighborhood. This paper gives the analysis and simulation results of the localized approach through reputation mechanism and proves that the localized access control with reputation is better than the localized approach without reputation.

1 Introduction

With recent performance advancements in computer and wireless communications technologies, advanced mobile wireless computing is expected to see increasingly widespread use and application. A mobile ad-hoc network (MANET) is a self-configuring network of mobile routers (and associated hosts) connected by wireless links, the union of which forms an arbitrary topology. This ad hoc networking technology provides "anytime" and "anywhere" services to the users, in a potentially large infrastructure less wireless network, based on the collaboration among individual network nodes. The routers are free to move randomly and organize themselves arbitrarily; thus, the network's wireless topology may change rapidly and unpredictably. A MANET is an autonomous system of mobile nodes. The system may operate in isolation, or may have gateways to and interface with a fixed network.

This paper deals with access control in ad-hoc networks. Access control is a mechanism of deciding which node can access the network i.e. who can use the network layer services like packet forwarding and routing. The ad-hoc network does not have any fixed base station as in cellular networks, or routers as in wired networks to perform this access control mechanism. All the nodes in an ad-hoc network acts as a router and cooperate among themselves for proper functioning of the network. It is

T. Kunz and S.S. Ravi (Eds.): ADHOC-NOW 2006, LNCS 4104, pp. 197–210, 2006.
© Springer-Verlag Berlin Heidelberg 2006

assumed that all the nodes that participate in the network will do forwarding and routing in favor of other nodes. But this assumption does not work in all cases. Sometimes the nodes agree to forward, but fail to do because they want to save their battery power and CPU cycles. They just keep receiving the data destined to them, and drop the data of other nodes without forwarding or routing them, which reduces the throughput of the network. These nodes are called as misbehaving nodes.

The main aim of access control is to allow only the well-behaving nodes to access the network and to isolate misbehaving nodes from the network. The misbehaving nodes are denied to use network services like packet forwarding and routing. This paper explains a localized approach for access control which focuses on packet forwarding and routing misbehaviors, and the use of reputation values in access control. Reputation refers to, the opinion of one node as seen by other nodes. It is a proper means of discovering misbehaving nodes in the network. This paper is organized into following sections. Section 2 reviews some of the existing approaches given in the literature and section 3 briefs the localized approach for access control. Section 4 gives details of using the reputation values in the localized approach for access control. Section 5 gives the simulation results for localized approach with reputation mechanism and for localized approach without reputation mechanism. Finally section 6 gives the conclusion.

2 Related Work

There are many approaches in the literature, which deals with access control in ad-hoc networks. But only few papers [1], [2], [3] deal with packet forwarding and routing misbehaviors.

L.Buttayan et al. [1] focuses on packet forwarding and they address the problem of stimulating co-operation in self-organizing Mobile Ad-hoc Networks for civilian applications. This approach uses a tamper resistant hardware module called "security module". This security module maintains a nuglet counter. The security module enforces two rules. Firstly, when the node wants to send one of its own packets, the number n of intermediate nodes that are needed to reach the destination is estimated. If the nuglet counter of the node is greater than or equal to n, then the node can send its packet, and the nuglet counter is decreased by n. Otherwise, the node cannot send its packet, and the nuglet counter is not modified. Secondly, when the node forwards a packet for the benefit of other nodes, the nuglet counter is increased by one. Every node has to maintain a +ve counter value in order to send its own data. The nuglet counter is protected from illegitimate manipulations by the tamper resistance of the security module. This approach ensures that the misbehavior is not beneficial and hence it should occur rarely only. But the availability of hardware module is not guaranteed in general.

S.Marti et al. [2] addresses the problem of nodes agreeing to forward packets of other nodes but fail to forward. This describes two mechanisms to improve the throughput of the network. One mechanism is the watchdog, which identifies the misbehaving node by monitoring the nearby nodes whether they forward the packets of other nodes in the network. The other mechanism is the path rater that defines the best route by avoiding those misbehaving nodes. Since this approach tries to avoid the misbehaving nodes for routing, there's less chance of dropping packets, thus providing a better throughput even in the presence of high number of misbehaving nodes. But this approach does not isolate the misbehaving nodes; they still utilize the network services, i.e. the nodes are not punished for misbehaving.

L.Zhou *et al.* [3] and G.Appenzeller *et al.* [4] proposed ticket based approaches. Tickets are provided for the nodes, which are well behaving, and network access is provided only to the nodes with a valid ticket. The ticket is obtained from a centralized authority [3] or from distributed servers [4]. The central server approach has several advantages and disadvantages. The central server approach can work well for a simple, less dynamic network. But for a dynamic network the delay will be more. The distributed approach has no much difference with central authority system except that here there are three or more central servers in the network. In both the approaches when the central server fails, the network functioning becomes vulnerable to attacks.

T.Michiardi and Molva [9] proposed a COllaborative Reputation (CORE) mechanism that also has a watchdog component for monitoring. Here the reputation value is used to make decisions about cooperation or gradual isolation of a node. Reputation values are obtained by regarding nodes as requesters and providers, and comparing the expected result to the actually obtained result of a request. In CORE the reputation value ranges from positive (+) through null (0) to negative (-). The advantage of this method is that having a positive to negative range allows good behavior to be rewarded and bad behavior to be punished. This method gives more importance to the past behavior and hence tolerable to sporadically bad behavior, e.g. battery failure. But the assumption that past behavior to be indicative of the future behavior may make the nodes to build up credit and then start behaving selfishly.

Sonja Buchegger *et al.* [8] proposed the reputation-based approach, CONFIDANT which consists of the following components: The Monitor, the Reputation System, the Path Manager, and the Trust Manager. These components are present in every node. CONFIDANT extends reactive routing protocols [10] with a reputation-based system in order to isolate misbehaving nodes. As a component within each node, the monitor registers these deviations from normal behavior. As soon as a given bad behavior occurs, the reputation system is called. ALARM messages are sent by the trust manager of a node to warn others about the malicious nodes. The reputation system in this protocol manages a table consisting of entries for nodes and their rating. The rating is changed only when there is sufficient evidence of malicious behavior that is significant for a node and that has occurred a number of times exceeding a threshold to rule out coincidences. If the rating of a node in the table has deteriorated so much as to fall out of a tolerable range, the path manager is called for action. The path manager performs the following functions: a) Path re-ranking according to security metric, e.g. reputation of the nodes in the path. b) Deletion of paths containing malicious nodes. c) Action on receiving a request for a route from malicious node (e.g. ignore, do not send any reply). The CONFIDENT method uses only negative reputation values. In CONFIDANT, the attacks involving building up of credits, before behaving selfishly, have less effect. But it is less tolerant to the failed nodes, which may exhibit failed behavior due to loss of power.

The localized approach for access control is proposed by Haiyoun Luo *et al.* [5]. Since this approach forms the basis of the proposed scheme it is explained in detail in the following section.

3 The Localized Approach

The localized approach [5] proposes a fully localized design paradigm to provide ubiquitous and robust access control for mobile ad hoc networks. This solution takes a ticket-based approach. Each well behaving node uses a certified ticket to participate in routing and packet forwarding. Nodes without valid tickets are classified as misbehaving. They will be denied from any network access, even though they move to other locations. Thus, misbehaving nodes are "isolated" and their damage to the mobile ad hoc network is confined to their locality. The access control operation emphasizes multiple node consensus and fully localized instantiation. Since any individual node is subject to misbehaviors, this approach does not rely on any single node. Instead, the nature of cooperative computing in an ad hoc network is leveraged and the approach depends on the collective behaviors of multiple local nodes. Here multiple nodes in a local network neighborhood, typically one or two hops away, collaborate to monitor a node's behavior and determine whether it is well-behaving or misbehaving using certain detection mechanism of their choice. These local monitoring neighbors will renew the expiring ticket of a well-behaving node collectively, while a misbehaving node will be denied from ticket renewal or be revoked of its ticket. In this way, the functionality of a conventional access control authority, which is typically centralized, is fully distributed into each node's locality. Every node contributes to the access control system through its local efforts and all nodes collectively secure the network.

The localized approach does not need any hardware module for security. It does not assume anything about the packet size or type of traffic or the type of data. It not only detects the misbehaving nodes but also isolates them from the network. Average delay for ticket renewal is tolerable, because the node gets its ticket from its locality rather than going to a central server. There's no necessity for the node to rely upon a single node for getting a ticket or for renewal. So this approach is highly robust and scalable.

The localized approach requires that each node should get k tickets from its local neighborhood. It is possible to get k number of tickets in a highly populated network. But it is not possible when the number of nodes in a network is less. Thus the localized approach cannot be used in a sparse network. Moreover the protocol used in localized approach broadcasts the ticket request to all its neighbors, which increases the communication overhead.

The efficiency of the localized approach depends upon the coalition size k. i.e. the number of partial tickets that the node should get to access the network. The parameters viz. average delay, overhead and success ratio, which are used for simulation in [5], vary depending upon the k value. The k value is fixed as 5 in [5] based on the network size. This value does not change when number of nodes in the network increases or decreases. But this value will not work for all the networks. It is applicable only to a large network. For a sparse network, collecting 5 tickets from the neighborhood will cause more delay, because the nodes may not have sufficient number of neighbors in their locality. So in order to reduce the number of tickets a node should receive before successful access of the network, reputation mechanism can be used.

4 Reputation Based Localized Access Control

The opinion of one node about another node is called as reputation [7]. For Mobile ad-hoc networks, reputation means participation of a node in routing and forwarding as seen by others. This reputation system can be used to make decisions about which nodes to cooperate and which nodes to exclude from the network. This system can be used with any misbehavior identifying schemes.

The goals of the reputation systems are 1) to provide information to distinguish between a trustworthy user in the network and an untrustworthy user and 2) to encourage users to act in a trustworthy manner and 3) to discourage untrustworthy user from participating in the network access.

Using the reputation mechanism in the localized approach for access control helps to reduce the number of tickets a node has to get to access the network. It is enough to get one ticket from a highly reputed node instead of waiting to get some k tickets from the neighbors.

This paper proposes a ticket-based approach, which uses reputation mechanism for evaluating the tickets. The nodes can access the network if they have a valid ticket. The tickets are obtained from the neighboring nodes, which have high reputation value. Initially the tickets are issued by a dealer. When the network is formed each node is assigned with a valid ticket but with a less expiration time. Once the expiration time reaches, the nodes have to renew their tickets. For renewing, the nodes will send the broadcast request to all its one-hop neighbors. On receiving the ticket renewal request, the neighbors have to d ecide whether to send a ticket or not by checking the reputation value of that node. Each node maintains the reputation value, by monitoring the behavior of the neighboring node using any monitoring mechanism. For simulations, watchdog [2] mechanism is used. When the requesting node receives a reply ticket, it checks the reputation value of the node, which has sent the reply. If the reputation value of the node is greater than a threshold value (this value is chosen based on the network behavior) then the requesting node accepts the ticket, otherwise it rejects the ticket from that node and looks for other replies. Once it receives a ticket from higher reputation node, the node uses that ticket to prove its behavior and access the network. This makes the ticket obtaining process simpler.

Whenever a node issues a network access request, its ticket and the reputation value of the node, which gave the ticket, is verified. This ensures that two nodes cannot collaborate with each other and generate false tickets. Moreover other nodes will also monitor the behavior of these nodes. Nodes may try to generate their own tickets for communication. But this will be identified because the tickets are signed and verified using RSA [11] algorithm. So this method is false proof and secure.

5 Simulation Results

Ns2 [6] is used for simulation. Ns2 is a discrete event simulator, which is widely used for simulation of both wired and wireless networks. An agent similar to UDP (User Datagram Protocol) is used for simulation. The average mobility of the nodes is set as 1-15m/s for the scenarios and for creating the scenario random waypoint model is used. The routing protocol is DSR (Dynamic Source Routing). The number of nodes is varied from 30 –100. The performance of *ticket renewal* service is measured using the parameters, success ratio, overhead, average number of retries and average delay. Success ratio is defined as the ratio of number of successful renewals to the maximum number of renewal requests sent by all the nodes within the simulation time. Overhead is the total number of bytes sent by all the nodes in the scenario. Average number of retries is the number of retries made by the nodes and average delay is the delay incurred for successful renewal of the tickets. In this paper each node calculates the reputation value of its neighbors using the formula,

$$R_{direct} = \frac{\sum\limits_{Pkts=0}^{\infty} F_{pkts}}{\sum\limits_{Pkts=0}^{\infty} S_{pkts}} \tag{1}$$

Where R_{direct} is the reputation value calculated by monitoring the neighbors directly and F_{pkts} is the number of packets forwarded by this nodes and S_{pkts} is the number of packets sent by this node. This formula is used to calculate the reputation value of a node by directly monitoring the neighboring node's behavior using a monitoring mechanism such as watchdog. Each node runs the monitoring mechanism and counts the number of packets forwarded by the neighboring nodes and number of packets originated from those nodes. It is also possible to pass this reputation value that is calculated directly by monitoring the neighbors, to 1 or 2 hop neighbors. But this exchange of reputation information will increase the communication overhead.

For simulations the threshold value for reputation is set as 0.8. This value is chosen based on the assumption that a value below this may lead a node to misbehave at any time, and it cannot be a node which forwards more packets of other nodes than sending its own packets. For example, a node with 0.5 reputation value means that it is forwarding as many packets as it is sending,(it is 50% well behaving node) and at any time the number of packets it is sending can exceed the number of packets it is forwarding. So it cannot be a *highly* reputed node.

The localized approach without reputation mechanism uses a k value of 5.i.e the nodes have to get 5 tickets to become a valid node.

Figs. 1-4 show the average delay, success ratio, number of retries and overhead respectively for the localized approach (LA) with reputation mechanism and for the localized approach without reputation mechanism. For these scenarios the average node speed is set as 3m/s and the channel error rate is 1%. The number of nodes is varied from 50 to 100. The localized approach with reputation mechanism exhibits both higher success ratio (Fig. 2) and lower delay (Fig. 1) than the LA without reputation system. This is due to the fact that in the LA without reputation system, the

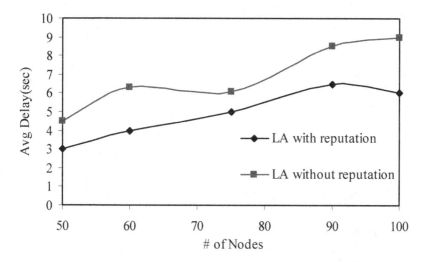

Fig. 1. Average Delay VS Number of nodes, node speed 3m/s

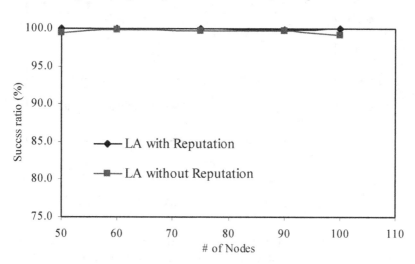

Fig. 2. Success ratio VS Number of Nodes, node speed 3m/s

nodes have to get k tickets from their neighbors for accessing the network, whereas in the reputation based localized control, the nodes need to get only one ticket from a node with high reputation in its locality.

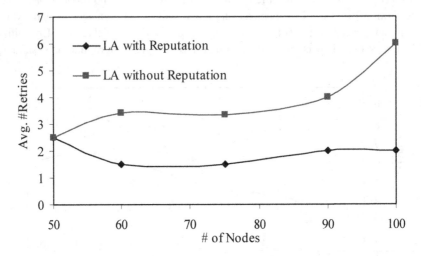

Fig. 3. Average retries VS Number of nodes, node speed 3m/s

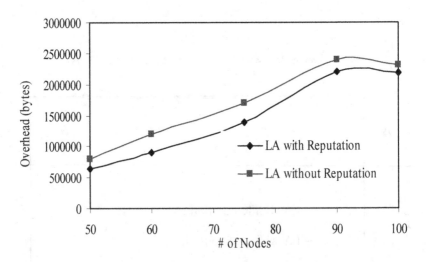

Fig. 4. Overhead VS Number of Nodes, node speed 3m/s

The average number of retries (Fig. 3) for the reputation-based system is also maintained at a minimum value compared to the localized approach. When the number of nodes increases, the communication overhead in the network (Fig. 4) increases more for the localized approach, compared to the reputation-based localized approach.

Figs. 5-9 compare the LA with reputation and the LA without reputation for the low channel error rate (1%, Figs. 5-7) and high channel error rate (10%, Figs. 8-9). For these scenarios the number of nodes is set as 100. The mobility of the nodes is varied from 1-15m/s. From the graphs it is found that the performance of LA without reputation mechanism degrades as the channel error rate increases. But in the LA with reputation mechanism, since the nodes need to get only one ticket, the performance is same even when the channel error rate increases from 1% to 10%.

Fig. 5 shows the average delay for various node speeds. For the reputation-based system the average delay is below 8sec even when the mobility increases.

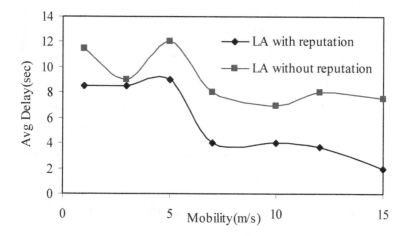

Fig. 5. Average Delay VS Mobility, 100 nodes, Channel Error Rate 1%

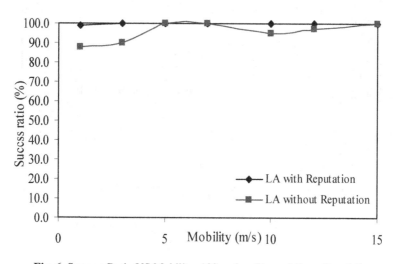

Fig. 6. Success Ratio VS Mobility, 100 nodes, Channel Error Rate 1%

But the delay for the localized approach without using the reputation mechanism is higher. The overhead (Fig. 7) in the reputation based localized approach is less compared to the localized approach without reputation. It is seen that the success ratio (Fig. 6) is 100 % in the reputation based localized approach even when the mobility increases.

The LA without reputation degrades in performance when the channel error rate increases to 10% (Figs. 8-9), but the reputation based LA is not much affected for variations in the channel error rate.

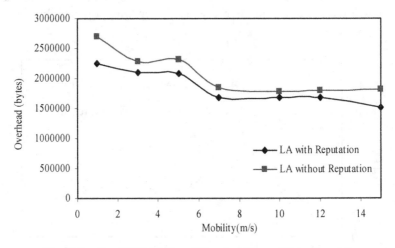

Fig. 7. Overhead VS Mobility, 100 nodes, Channel Error Rate 1%

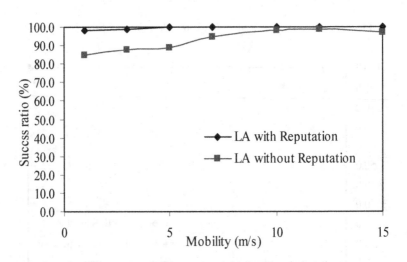

Fig. 8. Success Ratio VS Mobility, 100 nodes, Channel Error Rate 10%

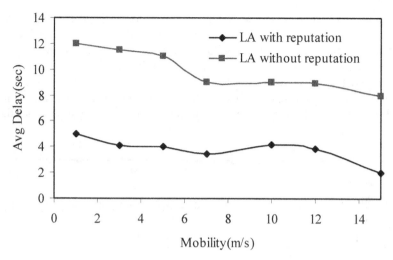

Fig. 9. Average Delay VS Mobility, 100 nodes, Channel Error Rate 10%

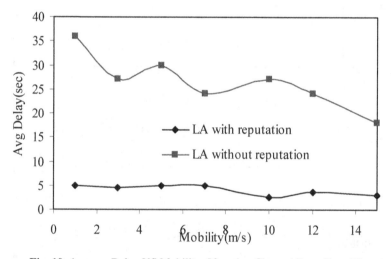

Fig. 10. Average Delay VS Mobility, 30 nodes, Channel Error Rate 1%

Figs. 10 -14 show the effect of sparse neighborhood on LA with reputation and on LA without reputation. For these scenarios the number of nodes is set as 30 and the channel error rate is set as 1% (low) for Figs. 10 -12 and 10% (high) for Fig. 13 and Fig. 14. The mobility of the nodes is varied from 1-15m/s. From the Figs 10-14 it is obvious that the performance of the LA without reputation degrades, as the nodes will not be able to get k tickets from the sparse network. Thus the LA without reputation system is not applicable for sparse neighborhood networks. But the reputation based LA system provides good performance by 100% availability, low delay, less number of retries and minimum overhead when compared to the localized approach. The parameters are not much affected even when the channel error rate increase to 10% in the LA with reputation system. Where as for the LA without reputation system, the increase in channel error rate has a bad impact on the parameters like success ratio and average delay (Figs. 13-14)

Thus the simulation results proves that reputation based localized approach outperforms the localized approach without reputation and also shows that the LA

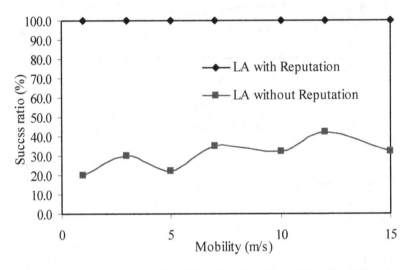

Fig. 11. Success Ratio VS Mobility, 30 nodes, Channel Error Rate 1%

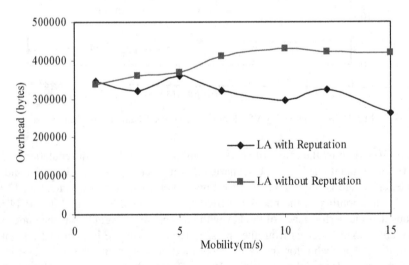

Fig. 12. Overhead Ratio VS Mobility, 30 nodes, Channel Error Rate 1%

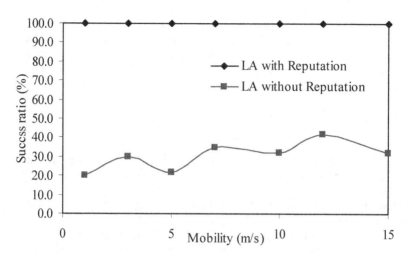

Fig. 13. Success Ratio VS Mobility, 30 nodes, Channel Error Rate 10%

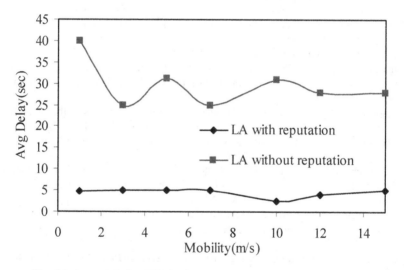

Fig. 14. Average Delay VS Mobility, 30 nodes, Channel Error Rate 10%

with reputation can used even in the sparse neighborhood networks. This approach is robust, since any node that needs a ticket can get from its own neighborhood, and it is also scalable.

6 Conclusion

The localized access control mechanism works well for a dynamic environment with high mobility. But it does not work for a sparse network and for networks with low mobility. The graphs from the simulation shows that, for localized approach with reputation mechanism the overhead is minimum, the delay is less, the average number of retries is minimum and also the success ratio is more. Moreover the probability of getting a ticket from a single, highly reputed neighboring node is more than getting tickets from 5 well behaving neighborhood nodes. So it is recommended to use the localized approach along with the reputation mechanism.

References

1. L.Buttayan and J.P.Hubaux, "Stimulating Cooperation in Self-organizing Mobile Ad-Hoc Networks," in *ACM/Kulwer Mobile Networks and Applications*, vol. 8, no.5, pp. 579-592, Oct 2003.
2. S.Marti, T.Giuli, K.Lai and M.Baker, "Mitigating Routing Misbehavior in Mobile Ad-Hoc Networks," in *Proc. ACM MOBICOM*, pp. 255-265, 2000.
3. L.Zhou, E.B.Schnder and R.Can Renese "COCA: a Secure Distributed on Line Certificate Authority," *ACM Tran, Computer Sys*, vol. 2, no. 4, pp. 329-368, Nov. 2002.
4. G.Appenzeller, M.Roussopoulous, and M.Baker, "User-friendly access control for public network ports," in *Proc. IEEE INFOCOM*, pp. 699-707, 1999.
5. Haiyoun Luo, P.Zerfos, Songwu Lu, L.Zhang "URSA-Ubiquitous and Robust Access Control for Mobile Ad hoc Networks," IEEE/ACM Transactions on Networking, vol. 12, no. 6, pg. 1049-1063, Dec. 2004.
6. www.isi.edu/nsnam/ns/ns-tutorial/tutorial-02.
7. Sonja Bucheggery, Jean-Yves Le Boudec, "Self Policing Mobile Ad-hoc Networks by reputation Systems," available online from http://lcawww.epfl.ch/Publications/LeBoudec/BucheggerL05.pdf
8. Sonja Buchegger and Jean-Yves Le Boudec, "Performance Analysis of the CONFIDANT Protocol: Cooperation Of Nodes — Fairness In Dynamic Ad-hoc NeTworks," In *Proceedings of IEEE/ACM Symposium on Mobile Ad-Hoc Networking and Computing (MobiHOC)*, Lausanne, CH, June 2002.
9. Pietro Michiardi and Refik Molva, "CORE: A collaborative reputation mechanism to enforce node cooperation in mobile ad hoc networks," Sixth IFIP conference on security communications, and multimedia (CMS 2002), Portoroz, Slovenia, 2002.
10. D. B. Johnson and D. A. Maltz, "Dynamic source routing in ad hoc wireless networks," in *Mobile Computing*, T. Imielinski and H. Korth, Eds. Norwell, MA: Kluwer, vol. 353, pp. 153–181, 1996.
11. Frankel, Gemmall, MacKenzie, Yung, "Proactive RSA," in *Proc. CRYPTO*, pp. 440–454, 1997.

Distributively Increasing the Percentage of Similarities Between Strings with Applications to Key Agreement*

Effie Makri[1] and Yannis C. Stamatiou[2]

[1] University of the Aegean, Department of Mathematics,
Karlovasi, 83200, Samos, Greece
effiem@aegean.gr

[2] University of Ioannina, Department of Mathematics, 451 10, Ioannina,
Greece and Research and Academic Computer Technology Institute,
N. Kazantzaki, University of Patras, 26500, Rio, Patras, Greece
istamat@cc.uoi.gr

Abstract. A central problem in distributed computing and telecommunications is the establishment of common knowledge between two computing entities. An immediate use of such common knowledge is in the initiation of a secure communication session between two entities since the two entities may use this common knowledge in order to produce a secret key for use with some symmetric cipher. The dynamic establishment of shared information (e.g. secret key) between two entities is particularly important in networks with no predetermined structure such as wireless mobile ad-hoc networks. In such networks, nodes establish and terminate communication sessions dynamically with other nodes which may have never been encountered before in order to somehow exchange information which will enable them to subsequently communicate in a secure manner. In this paper we give and theoretically analyze a protocol that enables two entities initially possessing a string each to securely eliminate inconsistent bit positions, obtaining strings with a larger percentage of similarities. This can help the nodes establish a shared set of bits and use it as a key with some shared key encryption scheme.

1 Introduction

A mobile ad-hoc network is a self-organizing, autonomous system of nodes with wireless communication capabilities which move around freely giving an unpredictable and changeable nature to the network topology. Security is a key concern in such networks due to the impromptu membership of nodes, which raises trust issues amongst them. The use of wireless links also poses a major security issue in that there is less effort involved in compromising a communication session than that involved in compromising wired links. This, coupled with computational and energy constraints of the devices, render the issue of security a most

---

* Partially supported by the IST Programme of the European Union under contact number IST-2005-15964 (AEOLUS) and the INTAS Programme under contract with Ref. No 04-77-7173 (Data Flow Systems: Algorithms and Complexity (DFS-AC)).

T. Kunz and S.S. Ravi (Eds.): ADHOC-NOW 2006, LNCS 4104, pp. 211–223, 2006.

taxing problem. In this paper we focus specifically on the use of symmetric key cryptosystems for mobile ad-hoc networks, since the computational and the communication overhead incurred with public key schemes renders them most inefficient for these types of networks, due to the resource constraints of mobile devices as well as bandwidth constraints. Furthermore, the use of symmetric key cryptosystems requires that all nodes in the network share a common key for encryption and decryption. Vital issues of security, such as authentication, integrity and confidentiality are dealt with using a symmetric key scheme; however the lack of a superior key management framework renders the whole security of the system ineffective since attacks are mainly made on the key management infrastructure. According to [2], there are three classes of key management: (i) key distribution, (ii) key agreement, and (iii) key pre-distribution. The authors of [2] propose a fourth class of key management: (iv) *distributed, key pre-distribution schemes*. Below, we briefly review these schemes.

A simple approach for the key distribution problem is to have a single (and trusted) member of the network - the leader - responsible for the generation of the group key and its distribution to all the members in the network via some secure channel [9]. This approach, however, essentially relies on a Trusted Third Party (TTP) being available for the network. If this TTP is compromised then network security is breached.

The *Group Key Management Protocol (GKMP)* described in [7] provides a simple key management scheme where the leader creates the group key and distributes it to the group members. A centralized key distribution service, however, creates a single point of failure in the whole security mechanism. Moreover, due to the mobility of the nodes of ad-hoc networks, the leader may be unavailable to some of the more distant nodes.

The authors of [4] use the concept of *Domain Key Distributor (DKD)* for the role of the leader and propose a system whereby the domain is divided into smaller areas each one containing an Area Key Distributor. The introduction of these "sub-leaders" results in minimizing the number of messages needed to be sent by the one leader to all the nodes in the network, and increases the availability of the key distribution service. An alternative scheme is proposed by [13], where the leader maintains a hierarchy of keys and aims mainly at improving scalability.

A possible fundamental weakness of the key distribution schemes discussed above is that they are costly in overhead (setting up and continuously operating the key distribution mechanism). This may pose a performance problem when these schemes are used in mobile ad-hoc networks where there are certain computing power and bandwidth allocation constraints. In addition, since the schemes rely on some centralized key management policy, there is also the issue of the existence of a single point of failure. In order to overcome the centralized nature of these key distribution schemes, the *key agreement* or *contributory* key management schemes (see [12]) have been proposed. The main feature of these schemes is that the key generation protocol dictates that the members of the network themselves contribute independently to the creation of the group key.

Distinction is made between a *Partially Contributory* protocol, where only a subset of the nodes contribute to the creation of the key and a *Fully Contributory* protocol where all the network nodes make contributions.

The majority of key agreement protocols, two of which are proposed by [8] and [10], extend the basic 2-party Diffie-Hellman key exchange protocol described in [5] to multiple parties. However, such schemes do not seem to be particularly suited for use in mobile ad-hoc networks either, due to the fact that the schemes are not robust against the changing nature of these networks (i.e. dynamic establishment and termination of connection links and irregular, dynamic communication topology).

In addition to the aforementioned schemes are the key pre-distribution schemes according to which the group key is distributed to the members of the network before it is actually formed by them. The authors of [6] propose a scheme for distributed sensor networks where sensors are pre-loaded with a random set of keys before being deployed. This ensures that only a small number of keys need to be saved on each sensor node to guarantee that any two nodes share a key with some chosen probability. Nodes can communicate with other nodes with which they do not share keys by establishing path-keys between them. This requires that two neighbouring sensor nodes require a single common key. A modification of this scheme is proposed by [5], where q common keys are needed instead of just the one single common key, which results in increased resilience of the network against node capture. The authors also propose the establishment of the path-key along multiple paths.

The above schemes assume that sensor nodes are deployed by a single party and therefore trusted. Also, keys need to be pre-loaded onto each node, thus not completely eliminating the need for trusted third parties, albeit off-line ones. In a mobile ad-hoc environment it is generally not the case that nodes have any prior contact with other nodes, thereby placing trust in a third party is not an advisable option. The authors of [2] suggest distributing the role of the TTP amongst all the nodes in the network, whereby each one contributes equally to the key pre-initialization process. This paradigm is termed the *Distributed Key Pre-distribution Scheme (DKPS)* by the authors. With this scheme, the nodes in the network jointly realize the key pre-distribution process with results identical to a key pre-distribution process performed by the TTP.

In this paper we present a simple, randomized key generation policy in the *key agreement* category, whereby two nodes wishing to initiate a communication session initially start with a sufficiently long string each and, through a number of bit-position elimination steps carefully chosen so as to favour the deletion of differing positions, compute strings in which the percentage of differences is much smaller than the percentage of differences in the initial strings. We, also, theoretically analyze the time evolution of the protocol using the differential equations method described by Wormald in [14] and show that it increases the initial percentage of similarities with the increase depending on the initial percentage. This increase can help locate a common set of bits that can serve as a secret key in the establishment of a secure communication session.

2 Statement of the Problem

Two entities wish to establish a shared key. They initially possess the strings A and B respectively, of size N, which agree on $X(A, B)$ positions. The entities' goal is to cooperatively locate a sufficiently long subset of bits common to both strings which can serve as a shared key for initiating secure exchange of messages using a secret key cryptoalgorithm. Let us consider the following trivial (but not successful) approach. The owners of A and B choose the same random set of k bits. Using the bits found at these positions, one of them encrypts something which when decrypted makes sense and sends it to the other. If the other decrypts it and find meaningful information, then he/she simply announces it to the sender and from this point on they can use these bits to securely exchange messages. However, it is highly unlikely that they will succeed if their strings initially differ substantially. For instance, if they agree on $X(A, B)$ positions then the probability that a randomly subset of positions will be a set of agreeing positions is equal to $\left(\frac{X(A,B)}{N}\right)^k$, where k will be the resulting key size. Unless $X(A, B)$ is sufficiently large, this probability will be very low. Thus, our approach is to, gradually, *increase* the quantity $\frac{X(A,B)}{N}$ by a bit elimination process that favours the deletion of bits at differing positions in the two strings. The end result of this elimination process will be a pair of strings having a greater percentage of similar position and, thus, will enable the two nodes to rapidly locate a subset of equal bits using the approach outlined above.

The bit-similarity is formalized by the following:

Definition 1. *The bit correlation $X(A, B)$ between strings A and B is the number of positions at which the two strings agree:*

$$X(A, B) = \frac{|\{i : A_i = B_i, 1 \leq i \leq N\}|}{N}.$$

When the bit correlation between two strings A, B is manipulated by a process (e.g. some protocol executed between the entities possessing the strings) we will find useful the notion of *time-dependent* bit-correlation:

Definition 2. *The time-dependent bit correlation $X(A, B, t)$ (or $X(t)$ when no confusion arises), $t \geq 0$, between strings A and B is defined as*

$$X(A, B, t) = \frac{|\{i : A_i = B_i, i \in N_t\}|}{|N_t|}$$

with N_t the remaining positions after t steps $(N_0 = \{1, \ldots, N\})$.

Obviously, $X(A, B, 0) = X(A, B)$.

3 Approximating the Evolution of Stochastic Processes

In order to track the density of positions where two strings agree, we will make use of Wormald's theorem (see [14]) to model the probabilistic evolution of the

protocol described in Section 4 using a deterministic function which stays prov-
ably close to the real evolution of the algorithm. The statement of the theorem
is as follows:

Definition 3. *A function f satisfies a Lipschitz condition on $D \subset \Re^j$ if there
exists some constant $L > 0$ such that*

$$|f(u_1, \ldots, u_j) - f(v_1, \ldots, v_j)| \leq L \sum_{i=1}^{j} |u_i - v_i|$$

for all $(u_1, \ldots, u_j)$ and $(v_1, \ldots, v_j)$ in D.

Definition 4. *Given a random variable X depending on n, denoted by $X^{(n)}$,
we say that $X^{(n)} = o(f(n))$ always if*

$$\max\{x | \mathbf{Pr}[X^{(n)} = x] \neq 0\} = o(f(n)).$$

Theorem 1. *Let $Y_i^{(n)}(t)$, $n \geq 1$, be a sequence of real-valued random variables,
$1 \leq i \leq k$ for some fixed k, such that for all i, all t and all n, $|Y_i^{(n)}(t)| \leq Bn$
$(n > 0)$ for some constant B. Let $\mathbf{H}(t)$ be the history of the sequence, i.e. the
matrix $\langle \vec{Y}(0), \ldots, \vec{Y}(t) \rangle$, where $\vec{Y}(t) = (Y_1^{(n)}(t), \ldots, Y_k^{(n)}(t))$.*
*Let $I = \{(y_1, \ldots, y_k) : \mathbf{Pr}[\vec{Y}(0) = (y_1 n, \ldots, y_k n)] \neq 0 \text{ for some } n\}$. Let D be
some bounded connected open set containing the intersection of $\{(s, y_1, \ldots, y_k) :
s \geq 0\}$ with a neighborhood of $\{(t/n, y_1, \ldots, y_k) : (y_1, \ldots, y_k) \in I\}$. (That is,
after taking a ball around the set I, D is required to contain the part of the ball
in the half-space corresponding to $s = t/n$, $s \geq 0$.)*
Let $f_i : \Re^{k+1} \to \Re, 1 \leq i \leq k$, and suppose that for some $m = m(n)$,

(i) for all i and uniformly over all $t < m$, always

$$\mathbf{E}[Y_i^{(n)}(t + 1) - Y_i^{(n)}(t) | \mathbf{H}(t)] = f_i(t/n, Y_0^{(n)}(t)/n, \ldots, Y_k^{(n)}(t)/n) + o(1),$$

(ii) for all i and uniformly over all $t < m$,

$$\mathbf{Pr}[|Y_i^{(n)}(t + 1) - Y_i^{(n)}(t)| > n^{1/5}] = o(n^{-3}), \text{ always,}$$

*(iii) for each i, the function f_i is continuous and satisfies a Lipschitz condition
on D.*

Then

(a) for $(0, \hat{z}^{(0)}, \ldots, \hat{z}^{(k)}) \in D$ the system of differential equations

$$\frac{dz_i}{ds} = f_i(s, z_0, \ldots, z_k), 1 \leq i \leq k$$

*has a unique solution in D for $z_i : \Re \to \Re$ passing through $z_i(0) = \hat{z}^{(i)}, 1 \leq
i \leq k$, and which extends to points arbitrarily close to the boundary of D;*

(b) almost surely $Y_i^{(n)}(t) = z_i(t/n) \cdot n + o(n)$, uniformly for $0 \leq t \leq \min\{\sigma, m\}$ and for each i, where $z_i(s)$ is the solution in (a) with $\hat{z}^{(i)} = Y_i^{(n)}(0)/n$, and $\sigma = \sigma(n)$ is the supremum of those s to which the solution can be extended.

What the theorem essentially states is that if we are confronted with a number of (possibly) interrelated random variables (associated with some random process) such that they satisfy a Lipschitz condition and their expected fluctuation at each time step is known, then the value of these variables can be approximated using the solution of a system of differential equations. Furthermore, the system of differential equations results directly from the expressions for the expected fluctuation of the random variables describing the random process.

4   Gradual Increase of the Bit-Correlation

Let U_0 and U_1 be two communicating entities possessing the strings V_0 and V_1 of initial bit-correlation $X(V_0, V_1, 0)$. Then the entities get involved in a process whereby each, in turn, selects a random subset of two positions from the set of available bit positions (initially this set is equal to $\{1, \ldots, N\}$) and sends the two positions as well as the bits stored at these positions to the other entity. If the other entity sees that the two positions differ by at least 1 bit, then it sends a message to the other entity to eliminate from further consideration one of the two positions (chosen at random) while eliminating the same position locally at its bit string. If no differences are encountered, then the entity sends a message not to eliminate any of the two positions. This process continues up to a certain, predetermined number of steps (defined in the next section).

The intuition behind this protocol is that when two 2-bit random substrings of two strings of bit-correlation $X(V_0, V_1)$ differ in at least one position, then it is more likely to eliminate a differing bit position than to eliminate a position where the strings agree. The probability of removing a differing bit position is equal to $1 - \frac{X(V_0, V_1, t)}{|N_t|}$. As long as $X(V_0, V_1, t)$ remains low, then the probability of eliminating a bit-position of disagreement is high. On the contrary, as $X(V_0, V_1, t)$ increases the probability of eliminating a differing position becomes low but this is due to the fact that since $X(V_0, V_1, t)$ is high, bit differences are not encountered often during the execution of the protocol.

Below, we describe in detail the general protocol that involves the examination of k places within the strings and not only 2 as in our discussion above. The protocol is based on two functions (see Section 7 for their implementation):
(i) $f(V(S))$: takes the substring of string V defined by the position set S and suitably encodes it so that it is easy to disclose without actually revealing it, whether the substring differs in at least half the positions with another substring.
(ii) $g(V(S), f(V'(S), r)$: takes as input a substring of V and an encoding of a substring of V' on the same set of positions S and returns 1 if the two substrings differ in at least half the positions. Otherwise it returns 0.

Protocol for user $U_c, c = 0, 1$
Protocol parameters known to both communicating parties: (i) k, the subset size, (ii) T, the number of protocol execution steps, (iii) the index (bit position) set N.

```
 1. i ← 1 /* The step counter. */
 2. N ← {1,...,N} /* The available string positions. */
 3. while i ≤ T and k ≥ N
 4.    begin /* while */
 5.       if odd(i + c) then /* Users 0 and 1 alternate moves. */
 6.          begin
 7.             S ← RANDOM(k, N) /* Return random subset of available positions. */
 8.             m ← (S, f(V_c(S))) /* Chosen positions and encoding of the substring. */
 9.             SEND(U_{c+1 mod 2}, m) /* Send the message ... */
10.             RECV(U_{c+1 mod 2}, p) /* ... and wait for the response. */
11.             N ← N − {p} /* Eliminate the position chosen by the other party. */
12.          end
13.       else
14.          begin
15.             RECV(U_{c+1 mod 2}, m) /* Receive positions and encoding of sent substring. */
16.             if g(V_c(m_1), m_2, ⌈k/2⌉) = 1 then/* If at least half the positions differ ... */
17.                p = RANDOM(1, {1,...,k}) /* ... eliminate a random position. */
18.             else
19.                p = 0 /* Otherwise, no position is eliminated. */
20.             SEND(U_{c+1 mod 2}, p) /* Send the chosen position to the other party ... */
21.             N ← N − {p} /* ... and eliminate it also locally. */
22.          end
23.       i ← i + 1
24. end /* while */
```

There are two random variables relevant to the computation of the density: $X(i)$ and the set N_i. The density at step i is equal to $\frac{X(i)}{|N_i|}$. The quantity $|N_i|$ decreases by one with *certainty* only at steps where the value subsets of V_0 and V_1 agree in all places. We would like to have it decrease by one at *each step* in order to facilitate the analysis. In order to allow this decrease at each step, in steps where the values agree (and, thus, no actual deletion of bit positions takes place) we will add to $X(i)$ a corrective quantity $s(i)$ such that the density remains constant in getting from step i to step $i + 1$:

$$\frac{X(i+1)}{|N_{i+1}|} = \frac{X(i)}{|N_i|} \implies \frac{X(i) + s(i)}{|N_i| - 1} = \frac{X(i)}{|N_i|} \implies s(i) = -\frac{X(i)}{|N_i|}. \tag{1}$$

Now $|N_i| = n - i$.

5   The General k-Place Elimination Protocol

Below we provide the deterministic differential equation that governs the evolution of the elimination protocol, using Wormald's theorem.

Theorem 2. *The differential equation that results from the application of Wormald's theorem on the quantity $X(i)$ (places of agreement at protocol step i) as it evolves in the k-place elimination protocol is the following:*

$$\frac{dx(t)}{dt} = - \sum_{j=\lceil \frac{k}{2} \rceil}^{k} \binom{k}{j} \left(1 - \frac{x(t)}{(1-t)}\right)^j \left(\frac{x(t)}{1-t}\right)^{k-j} \left(1 - \frac{j}{k}\right) +$$

$$\left[-\frac{x(t)}{(1-t)}\right] \sum_{j=0}^{\lceil \frac{k}{2} \rceil - 1} \binom{k}{j} \left(1 - \frac{x(t)}{(1-t)}\right)^j \left(\frac{x(t)}{1-t}\right)^{k-j}. \qquad (2)$$

Proof. Let $X(i)$ be the number of places of agreement at step i. The possible values for $X(i+1)$ are either $X(i)-1$ or $X(t)+s(i)$ (see (1)). Thus the expectation $\mathbf{E}[X(i+1) - X(i)]$ is equal to

$$(-1) \cdot \mathbf{Pr}[X(i+1) - X(i) = -1] + s(i)\mathbf{Pr}[X(i+1) - X(i) = s(i)]$$

$$= -\mathbf{Pr}[X(i+1) - X(i) = -1] - \frac{X(i)}{n-i}\mathbf{Pr}[X(i+1) - X(i) = s(i)]. \qquad (3)$$

Let E_1^j be the event *The chosen positions differ in j places* and E_2 be the event *Agreement position is deleted.* Then

$$\mathbf{Pr}[X(i+1) - X(i) = -1] = \sum_{j=\lceil \frac{k}{2} \rceil}^{k} \binom{k}{j} \mathbf{Pr}[E_1^j] \cdot \mathbf{Pr}[E_2|E_1^j] =$$

$$\sum_{j=\lceil \frac{k}{2} \rceil}^{k} \binom{k}{j} \left(1 - \frac{X(i)}{n-i}\right)^j \left(\frac{X(i)}{n-i}\right)^{k-j} \left(1 - \frac{j}{k}\right), \text{ and}$$

$$\mathbf{Pr}[X(i+1) - X(i) = s(i)] = \sum_{j=0}^{\lceil \frac{k}{2} \rceil - 1} \binom{k}{j} \left(1 - \frac{X(t)}{(n-i)}\right)^j \left(\frac{X(t)}{n-i}\right)^{k-j}.$$

Thus, (3) becomes

$$\mathbf{E}[X(i+1) - X(i)] = - \sum_{j=\lceil \frac{k}{2} \rceil}^{k} \binom{k}{j} \left(1 - \frac{X(i)}{n-i}\right)^j \left(\frac{X(i)}{n-i}\right)^{k-j} \left(1 - \frac{j}{k}\right)$$

$$- \frac{X(i)}{n-i} \sum_{j=0}^{\lceil \frac{k}{2} \rceil - 1} \binom{k}{j} \left(1 - \frac{X(t)}{(n-i)}\right)^j \left(\frac{X(t)}{n-i}\right)^{k-j}.$$

Performing the scaling $n \to 1, i \to t \cdot n$, and $X(i) \to x(t) \cdot n$ and applying Wormald's theorem we obtain the required differential equation.   □

6 The 2-Place Elimination Protocol

For $k = 1$ we obtain the trivial protocol that examines position subsets of cardinality 1. The elimination process is easily seen to obey the equation

$$x(t) = \frac{1-t}{1+c-ct}, c = \frac{1-x(0)}{x(0)} \tag{4}$$

with $x(0) = \frac{X(0)}{n}$ the initial percentage of agreement positions. When $k = 2$, from Theorem 2 we obtain the following differential equation:

$$\frac{dx(t)}{dt} = (-1)\left[\left(1 - \frac{x(t)}{(1-t)}\right)\frac{x(t)}{1-t}\right] + \left[-\frac{x(t)}{(1-t)}\right]\left[\frac{x(t)}{(1-t)}\right]^2. \tag{5}$$

Lemma 1. *The solution of the differential equation (5) is the function*

$$x(t) = \frac{1-t}{LambertW[c_0(1-t)]+1}, c_0 = \left[\frac{1-x(0)}{x(0)}\right]e^{\left[\frac{1-x(0)}{x(0)}\right]} \tag{6}$$

where $x(0) = \frac{X(0)}{n}$ is the percentage of the initial agreement positions.

For $x \in \Re$, the *Lambert W function* is defined as the principle branch of the function that satisfies the following equation (see, e.g., [3]):

$$LambertW(x)e^{LambertW(x)} = x.$$

Proof. The differential equation can be rewritten as follows:

$$\frac{dx(t)}{dt} = -\left(\frac{1}{1-t}\right)^3 x(t)^3 + \left(\frac{1}{1-t}\right)^2 x(t)^2 - \left(\frac{1}{1-t}\right)x(t).$$

This is an instance of the *Abel Equation of the First Kind* (see, e.g., [11]):

$$\frac{dx(t)}{dt} = f_3(t)x(t)^3 + f_2(t)x(t)^2 + f_1(t)x(t) + f_0(t) \tag{7}$$

with $f_3(t) = -\left(\frac{1}{1-t}\right)^3, f_2(t) = \left(\frac{1}{1-t}\right)^2, f_1(t) = \frac{1}{1-t}, f_0(t) = 0$. Since $f_0(t) = 0$ and $f_2(t) \neq 0$, we work as follows (see [11]). We set $v(t) = e^{\int f_1(t)dt} = 1-t, z(t) = \int vf_2(t)dt = -\ln(1-t), g(t) = v(t)\frac{f_3(t)}{f_2(t)} = -1$ and define a new differential equation

$$\frac{du(z)}{dt} = u(z)^2 + g(z)u(z)^3 = u(z)^2 - u(z)^3. \tag{8}$$

This differential equation has separate variables and the independent variable (z) is missing. It is, thus, of the form $\frac{u(z)}{dz} = ag(u(z))$ with a a constant which, in our case, is equal to 1 and $g(u) = u^2 - u^3$. Its solution is given by the integral equation ([11]) $\int \frac{dy}{g(u)} = az + c$, with c a constant. Substituting the values

$g(u) = u^2 - u^3$ and $a = 1$ we obtain, after integrating and exponentiating with base e, the equation $\left(1 - \frac{1}{u}\right)^u e = e^{-u(z+c)}$. We set $w = 1 - \frac{1}{u}$ and, thus, obtain the equation $-we^{-w} = e^{-z-c-1}$ from which we conclude that $w = -\text{LambertW}(e^{-z-c-1})$. Therefore,

$$x(z) = \frac{1-t}{1 + \text{LambertW}(e^{-z-c-1})}. \tag{9}$$

We now set $z = -\ln(1 - t)$. Setting, also, $t = 0$ and solving for c, we find $c - \ln\left[\frac{1-x(0)}{x(0)}\right] - \frac{1}{x(0)}$, obtaining (6). $\qquad\square$

7   Assessment of the Elimination Protocol

Let us first discuss each protocol's efficiency in reaching high agreement percentage states between the two initial strings. If we denote by $x_k(t)$ the solution of the differential equation of the s-place elimination protocol, then using Wormald's theorem, we can deduce that the the number of agreement positions, as a function of the elimination step i, is with high probability $X_k(i) = x_k(i/n)n + o(n)$, using the solutions $x_k(t)$ as given in Equations (4) and (1) for the 1-place and 2-place elimination protocols respectively. The agreement density $d_s(i)$ is thus equal to $X_s(i)/(n - i)$:

$$d_s(i) = \frac{X_s(i/n) + o(n)}{n - i} = \frac{x_s(i/n)n}{n - i} + o(1). \tag{10}$$

The 2-place elimination protocol is no better at eliminating disagreeing positions because at each step it deletes a single position despite the fact it has the opportunity to delete two positions. However, it can be efficient if used in an "aggressive" manner, whereby if the substrings disagree in at least 1 position, then both positions are deleted at once. Then it can be shown that the expected density at each step is bounded from below by $d_2(i/n)/(n - 2i)$, for $i = 0 \ldots \lfloor \frac{n}{2} - 1 \rfloor$. In Figures 1 and 2 we see the plot of the protocols we examined above (1-place elimination and 2-place elimination) plus the aggressive 2-place elimination protocol before half of the steps have been executed (we have used $n = 10000$). Notice that the time scale covers steps from 0 up to $n/2$ because the aggressive 2 place elimination protocol deletes 2 positions at a time. We observe that the density increases at a higher rate for the aggressive 2-place elimination protocol.

For instance, if we start with two strings of $n = 10000$ bits each which initially agree on half of their positions (see Figure 2) then after 4750 steps of the protocol the two remaining strings, each consisting (in the worst case) of about 250 bits, agree on 90% of their positions. This means that the two entities can form a shared 32-bit key as follows: they both repeatedly choose random 32-bit subsets of the 250-bit remaining strings, with one of the nodes encrypting (with the chosen 32-bit subset) some previously agreed upon token and sending it to the other. If the other node decrypts the message and discovers the token then the 32-bit string can serve as a shared key. This will take, on the average, $1/(0.9)^{32} \cong 29$

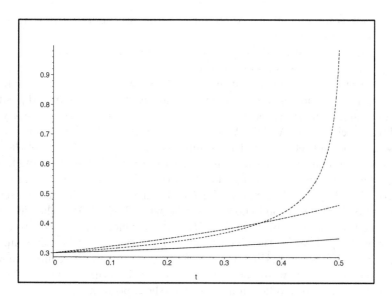

Fig. 1. Evolution of densities with initial agreement density equal to 0.3 (aggressive 2-place: dotted line, 2-place: continuous line, 1-place: dashed line)

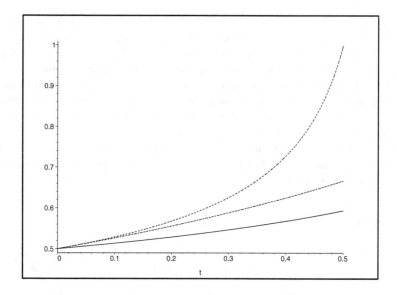

Fig. 2. Evolution of densities with initial agreement density equal to 0.5 (aggressive 2-place: dotted line, 2-place: continuous line, 1-place: dashed line)

steps. Alternatively, the two nodes may employ an error correction code and try to "correct" the chosen substrings until they discover agreement. Note that the remaining 250-bit substrings can be reused in discovering new 32-bit substrings for each new communication session without the two nodes needing to repeat the bit-elimination protocol.

With regard to security, it is based on the functions f and g (see Section 4). One way to define these functions is the following.

Our description of the elimination protocol assumes that the two communicating parties each possess a bit vector and thus, each of the compared positions contains a single bit. However, it is often the case (e.g. in deterministic key pre-distribution schemes in Mobile Ad-hoc Networks) that this bit vector actually represents a key subset of a predefined universal key set. Thus, each of the compared positions can be values chosen from within a predefined key set. In such a case, the functions f and g can be defined as follows, for the 2-elimination protocol: (i) $f(V(S)) = VE(S(1)) \oplus VE(S(2)$, where $S(1)$ is the first chosen position and $S(2)$ is the second chosen position while $VE(S(1))$ is the element of V corresponding to position $S(1)$ and $VE(S(2))$ the element of V corresponding to position $S(2)$. That is, the Exclusive-Or of the two key values is computed. (ii) $g(V(S), f(V'(S)) = f(V'(S)) \oplus VE(S1) \oplus VE(S2)$. If the computed value is 0 (i.e. no differences were encountered), then 0 is returned. Otherwise, 1 is returned (i.e. the values differ in at least 1 of the 2 positions).

8   Conclusions

In this paper we described a series of protocols, called k-place elimination protocols, that can be used in order to increase the percentage of agreement positions between two strings, held by two communicating parties without revealing the values. The protocols involve the examination of random position subsets of size k and the elimination of a random position if the two strings are found to disagree in more than half the examined positions. In this way, the random process governing the protocol is directed towards the elimination of disagreement positions more than the elimination of agreement positions. The protocol is general and may be used in any situation involving either wireless or conventional networks in which there is no trusted third party or key management authority among the network nodes.

References

1. H. Chan, A. Perrig, and D. Song, "Random Key Predistribution Schemes for Sensor Networks," Proceedings of the IEE Symposium of Privacy and Security", 11–14 May, 2003, pp. 197–213.
2. A. C-F. Chan and E.S. Rogers Sr., "Distributed Symmetric Key Management for Mobile Ad hoc Networks," INFOCOM 2004. 23rd Annual Conference of the IEEE Computer and Communications Societies, Vol. 4, 7–11 March, 2004. pp. 2414–2424.
3. R.M. Corless, G.H. Gonnet, D.E.G. Hare, D.J. Jeffrey, and D.E. Knuth. On The Lambert W Function. *Advances in Computational Mathematics* 5 (1996): 329–359.

4. B. DeCleene, L. Dondeti, S. Griffin, T. Hardhonio, D. Kiwior, J. Kurose, D. Towsley, S. Vasudevan, and C. Zhang, "Secure Group Communication for Wireless Networks," Military Communications Conference (MILCOM) 2001. Network-Centric Operations: Creating the Information Force. IEEE. Vol. 1, 28–31 Oct. 2001, pp. 113–117.

5. W. Diffie and M. Hellman, "New Directions in Cryptography," IEEE Transactions on Information Theory, Vol. 22, Issue 6. November 1976, pp. 644–654.

6. L. Eschenauer and V. D. Gligor, "A Key-Management Scheme for Distributed Sensor Networks," Proceedings of the 9th ACM Conference on Computer and Communication Security, November 2002, pp. 41–47.

7. H. Harney and C. Muchenhirn, 1997, "Group Key Management Protocol (GKMP) Specification," Internet Engineering Task Force - Network Working Group, Request for Comments No. 2093, July 1997.

8. Y. Kim, A. Perrig and G. Tsudik, "Communication-Efficient Group Key Agreement," IFIP SEC 2001, June 2001.

9. B. Lehane, "Ad Hoc Key Management," PhD Thesis, Department of Electronic and Electrical Engineering, Trinity College Dublin, 2004.

10. D. A. McGrew, A. T. Sherman, "Key Establishment in Large Dynamic Groups Using One-Way Function Trees," IEEE Transactions on Software Engineering, Vol. 29, Issue 5, May 2003, pp. 444–458.

11. G.M. Murphy, *Ordinary Differential Equations and their Solutions*, D. Van Nostrand Company Inc., 1960.

12. M. Steiner, G. Tsudik, and M. Waidner, "CLIQUES: A New Approach to Group Key Agreement," Proceedings of the 18th International Conference on Distributed Computing Systems (ICDCS '98), 26–29 May, 1998, pp. 380–387.

13. C.K. Wong, M. Gouda, and S.S. Lam, "Secure Group Communications Using Key Graphs," IEEE/ACM Transactions on Networking, Vol. 8, Issue. 1, February 2000, pp. 16–30.

14. N.C. Wormald, "The differential equation method for random graph processes and greedy algorithms," *Ann. Appl. Probab.* **5**, 1217–1235, 1995.

Key Revocation for Identity-Based Schemes in Mobile Ad Hoc Networks

Katrin Hoeper and Guang Gong

Department of Electrical and Computer Engineering
University of Waterloo
Waterloo, ON, N2L 3G1, Canada
khoeper@engmail.uwaterloo.ca, ggong@calliope.uwaterloo.ca

Abstract. Recently, identity-based cryptographic (IBC) schemes have been considered to secure mobile ad hoc networks (MANETs) due to their efficient key management properties. However, proposed schemes do not provide mechanisms for key revocation and key renewal. In this paper, we propose the first key revocation and key renewal mechanisms for IBC schemes that are especially designed for MANETs. In our fully self-organized revocation scheme, each node monitors nodes in communication range and securely propagates its observations. The public key of a node is revoked if a minimum number of nodes accused the node. To enable key renewal, we introduce a modified format for ID-based public keys, such that new keys can be issued for the same identity. The introduced revocation scheme is efficient because it uses pre-shared keys from the Weil pairing and messages are sent to an m-hop neighborhood instead to the entire network.

Keyword: Identity-Based Schemes, Revocation, Key Renewal, Mobile Ad Hoc Networks, Neighborhood Monitoring.

1   Introduction

A growing number of mobile wireless applications require that networks are spontaneously formed by the participating devices themselves. Such networks are referred to as *mobile ad hoc networks (MANETs)*. The idea behind MANETs is to enable connectivity among any arbitrary group of mobile devices everywhere, at any time. Slowly people realize that implementing security is of paramount importance in MANETs. However, the special properties of MANETs, such as the lack of infrastructure including the absence of trusted third parties (TTPs), as well as the constraints of the devices and the communication channels, make implementing security a very challenging task. Due to the problem of secure key distribution in symmetric schemes, the use of a public key infrastructure (PKI) is desirable in many MANET applications. The major challenges of implementing PKIs in MANETs are issuing and distributing certificates and enabling certificate revocation. Many PKI-based schemes have been proposed to secure MANETs,e.g. [3, 10, 12], whereas many of them do not provide mechanisms for certificate revocation at all, e.g. [12]. In general, one can distinguish two PKI

T. Kunz and S.S. Ravi (Eds.): ADHOC-NOW 2006, LNCS 4104, pp. 224–237, 2006.

implementations in MANETs. In the first one, a certification authority (CA) issues public key certificates to nodes before the nodes join the network, we call this an *off-line CA*. In the second case, nodes obtain their certificates from a group of network nodes that serve as *distributed on-line CA*. Distributed CAs can be implemented using (k, n)-threshold schemes [10,12]. Implementations using on-line TTPs have the advantage that they do not require the set-up of any infrastructure. However, the threshold schemes impose a lot of communication and computational load onto the network.

Recently, identity-based cryptographic (IBC) schemes have been considered as an alternative public key scheme to secure MANETs due to their efficient key management [4,6,9]. However, the proposed IBC schemes do not provide mechanisms for key revocation and key renewal. We believe that these mechanisms are of paramount importance and every node in a MANET should be able to instantly verify whether a public key has been revoked. Due to the weak physical protection of nodes, node compromises including key disclosures are very likely in MANETs. Frequent key renewals to prevent such compromises are either computationally challenging in solution with distributed on-line key generation center (KGC) or infeasible in solutions with off-line KGC. In this paper, we propose the first key revocation and key renewal schemes for IBC schemes that are especially designed to meet the requirements and constraints of MANETs. In our revocation scheme, each node uses a neighborhood watch scheme to monitor nodes in communication range for suspicious behavior. These observations are then securely propagated to an m-hop neighborhood. The public key of a node is revoked if at least δ nodes accused that node. Our key revocation scheme is scalable in parameters m and δ, i.e. the level of security can be chosen as performance trade-off. To enable key renewal in IBC schemes, we introduce a new format for ID-based public keys such new keys can be issued for the same identity after the previous key has been revoked. A detailed discussion about IBC schemes in MANETs including a first version of our revocation scheme can be found in [7].

The proposed key revocation scheme can be adapted to PKI-based solutions for MANETs to provide certificate revocation. Other than existing certificate revocation schemes for MANETs, e.g. [3,10], our scheme reduces the overall computational and communication network overhead by using pre-shared keys from the Weil pairing to secure accusations, and sending messages to an m-hop neighborhood rather than to the entire network. Other than in existing schemes, we discuss and efficiently solve the problems of nodes that wish to revoke their own keys and new nodes that join the network and wish to learn about past accusations and revocations.

The remainder of the paper is organized as follows. In the next section we discuss the system set-up for our key revocation and key renewal schemes for IBC-based MANETs. The revocation and renewal schemes are then introduced in Sect. 3. The security of the scheme is analyzed in Sect. 4 and compared to related work in Sect. 5. Finally, we discuss the contributions of the proposed schemes in Sect. 6.

2   System Set-Up

2.1   Preliminaries

In this paper, we present key revocation and key renewal schemes for IBC implementations that are based on IBC schemes from the Weil pairing [1]. In the remainder of the paper, we adopt many notations from [1], please refer to the original paper for details. IBC schemes provide a very efficient key management that helps reducing communication, computation, and memory costs. A summary of desirable properties of IBC scheme for implementations in MANETs can be found in [6, 7]. The main feature of IBC schemes is the use of self-authenticating public keys, which makes the use of public key certificates redundant. In IBC schemes, every network node is able to derive the public key Q_i of a communication partner ID_i in the network, e.g. $Q_i = H_1(ID_i)$ in [1] where H_1 is a publicly known hash function that maps an arbitrary string to an element in group $\mathbb{G}_1$, i.e. $H_1 : \{0,1\}^* \rightarrow \mathbb{G}_1^*$. Because the public key Q_i of a node ID_i is pre-determined, the private key d_i is derived from Q_i and a master secret key s that is only known to the KGC, i.e. $d_i = sQ_i$. The KGC generates and distribute the private keys during the initialization of all network nodes. All pairs of nodes ID_i and ID_j in a pairing-based IBC scheme are able to compute a pairwise pre-shared secret key $K_{i,j}$ [2] in a non-interactive fashion as given in (1).

$$K_{i,j} = \hat{e}(d_i, Q_j) = \hat{e}(Q_i, d_j) \tag{1}$$

For the key computation, both nodes compute the bilinear non-degenerate mapping $\hat{e}(\cdot)$, e.g. the Weil pairing, over their own private key d_i and the public key Q_j of the other node. The KGC is a key escrow because it knows all private and pre-shared keys in the network. Threshold schemes for distributed KGCs have been introduced for this matter [1]. Other preventions are known as well and the limited power of key escrows in MANETs has been analyzed in [5].

2.2   Choosing Identities

Identities ID_i must be *unique* for each entity i in the network. Furthermore, identities must be *unchangeably bound* to an entity for its entire *lifetime* and the identity is *not transferable*. The string of information that can be used as identity depends on the application. Generally, we can distinguish three cases of entities an identity can be bound to: (1) a user operating a network node, i.e. the ID string corresponds to the user, e.g. the user's email address; or (2) a device, i.e. the ID is bound to the hardware, e.g. the MAC address; or (3) a network interface, in that case the ID might be derived from the IP address.

For example, if an application enables two users to securely communicate with each other, user-dependent IDs seem desirable. In sensor networks or other MANETs in which user do not operate the devices, the MAC address seems to be a good choice. The third scenario might be of interest in some special applications. However, it is not feasible in many MANETs because network addresses such as IP addresses are dynamic or do not exist at all.

2.3   Public Key Format

To limit the validity period of an ID-based public key, an expiry date can be embedded in the key itself, e.g. $Q_i = H_1(ID_i||t_i)$, where the expiry date t is concatenated with the identity [1]. Only a node that is in possession of the private key d_i which corresponds to the date t_i can sign or decrypt messages. However, this key format is only sufficient in schemes without revocation. In schemes with explicit key revocation, i.e. keys can be revoked before they expire, nodes need to be able to request new keys even before the expiry date t. Since the identity ID_i is unchangeable in IBC schemes as discussed in Sect. 2.2, issuing a key for the same expiry date t would result into the same old compromised key. However, issuing new keys with a new expiry date t' might not be feasible, because a node ID_i is only eligible to possess keys until t. Furthermore, it is desirable in IBC schemes that expiry dates are chosen in a predictable manner, e.g. in fixed intervals, such that nodes do not need to exchange public keys after the previous keys expired. Hence, to provide immediate key renewal after key compromise, we need to add some additional data v to the public key that can be changed with every key renewal. We use the following format as given in (2) below, where v is the version number of the public key.

$$Q_i = H_1(ID_i||t_i||v_i) \tag{2}$$

For instance, upon revocation of $Q_i = H_1(ID_i||t_i||v_i)$, node ID_i can request a new key Q'_i before t_i, with $Q'_i = H_1(ID_i||t_i||v'_i)$ and $v'_i = v_i + 1$. Note that the version number v always starts with $v = 1$ for every new expiry date t and is incremented with each key renewal for the same date t. Key renewal and distribution with $v > 1$ are discussed in Sect. 3.3.

2.4   Assumptions

The necessary assumptions for the network and its nodes in our IBC key revocation and key renewal schemes can be summarized as follows: *1. bidirectional communication links*; *2. nodes are in promiscuous mode*; *3. each node has a unique identity*; *4. nodes know identities of their one-hop neighbors*; *5. nodes know distances to other nodes in m-hop neighborhood*; and *6. nodes obtain a private and public key pair from an off-line KGC prior joining the network.*

The first two assumptions are necessary to enable nodes to monitor their neighbor nodes in communication range. This is required in our revocation scheme. Bidirectional links are a common assumption in many lower-layer MANET protocols, e.g. in the AODV [11] and other AODV-based routing protocols. Promiscuous mode is assumed in dynamic routing protocols for MANETs, e.g. AODV and DSR [8]. Assumption 3 is necessary to unambiguously identify nodes. This kind of identifiers are required for many network tasks and protocols, such as routing and authentication. Assumption 4 is needed, because neighbor nodes need to be unambiguously identified before they can be marked as suspicious or trustworthy in the revocation scheme. This information is usually provided by routing and other lower-layer protocols, e.g. AODV. In case the identities of neighbors are not provided by lower layer protocols, users first explore their neighborhood by sending *hello* messages and waiting for the responses. Assumption 5 is necessary to enable nodes to decide which accusation values they need to consider for updating their revocation lists, e.g. accusations from nodes more than m hops away are discarded. This information is provided by the routing protocols, e.g. AODV and DSR. Assumption 6 is necessary because cryptographic keys are needed to provide message authentication in our revocation scheme. Here we assume an external off-line KGC that issues the keys. The assumption can be changed to a distributed on-line KGC that issues keys within the network as outlined in Sect. 5. We can summarize that the necessary assumptions are quite common, and in fact necessary in most ad hoc routing and security protocols. Hence, our assumptions do not impose an additional burden to the system.

3   Key Revocation and Renewal for IBC Schemes in MANETs

3.1   Key Revocation

Every node in a MANET needs to be able to verify whether a public key is revoked. Public key revocations need to be handled within the network, because nodes need to be able to immediately verify the status of a public key. So far in all IBC schemes, i.e. general schemes and schemes especially designed for MANETs, revocation referred to embedding an expiry date in the public key. As discussed earlier, this is not sufficient because nodes need to be able to revoke keys before they expire, e.g. in the case of key compromise or malicious behavior. In our scheme, keys are revoked either if a node notices that its own key has been compromised or if a group of at least δ nodes observes that another node behaves suspiciously.

In order to provide key revocation in IBC schemes for MANETs, we introduce three algorithms. First, nodes observe the nodes in their neighborhood for suspicious behavior using *Algorithm 1: Neighborhood watch*. Second, nodes need to be able to revoke their own public keys using *Algorithm 2: Harakiri*. Third, nodes securely inform each other about suspicious observations and generate key revocation lists in *Algorithm 3: Accusation scheme*.

Alg.1 Neighborhood watch: The neighborhood watch scheme is a local monitoring scheme, in which each node ID_i monitors its one-hop neighborhood $\mathcal{N}_i$ for suspicious behavior. Suspicious behavior can be frequent packet drops or a large number of sent messages. Therefore nodes observe their neighbors and check for instance whether the nodes forwarded messages that were addressed to another node. Suspicious behavior can have different causes, e.g. a node has been compromised and is now controlled by a malicious user, or a node is selfish and rather conserves its energy than forwarding messages.

For an easier representation and without loss of generality, we denote ID_i's one-hop neighbors as $ID_j \in \mathcal{N}_i$ with $j \in \{1, \ldots, N_i\}$, where N_i is the number of one-hop neighbors. User ID_i stores so-called accusation values $a_{k,j}^i$ for nodes ID_j together with the expiry date t_j^i and version number v_j^i of the current public key Q_j of ID_j. Note that in our notation the upper index i generally refers to the node ID_i that stores the value, where the accusation values $a_{k,j}^i$ stored by ID_i, indicate whether a node ID_k has accused a node ID_j (i.e. $a_{k,j}^i = 1$) or not ($a_{k,j}^i = 0$). A node ID_i sets its own accusation values $a_{i,j}^i = 1$, i.e. $k = i$, if ID_i observed ID_j to behave suspiciously, otherwise $a_{i,j}^i = 0$. The accusation values $a_{i,j}^i$ for all $ID_j \in \mathcal{N}_i$ that a node ID_i creates from its own neighborhood watch, can be represented as an accusation matrix

$$AM^i = \begin{pmatrix} ID_1 & (t_1^i, v_1^i) & a_{i,1}^i \\ \vdots & \vdots & \vdots \\ ID_{N_i} & (t_{N_i}^i, v_{N_i}^i) & a_{i,N_i}^i \end{pmatrix} \text{ with } a_{i,j}^i \in \{0,1\} \text{ and } j \in \{1,\dots,N_i\}.$$

Each row vector $\underline{r}^i(ID_j)$ in AM^i, we use $\underline{r}_j^i$ for short in the rest of the paper, corresponds to a neighbor $ID_j \in \mathcal{N}_i$ and the accusation values $a_{i,j}^i$ for the current public key Q_j with expiry date t_j^i and version number v_j^i. We refer to the third column in AM^i as column vector $\underline{c}_i^i$, which is the vector that contains all accusations made by ID_i. The accusation values are updated every time ID_i observes suspicious behavior in its own neighborhood $\mathcal{N}_i$. Once the flag $a_{i,j}^i$ is set, the value will not be reset to zero until a new public key Q_j' is received for node ID_j.

Alg.2 Harakiri: When a node ID_i realizes that its private key d_i has been compromised, it broadcasts a harakiri message hm_i, with

$$hm_i = (ID_i, d_i, Q_i, (t_i, v_i), \text{"revoke"}, hopcount),$$

to its m-hop neighborhood $m\text{-}\mathcal{N}_i$. Node ID_i initially sets $hopcount = m$ and sends the message to all one-hop neighbors $\mathcal{N}_i$. The receivers ID_j verify if the harakiri is authentic, by checking wether (3) is true.

$$K_{j,i} = \hat{e}(d_i, Q_j) \tag{3}$$

The check verifies whether the broadcasted private key d_i indeed corresponds to the public key Q_i and thus ID_i. Therefore, a recipient of hm_i, say node ID_j, looks up whether it is in possession of the public key Q_i and the pre-shared key $K_{j,i}$ and if so, uses the $K_{j,i}$ to check whether (3) is true. If ID_j is not in possession of these keys, ID_j first computes Q_i according to (2) and checks whether ID_i and Q_i correspond to each other. If this check is successful, ID_j derives $K_{j,i}$ according to (1) and then checks whether (3) is true. If (3) is true, the receiver ID_j updates its accusation value $a_{i,i}^j = 1$, decrements $hopcount$ and broadcasts the message again. Otherwise, hm_i is discarded. The broadcasting is repeated until $hopcount = 0$. This ensures that all nodes in an m-hop neighborhood of the compromised node ID_i receive the harakiri message and thus learn about the key compromise.

Alg.3 Accusation scheme: In this algorithm every node ID_i creates its own key revocation list $\mathcal{KRL}^i$ for its m-hop neighborhood. In this paragraph we will describe how revocation lists are created (Alg.3.1), securely propagated (Alg.3.2) and how nodes use received revocation lists and harakiri messages to update their own revocation lists (Alg.3.3).

Alg.3.1 Creating a key revocation list $\mathcal{KRL}$: Each node ID_i creates a key revocation list $\mathcal{KRL}^i$ of the following format:

$$
\mathcal{KRL}^i = \begin{pmatrix}
ID_1 & (t_1^i, v_1^i) & R_1^i & a_{1,1}^i & \cdots & a_{1,M_i}^i \\
\vdots & \vdots & \vdots & \vdots & \ddots & \vdots \\
ID_{M_i} & (t_{M_i}^i, v_{M_i}^i) & R_{M_i}^i & a_{M_i,1}^i & \cdots & a_{M_i,M_i}^i
\end{pmatrix},
$$

with $a_{j,k}^i \in \{0, 1, -\}$ and $j, k \in \{1, \ldots, M_i\}$, where M_i is the number of nodes in ID_i's m-hop neighborhood $m\text{-}\mathcal{N}_i$, including ID_i itself. The accusation value $a_{j,k}^i = -$ indicates that ID_j and ID_k are more than m hops away from each other, and thus cannot give a statement about each others trustworthiness. Each row vector $\underline{r}^i(ID_j)$, short $\underline{r}_j^i$, in $\mathcal{KRL}^i$ corresponds to a node $ID_j \in m\text{-}\mathcal{N}_i$, where the row contains the accusation values $a_{j,k}^i$ from all nodes $ID_k \in m\text{-}\mathcal{N}_i$ evaluating node ID_j. Each column vector $\underline{c}^i(ID_j)$, short $\underline{c}_j^i$, in $\mathcal{KRL}^i$ contains all the accusations $a_{k,j}^i$ made by node ID_j against all nodes $ID_k \in m\text{-}\mathcal{N}_i$. The upper index i denotes that the values are the current values in ID_i's $\mathcal{KRL}^i$, where other nodes might have different values. For example, $a_{j,k}^i \neq a_{j,k}^l$ for $i \neq l$ in some cases. Discrepancies in accusation values can exist, because accusation values may be more or less up to date, and nodes have different m-hop neighborhoods, and thus receive different accusation and harakiri messages.

The first field in each row $\underline{r}_j^i$ in $\mathcal{KRL}^i$ contains the identity of node ID_j, the second field the expiry date t_j^i and version number v_j^i of the most recent public key Q_j that ID_i knows of. The fields $4 - (M_i + 3)$ contain the accusation values $a_{j,1}^i$ - a_{j,M_i}^i, where value $a_{j,k}^i = 1$ indicates that node ID_k accused node ID_j, and $a_{j,k} = 0$ otherwise. The third field contains a 1-bit flag R_j^i that, when set, indicates that node ID_i considers the public key Q_j of node ID_j as revoked. The revocation flags R_j^i in ID_i's key revocation list $\mathcal{KRL}_i$ are set, i.e. $R_j^i = 1$, if one of the following four conditions is true:

(Cond.1): $a_{j,i}^i = 1$, i.e. node ID_i observed itself the malicious behavior of node ID_j during the neighborhood watch (Alg.1). This follows that ID_j and ID_i are 1-hop neighbors.

(Cond.2): t_j^i is expired, i.e. the current copy of the ID_j's public key Q_j is expired.

(Cond.3): $a_{j,j}^i = 1$, i.e. ID_i received an authentic harakiri message hm_j from ID_j.

(Cond.4): $A_j^i = \sum_{k=1}^{M_i} a_{j,k}^i > \delta$ for all k, s.t. $R_k^i = 0$, i.e. add all accusation values $a_{j,k}^i$ of row vector $\underline{r}_j$ from non-revoked nodes ID_k and check whether the sum is greater than δ. In other words, the public key Q_j is revoked if node ID_i received more than δ accusations from trustworthy nodes ID_k for a suspicious node ID_j. Note that "-" is treated as zero value in the sum.

If none of the four conditions applies, $R_j^i = 0$, i.e. ID_j and its current public key Q_i are considered to be trustworthy.

Alg.3.2 Propagating accusations: In this algorithm nodes securely propagate accusations through the network. Every time a node ID_i updates its accusation matrix AM^i because it observed some suspicious behavior in its neighborhood watch, ID_i also sends an accusation message to its one-hop neighbors. Similarly, every time ID_i updates its key revocation list $\mathcal{KRL}^i$ because it received accusations from other nodes, ID_i sends an accusation message to its neighbors. The accusation messages send by ID_i have the following format:

$$am_{i,j} = (f_{K_{i,j}}(ID_i, am_i), (ID_i, am_i)), \text{ for all } ID_j \in \mathcal{N}_i$$

where $am_i = AM^i$ for updates from ID_i's own neighborhood watch, and $am_i = \mathcal{KRL}^i$ for updates of the revocation list caused by accusation messages that ID_i received from other nodes. Optionally, am_i contains only the updated values to reduce bandwidth. The accusation messages $am_{i,j}$ are secured using the pre-shared keys $K_{i,j}$ for all $ID_j \in \mathcal{N}_i$ and then unicasted to each one-hop neighbor ID_j. The pre-shared keys serves as input in a secure hash function $f(\cdot)$ to authenticate the message. Upon receiving $am_{i,j}$, a neighboring node ID_j verifies the received message using its pre-shared key $K_{j,i}$. If the verification is successful, ID_j updates its key revocation list $\mathcal{KRL}^j$ accordingly, as we will explain in the next paragraph. Otherwise the message is discarded.

Alg.3.3 Updating key revocation lists: Every time a node ID_i receives a harakiri message hm_j from ID_j, the node verifies the message as described in Alg. 2 and if this verification is successful, node ID_i sets $a_{j,j}^i = 1$ and thus $R_j^i = 1$ in its revocation list $\mathcal{KRL}^i$ (see Cond. 3 in Alg. 3.2). If a node ID_i receives an accusation message $am_{j,i}$ of an one-hop neighbor ID_j, node ID_i performs the following steps to update its key revocation list $\mathcal{KRL}^i$:

(Step 1): check whether $R_j^i = 0$, i.e. whether ID_i considers ID_j to be trustworthy; if yes continue, else discard $am_{j,i}$ and stop.
(Step 2): verify authenticity of $am_{j,i}$ using the pre-shared key $K_{i,j}$ as described in Alg. 3.2; if verification is successful continue, else discard $am_{j,i}$ and stop.
(Step 3): extract column vector $\underline{c}_j^j$ of AM^j or $\mathcal{KRL}^j$ from $am_{j,i}$ to update column vector $\underline{c}_j^i$ in $\mathcal{KRL}^i$, i.e. adopt the accusation values from ID_j's neighborhood watch. Note that ID_i uses only accusation values that are addressed to nodes in ID_i's own m-hop neighborhood, other accusation values are discarded. If $am_{j,i} = AM^j$ stop, else continue.
(Step 4): discard all columns $\underline{c}_k^j$ from $\mathcal{KRL}^j$ for: $k = i$ because that is ID_i's own accusation vector; $k = j$ because that one was used in step 3; $k = l$ for all $R_l^i = 1$ with $l \in \{1, \ldots, M_i\}$ because ID_i does not trust nodes ID_l; $k = r$ for all $am_{r,i}$ that were accepted in step 2; $k = s$ for all $ID_s \notin m\text{-}\mathcal{N}_i$, i.e. nodes that are more than m hops away. Save all other columns $\underline{c}_k^j$.

Now ID_i repeats steps 1-4 for all received accusation messages $am_{j,i}$. Lets say ID_i saved d_k column vectors $\underline{c}_k^j$ from d_k different nodes ID_j for the same ID_k

in step 4. Now for every $d_k > \varepsilon$, with ε being a security threshold, ID_i performs step 5 below.

(Step 5): use all d_k saved column vectors $\underline{c}_k^j$ from step 4 to update ID_i's column vector $\underline{c}_k^i$ in $\mathcal{KRL}^i$. The update is done by using the majority vote for each element in the column vector, i.e. the majority for each accusation value $a_{l,k}^j$ with $l \in \{1, \ldots, M_i\}$ is computed. For simplicity, we assume ID_i saved d_k column vectors $\underline{c}_k^j$ from d_k neighbors ID_j with $j \in \{1, \ldots, d_k\}$ in step 4, with $d_k \geq \varepsilon$. Now ID_i computes

$$a_{l,k}^i = \begin{cases} 1 & \text{if } \sum_{j=1}^{d_k} a_{l,k}^j > \frac{d_k}{2}, \text{ with } l \in \{1, \ldots, M_i\} \text{ and } j \in \{1, \ldots, d_k\} \\ 0 & \text{if } \sum_{j=1}^{d_k} a_{l,k}^j < \frac{d_k}{2}, \text{ with } l \in \{1, \ldots, M_i\} \text{ and } j \in \{1, \ldots, d_k\} \\ a_{l,k}^i & \text{otherwise} \end{cases}$$

Again, only values $a_{l,k}^j$ for nodes ID_l that are within ID_i's m-hop range are considered, others are discarded. If no majority can be found, the accusation value in $\mathcal{KRL}^i$ remains unchanged. Node ID_i repeats this for all column vectors $\underline{c}_k^i$ for which the number of collected vectors $\underline{c}_k^j$ is $> \varepsilon$.

3.2   Example for $\mathcal{KRL}$ Update

We present an artificially small and simple network scenario to illustrate how Alg.3.3 works. In our example we consider six network nodes ID_i with $i \in \{1, \ldots, 6\}$ as shown in Fig. 1, where the nodes maintain key revocation lists for nodes in two hop distance, i.e. $m = 2$, and the security parameters are set to $\delta = 3$ and $\varepsilon = 2$. We now show how ID_1 updates its revocation list $\mathcal{KRL}^1$ upon receiving the accusation messages am_2, am_3, am_4 from its one-hop neighbors $\mathcal{N}_1 = \{ID_1, ID_2, ID_3, ID_4\}$. To do so, ID_1 executes Alg.3.3. For simplicity, we assume that the current expiry date is t for all nodes and the version number is $v = 1$ for all public keys. Hence, we neglect the values (t, v) in the revocation lists. The revocation lists from ID_3 and ID_4, and ID_1's list from before the update look as follows:

$$\mathcal{KRL}^1 = \begin{pmatrix} ID_1\,0\,0\,0\,0\,0\,0 \\ ID_2\,1\,1\,0\,0\,0\,- \\ ID_3\,0\,0\,0\,0\,0\,1 \\ ID_4\,0\,0\,0\,0\,0\,0 \\ ID_5\,0\,0\,0\,0\,0\,0 \end{pmatrix}, \quad \mathcal{KRL}^3 = \begin{pmatrix} ID_1\,0\,0\,1\,0\,0\,1\,- \\ ID_2\,1\,1\,0\,1\,1\,-\,- \\ ID_3\,0\,0\,1\,0\,0\,0\,- \\ ID_4\,0\,0\,0\,0\,0\,1\,1 \\ ID_5\,1\,0\,-\,1\,1\,0\,1 \\ ID_6\,1\,-\,-\,1\,1\,0\,0 \end{pmatrix}, \quad \mathcal{KRL}^4 = \begin{pmatrix} ID_1\,0\,0\,1\,0\,0\,0\,- \\ ID_2\,1\,1\,0\,1\,1\,-\,- \\ ID_3\,0\,0\,1\,0\,0\,1\,- \\ ID_4\,0\,0\,0\,0\,0\,0\,1 \\ ID_5\,1\,0\,-\,1\,1\,0\,1 \\ ID_6\,1\,-\,-\,1\,1\,1\,0 \end{pmatrix}.$$

We now go through the steps of Alg.3.3:

(Step 1): save am_3 and am_4, discard am_2 because $R_2^1 = 1$ in $\mathcal{KRL}^1$

(Step 2): am_3 and am_4 successfully authenticated using $K_{1,3}$ and $K_{1,4}$, resp.

(Step 3): use column vectors $\underline{c}_3^3$ from $\mathcal{KRL}^3$ and column vector $\underline{c}_4^4$ from $\mathcal{KRL}^4$ to update column vector $\underline{c}_3^1$ and $\underline{c}_4^1$, respectively , i.e.

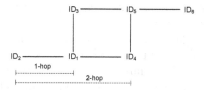

Fig. 1. Example of network setting

$$\underline{c}_3^1 = \begin{pmatrix} 0 \\ 1 \\ 0 \\ 0 \\ 1 \end{pmatrix}, \underline{c}_4^1 = \begin{pmatrix} 0 \\ 1 \\ 0 \\ 0 \\ 1 \end{pmatrix}.$$

(Step 4): from $\mathcal{KRL}^3$ discard $\underline{c}_1^3$ because $k = i = 1$, $\underline{c}_2^3$ because $R_2^1 = 1$, $\underline{c}_3^3$ because $k = j = 3$, $\underline{c}_4^3$ because am_4 was accepted in Step 2, and $\underline{c}_6^3$ because ID_6 is more than 2 hops away. Hence, only save $\underline{c}_5^3$ from $\mathcal{KRL}^3$. For similar arguments, save only $\underline{c}_5^4$ from $\mathcal{KRL}^4$.

(Step 5): since the number of saved columns for the same node ID_5 equals the required minimum, i.e. $d_5 = 2 \geq \varepsilon = 2$, continue with majority vote. The two saved column vectors $\underline{c}_5^3$ and $\underline{c}_5^4$ for ID_5 are used to update the corresponding column vector $\underline{c}_5^1$ and complete the update of $\mathcal{KRL}^1$, i.e.

$$\underline{c}_5^3 = \begin{pmatrix} 1 \\ - \\ 0 \\ 1 \\ 0 \\ 0 \end{pmatrix}, \underline{c}_5^4 = \begin{pmatrix} 0 \\ - \\ 1 \\ 1 \\ 0 \\ 1 \end{pmatrix}, \underline{c}_5^1 = \begin{pmatrix} 0 \\ - \\ 1 \\ 1 \\ 0 \end{pmatrix}, \mathcal{KRL}^1 = \begin{pmatrix} ID_1 \, 0\,0\,0\,0\,0\,0 \\ ID_2 \, 1\,1\,0\,1\,1\,- \\ ID_3 \, 0\,0\,0\,0\,0\,1 \\ ID_4 \, 0\,0\,0\,0\,0\,1 \\ ID_5 \, 0\,0\,0\,1\,1\,0 \end{pmatrix}.$$

3.3 Key Renewal

The presented IBC revocation scheme for MANETs needs to be complemented by a key renewal algorithm to enable a node ID_i to obtain a new key pair (Q_i, d_i) after its public key expired, or was revoked by a harakiri message or δ accusation messages. In any case, a node needs to access the off-line KGC for key renewal. When doing so, the node must re-authenticate itself to the KGC using some credentials that identify the node. An off-line KGC cannot distinguish between malicious nodes whose keys have been revoked because of bad behavior or honest nodes whose keys have been compromised. Therefore, malicious nodes can always request new keys once their old keys have been revoked due to malicious behavior. Note that malicious nodes are acting under their true identities and thus successfully authenticate themselves to the KGC. To restrict the power of such malicious nodes, we choose a maximum version number v_{max}, i.e. the number of key renewals for the same expiry date is restricted. Clearly, a node that requests more than v_{max} key renewals is either malicious or not able to appropriately protect its key data.

Upon receiving a new key pair and re-joining the network, a node ID_i only needs to broadcast its new public key Q_i' to the m-hop neighborhood, if ID_i received new keys with a version number $v_i' > 1$. The receivers of Q_i', update the version number in their revocation lists accordingly and set all accusation values for Q_i' to zero. In all other cases, the node does not need to inform other nodes about its new keys. This a based on the fact that at each new expiry interval t', all new public keys Q' are assumed to have $v' = 1$ and the accusation values of these new keys are all set to zero, until new accusations are received for Q'.

4   Security Analysis

We assume that the underlying IBC scheme including the pre-shared keys from (1) are secure [1,2] and limit our analysis to the introduced key revocation and key renewal schemes. In the revocation scheme, trust is based on monitoring one-hop neighbors. Node ID_i who trusts a neighbor ID_j, also trusts that this neighbor properly monitors its own neighbors and maintains a correct revocation list. The security of the key revocation scheme depends on security parameter δ, which is the threshold for key revocations. Hence, the system is secure for up to δ-1 colluding malicious nodes. Note that these malicious nodes need to remain undetected in the neighborhood scheme for a successful attack. The security parameter ε protects honest nodes from false accusations that were made by nodes that are more than one but maximally m hops away. In the proposed revocation scheme, nodes do not trust nodes that are not in their direct communication range and a majority vote is used for accusations from these distant nodes. Only if at least $d = \frac{\varepsilon}{2}$ different sources for an accusation agree in their values, an accusation made by distant nodes is accepted. We would like to point out that the majority vote is computed for each accusation value separately. Consequently, a group of at least $\frac{\varepsilon}{2}$ colluding nodes can manipulate one accusation value for a node ID_i. However, δ such manipulations are necessary to revoke ID_i's key. Hence, the majority vote with parameter ε significantly reduces the propagation of false accusations.

We now briefly discuss some effects of some typical attacks in MANETs on our scheme. A more detailed analysis is in the full version of this paper. We believe that node compromise and selfish nodes are very likely in MANETs. Both can be detected in our neighborhood watch scheme. In our scheme, keys from such malicious nodes are first locally revoked by one-hop neighbors and eventually revoked by all nodes in m-hop distance. In that way malicious users who control compromised nodes are excluded from the network, because they cannot request new keys since this requires authentication to the KGC. On the other hand, selfish nodes are encouraged to participate, because otherwise they are forced to frequently renew their keys which imposes even more costs than forwarding other nodes' messages. Malicious nodes cannot simply drop or manipulate accusations against themselves, because first would be detected by the neighborhood watch scheme, and second attempt would be prevented by using majority votes for accusations. A roaming adversary ID_i may move every time its number of accusation approaches δ. In another scenario, a roaming adversaries moves to a new neighborhood more than m hops away, such that nobody has any accusation values for ID_i. However, in both scenarios with roaming adversaries, the present one-hop neighbors quickly detect ID_i's malicious behavior and locally revoke its key. With those accusations propagating to all m-hop neighbors the adversary would need to move faster than the propagation of the accusations. Hence, the power of roaming adversary for launching attack is fairly limited since they need to move fairly fast and frequently and cannot remain at the same location for a longer period of time. Malicious nodes may try to bypass the security parameter δ by fabricating δ different identities. However, this is prevented by the KGC which checks the identity of every node before issuing keys, where the security of the scheme is based on the honesty of the KGC.

5   Related Work

5.1   IBC Schemes in MANETs

Recently, two IBC schemes have been introduced for securing MANETs [4, 9]. Both papers suggest emulating a distributed on-line KGC using (k, n)-threshold schemes. As mentioned earlier, the use of threshold schemes introduces a significant communication and computational overhead to the network. The key management in both solutions is entirely self-organized by the network nodes and the authors claim that their schemes are more efficient than fully self-organized PKIs due to the efficient key management of the underlying IBC schemes. However, both solutions do not introduce key revocation and key renewing mechanisms for their schemes.

Our proposed key revocation and renewal schemes can be applied to both IBC schemes in [4, 9] to provide key revocation and key renewal in MANETs. Our schemes can be easily modified for distributed on-line KGCs, such that revocation is done by all network nodes by executing Alg.1-3 from Sect. 3.1, where the distributed on-line KGC takes over the task of key renewal using the key format from (2). Since our revocation scheme works independent of the (k, n)-threshold schemes, the solution is very efficient.

5.2   PKI Schemes for MANETs

Many PKI-based schemes for MANETs have been introduced, e.g. [3, 10, 12], where some of them use (k, n)-threshold schemes to implement distributed on-line CAs [10, 12]. In [12], it is suggested to collaboratively revoke certificates, but no algorithm is introduced. In fact, a revocation scheme in this solution would require threshold signatures, which is computationally very demanding. In [10], an accusation scheme is briefly outlined, in which each node observes their neighboring nodes for malicious behavior. Based on their observations, nodes send their signed accusations to an m-hop neighborhood. All receivers verify the accusations and update their accusation lists accordingly. If the number of accusations for one node is greater than a threshold δ, the certificate is revoked. Compared to [10], our revocation scheme uses pre-shared keys to secure accusation messages instead of signatures. Hence, our scheme is more efficient, once the pre-shared keys have been computed for the first time. Furthermore, the problem of newly joining nodes is not discussed in [10]. In our solution, new nodes can simply start the revocation scheme (Alg.1-3), whereas in [10] joining nodes need to verify accusation tables from its neighbors to learn about past accusations and revocations. This requires the verification of all received accusation values, i.e. approximately N^2 verifications for N network nodes which is clearly too demanding. Furthermore, a harakiri algorithm for nodes that want to revoke their own keys is not discussed in [10].

In [3], a certificate revocation scheme for MANETs is presented that uses an accusation scheme with threshold δ. The scheme assumes an off-line CA that issues certificates to all network nodes before they join the network. All accusations are frequently broadcasted throughout the entire network. The revocation scheme uses a weighted accusation scheme to decide wether a certificate is revoked. Here, instead of just computing the sum of accusations, the accusations are weighted according to the number of accusations a node made, how many accusations were reported against this node, etc.. When a new node joins the network, the node receives the accusation tables from all network nodes. The accusation messages in [3] are not secured at all and the authors suggest to check for inconsistencies in received accusation tables and only trust accusations from senders with sufficient trust value. Compared to [3], our scheme secures accusation messages and we provide a detailed description how majority votes can be implemented to check for inconsistencies in accusation values. Furthermore, our scheme provides scalable performance by choosing an m-hop propagation range for accusations. A harakiri algorithm for nodes that want to revoke their own keys is not discussed in [3]. Note that the weighted accusation scheme from [3] can be applied to our revocation scheme, such that weighted accusation values are used in Alg.3.1.

6   Discussions and Conclusions

In this paper, we introduced the first key revocation and key renewal schemes for IBC schemes in MANETs. The proposed key revocation and key renewal schemes can be applied to the recently proposed IBC schemes for MANETs [4, 9] which do not provide these fundamental mechanisms. Furthermore, our solution is applicable to any kind of IBC scheme in MANETs and can be easily adapted to PKI schemes in MANETs with off-line or on-line CAs. Our neighborhood watch scheme helps to detect malicious, selfish, and any other misbehaving nodes in MANETs, where all observations are securely propagated to an m-hop neighborhood.

The proposed revocation scheme is scalable in its security and performance parameters m, δ, and ε. For instance, greater m decreases the chances of a roaming adversary to remain undetected, where smaller values increase the scheme's performance with respect to bandwidth and memory space. Security parameters δ and ε prevent up to δ-1 undetected colluding one-hop neighbors and at least that many m-hop neighbors from falsely revoking keys. Our solution is very efficient due to the use of pre-shared keys to secure accusation messages instead of signatures and propagating messages to an m-hop neighborhood instead of to the entire network. Other than existing PKI solutions for MANETs, our solution provides a very efficient way for nodes to revoke their own keys. Furthermore, newly joining nodes can simply join the network and start the revocation scheme without first verifying a large number of past accusations and revocations.

References

1. D. Boneh and M. Franklin. Identity-Based Encryption from the Weil Pairing, *Advances in Cryptology - CRYPTO '2001*, LNCS 2139, pp. 213-229, 2001.
2. C. Boyd, W. Mao, and K.G. Paterson. Key Agreemet Using Statically Keyed Authenticators, *Applied Cryptography and Network Security, ACNS 2004*, LNCS 3089, pp. 248-262, 2004.
3. C. Crépeau and C.R. Davis. A Certificate Revocation Scheme for Wireless Ad Hoc Networks. *Proceedings of ACM Workshop on Security of Ad Hoc and Sensor Networks (SASN '03)*, ACM Press, isbn 1-58113-783-4, pp.54-61, 2003.
4. H. Deng, A. Mukherjee, D.P. Agrawal. Threshold and Identity-based Key Management and Authentication for Wireless Ad Hoc Networks, *International Conference on Information Technology: Coding and Computing (ITCC'04)*, vol. 1, pp. 107-115, 2004.
5. K. Hoeper and G. Gong. Short paper: Limitations of Key Escrow in Identity-Based Schemes in Ad Hoc Networks, *Security and Privacy for Emerging Areas in Communication Networks (SecureComm 05)*, 2005.
6. K. Hoeper and G. Gong. Identity-Based Key Exchange Protocols for Ad Hoc Networks, *Canadian Workshop on Information Theory –CWIT'05*, 2005.
7. K. Hoeper and G. Gong. Bootstrapping Security in Mobile Ad Hoc Networks Using Identity-Based Schemes with Key Revocation, *Technical Report CACR 2006-04*, Centre for Applied Cryptographic Research, University of Waterloo, 2006.
8. D.B. Johnson and D.A. Maltz. Dynamic Source Routing in Ad Hoc Wireless Networks. *Mobile Computing*, vol. 353, chapter 5, pp. 153–181. Kluwer Academic Publishers, 1996.
9. A. Khalili, J. Katz, and W.A. Arbaugh. Toward Secure Key Distribution in Truly Ad-Hoc Networks, *Proceedings of the 2003 Symposium on Applications and the Internet Workshops (SAINT'03 Workshops)*, IEEE Computer Society, pp. 342-346, 2003.
10. H. Luo, P. Zerfos, J. Kong, S. Lu, and L. Zhang. Self-Securing Ad Hoc Wireless Networks, *Seventh IEEE Symposium on Computers and Communications (ISCC '02)*, 2002.
11. C.E. Perkins, E.M. Royer, and S.R. Das. Ad Hoc On Demand Distance Vector (AODV) Routing. IETF Internet draft, Internet Draft (draft-ietf-manet-aodv-09.txt), November 2001. Work in Progress.
12. L. Zhou and Z.J. Haas. Securing Ad Hoc Networks, *IEEE Network Journal*, vol. 13, no. 6, 1999, pp. 24-30.

Self-monitoring Security in Ad Hoc Routing

J.A. Ploskonka and A.R. Hurson

The Pennsylvania State University, Department of Computer Science and Engineering
University Park, Pennsylvania 16802
{ploskonk, hurson}@cse.psu.edu

Abstract. Ad hoc networks are useful in a variety of applications, particularly in areas where creating a network infrastructure is not cost effective or when timeliness is important, such as military or rescue operations. The sensitivity of these uses for ad hoc networks makes routing security an important consideration. Unlike traditional networks, in ad hoc networks each node participates in routing functions. A malicious node could introduce serious disruptions in the network by advertising false routing information or not participating in packet forwarding. We introduce modifications to the proposed AODV-S routing protocol to reduce energy consumption associated with the security solution. Both protocols are simulated and the performance is compared with AODV. These results are presented and analyzed. Additionally, the application of ad hoc networks and the proposed AODV-SMS protocol in an information sharing environment is discussed.

1 Introduction and Motivation

In this paper, we consider security in ad hoc networks. An ad hoc network is an autonomous system of mobile nodes connected by wireless links where a fixed infrastructure is not required. Rapid and unpredictable topology changes are characteristic of ad hoc networks. In an ad hoc network, each node acts as a router and the nodes must work together in order to communicate. Ad hoc networks are useful in a variety of applications, particularly in areas where creating a network infrastructure is not cost effective or when timeliness is important. For example, an ad hoc network could be used to establish communication in a military battlefield where no infrastructure exists or for rescue workers in an area where the infrastructure was destroyed by a terrorist attack or natural disaster [1]. It should be noted that two emerging computing environments – pervasive computing and sensor networks – share many common issues with ad hoc networks. Some of the challenges faced in the design of ad hoc networks include resource constrained devices, high error rates and other limitations of wireless communication media, security, and privacy. Power is the limiting factor as battery technology has relatively changed very little in the past decade.

Routing security in ad hoc networks is an important consideration. Unlike traditional networks, in ad hoc networks each node participates in routing functions. A malicious node could introduce serious disruptions in the network by advertising false routing information or not participating in packet forwarding. Since resources (especially bandwidth and power) are limited in ad hoc networks, intelligent attackers could exploit these limitations to drain a node's power supply.

T. Kunz and S.S. Ravi (Eds.): ADHOC-NOW 2006, LNCS 4104, pp. 238–251, 2006.

In this paper, we discuss proposed security solutions for ad hoc networks and present modifications to the proposed AODV-S protocol to reduce the power consumption associated with the security solution. We selected AODV-S because of its self-organized approach to security, which is unique among the proposed solutions. Additionally, this protocol focuses on both functions of the network layer, routing and packet forwarding; whereas many of the other proposed protocols address only routing. We then discuss simulation results comparing the performance of AODV, AODV-S, and the proposed AODV-SMS. The remainder of this paper is organized as follows. Section 2 presents background information on ad hoc networks, routing in ad hoc networks, and security. Section 3 introduces the proposed AODV-SMS protocol and discusses its application in an information sharing environment. Simulation results are discussed in Section 4. Finally, Section 5 presents conclusions and discusses future work.

2 Background

2.1 Ad Hoc Networks

In ad-hoc networks, a collection of wireless mobile nodes form a temporary network without the aid of centralized administration or standard support services. Information is available at many participating nodes and is no longer owned by trusted servers. The network consists of wireless mobile nodes grouped together to cooperatively perform network functions, such as packet forwarding. Each node acts as a router and must trust its neighbors to behave correctly. Mobile nodes within each other's radio range communicate directly via wireless links, while nodes that are physically far apart must rely on other nodes to relay data packets. Ad hoc networks provide a reasonable solution for applications where installing an infrastructure is not feasible [3].

Ad hoc networks are constrained by mobility and problems associated with the wireless medium. Topology changes are frequent in ad hoc networks and nodes have limited power supplies. Wireless links are less reliable than wired links and are subject to eavesdropping, signal interface, and jamming [4]. The capacity of wireless links is considerably less than wired links and the effects of noise, interference, and other propagation properties reduce the bandwidth further. Additionally, the effects of the medium access control (MAC) protocol must be considered.

2.2 Routing

The routing approaches used in wired networks are not appropriate in wireless ad hoc networks because of frequent topology changes, resource limitations of the mobile nodes, and limited bandwidth on wireless channels. Proposed ad hoc routing protocols fall into two main categories – table-driven and source-initiated on-demand [5]. Like routing protocols used in wired networks, table-driven routing protocols maintain current routing information for every other node in the network. Source-initiated on-demand routing protocols only discover and maintain entries in a routing table when a source node requires a route to some destination. Performance studies have demonstrated that the source-initiated on-demand protocols AODV [8][9] and DSR behave better than table-driven approaches, particularly in cases when the overall node mobility is very high [6] [7].

2.3 Security

Ad hoc networks are commonly used in applications where the information is particularly sensitive and timely, such as military operations. Therefore, security is crucial in these types of networks. Designing a secure ad hoc network is challenging because of the wireless link vulnerabilities, the distributed and dynamic nature of ad hoc networks, and the network size. Nodes in ad hoc networks rely on each other for packet forwarding and routing, but the environment is inherently insecure because no centralized server can be assumed. On the other hand, network redundancies can be advantageous and trust can be distributed among many nodes.

In general, security introduces processing, storage, and communication overhead. Both the devices and the wireless transmission media of ad hoc networks are resource constrained and unable to support the overhead required for traditional security solutions. Additionally, unlike wired networks, ad hoc networks lack a fixed infrastructure. If the ad hoc network is connected to a backbone network, the traditional mechanisms for authentication, authorization, and access control can be used for end-to-end communications. However, these approaches will require modification to operate effectively in a wireless environment. For example, key distribution schemes suitable in wired networks may not work well in a wireless environment [27] due to limitations introduced by the wireless transmission media, the relatively resource poor mobile devices, and the unstable network topology.

A variety of security extensions to AODV and DSR have been proposed [10][11], including Authenticated Routing for Ad hoc Networks (ARAN) [12], Security Aware Ad Hoc Routing (SAR) [4], Ariadne [13], Cooperation of nodes, Fairness in Dynamic Ad hoc NeTworks (CONFIDANT) [14], Watchdog and Pathrater [3], and Secure AODV (SAODV) [15].

AODV-S [16] is similar to the CONFIDANT and Watchdog and Pathrater approaches in that it uses neighbor monitoring to detect misbehaviors. This is a self-organized approach that does not require an *a priori* trust relationship between the nodes and has limited reliance on a centralized trusted server. In AODV-S legitimate nodes observe the actions of its neighbors with respect to routing updates and packet forwarding. When a node detects a misbehaving node, it broadcasts a Single Intrusion Detection (SID) packet. Relying on the observations of a single node is inaccurate and could allow an attacker to make false accusations against a legitimate node. An "m out of N" collaborative monitoring strategy improves the accuracy of neighbor monitoring and prevents a single attacker from effectively isolating other nodes. A node is only considered malicious if m of its N neighbors broadcast SID packets against it. When a node, v_x, receives m independent SID packets against a single node, it broadcasts a Group Intrusion Detection (GID) packet. A Token Revocation (TREV) packet is constructed by the first node that receives k GID packets against the same node. The TREV packet is signed by the system secret key and perceived attackers are isolated in the network as the TREV packet is propagated in the network. Like the Watchdog and Pathrater approach, nodes operate in promiscuous mode. We chose to study this approach further because the self-organized approach closely matches the nature of ad hoc networks. Reliance on a central server represents a single point of failure and creates additional overhead. Threshold cryptography requires multiple nodes to sign certificates, which provides some degree of protection against false accusations and colluding attackers. For further information on threshold cryptography and techniques useful in ad hoc networks, please see [17][18][19].

3   AODV-SMS

3.1   AODV-S Modifications

One of the stated goals of AODV-S is to decrease, over time, the security solution overhead when the network is in good condition [16]. This is accomplished by using an additive algorithm to determine the Expiration_Time field when re-issuing tokens. The neighbor monitoring mechanism requires each node to constantly monitor its neighbors, but operating in promiscuous mode has high energy costs [20]. To reduce this overhead, we propose a monitoring scheme that allows nodes to perform monitoring tasks for shorter periods of time if the estimated threat is low. Energy and security are generally tradeoffs – achieving a secure solution will not necessarily result in an energy efficient solution and will likely introduce energy overhead. Similarly, a highly energy efficient routing protocol may not produce the most secure system. Therefore, our goal is to maintain a high level of security while reducing the energy costs associated with neighbor monitoring. Our approach does not require additional control messages to determine when a node should monitor its neighbors. We rely on the TRL (Token Revocation List) to estimate the current threat and determine the monitoring level accordingly.

This scheme reduces the energy consumption required to perform neighbor monitoring when the network experiences limited attacks based on the number of tokens on the TRL. Let *NRT* be the number of tokens on the TRL. Time is divided into monitoring periods of length t_{mon}. If the perceived threat is HIGH, each node will monitor its neighbors for the entire period. This monitoring level corresponds to the approach adopted by AODV-S. At the beginning of each time period, each node determines how many tokens are currently on its TRL (*NRT*) and finds the appropriate monitoring level. We selected the threshold secret sharing parameter, k, as the parameter used to determine the monitoring level because every node in the system already has knowledge of this value. Once the monitoring level is determined, each node independently selects a uniformly distributed random time to begin monitoring and turns off promiscuous processing until the start of its monitoring period. Any time a node receives a TREV packet, it automatically switches to the highest monitoring level. Another potential approach for responding to attacks is to progress from the current monitoring level to the next highest monitoring level as the number on tokens on the TRL increases. The method we selected is the most paranoid, causing us to react quickly to any sign of an attack.

In small networks with fewer than 100 nodes, values of k will typically fall between 3 and 5 [19]. We will assume a small network for our discussion based on previous simulation studies of ad hoc networks. When the number of monitoring levels, *l*, is large relative to *k*, ambiguity between monitoring levels may occur. Additionally, the number of overlapping monitoring periods will decrease, thus increasing the potential for an attacker to remain undetected. Therefore, the number of monitoring levels must be selected with knowledge of *k*.

In our simulations, we consider three threat levels (HIGH, MEDIUM, and LOW) (Table 1). Using only two threat levels would result in two extremes (always monitoring or infrequently monitoring) while using more than three levels would unnecessarily increase the complexity of our approach.

Table 1. AODV-SMS Monitoring Levels

Monitoring Level	NRT	Monitoring Time (per node)	Start of Monitoring Time
HIGH	$NRT \geq \frac{2}{3}k$	t_{mon}	$rand(0, \frac{2}{3}t_{mon})$
MEDIUM	$\frac{1}{3}k \leq NRT < \frac{2}{3}k$	$\frac{2}{3}t_{mon}$	$rand(0, \frac{1}{3}t_{mon})$
LOW	$NRT < \frac{1}{3}k$	$\frac{1}{3}t_{mon}$	0

Our approach is a tradeoff between energy efficiency and accurate monitoring. With a LOW threat estimate, energy is saved by reducing the number of nodes monitoring at one time. When operating in a HIGH monitoring level, our solution has the same energy costs as AODV-S and has the greatest potential to effectively isolate attackers in the network. No additional control packets are required to achieve this tiered monitoring approach since each node makes monitoring decisions based on its own TRL

3.2 Application in an Information Sharing Environment

Portable computing devices are commonplace today and users demand access at all times, regardless of where they are. Within the context of an information sharing environment originally designed for traditional networks, there exists demand to expand user access to wireless networks. Here we present an application for AODV-SMS within such a paradigm.

The *Summary Schemas Model* (SSM) [21] is an adjunct to multidatabase language systems that supports automatic identification of semantically similar/dissimilar data entities. The model maintains a hierarchical meta-data based on access terms exported from underlying local databases. This meta-data is used to resolve name differences using word relationships defined in a standard dictionary such as Roget's Thesaurus. The SSM consists of three major components: a thesaurus, the local nodes, and the summary-schemas nodes. The thesaurus defines a set of access terms, namely global terms, the semantic categories they belong to, and their semantic relationships. A local node is a physical database containing real data. A summary-schemas node is a logical database that contains a meta-data called *summary schema*, which represents the concise and abstract contents of its children's schemas. Fewer terms are used to describe the information in a summary schema than the union of the terms in the input schemas while capturing the semantic information of the input schemas, and therefore, reducing the schema size at higher SSM levels. Relative to other multidatabase solutions, the SSM is a robust approach that preserves local autonomy, provides high performance, and has good scalability.

The SSM platform was extended to include mobility and wireless communication [22] and the authors of [23] proposed an extension of the work to meet user demands for "anywhere, anytime" information access. Broadening the scope of authorized SSM to accommodate mobility and wireless communication is challenging and adds to existing time and space overhead. The main goal is to allow the clients to be mobile in nature and allow the local databases to be either mobile or stationary in nature. Allowing clients and/or local databases to operate in an ad hoc manner is reasonable because of the limited range of wireless access points. Within the ad hoc portion of the network, AODV-SMS can be used to perform energy-efficient secured routing functions. Using AODV-SMS ensures that inappropriate network layer behavior (advertising incorrect routes, dropping packets, excessively duplicating packets, and occupying an unfair share of the bandwidth) is isolated from the network. An ad hoc node can join the network by either obtaining its initial token and secret share from a trusted server or by performing the localized self initialization procedure [24]. Once a node is able to communicate within the ad hoc network, the authorization mode from [23] can be used to establish secure communication channels for database queries. To tie the ad hoc network to the wireless LAN, the mobile nodes within the wireless LAN will require tokens so these nodes can send packets to ad hoc nodes. The server could issue these tokens to nodes within the wireless LAN with a period of validity determined based on the level of trust the server has in a given node. If a wireless LAN node were to move to the ad hoc network, it could renew its token like any other node in the ad hoc portion of the network. If the node remained in the wireless LAN, it would need to renew its token with the server. Using AODV-SMS to perform routing functions adds minimal overhead to the SSM structure since a central server is only required to issue initial tokens and secret shares as well as performing token renewals for nodes within the wireless LAN. If localized self-initialization is employed, the central server only needs to distribute k tokens and secret shares. To achieve a more secure solution, the central server should be used to issue initial tokens and secret shares for all nodes in the ad hoc network. The latency between issuing a request and receiving a reply in an SSM structure extended with an ad hoc network will increase compared to a wired SSM structure or a wired/wireless SSM structure. A request from a wired or traditional wireless client satisfied by an ad hoc local database will experience delays due to route discovery and communication latency if the database has moved far from the wireless access point (multiple communication hops). The delay experienced by an ad hoc client will vary greatly depending on the current topology of the ad hoc network. If the local databases are not ad hoc, this latency will only be experienced by nodes in the ad hoc network making requests since the remainder of the SSM structure remains unchanged.

4 Performance Evaluation

4.1 Simulation Environment

The ns2 [25] simulation package was used to evaluate the performance of AODV, AODV-S, and AODV-SMS. Ns2 is a discrete event simulator designed for network research that supports a wide range of protocols in both wired and wireless networks. The wireless extensions to ns2 were developed by the CMU Monarch project [6].

The implementation of AODV included in the ns2 package was used without modification in this study. HELLO messages were used for local connectivity, as opposed to using link layer detection, to allow a fair comparison among the three protocols. AODV-S used the included implementation of AODV as a base and incorporates the modifications described in [16] and AODV-SMS includes the extensions discussed in Section 3. The implementations of AODV-S and AODV-SMS were thoroughly tested to verify the correct behavior of sending and receiving each message type.

The simulation parameters (Table 2) were selected based on published studies of ad hoc routing protocols [3][26]. For each of the pause times, three movement patterns were generated. For AODV-SMS, the secret share was updated halfway during each simulation to allow the updated shares to be used. We also considered AODV-SMS when secret share updates were not performed as the simulation period was short compared to a realistic update frequency of several hours [17]. Note that running ns2 simulations for this increased time period is impractical due to the size/space limitations of ns2.

Two cases were considered. In Case I, we considered the performance of the protocols without attackers present in the network. In Case II, the impact of attackers on the network was evaluated. Case II attackers were defined as nodes that participate in the routing protocol but drop all data packets received for forwarding. The percentage of attackers ranged from 0% to 40% of the nodes in the network, in increments of 10%. Each group of attackers builds on the previous group; for example, to form a case with 20% attackers, the attackers from the 10% group and an additional set of attackers are used.

Table 2. Simulation Parameters

Simulation time	900 s
Number of nodes	50
Network dimensions	300m x 600m
Node pause times	0, 30, 60, 180, 300, 600, 900 s
Maximum node speed	20 m/s
Number of traffic sources	10
Packet size	512 bytes
Data rate	4 packets/s

4.2 Performance Metrics

The performance of the three protocols was compared using three metrics:

- **Throughput:** The percentage of data packets arriving at their destination,
- **Normalized routing load:** The number of routing packets sent per received data packet, and
- **Energy consumption:** The energy consumption per node due to routing protocol communication.

The first two metrics are similar to those used in previous simulation studies [26]; the third uses Equations 1 and 2 and the appropriate values for the coefficients m and b [20] to determine the energy costs associated with routing protocol communication. In these equations, m represents the incremental cost associating with sending or receiving a packet. This cost is proportional to the packet *size* in bytes. There is also a fixed cost associated with sending or receiving packets that depends on a number of factors and will vary based on the type of communication (broadcast, point to point, or promiscuous receipt).

$$E_{recv} = m_{recv} \times size + b_{recv\{bcast,p2p,non-dest_{(S,D)}\}} \tag{1}$$

$$E_{send} = m_{send} \times size + b_{send\{bcase,p2p\}} \tag{2}$$

Normalized routing load is an effective way to compare the efficiency of the routing protocols. In the case of AODV-S and AODV-SMS, routing packets include all packets required for token renewal and neighbor monitoring. Energy consumption takes into account the cost of sending and receiving routing protocol packets. The cost of neighbor monitoring (receiving packets promiscuously) is also considered since this is an integral part of both security-enhanced routing protocols.

4.3 Simulation Results

4.3.1 Case I – No Attackers

In this case, we evaulate the performance of AODV-S and AODV-SMS relative to AODV without the presence of attackers in the network. In general, our AODV results are similar to the findings in [26], although we used HELLO messages for local connectivity resulting in slightly lower throughput and higher routing load than the published study.

Each of the protocols experiences improved throughput (Figure 1) as mobility decreases because fewer routing changes are occurring and therefore less packets are dropped while the network responds to the topology change. Note that the rise in throughput corresponds with a decrease in routing load (Figure 2). At high mobility levels, the security enhanced protocols see a 17-27% drop in throughput compared to AODV. This performance degradation is due to false accusations occurring more often at higher mobility levels because the topology is changing frequently and neighbor monitoring is not as accurate. Additionally, the overhead of the security solution exaggerates the impact mobility has on the underlying AODV protocol. As node movement decreases, the security enhanced protocols performance improves in terms of throughput and fewer nodes are accused of misbehavior. With respect to

AODV-SMS, when secret share updates are performed the throughput is impacted (approximately a 5% reduction). If secret share updates were performed once halfway through a live implementation with a duration on the order of hours [17], we expect that the effect of the updates on throughput would be less pronounced.

For the security enhanced protocols, we observe that more neighbor monitoring packets (i.e. false accusations) are sent at the higher mobility levels indicating that the neighbor monitoring scheme is impractical in the presence of a highly mobile network. Each of the protocols appears to have some difficulty with the high mobility cases with the increase and decrease in routing packets. Further examination reveals that this increase directly correlates with the number of RREQ packets. At pause times of 180 seconds and greater, the percentage of routing packets consistently decreases for each of the protocols, indicating that the behavior in the high mobility cases is related to the rapidly changing network topology, false node accusations, and other limitations of the wireless transmission medium. As the network topology changes less frequently, neighbor monitoring results in fewer false accusations since there is less network congestion (fewer routing packets being sent) and more stability overall. At the lower mobility levels, the security enhanced protocols still require more routing packets than AODV because of the token renewal process. Token renewal represents a significant portion of the routing packets sent by the security enhanced protocols. Note that AODV-SMS without the use of Secret Share updates has the same routing load as AODV-S. When Secret Share updates are used with AODV-SMS, the routing load is higher because of the additional communication required to update the shares.

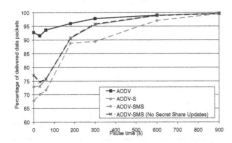

Fig. 1. Throughput – no attacks

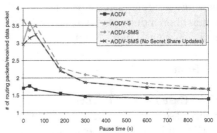

Fig. 2. Normalized Routing Load – no attacks

The energy consumption of AODV-SMS is significantly (on average 40%) better than AODV-S (Figure 3). For each of the protocols, the energy consumption decreases with decreasing mobility. AODV and AODV-S experience this decrease because fewer routing packets are being sent. Both models of AODV-SMS experience a further decrease in energy consumption after the 60 second pause time case because the nodes are in a lower monitoring level (either LOW or MEDIUM) for a longer period of time since fewer false accusations (i.e. Neighbor Monitoring packets) are sent.

Based on these findings, we will only consider the cases of 300, 600, and 900 seconds pause times in Case II – we have shown that AODV-S and AODV-SMS perform poorly in the presence of high mobility. Also, we will only consider AODV-SMS

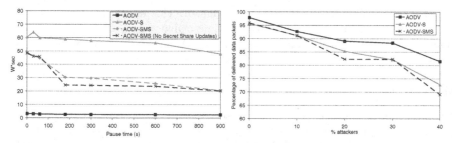

Fig. 3. Energy Consumption – no attacks   **Fig. 4.** Throughput – packet drop attackers – pause time: 300s

without secret share updates as we have shown that introducing the share updates results in additional overhead over the simulation period but the behavior follows the same trend as the protocol without the use of share updates.

4.3.2 Case II – Packet Drop Attackers

4.3.2.1 300 second pause time. In this case, AODV saw an approximately 20% decrease in packet throughput (Figure 4) as the number of attackers in the network increased, compared to a roughly 25% decrease experienced by the security enhanced protocols. In terms of routing load, AODV experienced a 22% increase with increasing attacks (Figure 5) while the security enhanced protocols saw 44% (AODV-S) and 51% (AODV-SMS) increases. The security enhanced protocols experienced this additional overhead as a result of alerting the network upon detection of a misbehaving node and attempting to route around the attacker. In the 20% and 40% attacker cases, AODV-SMS had a 3% lower throughput than AODV-S due to a higher routing load at these points.

The energy required (Figure 6) by AODV-S decreases with increasing attacks because AODV-S monitors all overheard packets and fewer packets are sent as the number of attackers increases. On the other hand, AODV-SMS requires more energy as the number of attackers increases because the protocol operates using a tiered monitoring approach, which causes nodes to monitor overheard packets for a greater

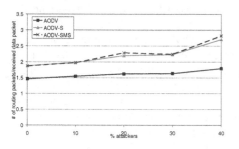

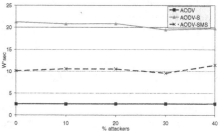

Fig. 5. Normalized Routing Load – packet drop attackers – pause time: 300s

Fig. 6. Energy Consumption – packet drop attacks – pause time: 300s

period of time when the perceived threat is higher. This monitoring approach uses 49% less energy on average than AODV-S while maintaining a level of performance in terms of throughput and routing load similar to AODV-S. The energy required by AODV stays relatively constant as the attacks increase.

Our results for the 600 second pause time case were as expected; for brevity, we will not include them here.

4.3.2.2   900 second pause time. In the 900 second pause time case, all three protocols experience a decrease in throughput (Figure 7) with increasing attacks, with AODV-S performing best, followed by AODV-SMS and AODV. In the 0-30% attacker cases, AODV-SMS performs well – exceeding the throughput of AODV and beating AODV-S in the 20% and 30% attacker cases. In the 40% attacker case, AODV-SMS experiences an increase in routing load (Figure 8), causing the drop in throughput observed in Figure 7.

We observe in Figure 8 again that normalized routing load is a good indicator of behavior in terms of throughput – when AODV-SMS has a lower routing load than AODV-S, it also sees a higher throughput.

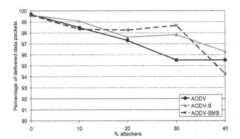

Fig. 7. Throughput – packet drop attackers pause time: 900s

Fig. 8. Normalized Routing Load – packet drop attacks – pause time: 900s

In terms of energy consumption (Figure 9), AODV-SMS uses on average 50% less energy than AODV-S. With increasing attacks, AODV-SMS uses more energy for reasons discussed earlier. AODV uses a constant amount of energy with increasing attacks. AODV-S requires slightly more energy in the 10%-30% attacker cases when the throughput remains high and less in the 40% case where AODV-S experienced a drop in throughput.

AODV-SMS exhibits similar performance to AODV-S in terms of throughput and routing load while consuming 50% less energy than AODV-S. This shows that the tiered monitoring approach is a reasonable way to reduce the energy consumption associated with the security solution without sacrificing performance otherwise. The approach to detecting an attacker employed by the security enhanced protocols may not be optimal, as evidenced by the throughput being worse than AODV in several cases (most notably in cases where the network topology is changing frequently). We also observed that the increase in normalized routing load with increasing attacks associated with the security enhanced protocols as compared with the increase seen in AODV is a good indicator of the protocol's throughput behavior. If the security enhanced protocol(s) had an increase in routing load approximately similar to AODV,

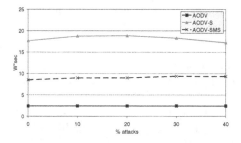

Fig. 9. Energy Consumption – packet drop attacks – pause time: 900s

the throughput was similar. In the 300 second pause time case, the security enhanced protocols experienced a much larger increase in routing load compared to AODV and were not able to maintain a throughput similar to AODV. However, in the 600 and 900 second pause time cases, AODV-S and AODV-SMS required an increase in routing load similar to AODV and the security enhanced protocols saw throughputs that met or exceeded that of AODV.

5   Conclusions and Future Work

Resource limitations and complications introduced by the mobile nature of ad hoc networks make security a difficult goal to achieve. To improve the energy efficiency of AODV-S, we introduced several protocol changes. Our proposed AODV-SMS protocol used a tiered monitoring approach – when attackers are detected in the network, each node monitors constantly; as attackers are isolated, the time spent monitoring decreases. To improve the security of the network over time, we employ secret share updates.

The security enhanced protocols performed poorly in terms of routing load and throughput in the presence of high mobility. We discovered the percentage of routing packets required for token renewals was significant and believe this is the major reason for the poor performance. At lower mobility levels, the security enhanced protocols performance is more inline with AODV with an acceptable level of overhead.

In the presence of packet drop attackers, the security enhanced protocols performance met or exceeded the throughput of AODV in cases where the topology changes were minimal. We also made the following observations:

- The number of neighbor monitoring packets (SID, GID, and TREV) increased with the percentage of attackers.
- AODV-S required less energy as the attacks increased since fewer packets were being sent. AODV-SMS used more energy as the percentage of attackers increased.
- On average, AODV-SMS required 50% less energy than AODV-S.
- AODV-SMS exhibited similar performance to AODV-S in terms of throughput and normalized routing load while using significantly less energy.

This effort has shown that AODV-S can be improved in terms of energy efficiency by employing a tiered monitoring approach. The performance of AODV-SMS could certainly be improved to make it a reasonable solution to the problem of security in ad hoc networks. Here we offer several suggestions on areas of improvement to the protocol:

- To improve the behavior of the security enhanced protocols in the presence of packet drop attackers, a different threshold detection mechanism should be employed.
- A major source of routing overhead in AODV-S and AODV-SMS came from token renewals. Reducing this overhead would likely have a large impact on the performance of the security enhanced protocols.
- AODV-S and AODV-SMS require the use of periodic HELLO messages (rather than link layer connectivity) to maintain a list of trusted neighbors. HELLO messages require significant overhead so developing an approach that eliminates the usage of HELLO messages and finds a different way to establish a trust relationship between neighbors (by verifying a node's token) could also dramatically improve the performance of the secure protocols.
- Enhance SSM by adding support for mobile ad hoc nodes using AODV-SMS.

Acknowledgments

This work in part is supported by Office of Naval research under the contract N00014-02-1-0282 and the National Science Foundation under the contract IIS-0324835.

References

1. Tanenbaum, A.S. *Computer Networks*. 4th ed. Prentice Hall, 2003.
2. Corson, S. and J. Macker. "Mobile Ad Hoc Networking (MANET): Routing Protocol Performance Issues and Evaluation Considerations." IETF MANET RFC2501, January 1999.
3. Marti, S., T. Giuli, K. Lai, and M. Baker. "Mitigating routing misbehavior in mobile ad hoc networks." In *Proceedings of the Sixth annual ACM/IEEE International Conference on Mobile Computing and Networking*, pages 255--265, 2000.
4. Yi, S. P. Naldurg, R. Kravets. "Security-Aware Ad Hoc Routing for Wireless Networks." Poster presentation at ACM Symposium on Mobile Ad Hoc Networking and Computing, October 2001. University of Illinois at Urbana-Champaign Technical Report UIUCDCS-R-2001-2241, August 2001.
5. Royer, E.M. and C-K Toh. "A Review of Current Routing Protocols for Ad Hoc Mobile Wireless Networks." *IEEE Personal Communications*, April 1999.
6. Broch, J. D. A. Maltz, D. B. Johnson, Y. C. Hu, and J. Jetcheva. "A Performance Comparison of Multi-Hop Wireless Ad Hoc Network Routing Protocols." In Proceedings of the ACM/IEEE MobiCom, October 1998, pp. 85-97.
7. Manzoni, P. and J.-C. Cano. "A Performance Comparison of Energy Consumption for Mobile Ad Hoc Network Routing Protocols." In *Proceedings of the 8th International Symposium on Modeling, Analysis and Simulation of Computer and Telecommunication Systems*, 2000, pp. 57-64.

8. Perkins, C.E. and E.M. Royer. "Ad hoc On-Demand Distance Vector Routing." In *Proceedings of the 2nd IEEE Workshop on Mobile Computing Systems and Applications*, February 1999, pp. 90-100.
9. Perkins, C.E., E. Belding-Royer, and S. Das. "Ad Hoc On-Demand Distance Vector (AODV) Routing." IETF MENET RFC3561, July 2003.
10. Molva, R. and P. Michiardi. "Security in Ad Hoc Networks." In *Proceedings of Personal Wireless Communications*, September 2003, Venice, Italy.
11. Law, Y.W., S. Dulman, S. Etalle and P. Havinga. "Assessing Security-Critical Energy-Efficient Sensor Networks." TR-CTIT-02-18, Department of Computer Science, University of Twente, 2002.
12. Sanzgiri, K., B. Dahill, B.N. Levine, C. Shields, and E. M. Belding-Royer. "A Secure Routing Protocol for Ad Hoc Networks." In *Proceedings of the 10th IEEE International Conference on Network Protocols*, 2002.
13. Hu, Y.-C., A. Perrig, and D. B. Johnson, "Ariadne: A secure on-demand routing protocol for ad hoc networks," In The 8th ACM International Conference on Mobile Computing and Networking, September 2002.
14. Buchegger , S. and J.-Y. Le Boudec. "Nodes bearing grudges: Towards routing security, fairness, and robustness in mobile ad hoc networks." In 10th *Euromicro Workshop on Parallel, Distributed and Network-based Processing*.
15. Zapata, M.G. and N. Asokan. "Securing Ad Hoc Routing Protocols." In *Proceedings of the ACM Workshop on Wireless Security*. September 2002.
16. Yang, H., X. Meng, and S. Lu. "Self-Organized Network-Layer Security in Mobile Ad Hoc Networks." In *Proceedings of the ACM Workshop on Wireless Security*. September 2002.
17. Luo, H., P. Zerfos, J. Kong, S. Lu, and L. Zhang. "Self-securing Ad Hoc Wireless Networks." In *The 7th IEEE Symposium on Computers and Communications*, 2002.
18. Shamir, A. "How to Share A Secret." Communications of the ACM, November 1979.
19. Kong, J., P. Zerfos, H. Luo, S. Lu, and L. Zhang. "Providing Robust and Ubiquitous Security Support for Mobile Ad Hoc Networks." In *Proceedings of IEEE ICNP*, 2001.
20. Feeney, L., and M. Nilsson. "Investigating the Energy Consumption of a Wireless Network Interface in an Ad Hoc Networking Environment." In *Proceedings of IEEE INFOCOM* Anchorage, AK, 2001.
21. Hurson, A. R. and M.W. Bright. Multidatabase Systems: An advanced Concept in Handling Distributed Data. Advances in Computers, vol. 32, 1991.
22. Lim, J. B. and A. R. Hurson. "Transaction processing in a mobile, multi-database environment." Multimedia Tools and Applications. 15(2): 161-185, 2001.
23. Haridas, H. A.R. Hurson, and Y. Jiao. "Security Aspects of Wireless Heterogeneous Databases – Protocol, Performance, and Energy Analysis." In *Proceedings of the 2003 International Conference on Parallel Processing Workshops*, pp. 417-424, October 2003.
24. Kong, J., P. Zerfos, H. Luo, S. Lu, and L. Zhang. "Providing Robust and Ubiquitous Security Support for Mobile Ad Hoc Networks." In *Proceedings of IEEE ICNP*, 2001.
25. <http://www.isi.edu/nsnam/ns/>
26. Das, S.R., C.E. Perkins, and E.M. Royer. "Performance Comparison of Two On-demand Routing Protocols for Ad Hoc Networks." In *Proceedings of the IEEE Conference on Computer Communications* (INFOCOM), Tel Aviv, Israel, March 2000, pp. 3-12.
27. A. Khalili, J. Katz, and W.A. Arbaugh. Toward Secure Key Distribution in Truly Ad-Hoc Networks. In IEEE Workshop on Security and Assurance in Ad-Hoc Networks, 2003.

Improved Pairing Protocol for Bluetooth

Dave Singelée and Bart Preneel

ESAT-COSIC, K.U. Leuven, Belgium
Dave.Singelee@esat.kuleuven.be

Abstract. The Bluetooth wireless technology realizes a low-cost short-range wireless voice- and data-connection through radio propagation. Bluetooth also has a security architecture. In this paper, we focus on the key agreement protocol, which is the most critical part of this security architecture. Several security flaws have been identified within the Bluetooth protocols: an attacker can track users by monitoring the Bluetooth hardware address, all keys depend on a low-entropy shared secret (the PIN), there are some very easy to perform Denial of Service attacks. We propose a new initialization mechanism for the key agreement protocol of Bluetooth. This improved pairing protocol can be easily extended so that it will not only solve the dependency of the keys on the PIN, but also the location privacy problem and an important Denial of Service attack. Our solution is user friendly and energy-efficient, two essential features for Wireless Personal Area Networks (WPAN).

Keywords: Bluetooth, security, pairing.

1   Introduction

In February 1998, the *Bluetooth Special Interest Group (SIG)* [1] was founded by some major players in the telecommunications and network industries: Ericsson, IBM, Intel, Nokia and Toshiba. In the next 6 years, several other companies joined the SIG and now there are already more than 3000 members. The major task of this organization was the creation of the Bluetooth specification which describes how mobile phones, computers, PDAs, headsets and other mobile devices can communicate over a wireless link with each other. In June 2000, the Bluetooth standard was included in *IEEE 802.15* [2], the *Wireless Personal Area Network (WPAN) Working Group*.

The Bluetooth wireless technology [3,4] realizes a low-cost short-range wireless voice- and data-connection through radio propagation. With a normal antenna, the maximal range is about 10 m. The Bluetooth wireless technology uses the 2.4 GHz band, which is unlicensed, and can be used by many other types of devices. Any device designed for use in an unlicensed band should provide robustness in the presence of interference, and the Bluetooth wireless technology has many features to achieve this, including *spread spectrum and frequency hopping*. Every time a Bluetooth wireless link is formed, it is within the context of a piconet. A piconet consists of maximally 8 devices that occupy the same physical channel.

T. Kunz and S.S. Ravi (Eds.): ADHOC-NOW 2006, LNCS 4104, pp. 252–265, 2006.

In each piconet, there is one *master*, the other devices are called *slaves*. The theoretical maximum bandwidth is 1 Mbit/s. The real bandwidth is lower because of forward error correction. One of the main differences between Bluetooth and some other wireless technologies is the ability to connect different types of devices (e.g., a mobile phone with a PDA).

We now briefly discuss the security architecture of Bluetooth and give an overview of its main security weaknesses. Next, a new initialization mechanism (also called *pairing* in the Bluetooth standard) is proposed. This improved pairing protocol solves the dependency of the keys on the PIN. Our main contribution is the extension of the protocol, which will make Bluetooth robust to a critical Denial of Service attack and which addresses the location privacy problem.

2   The Bluetooth Key Agreement Protocol

The key agreement protocol [5] is a crucial part of the security architecture of Bluetooth [6]. Suppose that two Bluetooth devices, called A and B, want to communicate securely (in the rest of this paper, we will assume that A initiates the communication). Initially these devices do not share a secret. They perform a key agreement protocol to generate a *link key* and an *encryption key*. The latter is fed to the stream cipher E_0. The process of generating a shared secret is called *pairing* (two Bluetooth devices are paired when they share a key which can be used to communicate securely).

2.1   Generation of the Unit Key

When a Bluetooth device is turned on for the first time, it calculates a *unit key*. This is a key that is unique for every device and that is almost never changed. It is stored in non-volatile memory. The unit key is only used if one of the devices does not have enough memory to store session keys (see also Sect. 2.4 for more details). It is generated as follows: first, the device computes a random number *RAND*. The unit key is based upon this random number and the *Bluetooth hardware address* (which is a factory-established parameter unique for every device).

2.2   Generation of the Initialization Key

At the start of a communication session, the Bluetooth devices do not yet share a session key. This will be achieved in different steps. First, an *initialization key* is generated. This temporary key is a function of a random number *IN_RAND* (generated by A and sent to B in clear), a shared PIN and the length L of this PIN. The PIN should be entered in both devices. The length of the PIN can be chosen between 8 and 128 bits. Typically, it consists of 4 decimal digits. If one of the devices does not have an input interface, a fixed PIN is used (often, the default value is 0000). This procedure is shown in Fig. 1. The result is a temporary shared key (the *initialization key*). Note that a low-entropy shared secret (the PIN) is used to generate the (initialization) key. An eavesdropper which is present during initialization, will know the random number *IN_RAND*.

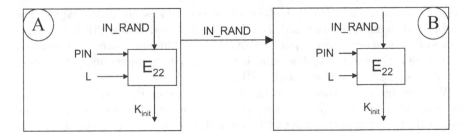

Fig. 1. Generation of the initialization key

2.3   Mutual Entity Authentication

Each time a new shared key is generated (an initialization key (§2.2) or a link key (§2.4)), both devices perform a mutual authentication protocol. The authentication scheme is based on a challenge-response protocol. This protocol is performed twice. First, B authenticates itself to A, as shown in Fig. 2. If this authentication is successful, the roles are switched (B becomes the verifier and A the prover). The authentication goes as follows. A generates a random number AU_RAND and sends this to B. This random number is called the *challenge*. Both devices now calculate a *response* $SRES = E_1(ADDR_B, K_{link}, AU\_RAND)$. $ADDR_B$ is the Bluetooth hardware address of B and K_{link} is the shared key (initialization key or link key). B sends its response to A. If this response corresponds to the value that A has calculated, then the authentication is successful. The value ACO (*Authenticated Ciphering Offset*) is used for the generation of the encryption key. Algorithm E_1 is based on the SAFER+ block cipher, with some small modifications.

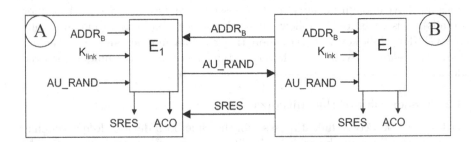

Fig. 2. Mutual entity authentication protocol

2.4   Generation of the Link Key

Both devices now share an initialization key. This key will be used to agree on a new, semi-permanent key (called the *link key*). The link key will be stored on both devices for future communication. Depending on the memory constraints of both devices, the link key can be the unit key of the memory-constrained device or a combination key derived from the input of both devices (Fig. 3).

If the unit key of device A is the link key, it is transmitted encrypted from A to B. This encryption is done by XOR'ing the unit key of A with the initialization key.

If the link key is a combination key, then both devices first generate a random number LK_RAND. These random numbers are encrypted with the initialization key and sent to the other device. Now they both calculate $LK\_K_A = E_{21}(LK\_RAND_A, ADDR_A)$ and $LK\_K_B = E_{21}(LK\_RAND_B, ADDR_B)$. The combination key K_{AB} is the XOR of LK_K_A and LK_K_B. This is shown in Fig. 3. Algorithm E_{21} is based on the SAFER+ block cipher, with some small modifications. After the generation of the link key, the (old) initialization key is definitively discarded and a mutual authentication is started with the exchanged link key that is shared between both devices (this has already been discussed in Sect. 2.3). The procedure shown in Fig. 3 is also carried out when a new link key is computed. The only difference is that the random numbers LK_RAND are encrypted with the old link key. After the generation of the new link key, the old one will be discarded.

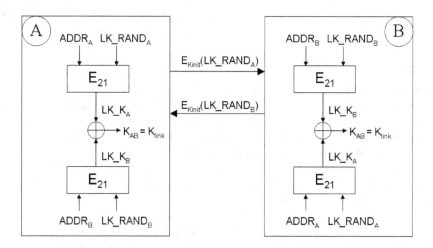

Fig. 3. The link key is a combination key

2.5 Generation of the Encryption Key and the Key Stream

After a successful generation of the link key and execution of the mutual authentication protocol, the encryption key can be generated. A generates a random number EN_RAND_A and sends this to B. Both devices generate the encryption key $K_C = E_3(EN\_RAND_A, K_{link}, COF)$. The COF value (*Ciphering Offset Number*) is the ACO value which was generated during the mutual authentication protocol. However, if the encryption key is used for broadcast, then the COF is the concatenation (denoted by $||$) of the Bluetooth hardware address of the sender with itself (so $COF = (ADDR||ADDR)$). The encryption key K_C has a length of 128 bits, but its length can be reduced to an encryption key K_C'

if necessary. Finally, the encryption key K_c (or the length-reduced key K'_C) is fed to the encryption scheme E_0 together with the Bluetooth hardware address and the clock of the master. The result is the key stream K_{cipher}. The master clock is used in order to make the key stream harder to guess.

3   Security Weaknesses

There are several security weaknesses in the Bluetooth standard [7,6]. We now give an overview of the most important security problems.

3.1   Unit Key

The unit key is used if one of the Bluetooth devices does not have enough memory to store session keys. This key is stored in non-volatile memory and almost never changed. As already described in Sect. 2.4, the unit key is sent encrypted (with the initialization key) to the other device. The result is the following weakness: if A has sent its unit key to device B, then B knows the key of A and can impersonate itself as A to a device C. This impersonation attack is impossible to detect. It is strongly recommended to avoid the use of unit keys.

3.2   Location Privacy

When two or more Bluetooth devices are communicating, the transmitted packets always contain the Bluetooth hardware address of the sender and the receiver. When an attacker eavesdrops on the transmitted data, he knows the Bluetooth addresses of these devices. The attacker does not have to be physically close to the communicating devices, (s)he can use a device with a stronger antenna (e.g., it is very easy to construct an antenna which can intercept Bluetooth communication from more than one mile away [8,9]) or just place a small tracking device near the two Bluetooth devices.

This way, the attacker can keep track of the place and time these two devices were communicating. This is a violation of the privacy of the user. The location information can be sold to other persons and used for location dependent commercial advertisements (e.g., a shop can send advertisements to everybody that is near the shop). It should be possible for the user to decide when his location is revealed and when not.

3.3   Security Depends on Security of PIN

The initialization key is a function of a random number IN_RAND, a shared PIN and the length L of the PIN. The random number is sent in clear and hence known by an attacker that is eavesdropping during the initialization phase. This means that only the PIN is a secret value. If an attacker obtains the PIN, (s)he knows the initialization key. It even gets worse: since all the other keys are derived from the initialization key, they will also be known by the attacker. Hence the security of the keys used in Bluetooth depends on the security of the PIN. If this value is too short or weak (e.g., 0000), it is very easy for an attacker to guess the PIN (and hence the initialization key).

Note that it is possible to verify a guess of the PIN. The reason is that a mutual authentication protocol is executed after the generation of the initialization key. If an attacker observes this protocol, (s)he obtains a challenge and the corresponding response. The attacker calculates for every guess of the PIN the corresponding response and when this is equal to the observed response, the guess of the PIN was correct. The shorter the PIN, the faster this brute force attack can be carried out. Shaked and Wool showed that this attack can be optimized by employing an algebraic representation of SAFER+, the cryptographic primitive used in the mutual authentication protocol [10]. The authors state that a PIN of 4 digits can be cracked in less than 0.06 seconds on a standard PC.

This is a very critical security problem. To solve it, we have to change the initialization procedure. The combination of a mutual authentication protocol and the generation of keys from low-entropy secrets will always be insecure. In the next section, we will suggest to use another pairing protocol. The goal is to improve the security, without changing the Bluetooth standard too much.

3.4   Denial of Service Attacks

Mobile networks are always vulnerable to Denial of Service (DoS) attacks. They consist of mobile devices and these devices are often battery powered. Bluetooth is no exception. An attacker can send dummy messages to a mobile device. When this device receives a message, it performs some computation, which consumes battery power [11]. After some time, all battery power will be consumed. This exhaustion of the battery power is called the *sleep deprivation attack* [12]. This attack is almost impossible to prevent.

There are also some Denial of Service attacks caused by implementation decisions. A nice example is the *black list* which is used during the mutual authentication protocol. To avoid that a device would start the authentication protocol over and over again (and eventually guess the correct PIN), each device has a black list of the Bluetooth addresses of the devices which failed to authenticate themselves correctly. These devices can not start an authentication procedure during some period. Candolin discovered that this mechanism can be exploited in several Denial of Service attacks [12]. An attacker can try to authenticate itself to device A, but change every time its Bluetooth hardware address. All these authentication attempts will fail and the black list of A will become quite large. If there is no upper limit on this black list, the entire memory of A will be filled with the entries of the black list and device A will crash. This is not the only DoS attack. Suppose device B wants to authenticate itself to A. After A has sent a challenge to B, the attacker sends a wrong response to A using the Bluetooth address of B. The authentication will fail, B will be put on the black list of A and the (correct) response of B will be ignored by A. The attacker keeps repeating this attack and B will never be able to authenticate itself successfully to A. Note that the same result could be obtained by jamming the radio signal, but the Denial of Service attacks described above are much easier to perform.

3.5   Other Security Problems

Bluetooth uses the encryption algorithm E_0. This stream cipher has some security flaws [13, 14, 15, 16, 17, 18]. A lot of attacks on E_0 are published, but most of these attacks do not work on the algorithm which implements E_0 in Bluetooth. There are however some exceptions. The fastest and most practical known-plaintext attack on the Bluetooth encryption scheme is recently found by Vaudenay [19]. The attack is based on a recently detected flaw in the resynchronization of E_0, as well as the investigation of conditional correlations in the Finite State Machine (FSM) governing the keystream output of E_0. This attack finds the original encryption key for two-level E_0 using the first 24 bits of $2^{23.8}$ frames and with 2^{38} computations.

Another security flaw is that there are no integrity checks. An attacker can always modify a transmitted Bluetooth packet. There are also some security problems in the challenge-response protocol, but this is a theoretical issue and not within the scope of this paper.

4   A Secure Initialization Mechanism for Bluetooth

One of the most important security problems of Bluetooth is the generation of the initialization key. The security depends on the security of the PIN. One could replace this weak initialization mechanism by an advanced pairing protocol. J.H. Hoepman has designed several pairing protocols that make use of an extra secure communication channel that is shared between both (Bluetooth) devices [20, 21]. This extra communication channel can be private and/or authentic. A private channel is a confidential communication channel and when two devices share an authentic channel, they know that only the other device could have sent a certain message. There are however some small practical problems. Bluetooth devices are typically energy-constrained, and cannot perform the modular exponentiations used in these pairing protocols. Moreover, in the most general scenario, the two Bluetooth devices only share one secure communication channel, being the user itself. The user can be considered as a very low-bandwidth secure channel. The secure communication channels used in the pairing protocols of Hoepman need a much higher bandwidth.

So we need a more efficient solution. Our improved pairing protocol combines the ideas of Hoepman with Elliptic Curve Cryptography (ECC) [22], temporary random addresses and Manual Authentication (MANA) protocols. In a MANA protocol, the user is considered to be a private and authentic communication channel, used to authenticate some data [23,24]. Our improved pairing protocol is efficient, user-friendly and it solves some critical security problems of Bluetooth. It is particularly designed to improve security, without renouncing the user-friendliness.

4.1   Improved Pairing Protocol

Definitions and assumptions: We use the following notation in the description of the protocol. A and B are the two Bluetooth devices that execute the pairing protocol. $\longrightarrow$ denotes sending a message using the Bluetooth technology. After the execution of the pairing protocol, the devices will share a temporary, random identifier R. When the devices are working in anonymous mode (which should be the default security mode), the headers of the packets sent via Bluetooth will not contain the Bluetooth hardware addresses. All messages (also after the execution of the pairing protocol) are broadcasted, but will contain the temporary identifier R. $\dashrightarrow$ denotes an action performed by the user (pressing a button or entering a number on the input interface of the device). The user is modeled as a private and authentic low-bandwidth communication channel. A and B will carry out the Diffie-Hellman protocol in the group of points defined by an elliptic curve $\mathcal{E}$ over a finite field [22, 25]. We will assume that the Bluetooth devices already share this elliptic curve $\mathcal{E}$ and a point P on this curve (e.g., all devices in the same piconet will share the same parameters). We recommend an elliptic curve key size of 160 bits.

$h_2(p), h_3(p)$: $\mathcal{E} \to \{0,1\}^n$ are two pairwise independent cryptographic hash functions that map a point p on the elliptic curve $\mathcal{E}$ to a n bit string. Typically, n will be equal to 128 (to be compatible with the current Bluetooth standard). In practice, both hash functions can be derived from a single hash function h using the equation $h_i(m) = h(m \parallel i)$ (where $\parallel$ denotes the concatenation of bits) [20].

$CV_k(m)$: $M \times \mathcal{E} \times K \to \{0,1\}^q$ is a check-value function that maps a message m from a message space M, a point p on an elliptic curve $\mathcal{E}$, and a key k from a key space K, to a q bit string (which is called the *check-value*). In order to provide a high level of security, the collision probability of the output of the check-value function (for a fixed key k) must be low. A key length of 16-20 bits and a check-value length of 16-20 bits is sufficient [23, 24].

$MAC_k(m)$: $\mathcal{E} \times M \times K \to \{0,1\}^n$ is a MAC (*message authentication code*) function that maps a point p on an elliptic curve $\mathcal{E}$, a message m from a message space M, and a key k from a key space K, to an n bit string. The key length of the MAC function should be 128 bits (or larger) to provide a high level of security. The NESSIE project [26] has published a list of several secure MAC algorithms.

Our improved pairing protocol is shown in Fig. 4. It consists of the following steps:

1. A and B start the pairing protocol. This is normally achieved by the user pressing a button on both devices. From this moment on, both devices are in "pairing-mode" and will not ignore communication from unknown devices anymore.
2. A chooses a random secret a and computes the point aP on the elliptic curve $\mathcal{E}$. B chooses the secret b and computes bP. A also generates a random, temporary identifier R which will replace the Bluetooth hardware addresses.

Finally, A generates a random, one-time key k and uses this key k to compute the check-value s of (aP, R). The length of k and s is around 16-20 bits (so only 4 or 5 hexadecimal digits).

3. The MANA 1 protocol is executed to authenticate the data $D = (aP, R)$ (other MANA protocols could also be used in our pairing protocol). This requires the following steps: k and s are shown on the output interface of device A and the user enters them on the input interface of device B. As both numbers consist of only 4 or 5 hexadecimal digits, this is not very cumbersome.

4. A broadcasts the data $D = (aP, R)$. All devices will receive this message, but only the devices which are in "pairing-mode" will store D in temporary memory. The other ones will just ignore this message. Note that this step could also be executed before step 3.

5. For every message $D = (aP, R)$ it receives, B computes the check-value and compares it with the value s that the user has entered in step 3. If the check-value is correct, B computes the point abP on the elliptic curve $\mathcal{E}$. This Diffie-Hellman key is mapped to an initialization key (called key in Fig. 4) and a MAC-key x. B stores the entry (R, key) in a table. This table contains the temporary identifiers and the keys used for communication with trusted (paired) devices. Next, it generates the check-value t and the MAC-value u (see Fig. 4). Finally, it broadcasts the quadruple (R, bP, t, u) using Bluetooth.

6. A waits until it receives a quadruple that contains the identifier R (all other messages are ignored) and first verifies the check-value t. When t is correct, A computes the point abP on the elliptic curve $\mathcal{E}$ and maps it to an

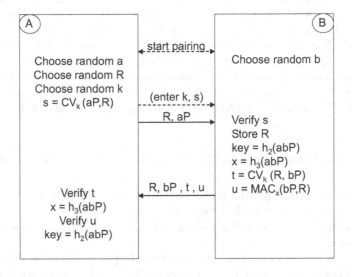

Fig. 4. Our improved pairing protocol

initialization key (called *key* in Fig. 4) and a MAC-key x. Finally, after successful verification of the MAC-value u (using the key x), the entry (R, key) is stored in a table used for communication with trusted (paired) devices. Note that the verification of the check-value t is not really necessary. However, it prevents a Denial of Service attack. An attacker could send random messages containing the identifier R and force A to perform many point-multiplications on the elliptic curve $\mathcal{E}$. Such a point-multiplication has an average energy cost of 300 mJoules [11]. This attack would hence exhaust the battery power of A very fast. That is why the Diffie-Hellman protocol is only executed after the verification of the check-value t.

7. The devices A and B are now paired and share a temporary identifier R and an initialization key. Next, they will perform the mutual authentication protocol and generate a link and encryption key (see Sect. 2). Communication going from A to B (or vice versa) will contain R (and no Bluetooth hardware addresses). The identifier R is meant to be used permanently as the new hardware address (i.e. not just for key agreement but also for all data packets).

4.2   Evaluation

We will now evaluate the (security) properties of our pairing protocol and show how it solves the aforementioned security problems of Bluetooth.

When the Bluetooth devices are working in anonymous mode (which should be the default security mode), **location privacy** is not an issue anymore. After a successful execution of the pairing protocol, the two (paired) Bluetooth devices share a temporary, random identifier R. All transmitted data packets contain this identifier R, and no other identification data. It is impossible for an attacker to link R to the Bluetooth hardware addresses of the sender and/or receiver of a Bluetooth packet. To avoid tracking, R is updated regularly (e.g., every time the two devices generate a new link key). The length of R should be 48 bits (to be compliant with the Bluetooth standard [4]). As R is chosen randomly, the probability of a collision is very small. Replacing the fixed Bluetooth hardware address by a temporary random identifier R has however some important consequences. The generation of the hopping sequence, which normally uses the Bluetooth hardware address of the master device as input, has to be modified. Any input can be used, as long as it is publicly known by all devices in the neighborhood (e.g., something that is broadcast during the inquiry phase). As this problem is not security related, we will not focus on it anymore.

The **security of the keys no longer depends on the security of the PIN**. The two devices carry out a Diffie-Hellman protocol on an elliptic curve and map this point to an n bit string (which will be the initialization key). The consequence is that the strength of the initialization key depends on the security properties of the Diffie-Hellman protocol on an elliptic curve and the hash function that maps a point on this curve to a bitstring (e.g., one could take the x-coordinate of the point and then apply a hash function). By carefully selecting a collision resistant hash function and an elliptic curve with good security properties, it will be computationally very hard for an attacker to obtain the initialization key (and hence the other keys). The reason is that the elliptic curve discrete logarithm problem (on a "good" curve) is a computationally difficult problem [22, 25]. Therefore, our pairing protocol improves the overall security dramatically. The same level of security can not be obtained by using a PIN to generate a session key, even if the PIN is updated frequently.

In our scheme, the PIN is replaced by a short key k and a check-value s. Both are relatively short (4 or 5 hexadecimal digits). So our pairing scheme is almost as **user-friendly** as the original pairing protocol used in Bluetooth.

An attacker obtaining k and s can perform a *substitution attack*. This **man-in-the-middle-attack** is done by first searching for a data string $D' = (a'P, R')$ that satisfies the property $CV_k(a'P, R') = CV_k(aP, R)$. The attacker then substitutes the message $D = (aP, R)$ (sent in step 4 of the pairing protocol) by $D' = (a'P, R')$ and hence forces B to compute the point $a'bP$. Finally, the attacker generates a random value b', calculates $t' = CV_k(R, b'P)$ and $u' = MAC_{x'}(b'P, R)$, and sends the quadruple $(R, b'P, t', u')$ to A, which will compute the point $ab'P$. As a result of this substitution attack, the attacker shares a key $h_2(ab'P)$ with A and a key $h_2(a'bP)$ with B. However, both devices think that they share an initialization key with each other. When k and s have a length of 16 bits, the probability of such a substitution attack to succeed is less than 2^{-12}. A length of 20 bits decreases the probability of a successful attack to 2^{-17}. Note that an attacker cannot verify a correct guess of k and s. These values are only used to authenticate the date D. So this substitution attack is less critical than the PIN guessing attack of the original pairing protocol of Bluetooth (see Sect. 3)!

The use of a blacklist was included in the Bluetooth security architecture to avoid repetitive authentication attempts in which an attacker can verify a guess of the PIN. After a relatively short amount of time, the attacker would find the correct PIN. This was avoided by the use of a blacklist. The problem is that this list can be abused in a **Denial of Service attack**. In our pairing protocol, the attacker can not verify a guess of the short key k and the check-value s. So we no longer need the blacklist. Other Denial of Service attacks are still possible (e.g., the sleep deprivation attack), but this can not be prevented in mobile ad-hoc networks.

Our pairing protocol is not only user-friendly, it is also quite **energy efficient**. The most energy consuming cryptographic operation in the pairing protocol, is carrying out the Diffie-Hellman protocol in the group of points defined by an elliptic curve. One of the advantages of working in this group, is the fact that it is a factor 3 more efficient than performing (standard) Diffie-Hellman in the multiplicative group $\mathbb{Z}_p^*$ [11, 27] ($\mathbb{Z}_p^* = \{a \mid 1 \leq a \leq p - 1\}$ with p a large prime). If one increases the key length, elliptic curve cryptography becomes even more efficient compared to standard Diffie-Hellman. Another advantage of elliptic curve cryptography is a shorter key length (160 bits, which is equivalent to standard Diffie-Hellman [28] with a key length of 1024 bits [29]). However, it can still be a very demanding operation for mobile devices with limited resources. Fortunately, the pairing protocol is only executed the first time two Bluetooth devices want to communicate securely. So if the mobile device with limited capacities does not have to make too many new (secure) connections, this will not form a problem. Besides, the cost of computation power decreases every year (devices get more powerful). Symmetric cryptography would be a more efficient solution, but the price one would have to pay, is user-friendliness (if one wants to achieve a high level of security), unless there would be extra communication channels available with medium or high bandwidth. In that case, other pairing protocols could be used to securely initialize a Bluetooth connection [30, 20, 21]. Our protocol assumes that at least one device has an input interface (device B) and one device an output interface (device A). However, some Bluetooth devices do not have an input/output interface (e.g., wireless headset). In this case, one can not use the user as a secure communication channel and other pairing protocols have to be used. Unfortunately, this probably requires some hardware changes (e.g., the presence of an infrared port on the device).

5   Conclusion

We have given a short overview of the key agreement protocol in Bluetooth. This is the most important part of the security architecture. Unfortunately, there are several security flaws in the Bluetooth standard. We focussed on three critical problems: the ability of an attacker to track users by monitoring the Bluetooth hardware address, the fact that all keys depend on a low-entropy shared secret (the PIN) and the Denial of Service attack that exploits the use of a black list. Recently, the new Bluetooth standard (version 2.0) was published, but unfortunately all security problems are still present in this new standard.

In this paper, we proposed a new initialization mechanism for the key agreement protocol of Bluetooth. This improved pairing protocol can be easily extended so that it will not only solve the dependency of the keys on the PIN, but also the location privacy problem and the *black list attack*. Our solution takes into account that Bluetooth devices are often memory- and energy-constrained. It combines the ideas of other pairing protocols, Elliptic Curve Cryptography, temporary random addresses and Manual Authentication protocols. Another important feature of our protocol is its user-friendliness. The user only has to enter two small numbers (4 or 5 hexadecimal digits). Our solution is particularly designed to improve security, without renouncing the user-friendliness. If there are extra communication channels with medium or high bandwidth available, or if the devices do not have an input/output interface, other pairing protocols could be used as well to securely initialize a Bluetooth connection.

Acknowledgments

Dave Singelée is funded by a research grant of the Institute for the Promotion of Innovation by Science and Technology in Flanders (IWT). This work was also supported by the Concerted Research Action (GOA) Ambiorics 2005/11 of the Flemish Government.

References

1. Bluetooth Special Interest Group. http://www.bluetooth.com/.
2. IEEE 802.15, the Wireless Personal Area Network Working Group. http://www.ieee802.org/15/.
3. J. Haartsen, M. Naghshineh, J. Inouye, O. Joeressen, and W. Allen. Bluetooth: Visions, Goals and Architecture. In *ACM Mobile Computing and Communications Review*, pages 38–45, 1998.
4. Bluetooth Specification. https://www.bluetooth.org/spec/.
5. G. Lamm, G. Falauto, J. Estrada, and J. Gadiyaram. Security Attacks against Bluetooth Wireless Networks. In *Proceedings of the 2001 IEEE Workshop on Information Assurance and Security*, pages 265–272. U.S. Military Academy, West Point, NY, June 2001.
6. D. Singelée and B. Preneel. Security Overview of Bluetooth. COSIC Internal Report, 2004.
7. M. Jakobsson and S. Wetzel. Security Weaknesses in Bluetooth. In *Proceedings of the Cryptographer's Track at the RSA Conference (CT–RSA '01)*, Lecture Notes in Computer Science, LNCS 2020, pages 176–191. Springer-Verlag, 2001.
8. DEF CON. Computer Underground Hackers Convention. http://www.defcon.org.
9. Humphrey Cheung.s The Bluesniper Rifle, 2004. http://www.tomsnetworking.com/Sections-article106.php.
10. Y. Shaked and A. Wool. Cracking the Bluetooth PIN. In *Proceedings of the 3rd International Conference on Mobile Systems, Applications, and Services (MobiSys '05)*, pages 39–50. ACM, June 2005.
11. A. Hodjat and I. Verbauwhede. The Energy Cost of Secrets in Ad–Hoc Networks. In *Proceedings of the IEEE Workshop on Wireless Communications and Networking (CAS '02)*, 2002.
12. C. Candolin. Security Issues for Wearable Computing and Bluetooth Technology. http://www.tml.hut.fi/~candolin/Publications/BT/, 2000.
13. C. De Cannière, T. Johansson, and B. Preneel. Cryptanalysis of the Bluetooth Stream Cipher. COSIC Internal Report, 2001.
14. N. Courtois and W. Meier. Algebraic Attacks on Stream Ciphers with Linear Feedback. In *Advances in Cryptology - EUROCRYPT 2003*, Lecture Notes in Computer Science, LNCS 2656, pages 345–359. Springer-Verlag, 2003.
15. S. Fluhrer and S. Lucks. Analysis of the E0 Encryption System. In *8th Annual International Workshop of Selected Areas in Cryptography (SAC 2001)*, Lecture Notes in Computer Science, LNCS 2259, pages 38–48. Springer-Verlag, 2001.
16. J. Golic, V. Bagini, and G. Morgari. Linear Cryptanalysis of Bluetooth Stream Cipher. In *Advances in Cryptology - EUROCRYPT 2002*, Lecture Notes in Computer Science, LNCS 2332, pages 238–255. Springer-Verlag, 2002.
17. M. Hermelin and K. Nyberg. Correlation Properties of the Bluetooth Combiner Generator. In *Proceedings of the 2nd International Conference on Information Security and Cryptology (ICISC '99)*, Lecture Notes in Computer Science, LNCS 1787, pages 17–29. Springer-Verlag, 1999.
18. F. Armknecht and J. Lano and B. Preneel. Extending the Resynchronization Attack. In *Proceedings of the 11th Annual International Workshop of Selected Areas in Cryptography (SAC 2004)*, Lecture Notes in Computer Science, LNCS 3357, pages 19–38. Springer-Verlag, 2004.

19. Y. Lu, W. Meier, and S. Vaudenay. The Conditional Correlation Attack: A Practical Attack on Bluetooth Encryption. In *Advances in Cryptology - CRYPTO 2005*, Lecture Notes in Computer Science, LNCS 3621, pages 97–117. Springer-Verlag, 2005.
20. J. H. Hoepman. The Ephemeral Pairing Problem. In *Financial Cryptography*, Lecture Notes in Computer Science, LNCS 3110, pages 212–226. Springer-Verlag, 2004.
21. J. H. Hoepman. Ephemeral Pairing on Anonymous Networks. In *Proceedings of the Second International Conference on Security in Pervasive Computing (SPC05)*, Lecture Notes in Computer Science, LNCS 3450, pages 101–116. Springer-Verlag, 2005.
22. A. J. Menezes. *Elliptic Curve Public Key Cryptosystems*. Springer, July 1993.
23. C. Gehrmann and K. Nyberg. Security in Personal Area Networks. In *Security for Mobility*, pages 191–230. IEEE, 2004.
24. C. Gehrmann, C. Mitchell, and K. Nyberg. Manual Authentication for Wireless Devices. *RSA Cryptobytes*, 7(1):29–37, 2004.
25. A. J. Menezes, P. C. van Oorschot, and S. A. Vanstone. *Handbook of Applied Cryptography*. CRC Press, October 1996.
26. New European Schemes for Signatures, Integrity, and Encryption. http://www.cryptonessie.org.
27. N. Potlapally, S. Ravi, A. Raghunathan, and N. Jha. Analyzing the Energy Consumption of Security Protocols. In *Proceedings of the 2003 International Symposium on Low Power Electronics and Design (ISLPED '03)*, pages 30–35. ACM, August 2003.
28. W. Diffie and M. Hellman. New Directions in Cryptography. In *IEEE Transactions on Information Theory*, pages 644–654, 1976.
29. National Security Agency. The Case for Elliptic Curve Cryptography. http://www.nsa.gov/ia/industry/crypto_elliptic_curve.cfm.
30. D. Balfanz, D. Smetters, P. Stewart, and H. Wong. Talking to Strangers: Authentication in Adhoc Wireless Networks. In *Proceedings of the Network and Distributed System Security Symposium (NDSS 2002)*. The Internet Society, 2002.

Threats to Security in DSRC/WAVE

Christine Laurendeau and Michel Barbeau

School of Computer Science, Carleton University
1125 Colonel By Drive, Ottawa, ON Canada K1S 5B6
Tel.: 613-520-2600; Fax: 613-520-4334
{claurend, barbeau}@scs.carleton.ca

Abstract. Dedicated Short Range Communications (DSRC) enabled
road vehicles are on the brink of actualizing an important application
of mobile ad hoc networks. It is crucial that the messages exchanged
between the vehicles and between the vehicles and specialized infrastructure be reliable, accurate and confidential. To this end, we propose
to identify the security threats inherent in the emerging DSRC Wireless
Access in Vehicular Environments (WAVE) architecture. We rank the
identified threats according to the European Telecommunications Standards Institute's (ETSI) threat analysis methodology. We also discuss
possible countermeasures to the most critical threats.

1 Introduction

As the emerging field of vehicle communications is poised to become the technology of the next decade, it becomes more imperative that the architecture and
infrastructure upon which vehicular networks are based be reliable and secure.
With new vehicles being outfitted with on-board equipment which essentially
renders each one into a mobile device capable of communicating with other vehicles' equipment and with fixed roadside stations, drivers are able to call upon
a veritable panoply of new and exciting applications. In addition to gaining access to an unending stream of navigational and localized information available
through Location-Based Services (LBSs), vehicles can benefit from a tremendous
increase in traffic safety when they receive real-time notification of upcoming
road hazards, imminent collisions and movements of emergency vehicles. Dedicated Short Range Communications (DSRC) [1] and its wireless component,
Wireless Access in Vehicular Environments (WAVE) [2] [3] [4] [5], provide an
architecture for vehicular networks.

At present, there is a dearth of discussion on security issues pertaining to
vehicle communications, although some issues have been addressed. In [6], Raya
and Hubaux identify threats to the traffic messages exchanged between vehicles
in Vehicular Adhoc NETworks (VANETs). The authors propose a security architecture featuring the use of digital signatures and multiple anonymous key
pairs to deal with those threats. As well, Blum and Eskandarian [7] offer an
outline of some security threats, including jamming, impersonation and fabrica-
tion, as they pertain to traffic messages. The authors outline an architecture for

T. Kunz and S.S. Ravi (Eds.): ADHOC-NOW 2006, LNCS 4104, pp. 266–279, 2006.

maintaining a VANET's integrity and functionality through the use of vehicular clusterheads managing access to the network.

Given the private nature of the application information exchanged and potential for attackers to wreak havoc in the vehicular network, there is a compelling need to identify and address the most severe security threats specific to the DSRC/WAVE architecture. We endeavour to present here an analysis of those threats using the European Telecommunications Standards Institute's (ETSI's) [8] methodology, where identified threats can be ranked as critical, major or minor depending on their likelihood of occurrence and impact on the user or the network.

Section 2 presents an overview of DSRC/WAVE's architecture as put forth in IEEE draft standards. Section 3 outlines the methodology used to rank the threats we uncover. Section 4 provides an analysis of the identified threats along with their risk assessments. Section 5 discusses possible countermeasures to the critical threats identified in Section 4, and Section 6 concludes the paper.

2   DSRC/WAVE Architecture

DSRC was conceived to provide an architecture for nodes within a vehicular network to communicate with each other and with the infrastructure in a secure and efficient manner. In DSRC, subsequently specialized as WAVE, GPS-enabled vehicles are equipped with On-Board Units (OBUs) which can communicate with each other to propagate safety warnings through Vehicle-to-Vehicle (V2V) communications. As well, OBUs have access to infrastructure devices called Road Side Units (RSUs) which are located intermittently along city streets and highways to provide access to a variety of services and applications, such as LBSs and wireline access, through Vehicle-to-Infrastructure (V2I) communications. Some OBUs may also be configured to provide service access to other OBUs through the *OBU to Vehicle Host Interface* (OVHI). Additionally, a special class of OBUs, called Public Safety OBUs (PSOBUs), are entrusted with special capabilities. These vehicles, which include police cars, fire trucks and ambulances, may operate as OBUs or RSUs, as circumstances dictate.

DSRC/WAVE operates in the 5.9 GHz band and has 75 MHz of bandwidth allocated for vehicle communications, which are based on line of sight with a range of up to 1 km and vehicle speeds of up to 140 km/h. While the standards upon which the WAVE architecture is based are still in development, a basic framework appears to have stabilized as outlined below.

2.1   Protocol Layers

The general WAVE architecture, along with the standards on which its different layers are based, can be found in Figure 1. The physical (PHY) and Medium Access Control (MAC) layers implement the IEEE P802.11p [2] standard. One of the additions in this standard consists of the use of dynamic MAC addresses to identify the OBUs in a somewhat anonymous manner. These MAC addresses are initialized upon startup, then set to a new designation whenever a message is received from a nearby node with a duplicate address.

The IEEE P1609.4 Multi-Channel Operation standard [5] describes how WAVE's MAC layer operates with the available spectrum allocated as one control channel (CCH) and four to six service channels (SCHs). Traffic safety messages are broadcast on the CCH, as are announcements of available services. Subsequent communications between an OBU and a service provider occur on a SCH. However, the OBU is required to periodically monitor the CCH for incoming announcements.

The IEEE P1609.3 Networking Services standard [4] outlines the WAVE architecture at the network layer. Messages may be conveyed using one of two protocols: the Internet Protocol version 6 (IPv6) for transaction applications involving LBSs for example, or the custom WAVE Short Messages Protocol (WSMP) for broadcast applications which require low latency for delivering crucial safety information.

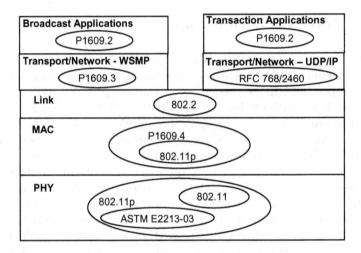

Fig. 1. DSRC/WAVE Protocol Layers and Standards

2.2   Security Measures

The IEEE P1609.2 standard [3] describes the measures designed to ensure that the messages exchanged in WAVE are secure.

Broadcast Messages are untargeted and usually related to traffic safety applications. They are broadcast using WAVE Short Messages (WSMs). Given the non-sensitive nature of the information carried in WSMs, these messages are not encrypted but merely signed with the sender's certificate. Every signed message includes the current timestamp, obtained from the OBU's internal clock which is synchronized with the GPS time. This timestamp is used by the receiver to verify the message against a cache of recently received messages to ensure against replay attacks. The signature algorithm specified in the standard is the Elliptic

Curve Digital Signature Algorithm (ECDSA) [9]. An example of a traffic application using broadcast messages is *platooning*, where vehicles can be organized as a convoy to increase capacity on major roads by having a lead OBU broadcast its changes in vehicle speed so that the OBUs behind it match its speed.

Transaction Messages are generally unicast and sent using the IP stack to an application running on a service provider host via either a RSU or another OBU over an OVHI. Because these messages may be used to access LBSs and personal data may be exchanged, they are encrypted with a symmetric encryption algorithm, such as the Advanced Encryption Standard in CCM mode (AES-CCM) [10], using a random key. This random key is in turn encrypted using the asymmetric encryption algorithm Elliptic Curve Integrated Encryption Scheme (ECIES) [11].

In order for vehicles to sign messages, they must possess a digital certificate. The standard recommends that procedures for PSOBU and RSU certificate enrollment include a manual component, but OBU certificate enrollment procedures have not yet been put forth. When signing a message, a vehicle may also include a *certificate chain*, i.e. the certificate that authorized the vehicle's certificate, as well as the certificate that authorized the certifying certificate, and so on, up to a *root certificate* which is issued by a higher authority such as a governmental agency responsible for licensing vehicles. Upon receipt of a signed message, each node must verify that it recognizes the root certificate used to authorize the sender's certificate. The node must also ensure that none of the certificate chain's members have been revoked by comparing each one against a Certificate Revocation List (CRL). These CRLs and the collection of valid root certificates are stored locally on each node and must be periodically updated.

3   Analysis Methodology

In 2003, the European Telecommunications Standards Institute (ETSI) developed a methodology for analyzing security threats to its meta-protocol, Telecommunications and Internet Protocol Harmonization Over Networks (TIPHON) [8]. This methodology allows for identified threats to be ranked in terms of risk, using estimated values for the likelihood of occurrence and impact upon the user or system.

The *likelihood* of occurrence of the threat indicates whether theoretical and practical knowledge is available for attackers to carry out an attack. Three levels of likelihood are defined with an associated numerical value: *Likely* (3) – all elements in place; *Possible* (2) – some elements in place; *Unlikely* (1) – important elements missing. Although the *impact* of a threat has no bearing on whether an attack occurs, it can indicate if the threat is serious enough to warrant further research into possible countermeasures. The values associated with the impact are the following: *High* (3) – serious consequences for the user or network; *Medium* (2) – short-term outages; *Low* (1) – minor consequences for the user or network. The *risk* is calculated as the product of the numerical values of the likelihood and impact. The categories in which the risk is deemed to fall are defined as:

Critical (9,6) – countermeasures must be devised without delay; *Major* (4) – the threat will eventually require attention; *Minor* (3,2,1) – the threat can be ignored in the short term.

In [12], we used the definitions provided in [8] to further break down the likelihood component into its two natural components: the technical difficulty in carrying out the threat and the motivation or potential gain on the part of the attacker for him or her to proceed. We make use of this idea here in order to fine-tune the risk assessment process. The values for technical *difficulty* can be defined in terms of whether or not the threat has previously been considered in theory or in practice: *None* – a precedent for the attack exists; *Solvable* – the attack is theoretically possible; *Strong* – theoretical elements missing. The levels for *motivation* include: *High* – significant gains for attacker; *Moderate* – service disruption only; *Low* – no significant gains. The technical difficulty and motivation associated with a given threat can be used with its impact to determine the risk assessment, as depicted in Table 1.

Table 1. Risk Assessment

Motivation	Difficulty	Likelihood	IMPACT		
			High	Medium	Low
High	None	Likely	Critical		Minor
	Solvable				
Moderate	None				
	Solvable	Possible		Major	
Low	Any	Unlikely	Minor		
Any	Strong				

4   Threat Analysis

In our analysis, we focus on the most basic security attributes to be preserved in vehicular networks: availability, authenticity and confidentiality. The collated list of threats, organized by risk category, can be found in Table 2.

4.1   Threats to Availability

Threats to the availability and consistent behaviour of the vehicular network include denial of service (DoS) attacks, the introduction of malicious software (malware) and the potentially high volume of messages introduced through spamming.

DoS. DoS attacks render the network unavailable to its users, for example by flooding the nodes with messages or by jamming signals at the physical layer. These attacks can be carried out either by network insiders turned rogue or by outsiders to the network.

Flooding. One way to incapacitate the vehicular network is to artificially generate such a high volume of false messages on the CCH that the network's nodes,

Table 2. Threat Analysis

Threat	Motivation	Difficulty	Likelihood	Impact	Risk
Black Hole, GPS Spoofing, Location Tracking	High	Solvable	Likely	High	Critical
Malware	Moderate	Solvable	Possible	High	Critical
DoS	Moderate	Solvable	Possible	Medium	Major
Masquerading, Replay, Transaction Tampering, Outsider Eavesdropping	High	Strong	Unlikely	High	Minor
Broadcast Tampering	Moderate	Strong	Unlikely	High	Minor
Spamming, Insider Eavesdropping	Moderate	None	Likely	Low	Minor

OBUs and RSUs alike, cannot sufficiently process the superfluous data. Important messages are lost as a result. Consequences may include accidents if collision warnings or platoon directives are not delivered.

Jamming. By creating interference on the CCH, an attacker can hamper message delivery, thereby compromising the traffic safety applications which depend upon it. Alternately, jamming can be used to cloak an attacker so that he or she cannot be identified.

Given that DoS represents a disruption rather than an opportunity for gain, the motivation required on the part of an attacker is rated as moderate according to the criteria provided in Section 3. The technical difficulty involved is solvable, given that it is theoretically possible. Since DoS would result in temporary outages, the impact on the network is ranked as medium, and according to Table 1, the threat is assessed as major.

Malware. The introduction of malware, such as viruses or worms, into the vehicular network has the potential to cause serious disruptions to its operation. Since the OBUs and RSUs are expected to receive periodic software and firmware updates, this threat is more likely to be carried out by a rogue insider than by an outsider. The associated motivation is ranked as moderate because it consists of a disruption in service. Since the threat is theoretically possible, the technical difficulty is a solvable one if countermeasures are not in place. The impact on the user is considered high due to the resulting long-lasting outages. As a result, the malware threat is ranked as critical.

Spamming. There is a risk in increased transmission latency due to the presence of spamming messages. The motivation for marketers to acquire a RSU or an OVHI-enabled OBU for this purpose is best rated as moderate. On one hand, it is likely to be very lucrative, but on the other hand, the business is ultimately accountable to its customers who typically resent such a waste of their time and bandwidth. With the technical difficulty rated as low since the marketer is an insider, and with the impact on the user also low because it represents little more than an annoyance, the threat is ranked as minor.

4.2   Threats to Authenticity

Ensuring the authenticity of the network includes protecting legitimate nodes from rogue insiders and outsiders infiltrating the network under an assumed identity, detecting the presence of black holes, identifying attacks replaying legitimate interactions, exposing spoofed GPS signals, and thwarting the introduction of misinformation into the network.

Masquerading. By posing as legitimate nodes in the vehicular network, outsiders can proceed to conduct more types of attacks than they otherwise could, for example forming black holes or fabricating false messages. However, given how easy it is to become part of the network by simply joining it with a working OBU, the masquerading exercise for an outsider becomes analogous to breaking a window to get into a house when the front door is wide open. There is, however, much to be gained by a rogue insider masquerading as another OBU or a RSU. By assuming a false identity, an attacker can create mischief with impunity, such as injecting false messages into the network and deceiving authorities into believing that another node was responsible. With PSOBUs possessing special privileges within the network, and RSUs providing wireline access and LBS information, spoofing such nodes can be the first step in accessing personal user information and possibly compromising privacy. However, because OBUs and RSUs can be identified by their certificate which can be distributed in Certificate Revocation Lists (CRLs) if a node turns rogue, such a deception would be difficult to successfully carry out. With the strong technical difficulty in conducting this attack, despite its high impact on the user and the network due to compromised integrity, the threat is ranked as minor.

Black Hole. A black hole is formed by nodes which fail to propagate messages. Such an attack can only be carried out by rogue insiders, since network outsiders are not expected to repeat messages. The consequences of having a black hole in the network include dropped traffic messages, service requests and replies. With sufficient numbers of rogue nodes colluding to form a black hole past which no messages are propagated, it may be possible for attackers to partition the vehicular network in such a way that legitimate nodes never receive messages. If this scenario succeeds, nodes may be prevented from receiving critical updates to their root certificate lists and CRLs, leaving them vulnerable to masquerading attacks from nodes using expired, revoked or falsified certificates. With significant gains to be made from this attack, its technical difficulty solvable and its tremendous impact on the security of the network, the threat is ranked as critical.

Replay Attack. Vehicular networks operating in the WAVE framework are protected from replay attacks by having each node maintain a cache of recently received messages against which new messages are compared. Messages older than a configurable time are discarded. The others are compared against this cache to ensure that they have not previously been received. This scheme assumes that an accurate source of time is available. The case where the clock accuracy is compromised is considered below in the section on GPS spoofing. Therefore, despite the potential gains to be made in terms of network manipulation, there are strong technical difficulties in carrying out a replay attack, rendering this a minor threat.

GPS Spoofing. By using a GPS satellite simulator to generate radio signals stronger than those received from the genuine GPS satellite, an attacker can lead nodes to believe they are in a different location than they actually are [13], potentially causing collisions. Also, if GPS time is used to timestamp messages, a spoofing of the GPS clock could result in nodes accepting expired messages as new ones and could thus lead to a successful replay attack. Given the potential gains for an attacker, the solvable technical difficulties involved in this type of attack and its high impact on the network and the users, the threat is ranked as critical.

Broadcast Tampering. It is possible that a rogue insider may attempt to inject false traffic safety messages into the network for the purpose of creating mischief, for example causing accidents by suppressing traffic warnings or manipulating the flow of traffic to clear a chosen route. Broadcast messages are meant for general consumption, but they need to be signed in order to deter attackers from generating false messages. Although a rogue insider, as a genuine, authenticated member of the network, could use its digital certificate to sign any number of false messages, it would soon find itself included on CRLs meant to black-list such nodes. The strong technical difficulty involved in carrying out this threat indicates that it is a minor one.

Transaction Tampering. Another possible threat to message integrity consists of an attacker modifying the messages exchanged in V2I in order to falsify transaction application requests or forge the associated replies. With a little imagination, a malicious agent could create an entire alternate life or lifestyle for an unsuspecting user. However, the technical difficulty involved is quite strong given that transaction messages must be encrypted as described in Subsection 2.2. As a result, the threat is minor.

4.3   Threats to Confidentiality

With the messages exchanged between the nodes of a vehicular network being accessible over the air, the threats to confidentiality include the illegitimate collection of transaction information through eavesdropping and the collection of location information available through the retransmission of broadcast messages.

Outsider Eavesdropping. Broadcast messages generally pertain to traffic safety information and are therefore uninteresting for the purposes of eavesdropping. The same cannot be said of LBS and other types of transaction application information. In accumulating a series of such service requests, a malicious agent can construct a profile of a given user by observing which services are used regularly, when, from which location and how much is spent. However, even with the potential gains for the attacker, the associated technical difficulty is quite strong, given the level of encryption required to protect this type of information, as described in Subsection 2.2. The threat is therefore ranked as minor.

Insider Eavesdropping. As long as insider nodes collect information in keeping with the terms of an agreement with the user, there is no problem. However, there is the possibility that an insider may collect information at a time when the user is unaware of the collection. For example, a traveling sales agent may have an agreement with an employer to have his or her movements tracked during business hours and not afterward. Because the impact on the user constitutes no more than an annoyance, the threat is considered minor.

Location Tracking. With the emerging potential of vehicle locations being constantly tracked, it is not difficult to imagine the temptation for attackers to exploit this new opportunity. By collecting an unsuspecting individual's location trace over time, such information can be used in tracking, stalking, or in building a potentially damaging profile of the user. Every time an OBU propagates a broadcast message to alert other vehicles to traffic safety updates, it digitally signs the repeated message with its own certificate which can identify the OBU and its current position to the receiving nodes. Given the attacker's gain in personal location information, the solvable technical difficulties associated with this threat and its tremendous impact on the user, it is ranked as critical.

5   Countermeasures and Open Issues

In our analysis in Section 4, we uncovered four critical security threats inherent in the WAVE architecture: malware, black hole, GPS spoofing and location tracking, as well as one major threat, DoS. These problems need to be addressed before a WAVE implementation can be considered secure.

Malware. According to the security considerations outlined in the informative portion of the WAVE security standard [3], received software and firmware upgrades should only be permitted if they are sent and digitally signed by nodes which have the required permission. In [14], a new security standard proposed by a large consortium of smart phone vendors is discussed. Although the details are sketchy, it appears that this standard will be based on the Trusted Platform Module (TPM) chip used in PCs and laptops [15]. This chip protects a device from malware by providing authentication and cryptography services, allowing other trusted devices to selectively access the functions available on the device. This is in essence the same approach recommended (although not mandated) in the WAVE security standard. This countermeasure is considered to be state of the art for smart phones and other mobile devices. It may be equally effective in vehicular networks, provided that it is implemented. Otherwise, entry points for malware will need to be closed through future upgrades as they are discovered.

Black Hole. Broadcast messages in WAVE are propagated by flooding. All OBUs have the same range, and we may assume that the links are symmetrical. It may therefore be feasible to detect a black hole by having the sending nodes listen to their neighbors' retransmissions to ensure that they repeat the messages. This concept is known as passive acknowledgement [16]. According to the IEEE P802.11p [2] standard, nodes periodically send Nearby Station request messages to ensure that none of their neighbors have a duplicate MAC address. If a given node sends a broadcast message and knows who its neighbors are, it can flag a neighboring node which fails to repeat messages. It can then follow the appropriate repudiation procedure, as discussed in *Rogue Repudiation* in the sequel.

More conventional black hole prevention techniques, including multipath routing and the use of backup routes, are outlined in [17]. However, WAVE has a measure of resistance to black holes by design, due to the fact that it propagates messages by flooding without any route optimization mechanisms. With messages sent to every node within range, each of which repeats to all the nodes within its range, redundancy provides resilience to black holes. Alternately, it may be feasible to have RSUs repeat traffic messages as well. Although these infrastructure nodes are vulnerable to physical tampering, the non-responsiveness of known RSUs can raise an alarm.

GPS Spoofing. The WAVE security standard [3] recommends the implementation of plausibility rules regarding changes in vehicle location, as well as the use of special calibration measures on the OBU clock so that updates to the time are performed by accelerating or decelerating the clock in a continuous manner, rather than in an abrupt, discrete fashion. Such rules offer a good basis for a countermeasure to the GPS spoofing threat. Similar plausibility-based countermeasures are offered in [13], such as profiling satellite availability, locations and signal strength in order to detect unaccountable changes to the configuration. If an OBU knows that at a given set of coordinates it can usually receive signals from four satellites, each with a specific signal strength, and it suddenly receives much stronger signals from two or seven satellites, there is a chance that a GPS satellite simulator has been set up nearby. A problem arises with this countermeasure when environmental conditions, such as the weather or ionosphere, affect the quality of the received signal. Changes in signal strength may be due to the presence of a GPS simulator or to natural phenomena.

However, a countermeasure to the GPS spoofing threat already exists. It consists of using the encrypted Precise Positioning System (PPS) military signal rather than the Standard Positioning System (SPS) civilian signal. The state-of-the-art PPS-based GPS receiver known as Selective Availability Anti-Spoofing Module (SAASM) [18] uses a combination of symmetric and asymmetric encryption, where an asymmetric scheme is used to distribute a symmetric key, as outlined in [19]. As a result, if transportation authorities are granted permission from the U.S. military to use the SAASM receiver, as some other governmental agencies have been, vehicular networks can be well protected from GPS spoofing attacks.

Location Tracking. Location tracking is an even more complex issue. An important conundrum is posed by the very nature of ad hoc vehicular networks: the balance of user privacy versus accountability. On one hand, users have the right to expect a certain level of confidentiality with regards to their identity, location and application use. The vehicular network must ultimately allow them to anonymously repeat broadcast traffic messages. On the other hand, it is imperative that malicious nodes be identified and neutralized in order to prevent serious harm to other users. This means that nodes must be uniquely and persistently identified so that attackers can be flagged over a significant distance and period of time. There is currently no mechanism in WAVE to support anonymous broadcast messages.

Much of the existing research into location privacy involves protecting users from LBS providers by blurring the user's exact location in space and time. By cloaking a user's LBS request so that it is indistinguishable from $k-1$ other requests, *k-anonymity* [20] [21] is achieved, with higher values of k resulting in greater anonymity and thus decreased granularity. One such implementation for vehicular networks is described in [22]. Unfortunately, such schemes are not applicable in traffic safety applications such as platooning and collision avoidance systems where each vehicle's precise location is of paramount importance.

DoS. DoS attacks in Internet applications and e-commerce can be mitigated by requiring sending nodes to perform a task such as solving a puzzle [23] before receiving services from a server. This idea ensures that legitimate nodes only have a simple computational chore to perform while attackers find themselves bogged down with the same task due to the sheer volume of messages sent. Ideally, DoS attacks should be mitigated at the lowest possible layer in the protocol stack. By providing link layer authentication, for example through the use of cryptographically generated MAC addresses, outsider attacks may be deterred. The concept of cryptographically generated IPv6 addresses, described in [24], may be applicable to MAC addresses in vehicular networks, although rogue insider nodes remain the greater threat. Further experiments need to be conducted to determine whether similar or other DoS prevention schemes can be applied to VANETs without adversely affecting transmission latency. It should be noted, however, that none of these countermeasures can prevent a wideband jammer from disrupting the VANET signals. Such an attack may be countered with the use of directional antennas which could allow vehicles to circumvent the jammed area.

In addition, a vehicular network bogged down with a significant load of legitimate messages becomes more vulnerable to DoS attacks. A further measure to increase WAVE's resilience to this threat would be the introduction of a routing protocol minimizing the number of broadcast messages.

Rogue Repudiation. Rogue nodes must be identified and repudiated as soon as they are detected. The WAVE security standard proposes to deal with such nodes by including their certificate identifiers on CRLs which are distributed to all other nodes in the vehicular network so that none will accept messages from a repudiated node. However, it is unclear how rogue nodes can be black-listed by the individual nodes detecting them.

It may be possible to use reputation systems such as [25] for each node to accrue good will or ill will, depending on how other nodes perceive its trustworthiness. Once a node has exceeded a pre-determined threshold for unreliability, it can be referred to some higher authority who can then black-list the sender by placing its certificate identifier on a CRL. Furthermore, once a rogue node moves out of the range of a given group of nodes, its reputation needs to follow it so that it cannot provoke another attack on a new group of unsuspecting nodes. The implementation of a rogue repudiation scheme, as well as the efficient dissemination of up-to-date CRLs to all the nodes, remain open issues.

6   Conclusions

We have identified some important threats to the security of DSRC/WAVE vehicular networks. Using the ETSI's threat analysis methodology, we have identified malware, black hole, GPS spoofing and location tracking as critical threats. The DoS attack is considered to be major.

We have outlined possible countermeasures to the critical and major threats. Malware threats can be dealt with by requiring that software and firmware upgrades be received only from authenticated nodes which possess the appropriate permissions, as outlined in the WAVE security standard. GPS simulators can be countered through the use of specialized GPS receivers accessing an authenticated signal. Passive acknowledgements may be used by sending nodes to detect black holes. Location tracking remains an open issue. DoS threats may be mitigated with the use of puzzles and directional antennas. Further work is required to simulate catastrophe situations in order to adequately assess the impact of the aforementioned attacks, most notably DoS.

It is clear that with some of these possible countermeasures and a rogue repudiation scheme still in the embryonic stages, the remaining security issues must be addressed before DSRC/WAVE vehicular networks and their applications can be fully realized.

Acknowledgements

The authors gratefully acknowledge the financial support received for this research from the Automobile of the 21st Century (AUTO21) Network of Centres of Excellence (NCE).

References

1. IEEE Vehicular Technology Society: 5.9 GHz Dedicated Short Range Communications (DSRC) - Overview. [Online]. Available: http://grouper.ieee.org.groups/scc32/dsrc/
2. IEEE 802 Committee of the IEEE Computer Society: Draft Amendment to Standard for Information Technology - Telecommunications and Information Exchange Between Systems - LAN/MAN Specific Requirements - Part 11: Wireless LAN Medium Access Control (MAC) and Physical Layer (PHY) Specifications: Wireless Access in Vehicular Environments (WAVE). Draft IEEE Standard, IEEE P802.11p/D1.1, January 2005.
3. SCC32 Committee of the IEEE Intelligent Transportation Systems Society: Draft Standard for Wireless Access in Vehicular Environments - Security Services for Applications and Management Messages. Draft IEEE Standard, IEEE P1609.2/D1.3, October 2005.
4. SCC32 Committee of the IEEE Intelligent Transportation Systems Society: Wireless Access in Vehicular Environments (WAVE) Networking Services. Draft IEEE Standard, IEEE P1609.3/D14, September 2005.
5. SCC32 Committee of the IEEE Intelligent Transportation Systems Society: Wireless Access in Vehicular Environments (WAVE) Multi-Channel Operation. Draft IEEE Standard, IEEE P1609.4/D03, April 2005.
6. M. Raya and J.-P. Hubaux: The Security of Vehicular Ad Hoc Networks. Proceedings of the 3rd ACM Workshop on Security of Ad Hoc and Sensor Networks, 2005.
7. J. Blum and A. Eskandarian: The Threat of Intelligent Collisions. IT Professional, vol. 6, no. 1, January/February 2004, 24–29.
8. ETSI: Telecommunications and Internet Protocol Harmonization Over Networks (TIPHON) Release 4; Protocol Framework Definition; Methods and Protocols for Security; Part 1: Threat Analysis. Technical Specification ETSI TS 102 165-1 V4.1.1, 2003.
9. American National Standards Institute: Public Key Cryptography for the Financial Services Industry, The Elliptic Curve Digital Signature Algorithm (ECDSA). ANSI Standard, X9.62-2005, 2005.
10. Internet Engineering Task Force: IETF Request for Comments: 3565, Use of Advanced Encryption Standard (AES) Encryption Algorithm in Cryptographic Message Syntax (CMS). IETF RFC 3565, 2003.
11. Certicom Research: Standards for Efficient Cryptography, SEC 1: Elliptic Curve Cryptography, Version 1.0. [Online]. Available: http://www.secg.org/download/aid-385/sec1_final.pdf
12. M. Barbeau: WiMax/802.16 Threat Analysis. Proceedings of the 1st ACM International Workshop on QoS and Security for Wireless and Mobile Networks (Q2SWinet), 2005.
13. J.S. Warner and R.G. Johnston: GPS Spoofing Countermeasures. The Weekly Homeland Security Newsletter, 12 December 2003. [Online]. Available: http://www.homelandsecurity.org/bulletin/121203.htm
14. N. Leavitt: Will Proposed Standard Make Mobile Phones More Secure? Computer, vol. 38, no. 12, December 2005, 20–22.
15. Trusted Computing Group: TCG TPM Specification. [Online]. Available: https://www.trustedcomputinggroup.org/groups/tpm/

16. J. Jubin and J.D. Tornow: The DARPA packet radio network protocols. Proceedings of the IEEE, vol. 75, no. 1, 1987, 21–32.
17. I. Aad, J.-P. Hubaux and E.W. Knightly: Denial of Service Resilience in Ad Hoc Networks. MobiCom '04: Proceedings of the 10th Annual International Conference on Mobile Computing and Networking, 2004.
18. J. Nielson, J. Keefer and B. McCullough: SAASM: Rockwell Collins' Next Generation GPS Receiver Design. Position Location and Navigation Symposium, IEEE 2000, March 2000, 98–105.
19. S. Callaghan and H. Fruehauf: SAASM and Direct P(Y) Signal Acquisition. GPS World, July 2002.
20. A. McDiarmid and J. Irvine: Achieving Anonymous Location-Based Services. Proceedings of the 60th International Conference on Vehicular Technology (VCT2004), vol.4, 2004.
21. B. Gedik and L. Liu: Location Privacy in Mobile Systems: A Personalized Anonymization Model. Proceedings of the 25th IEEE International Conference on Distributed Computing Systems, 2005.
22. M. Gruteser and D. Grunwald: Anonymous Usage of Location-Based Services Through Spatial and Temporal Cloaking. Proceedings of the First International Conference on Mobile Systems, Applications and Services, 2003.
23. A. Juels and J. Brainard: Client Puzzles: A Cryptographic Countermeasure Against Connection Depletion Attacks. Proceedings of the 1999 Network and Distributed Systems Security Symposium (NDSS), February 1999, 151–165.
24. T. Aura: IETF Request for Comments 3972, Cryptographically Generated Addresses (CGA), 2005. [Online]. Available: http://www.rfc-archive.org/getrfc.php?rfc=3972
25. P. Michiardi and R. Molva: CORE: A Collaborative Reputation Mechanism to Enforce Node Cooperation in Mobile Ad Hoc Networks. Proceedings of the IFIP TC6/TC11 Sixth Joint Working Conference on Communications and Multimedia Security: Advanced Communications and Multimedia Security, September 2002.

LSTOP: A Light-Weight Scalable Truthful Routing Protocol in MANETs with Selfish Nodes*

Yongwei Wang and Mukesh Singhal

Department of Computer Science, University of Kentucky
Lexington, KY 40506, USA
{ywang7, singhal}@cs.uky.edu

Abstract. We introduce Light-weight Scalable Truthful rOuting Protocol (*LSTOP*) for selfish nodes problem in mobile ad hoc networks where a node may use different cost to send packets to different neighbors. *LSTOP* is a truthful and scalable routing protocol. It incurs an overhead of $O(n^2)$ in the worst case and only $O(n)$ on the average, a substantial improvement over a recently proposed truthful routing protocol, which incurs an overhead of $O(n^3)$. *LSTOP* encourages nodes' cooperation by rewarding nodes for their forwarding service according to their cost and provides near-least-cost paths and even least paths with high probability in dense networks. We show the truthfulness of *LSTOP* and present results from an extensive simulation study to evaluate the performance of our protocol. Simulation results show that *LSTOP* approaches optimal (cost) routing, provides significantly higher packet delivery ratio, generates significantly lower control overhead, lower end-to-end delay and has better overpayment ratio compared to ad hoc-VCG.

Keywords: Routing Protocol, Mobile Ad Hoc Networks (MANETs), Selfish Nodes, Truthfulness, the VCG Mechanism.

1 Introduction

In mobile ad hoc networks (MANETs), nodes have limited transmission power and thus communication beyond radio range must rely on the forwarding help of other nodes. However, in such networks, nodes have limited resources, especially the battery power. As forwarding packets consumes more power than receiving packets or staying in idle status, a node may have no interest to spend its resource to forward packets for other nodes, especially if nodes in a network belong to different authorities. Such *selfish* behavior will result in failure of multi-hop data communication and thus it poses real threats to the proper functioning of mobile ad hoc networks.

To enforce nodes' cooperation, incentive as virtual money is introduced into MANETs. Nodes forwarding packets can get reimbursed for their cost and some bonus. Due to diversity of emitting power and that of cost of unit power, which may vary over time according to its battery status, nodes incur different cost to transmit a packet. Thus it is desirable to reimburse nodes according to their cost so that nodes get enough incentive, whereas the nodes getting forwarding services can pay these services with lowest possible price. However, to maximize their utility, selfish nodes may not expose their

---

* This research was partially supported by NSF grants IIS-0242384 and IIS-0324836.

T. Kunz and S.S. Ravi (Eds.): ADHOC-NOW 2006, LNCS 4104, pp. 280–293, 2006.

true cost. This poses the requirement for truthful protocols (strategies) to prevent such a cheating.

A protocol is *truthful* (or strategy-proof) if it maximizes the utility of nodes only when they reveal their true cost. Anderegg and Eidenbenz [1] recently proposed ad hoc VCG, the only truthful routing protocol that pays node according to their cost, for mobile ad hoc networks with selfish nodes. To collect complete existing links information, their protocol incurs a high route discovery overhead of $O(n^3)$, where n is the number of nodes in the network. Such an overhead may be prohibitively large in some networks, especially as the network size grows. Since the cost of overhead in term of control messages must also be reimbursed, controlling overhead is as important as guaranteeing truthfulness for any practical protocol. On the other hand, such routing protocol relies on broadcast approach to disseminate routing information. However, it is not practical to collect complete existing links information in the network based on broadcast messages since they are not reliable.

In this paper, we focus on finding a practical solution which rewards selfish nodes according to their cost in MANETs where a node may use different cost to send packets to different neighbors. Our protocol, named *LSTOP* (Light-weight Scalable Truthful rOuting Protocol), can guarantee the truthfulness and provide near-optimal routes (optimal routes with high probability) for data traffic in a network with sufficient high density. The most prominent feature of *LSTOP* is that it causes a low message overhead of $O(n)$ on the average and thus it is scalable. Specifically, we make the following contributions: 1) We present an algorithm that collects topology information using much lower overhead (from $O(n^3)$ to $O(n^2)$) compared to ad hoc-VCG [1]. 2) We present a scalable method to find near-optimal routing with a low overhead of $O(n)$ on the average. 3) Our protocol guarantees the truthfulness of routing in a network with sufficient high density. 4) We present an easy and accurate way to calculate payment for nodes involving in route discoveries. 5) To the best of our knowledge, we are the first to conduct extensive simulation study to evaluate important generic network metrics such as packet delivery ratio, overhead and end-to-end delay for protocols using mechanism design for selfish nodes problem. Simulation results show that *LSTOP* performs significantly better than ad hoc-VCG. It also approaches the optimal routing solution and provides better overpayment ratio over ad hoc-VCG.

The remainder of the paper is organized as follows. Section 2 reviews the related work. In Section 3, we present *LSTOP* in details, including the algorithm and analysis. In Section 4 we analyze the performance of our protocol using simulations and compare it with ad hoc-VCG. Section 5 concludes the paper.

2   Related Work

The problem of nodes cooperation in the MANETs has been an active research area [1]-[7][10]-[20]. Marti et al. [16] proposed two tools, watchdog and pathrater, to mitigate routing misbehavior (including selfish nodes) in ad hoc networks. Watchdog is used to identify malicious and selfish nodes, while pathrater is used to select a route to avoid them. Buchegger et al. [14] [15] proposed CONFIDANT protocol to monitor the behavior of nodes, evaluate the reputation of corresponding nodes and punish selfish nodes. Other schemes based on neighbor watching include [13] [17] [18] [19].

The use of virtual money (specifically called nuglet or credit) to simulate nodes cooperation has been suggested in [6][10]. Security modules independent of nodes are used to protect the nuglets or credit value from modification and other attacks. Salem et al. [11] proposed a charging and rewarding scheme in multi-hop cellular networks. Fratkin et al. [12] proposed APE (Ad Hoc Participation Economy) system, using a banker node to assure the payment consolidation and its integrity. Zhong et al. [7] proposed Sprite to stimulate cooperation among nodes. Each node reports the digest of received or forwarded packets to a central Credit Clearance Service (CCS), which determines the charge and credit to nodes involving in the packets forwarding.

Mechanism design was introduced to address selfish nodes problem recently. Anderegg and Eidenbenz [1] provide ad hoc-VCG routing protocol for ad hoc networks with selfish agents. This protocol is truthful and cost-efficient for data transmissions by using the VCG mechanism (named after Vickey [21], Clarke [22] and Grove [23]). However, it has some shortcomings. First, it needs $O(n^3)$ control messages for a route discovery. It is prohibitive for the source to pay this cost for its traffic. Second, to collect complete topology based on broadcast approach is not practical. Third, with $O(n^3)$ overhead, this protocol cannot provide good performance on metrics such as packet delivery ratio and end-to-end delay. Fourth, the payment calculation for route discoveries is very complicated. All these problems make this protocol difficult to use in practice. To overcome the problems of this protocol and keep the truthfulness are the motivations of our protocol. Zhong et al. [20] combined cryptography and VCG mechanism, as discussed in Section 3.5. Chen and Nahrstedt [4] proposed iPass where a node sets an auction in itself and gives its power and bandwidth to the winner of bidding flows. To set the auction, iPass requires at least two flows passing through any intermediate node concurrently. Besides, the source awards each intermediate node the same amount of money for packet forwarding, taking no consideration of their cost diversity. Cai and Pooch [3] proposed a protocol named $TEAM$. In this scheme, the payment received by an intermediate node may not cover its cost and thus nodes may have no incentives to forward packets. Also, the sender will not save energy and may get poor service because more hops are used. Wang and Li [5] proposed a time optimal algorithm to calculate payment to nodes for truthful routing in selfish wireless networks. Their model assumes that a node v_i has a fixed cost c_i of sending a packet to any of its outgoing neighbor. Srinivasan et al. [2] proposed a distributed and scalable acceptance algorithm called GTFT where a node decides to accept a relay request or not based on if it receives enough help from other nodes. This algorithm does not use the concept of virtual money.

3   LSTOP - A Light-Weight Scalable Truthful Routing Protocol

3.1   The System Model

A mobile ad hoc network composed of n nodes can be modeled as a directed weighted graph $G = (V, E, W)$. V is the set of nodes. E is the set of links between nodes. A node v_j within radio range of node v_i is represented as link (v_i, v_j). W is the set of weights for each link, indicating the cost to forward a packet along that link. *Each link may have a different weight*, i.e., a node has different cost to forward a packet to different neighbors. The power consumption is used as the cost metric. Denote $P_{i,j}^{min}$ as the minimum emitting power of node i so that node j can receive the packet from i. The weight w_{ij} of link (v_i, v_j) is equal to the product of $P_{i,j}^{min}$ multiplying i's cost of unit power c_i. Technically, $P_{i,j}^{min} = \frac{P_i^{emit} \cdot P_{min}^{rec}}{P_{i,j}^{rec}}$, where P_i^{emit} and $P_{i,j}^{rec}$ are the emitting power of i and the received signal of j, respectively. P_{min}^{rec} is the threshold over which j can receive the packet correctly. The network is assumed to be biconnected, i.e., the graph is still connected on removing any node and its incident links.

A source pays other nodes for their data forwarding service, and gives those nodes some marginal utility as an incentive. The payment is on per packet basis. Note that cost of control messages should also be reimbursed. Our goal is to reduce the total cost by providing sub-optimal path for data forwarding with very low overhead. We model routing as a game where nodes are players. Nodes have their private information (emitting power and cost of unit power). They are selfish and rational in that they take strategies to maximize their utility , and thus they may not reveal their true cost[1]. We assume no collusion between nodes. We also assume a payment mechanism [7] that takes care of accounting and transferring of payment between nodes, while insuring that nodes receive proper payment for their forwarding service.

3.2 Basic Idea

It is desirable to find a *truthful* and optimal (*least-cost*) path between a source and a destination, and VCG is the only mechanism that provides both these properties [8]. The VCG mechanism requires multiple optimal routes and thus the complete underlying weighted graph should be found. It costs ad hoc VCG [1] an overhead of $(O(n^3))$ to achieve this, which may not be practical specially in ad-hoc environments. Moreover, finding the complete underlying graph based on unreliable broadcast approach is not practical. A near-optimal (near-least cost) path for data traffic with very low overhead may reduce the total cost and may be more scalable. This is the stimulus for our *LSTOP* protocol. Compared to ad hoc-VCG, *LSTOP* provides near-optimal paths for data traffic with significantly low overhead of $O(n^2)$ in the worst case and $O(n)$ on the average. The basic idea of *LSTOP* works as follows: the source invokes route discovery on demand. Based on route request messages, nodes construct their neighbor lists. One route is selected by the destination as the base to construct a subgraph G^p. Only the nodes around this selected route report their neighbor lists to the source. Using these neighbor lists, the source constructs a subgraph G^p and finds the least cost path in G^p using the Dijkstra algorithm. Then VCG mechanism is applied to calculate the payment to nodes so that a rational node has no incentive to cheat over its cost.

3.3 Route Discovery

Route discovery intends to construct a subgraph G^p of the whole network. The subgraph G^p includes the source, destination and some intermediate nodes. To apply the VCG mechanism, this subgraph should be biconnected.

---

[1] Note that selfish nodes are different from malicious nodes, which intends to harm other nodes or destroy the network.

Whenever a source s want to communicate with a destination d, it initiates a $RREQ$ message. The $RREQ$ message includes such fields as $< s, d, seqNo, c_s, P_s^{emit} >$, where s and d are the node IDs of the source and the destination, respectively; $seqNo$ is the sequence number of this request; c_s is the sender's cost of unit power; P_s^{emit} is the emitting power of the sender.

Upon receiving a $RREQ$ message from a node i, a node j takes the following actions:

1. Determine the power P_j^{rec} at which j received the message.
2. Estimate the minimum power from node i to node j as $P_{i,j}^{min} = \frac{P_i^{emit} \cdot P_{min}^{rec}}{P_{i,j}^{rec}}$.
3. Append node i, c_i and $P_{i,j}^{min}$ to its neighbor list. Neighbor list is a list of structure consisting of node ID and the minimum emitting power from the neighbors of node j to node j and cost of unit power of these neighbors.
4. Check the freshness of $RREQ$ message by its sequence number. If it's a new message, node j appends its ID j, its cost of unit power c_j and its emitting power P_j^{emit} to $RREQ$ message and rebroadcasts it.

After receiving the first $RREQ$, the destination waits for a period τ and then sends a $RREP$ to the source. Denote the path along which the first $RREQ$ arrived as SP^{hop}. The $RREP$ is sent along SP^{hop} in reverse direction and includes the node list of that path. Waiting period τ is used to let $RREQ$ messages reach as many nodes as possible so that nodes can get as complete neighbor lists as possible.

After forwarding $RREP$ to the source, nodes on the path SP^{hop} report their neighbor lists to the source. Denote the set of nodes on the path SP^{hop} as V_{sp}^{on} and the set of nodes not on path SP^{hop} but within radio range of any node in V_{sp}^{on} as V_{sp}^{near}. Nodes in the network work in the promiscuous mode. Upon overhearing $RREP$ messages, nodes in V_{sp}^{near} send their neighbor lists to the source using $Neighbor$ messages. A $Neighbor$ message from node i includes all entries in i's neighbor list, consisting of neighbor's id, neighbor's cost of unit power and the minimum power needed from that neighbor to i. These $Neighbor$ messages use source routing. The routes to the source are established as the reverse routes of the routes along which nodes received the $RREQ$ messages. Nodes not overhearing $RREP$ messages don't send the $Neighbor$ messages.

Upon receiving all these neighbor lists from nodes in V_{sp}^{near} and V_{sp}^{on}, the source constructs a subgraph G^p which includes the source, the destination and all nodes within the radio range of nodes on path SP^{hop}. In a network with sufficient high density, this subgraph G^p is biconnected. In case that the acquired subgraph is not biconnected, the source invokes a new route discovery. By using Dijkstra or Bellman-Ford algorithm, the source can find a least cost path in G^p, denoted as SP^p.

Figure 1 illustrates this route discovery. s and d denote the source and the destination, respectively; the thin solid line denotes the SP^{hop}; the thick dash-dot line denotes the SP^p; the area enclosed by the dotted line is the found subgraph G^p. Only nodes in this enclosed area send $Neighbor$ messages to the source node. An observation is that, with high probability, the least cost path is present in the area enclosed by the dotted line. Thus SP^p may be a good approximation of the least-cost path for the whole network.

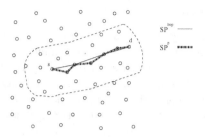

Fig. 1. Illustration of LSTOP

Data Forwarding: After finding the least cost path, the source can send data packets along this path. The header of data packets includes the node list $< s, v_1, v_2, \cdots, v_k, d >$, the minimum power $P_i^{min}, i \in \{1, 2, \cdots, k\}$ needed for every intermediate node v_i to forward data to its downstream node, and payment P_i to every intermediate node v_i. The payment to each intermediate node can also be accounted and reimbursed accordingly by a central server [7].

Upon link breakage due to mobility or cost change, the corresponding node sends a route error ($RERR$) message to the source. The source then invokes a new route discovery.

Message Complexity: In *LSTOP*, route discovery consists of two stages. In the first stage, $RREQ$ is flooding to the network. Every node forwards one $RREQ$ message. Thus a total of n $RREQ$ messages are sent. In the second stage, nodes in V_{sp}^{on} and V_{sp}^{near} send one $Neighbor$ message each to the source using source routing. The total number of $Neighbor$ messages forwarded is $\sum_{i \in V_{sp}^{on}} hop_s^i + \sum_{i \in V_{sp}^{near}} hop_s^i$, where hop_s^i denotes the hop counts from node v_i to the source s. Typically, the diameter of a network is $\sqrt{n}$, i.e., the number of nodes on a path is $\sqrt{n}$. The number of neighbors of a node, denoted as μ, is determined by the node density. Thus the neighboring nodes along the path SP^p is $\mu\sqrt{n}$, where μ is typically $6 \sim 8$. Since some nodes are common neighbors of two consecutive nodes on SP^p, i.e, the neighbor sets of two nodes will intersect with good likelihood, the actual value of μ is smaller. Under the assumption that typical network diameter is $\sqrt{n}$, a total of $O(\mu\sqrt{n} \cdot \sqrt{n}) = O(n)$ $Neighbor$ messages are forwarded. In the worst case, every node in the network is within the radio range of at least one of the nodes on the path SP^p. In this case all nodes report their neighbor lists and thus the overhead could be $O(n^2)$ in the worst case and $O(n^{3/2})$ on the average. However, in such a case, a least cost path is guaranteed to be found since the information on the whole graph is available to the source node. To sum up the two stages, a single discovery incurs message overhead of $O(n^2)$ in the worst case and $O(n)$ on the average.

In ad hoc-VCG, a $RREQ$ message is initiated by the source node and carries the information of links it passed by. A node forwards (broadcasts) every $RREQ$ message containing *any* link unknown to it. Since the number of links in a network is $O(n^2)$, each node may send as many as $O(n^2)$ $RREQ$ messages containing at least one new link weight. Thus ad hoc-VCG incurs an overhead of $O(n^3)$ control messages.

3.4   Payment Calculation

Payment to Route Discovery. We pay a unit amount for each node to forward a routing message. This unit payment can cover the maximum cost of a node for a routing message. Payment for a route discovery consists of two parts. The first part pays nodes involving in flooding $RREQ$. Every node broadcasts $RREQ$ exactly once. The second part pays nodes involving in the $Neighbor$ message reporting. Every node that overhears $RREP$ message along SP^{hop} sends its neighbor list to the source using a $Neighbor$ message. Since a $Neighbor$ message uses source routing, the source node can know all the intermediate nodes relaying the message by checking the header of the $Neighbor$ message. Upon receiving a $Neighbor$ message, the source increases the corresponding credits for the nodes forwarding the packet, and for the initiator of a $Neighbor$ message.

Payment for Data Forwarding. Payment calculation for data forwarding is based on subgraph G^p only. Nodes not in G^p or not on the least cost path get no payment for data forwarding. The payment to the nodes on the least cost path is calculated according to the VCG mechanism. Specifically, suppose the least cost path is $< s, v_1, v_2, \cdots, v_k, d >$, denoted as LP, and the cost of LP is C_{lp}. The declared cost of a node v_i, $i \in \{1, \cdots, k\}$, is T_i. Denote the least cost path from s to d excluding node v_i as LP^{-v_i} and the cost of this path as $C_{lp}^{-v_i}$. Then, P_{v_i}, the payment to a node v_i, $i \in \{1, \cdots, k\}$, is calculated as

$$P_{v_i} = C_{lp}^{-v_i} - (C_{lp} - T_i) \tag{1}$$

That is, the payment to a node v_i is the difference of the cost of the least cost path from the source to the destination which does not contain node v_i, and the cost of the least cost path from the source to the destination when the cost of node v_i is assumed to be zero.

Figure 2 shows an example of a weighted graph. The least cost path is $LP =< s, v_2, v_3, d >$ and $C_{lp} = 2 + 3 + 2 = 7$. The least cost path without v_2 is $LP^{-v_2} =< s, v_1, v_3, d >$ and its cost is $C_{lp}^{-v_2} = 6 + 2 + 2 = 10$. Thus, payment to v_2 is $P_{v_2} = C_{lp}^{-v_2} - C_{lp} + T_{v_2} = 10 - 7 + 3 = 6$. Similarly, the least cost path without v_3 is $LP^{-v_3} =< s, v_1, d >$ and its cost is $C_{lp}^{-v_3} = 6 + 5 = 11$. Thus payment to v_3 is $P_{v_3} = C_{lp}^{-v_3} - C_{lp} + T_{v_3} = 11 - 7 + 2 = 6$. Note that the total payment to the nodes

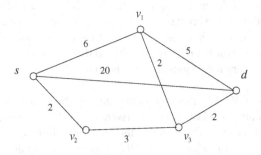

Fig. 2. A weighted graph

on the least cost path is greater than the cost of that path since the intermediate nodes are awarded more than their own cost. These marginal utilities are what those nodes pursue.

An alternative is to distribute the overpayment evenly among all intermediate nodes in a path [1]. In that case, every node gets the reimbursement equal to its cost. Bonus (payment subtracted by cost) of all intermediate nodes on a path will be summed up and be distributed evenly among them.

3.5   Truthfulness Analysis

We show that node's telling truth about its cost is its dominant strategy. Path selection is determined in two stages. In the first stage, SP^{hop} is chosen. SP^{hop} is the path along which the first $RREQ$ arrives at the destination. It is not dependent on the declared emitting power or cost of unit power of any node. Thus a node's under-declaration or over-declaration of its cost makes no sense for this stage. In the second stage, SP^p is found. After SP^{hop} is determined, subgraph G^p is constructed based on neighbor list information of nodes along and around SP^{hop}. The payment to nodes in the subgraph G^p is determined by applying the VCG mechanism. Only links in the subgraph are used. A link l_{ij}'s cost is determined by the sender i and the receiver j. Next, we show that nodes in the subgraph have no incentive to lie about their cost.

A sender v_i may under-declare its cost T_i by under-declaring its emitting power or its cost of unit power c_i in an attempt to get extra utility (margin). There are two cases. In the first case, node v_i is not on the least cost path initially. After under-declaration of its cost, if v_i is still not on the least cost path, then it gets no payment for data forwarding. On the other hand, v_i may be selected to be on the least cost path (LP). Denote the actual least cost path before v_i's under-declaration of its cost as LP_t and its cost as C_{lp}^t. Note that LP_t will be the least cost path after excluding v_i from the network, i.e., $C_{lp}^{-v_i} = C_{lp}^t$. Denote the actual cost of v_i as T_i^t. Then $\sum_{j \in LP \wedge j \neq i} T_j + T_i^t > C_{lp}^t = C_{lp}^{-v_i}$ and thus $C_{lp}^{-v_i} - \sum_{j \in LCP \wedge j \neq i} T_j < T_i^t$. From Equation 1, we have

$$P_{v_i} = C_{lp}^{-v_i} - C_{lp} + T_i = C_{lp}^{-v_i} - \sum_{j \in LP} T_j + T_i$$

$$= C_{lp}^{-v_i} - \sum_{j \in LP \wedge j \neq i} T_j. \tag{2}$$

Thus $P_{v_i} < T_i^t$, i.e., it implies that v_i's payment can't cover its cost and thus v_i gets negative utility. In the second case, v_i is already on the least cost path initially. In this case items in Equation 2 do not change even if v_i under-declares its cost. Thus v_i gets the same payment as when it tells the truth. Therefore, we conclude that v_i has no incentive to under-declare its cost.

Alternatively, v_i may over-declare its cost by over-declaring its emitting power or cost of unit power in hope of getting higher reimbursement. The over-declaration of cost makes the path through v_i more expensive. On one hand, this may cause the least cost path not go through v_i and thus v_i losses the chance to get the utility. On the other hand, if v_i is still on the least cost path, then items in Equation 2 do not change due to

v_i's over-declaration and thus v_i's payment will be still the same as the payment when it tells the truth. Therefore, v_i has no incentive to over-declare its cost.

Similarly, a receiver v_j's best strategy is to declare the true estimated minimum emitting power $P_{i,j}^{min}$. If v_j under-declares $P_{i,j}^{min}$, it may let the least cost path pass through it. However, the sender v_i will use a lower power to send packets and thus v_j will be out of the radio range of v_i. v_j can forward nothing and gets no payment. If v_j over-declares $P_{i,j}^{min}$ and thus making the path through v_j more expensive, the least cost path may not go through link l_{ij}. Even if link l_{ij} is still on the least cost path (after the over-declaration), v_j's over-declaration of cost increases the cost of the path C_{lcp}. From Equation 1, since $C_{lp}^{-v_j}$ and T_j do not change while C_{lp} is increased, v_j will get less payment than it gets when it tells truth about $P_{i,j}^{min}$. Thus v_j has no incentive to tell lie about its estimated minimum emitting power $P_{i,j}^{min}$.

Zhong et al. [20] had an argument about v_j's strategy in case of v_i's over-declaring P_i^{emit} as follows: If v_i over-declares P_i^{emit} by α times and v_j declares cost truthfully, then the estimated minimum power to cover link l_{ij} will be $\alpha P_{i,j}^{min}$. Thus l_{ij} may not be on the least cost path. If v_j under-declares the cost by β times ($1 < \beta \leq \alpha$), the estimated minimum power will be $\alpha P_{i,j}^{min}/\beta < \alpha P_{i,j}^{min}$, thus l_{ij} has more chance to be on the least cost path. *Since $\alpha P_{i,j}^{min}/\beta > P_{i,j}^{min}$, v_i can reach v_j.* Thus v_j get positive payoff than declaring the true minimum power. *However, note that v_i will not use this estimated minimum power $\alpha P_{i,j}^{min}/\beta$ to send packets to v_j. Instead, it will factor out its over-declaring ratio α and uses $P_{i,j}^{min}/\beta < P_{i,j}^{min}$ to send packet. Thus v_i can not reach v_j. v_j get no payoff, even if l_{ij} is still on the least cost path after the over-declaration of P_i^{emit}. Thus v_j's best strategy is still to declare the true estimated minimum power $P_{i,j}^{min}$.*

Base on their argument, Zhong et al. provided a protocol to prevent a node from under-declaring $P_{i,j}^{min}$ by combining the VCG mechanism with the cryptography. Their protocol results in even higher overhead than ad-hoc VCG. LSTOP's route discovery can be used in their protocol and can reduce their message overhead to $O(n^2)$ in the worst case and $O(n)$ on the average.

4   Simulation Study

4.1   Simulation Setup

We conducted an extensive simulation study to evaluate the performance of our protocol on a network with selfish nodes. We compared the proposed protocol with the ad hoc-VCG protocol [1] since both the protocols are aimed at networks where nodes are selfish and need some incentives to forward packets, and both achieve the truthfulness of the selected routes. We did not compare our protocol with generic routing protocols such as DSR and AODV because those protocols work under a different assumption that nodes follow the protocol and are willing to cooperate.

In our simulation study, we evaluate the following performance metrics:

- Approximation ratio. The approximation ratio of cost is the ratio of total cost of nodes based on *lstop* to that based on the whole graph (which gives the optimal solution, i.e., least cost paths). Similarly, the approximation ratio of payment is the ratio of total payment plus cost of source nodes based on *lstop* to that based on the whole graph. Approximation ratio shows how good *lstop* approaches the optimal solution.

- Packet delivery ratio. It is the percentage of the total number of packets received by the intended receivers to the total number of packets originated by all nodes, taking into consideration of factors such as packet collisions.
- Overhead. It is the total protocol control messages exchanged by nodes in the network, including routing messages as *RREQ*, *RREP* and *RERR*.
- Delay. It is the average end-to-end delay of data packets from senders to receivers.
- Overpayment ratio. This is the ratio of total payment to the intermediate nodes paid by all the source nodes plus the cost of the sources to total cost all nodes incurred for data transmission.
- Energy consumption. This is the total energy (power) consumed to send all packets, including route discoveries and data transmissions.

In the following discussion, the curves for the ad hoc-VCG protocol [1] is labelled as *vcg*. The curves for the proposed protocol *LSTOP* is labelled as *lstop*. Due to the space limitation, only the impact of network size is presented.

We used GloMoSim [9] for simulations. Unless specified otherwise, the following parameters were used in simulations. Nodes were placed uniformly in a square area with a density of 6000 square meter per node. Four levels of power emission for nodes, i.e., 1, 3, 5, 7 dBm, were used, corresponding to the radio range of 125, 158, 198 and 250 meters, respectively. Routing messages were always sent with the highest power level. For simplicity, we assume that nodes have same cost of unit power. We used 802.11 protocol with DCF as the MAC protocol and modified it to send packets with different power levels. All nodes followed Random Waypoint mobility model with a maximum speed of 5 m/s and a pause time of 30 seconds. Each simulation lasted for 900 seconds of simulated time. 10 CBR flows were simulated, starting at 100 seconds and ending at 880 seconds. Each CBR flow sent four 512-byte data packets per second. Data points represented in the graphs were averaged over 10 simulation runs, each with a different seed.

4.2 Simulation Results

Approximation Ratio. To study the approximation ratio of *lstop*, we collect the whole topology. Based on it, we apply VCG mechanism to calculate the total cost on the least cost paths and total payment to all intermediate nodes for data traffic. Since the paths found based on whole graph and those based on the sub graph may be different, mobility may cause different link breakages and thus results in different route discoveries. To be comparable, we set the network to be static.

Figure 3 shows the approximation ratio of cost and payment with respect to node density keeping 60 nodes. Both approximation ratio of cost and approximation ratio of payment are almost equal to 1. As the network area increases, i.e., node density decreases, the approximation ratio changes very little. The highest approximation ratio for either cost or payment is less than 1.007. It shows that selecting path using *lstop*

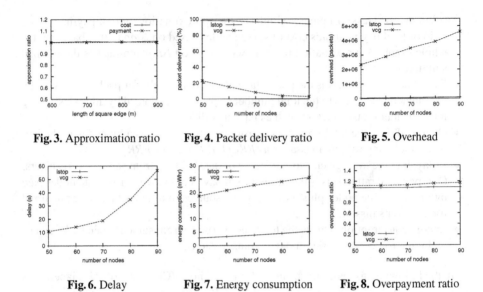

Fig. 3. Approximation ratio   **Fig. 4.** Packet delivery ratio   **Fig. 5.** Overhead

Fig. 6. Delay   **Fig. 7.** Energy consumption   **Fig. 8.** Overpayment ratio

approaches the optimal solution of selecting path based on whole topology in a network with sufficient high density.

Impact of Network Size

Packet Delivery Ratio. Fig. 4 shows the packet delivery ratio with respect to the number of nodes keeping the density fixed. We observe that the packet delivery ratio of *lstop* is significantly higher than that of *vcg*. *vcg*'s poor performance is due to three problems, all resulted from its $O(n^3)$ overhead.

First, *vcg*'s large number of $RREQ$s causes heavy congestion in MAC layer and thus results in high delay of message propagation. Results of our simulation study show that, in a static network, a destination node needs several seconds to collect *most* links information, which are carried by these $RREQ$s, and even more time to collect *all* links information. In a dynamic network, links collecting may not converge since the topology keeps changing during the $RREQ$ flooding period. Note that *vcg* requires the complete topology information to guarantee that the selected path is the least cost and truthful. After collecting links information, the destination node sends a $RREP$ to the source, including the least cost path and related information. However, the propagation of this $RREP$ message is time consuming due to heavy congestion in MAC layer. Results of our simulations show that it may take several seconds to reach the source node. Therefore, the whole route discovery may take long time. In a dynamic network, when the source node receives the $RREP$, the selected route (the least cost path) may be outdated. Thus data transmissions along the selected route may encounter link breakages, resulting in new route discoveries. New route discoveries aggravate the network traffic and further slow down route discoveries, which result in the selected paths to be even more outdated.

Second reason is that large number of $RREQ$ messages cause severe overflows of the network output queues and thus a lot of packets are dropped, including data packets and $RREP$ messages. The drop of $RREP$ messages results in increased route discoveries, which generates even more $RREQ$ messages and thus worsens the situation.

Third reason is that large number of $RREQ$ messages cause severe radio interference, resulting in data packets drops. Nodes outside the radio range but within the interference range of a communicating node pair may interfere with that communication and result in packets drops. The data packets drops result in increased route discoveries, which results in more $RREQ$ messages and thus worsens the situation. Even in a static network, many packets are dropped and multiple route discoveries are invoked for a single data session.

As the number of nodes increases, the network area increases and the average hop count of a path increases, and thus node mobility is more likely to cause link breakages for both the protocols. Besides, an increase in the number of nodes results in an increase in the number of links, thus route discoveries in vcg need more time to collect links information. The selected paths are more likely to be outdated, causing more link breakages. An increase in link breakages increases the packets drops and route discoveries, which generate more control messages especially for vcg and thus worsens the situation.

Overhead. Fig. 5 shows the overhead with respect to the number of nodes keeping the density fixed. We observe that the overhead of *lstop* is far (2 magnitude of order) lower than that of vcg. This is because vcg generates $O(n^3)$ overhead while *lstop* generates an overhead of $O(n)$ on the average. As the number of nodes increases, the overhead increases for both protocols due to two reasons. First, an increase in the number of nodes results in an increase in control messages of a route discovery. Second, an increase in the number of nodes results in more link breakages, as discussed above, and thus results in more route discoveries.

End-to-end Delay. Fig. 6 shows the end-to-end delay with respect to the number of nodes keeping the density fixed. The delay of *lstop* is around 0.1 second and is on the horizontal axis. We observe that *lstop* incurs far (2 magnitude of order) lower delay than vcg. This is due to the reasons described for packet delivery ratio in Section 4.2. First, vcg takes a long time for a route discovery. Second, multiple route discoveries for a single route request increase the delay. Route discoveries may fail due to $RREP$ messages drops resulting from the output queue overflows. It may also fail due to late arrival of $RREP$ messages, i.e., when $RREP$ messages reach the source nodes, the waiting timer for the route discoveries might have timed out already, and the source nodes might have already invoked the route discoveries again. Third, congestion and long queue in MAC layer increase the delay. As the number of nodes increases, the delay increases for vcg. An increase in the number of nodes increases the control messages, which results in longer output queues and aggravates the queue overflow and congestion in MAC layer and thus increases data packets' waiting time. Queue overflow causes the failure of a route discovery. Besides, route discoveries need more time to collect links information. Both result in data packets' waiting in the source nodes' buffers for longer time.

Energy Consumption. Fig. 7 shows the energy consumed for sending packets with respect to the number of nodes keeping the density fixed. We observe that *lstop* consumes much less energy than *vcg* because it sends far less control packets than *vcg*. As the number of nodes increases, more nodes consume energy and thus the energy consumption increases for both the protocols.

Overpayment Ratio. Fig. 8 shows the overpayment ratio with respect to the number of nodes keeping the density fixed. *vcg*'s overpayment ratio is a little higher than that of *lstop*. It implies that *lstop* can find better paths than *vcg*. The main reason is that in *vcg* with node mobility, it is difficult to collect the complete links information to get the most effective routes. Thus it doesn't achieve the most effective overpayment ratio. As the number of node increases, it's more difficult for *vcg* to collect the complete links information, and thus the overpayment ratio of *vcg* increases.

5   Conclusion

In this paper, we presented *LSTOP*, a scalable truthful routing protocol for MANETs with selfish nodes, where a node may use different cost to send packets to different neighbors. *LSTOP* approaches optimal routing solution in a dense network and incurs extremely low overhead of $O(n)$ on the average. It presents an efficient and simple method to collect topology information and constructs a selected subgraph. By applying VCG mechanism, *LSTOP* guarantees that nodes have no incentive to cheat over their cost. Besides, we presented a simple way to count the payment to nodes involving in route discovery. The most prominent feature of *LSTOP* is that it reduces the overhead from $O(n^3)$ of ad hoc-VCG [1] to $O(n^2)$ in the worst case and to $O(n)$ on the average, and thus it provides far better performance and scalability.

We conducted an extensive simulation study to evaluate our protocol and compare it with the ad hoc-VCG protocol [1]. By far, most papers using mechanism design method for selfish nodes problem evaluate overpayment ratio as the main performance metric. In addition to overpayment ratio, we focused on important network metrics such as packet delivery ratio, overhead and end-to-end delay. To the best of our knowledge, we are the first to evaluate these metrics for mechanism design methods. Simulation results show that our protocol approaches the optimal routing solution, achieves significantly higher packet delivery ratio, generates far (2 magnitude of order) lower overhead, incurs far (2 magnitude of order) lower end-to-end delay and has better overpayment ratio.

References

1. Luzi Anderegg, Stephan Eidenbenz, *Ad hoc-VCG: A truthful and Cost-Efficient Routing Protocol for Mobile Ad Hoc Networks with Selfish Agents*, in Proceedings of MobiCom'03, pp. 245-259, September, 2003
2. V. Srinivasan, P. Nuggehalli, F. Chiasserini, R. R. Rao, *Cooperation in wireless ad hoc networks*, in Proceeding of Infocom'03
3. Jianfeng Cai, Udo Pooch, *Play Alone or Together-Truthful and Efficient Routing in Wireless Ad Hoc Networks with Selfish Nodes"*, in Proceeding of MASS'04

4. Kai Chen and Klara Nahrstedt, *iPass: an Incentive Compatible Auction Scheme to Enable Packet Forwarding Service in MANET*, in Proceedings of ICDCS'04, pp. 534-542

5. Weizhao Wang and Xiang Yang Li, *Truthful Low Cost Unicast in Selfish Wireless Networks*, in Proceedings of IPDPS'04-Workshop 12

6. L. Buttyan and J. Hubaux, *Stimulating Cooperation in Self-Organizing Mobile Ad Hoc Networks*, ACM/Kluwer Mobile Networks and Applications, Vol. 8, No. 5, pp.579-592, 2003

7. S. Zhong, J. Chen and Y. R. Yang, *Sprite: A Simple, Cheat-Proof, Credit-Based System for Mobile Ad-Hoc Networks*, in Proceedings of Infocom 2003, Mar 2003

8. J.Feigenbaum, C. Papadimitriou, R. Sami and S.Shenker, *A BGP-based mechanism for lowest-cost routing*, in Proceedings of PODC 2002, pp.173-182

9. http://pcl.cs.ucla.edu/projects/glomosim/

10. Levente Buttyan and Jean-Pierre Hubaux, *Enforcing Service Availability in Mobile Ad-Hoc WANs*, in Proceedings of MobiHoc'00, 2000

11. N. B. Salem, L. Buttyan, J. Hubaus, M. Jakobsson, *A Charging and Rewarding Scheme for Packet Forwarding in Multi-hop Cellular Networks*, in Proceedings of MobiHoc'03, pp. 13-24

12. E. Fratkin, V. Vijayaraghavan, Y. Liu, D. Gutierrez, TM Li, and M. Baker. *Participation Incentives for Ad Hoc Networks*, http://www.stanford.edu/ yl314/ape/ paper.ps

13. Pietro Michiardi and Refik Molva, *Prevention of Denial of Service Attacks and Selfishness in Mobile Ad Hoc Networks*, Research Report RR-02-063 - January 2002

14. S. Buchegger, J.Y. Le-Boudec, *Nodes Bearing Grudges: Towards Routing Security, Fairness, and Robustness in Mobile Ad Hoc networks*, in Proceedings of EUROMICRO-PDP 2002

15. S. Buchegger, J.Y. Le-Boudec, *Performance Analysis of the Confidant Protocol (Cooperation of Nodes: Fairness in Dynamic Ad-hoc Networks)*, in Proceedings of MobiHoc'02, pp.226-236, June 2002

16. S. Marti, T. J. Giuli, K. Lai and M. Baker, *Mitigating Routing Misbehavior in Mobile Ad Hoc Networks*, in Proceedings of MobiCom'00, pp. 255-265, August 2000

17. Hugo Miranda and Luis Rodrigues, *Preventing Selfishness in Open Mobile Ad Hoc Networks*, in Proceedings of 7th CaberNet Radicals Workshop, October 2002

18. K. Paul, D. Westhoff, *Context Aware Detection of Selfish Nodes in DSR based Ad-Hoc Networks*, in Proceedings of VTC'02

19. Yongwei Wang, Venkata C. Giruka, Mukesh Singhal, *A Fair Distributed Solution for Selfish Node Problem in Mobile Ad Hoc Networks*, in Proceedings of ADHOCNOW'04, pp. 211-224

20. S. Zhong, L. Li, Y. G. Liu and Y. R. Yang, *On Designing Incentive-compatible Routing and Forwarding Protocols in Wireless Ad-Hoc Networks - An Integrated Approach Using Game Theoretical and Cryptographic Techniques*, in Proceedings of MobiCom'05

21. W. Vickrey, *Counterspeculation, auctions and competitive sealed tenders*, Journal of Finance 16, pp. 8-37, 1960

22. E. Clarke, *Multipart Pricing of Public Goods*, Public Choice 11, pp. 17-33, 1971

23. T. Groves, *Incentives in Teams*, Econometrica 41, pp. 617-663, 1973

Modelling and Analysis of Attacks on the MANET Routing in AODV

Peter Ebinger and Tobias Bucher

Fraunhofer Institute for Computer Graphics Research IGD
Security Technology Department
Fraunhoferstr. 5
64283 Darmstadt, Germany
{peter.ebinger, tobias.bucher}@igd.fraunhofer.de

Abstract. Mobile ad-hoc networks (MANETs) are especially vulnerable to attacks because of lacking infrastructure and data transfer using radio communication. In the past various types of MANET attacks have been detected and analyzed. In this article primary attack methods (black hole, wormhole, rushing, and Sybil) to MANET routing using the Ad-hoc On-demand Distance Vector (AODV) protocol are modelled and analyzed in a consistent way. This process comprises a semi-formal analysis of pre-conditions (costs, probability of success, required skills) and resulting damage from each attack by means of attack trees. We thereby get a uniform way of modelling and analyzing various attacks to MANET routing that allows to directly compare the different attacks and to assess the risks.

1 Introduction

Mobile ad-hoc networks (MANETs) are especially vulnerable to attack due to the lack of infrastructure. In addition, because data is transferred wirelessly it can easily be intercepted or interfered with. Network topology is permanently changing due to the mobility of the involved nodes. Finally, restricted energy supply and limited computational power narrow down the use of complex security mechanisms.

In the past various types of MANET attacks have been detected and analyzed. In each case, an identified attack tied to a particular protocol was examined and treated with individualized responses (e.g. black hole [1,2,8], wormhole [11] or Sybil [6] [4]). Until now no comprehensive conspectus of the different attacks has been carried out with a comparative evaluation for a routing protocol.

It is the aim of this article to model and analyze primary attacks to MANET routing in a consistent way. By means of attack trees [9] the different attack types are split into segments and modelled in a (semi-)formal way. Attack trees enable an analysis of each aspect using specific parameter values while attaching the necessary pre-conditions (including effort, the probability of success and the necessary skills) each attack requires to be successful. Damages that can result from of a successful attack are also examined.

T. Kunz and S.S. Ravi (Eds.): ADHOC-NOW 2006, LNCS 4104, pp. 294–307, 2006.

The MANET routing protocol AODV (Ad-hoc On-demand Distance Vector Routing Protocol) is investigated, which has been adopted by the IETF as RFC [7] (with active further development).

The analysis starts with a detailed overview of the different attacks and allows the reader to directly compare and evaluate possible risks to the system. This article is based on the analysis and modelling of routing methods in [3] using evaluation criteria from [5].

In the next section the basic concepts and ideas of primary attacks to MANET routing are introduced. Section 3 gives an overview and a definition of the attack tree analysis method.

In section 4 the specific examination criteria and scenarios are defined. A detailed analysis of the AODV protocol in relation to each specific attack type is carried out. Each attack is subdivided into individual steps and analysed using the previously defined evaluation criteria. Section 6 presents an evaluation of these analysis results and a comparison of the different attacks.

2 Attacks to MANET Routing

The following presents the basic concepts of some important attacks to MANET routing which are then modelled and analyzed with attack trees for the AODV routing protocol.

2.1 Black Hole Attack

The black hole attack [1,2] uses the idea of purposefully generating incorrect routes so that packages are no longer forwarded to the proper recipient D but instead get lost or sent to an attacker. Thus the name derives from the physical phenomenon black hole since something similar to a black hole is created in order to "swallow" the data packages. Fig. 1 shows an example of normal data traffic transferred via adjacent nodes to node D on the left and the effects of a successful attack on the right. Messages intended for node D do not reach their actual target but are intercepted by the attacker.

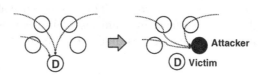

Fig. 1. Data flow to target D before and during a black hole attack

The attacker may also distribute fake routing information in order to become included in as many valid routes of the network as possible. This type of attack is always used during route finding or routing information update phases of the process. A black hole attack can also serve as a precursor for the execution of further attacks.

2.2 Wormhole Attack

A wormhole attack [11] uses two cooperating nodes of a network to re-route data traffic. In order for this to be successful, the two nodes must "ally" themselves and establish an additional channel outside normal network communications which serves as a tunnel. This shortcut is named after a wormhole as it mimics this hypothetical physical phenomenon.

In this type of attack the two nodes mask that they are not directly adjacent nodes and pretend to be neighbours (therefore disposing a fast connection to each other and their neighbours). As these paths are used for sending data that is not part of the proper network, wormholes are very difficult to detect.

Wormholes themselves are not necessarily only negative for a network as such a shortcut can have positive benefits such as relief for the network or shorter transfer times for packages on the routes containing the wormhole. Attackers use wormholes in the network to make their nodes appear more attractive (with perceived faster transfer times) so that more data is routed through their nodes. Similar to the black hole attack, the wormhole attack can also be used as a basis for further attacks.

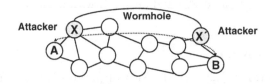

Fig. 2. Data flow during a wormhole attack of X and X'

2.3 Rushing Attack

This attack is based on the idea of transferring messages as soon as possible so that they forestall other messages in finding a route. In this way an attacker can exert a considerably greater influence on route generation (ensuring the inclusion of the attackers node on the route). This works especially well as many routing protocols dispose of security mechanisms against copies with the result that only the data package that arrives first is evaluated while all others are discarded.

2.4 Sybil Attack

A Sybil attack[1] [6] [4] occurs when a node in the network tries to masquerade as several identities. This can be achieved in two ways, by feigning the existence of an additional node or by stealing the identity of an existing node. The advantage

---

[1] Sybil is the name of a book [10] with the authentic report of the first psychoanalysis of a multiple split personality.

of controlling several identities to an attacker is that he can extensively conceal his activities within the network (e.g. a black hole attack).

3   Attack Trees

Attack trees are a semi-formal illustrative method of analysis which documents the security of a system in relation to various attack targets. First suggested by Bruce Schneier in [9] in December 1999, the principle is based on fault trees, used in business economics (e.g. for the examination of the reliability and failure frequency of machines).

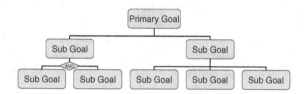

Fig. 3. Basic scheme of an attack tree

An attack is graphically represented as a tree, the root representing the primary goal of the attack. The branches represent the secondary goals of the attack or subattacks which are necessary preconditions for achieving the primary goal. Branching can represent either a logical "OR" or "AND" connection. If there is no further branching possible or intended (i.e. if it is an elementary action) you have reached a leaf of the tree.

In Fig. 3 the schematic structure of attack trees is shown. Only the "AND" nodes are emphasized within the figures, all other nodes are considered "OR" nodes.

3.1   Formal Definition

The differentiation between "OR" and "AND" nodes is defined below. Let $B = \{b_0, b_1, ..., b_n\}, n \geq 2$ be the set of child nodes b_i of a node a and A the state of the system after the sub goal represented by a was successfully achieved.

- If node a is marked with "OR", A is associated with the state the system can assume after the execution of one or more actions associated with b_0 to b_n (if b_i is a leaf) or the achievement of the respective sub goal (if b_i is an inner node).
- If node a is marked with "AND". A is associated with the state the system can assume after the execution of all actions associated with b_0 to b_n (if b_i is a leaf) or the achievement of the respective sub goal (if b_i is an inner node).

3.2   Parameter and Value Calculation

Values can be assigned to each leave of the attack tree to allow the calculation of the overall value of an examined parameter, e.g. the overall damage caused by a particular attack. The value of each node is calculated from the values of their child nodes and may include such value parameters as effort required, costs involved, the probability of success, the needed technical skills, or the expected damage. Formulas for node value calculations are selected subject to the type of parameter and node.

4   General Conditions and Evaluation Criteria

The scenario examined in the analysis is an ad-hoc network consisting of about 5 to 50 mobile nodes representing diverse devices such as PDAs and laptops. Furthermore we assume that the nodes that are part of the network can mutually authenticate each other e.g. by a previous offline key exchange. The following gives a description of parameter values used in the attack trees, which is based on a report [5] of the project (MANET Intrusion Detection in Tactical Environments) MITE.

4.1   Selection of the Evaluation Criteria and Quantification

Parameters are used to analyze, compare and evaluate each attack to MANET routing. Three parameters characterizing the preconditions and requirements for a successful attack were chosen (costs, probability of success, skills) along with an additional parameter describing the possible resulting damage. The co-domain of all four parameters is the closed interval $[0, 1]$.

Detailed specifications for each particular parameter values are very difficult to quantify. The following provides a rough assessment based on examples and categorization techniques. Another difficulty in assigning values is that parameters are not independent to one another (they overlap). Assessed values for elementary actions of each leave and resulting values for root nodes can therefore only be taken as a rough classification guide for each attack. Parameters were further co-classified into five categories: "very small", "small", "medium", "high" and "very high".

4.2   Costs

The parameter *costs* encompasses time cost to the attacker not monetary costs as the latter could soon be no longer valid due to the fast progress of information technology. Specifically it refers to the time required to execute the attack, not the preparation and development of attack methods and tools as these costs are included in the parameter skills described below.

An example of a low *costs* attack (category 1, duration < 1 minute, *costs* between 0 and 0.1) is a denial-of-service (DoS) attack which sends a *ping of death* package. In contrast a high *costs* attack would include "Brute-Force" cryptographic key attacks which fall into category 5 (duration: hundreds of hours and more, *costs* = 1).

4.3   Probability

The *probability* of a successful attack is a direct result of the product of the *probabilities* of each sub-task. This is exactly the case if all sub-tasks are stochastically independent, which can not generally be assured. A precise analysis should be based on a more elaborate probability model which takes conditional probabilities into account. The probabilities presented in this paper are therefore only approximate values which give an estimate of the size of the real probabilities.

An example of a low *probability* attack would be the random discovery of cryptographic key material (category 1, *probability* p=0.1). Boundary conditions can make exact quantification of these *probabilities* extremely difficult.

4.4   Skills

The technical skill and experience an attacker requires to successfully perform an attack is accounted for in the *skills* parameter. There is a negative correlation to the *costs* parameter since an attacker can possibly compensate lacking skills with increased resources and/or time (and vice versa).

Jamming (category 1, *skills* between 0 and 0.1) is an example of an attack requiring a low *skill* input level while use of the Differential-Power Analysis method requires "high" to "very high" *skill* level (categories 4 or 5, *skills* between 0.75 und 1).

4.5   Damage

In contrast to the other parameters *damage* does not address preconditions for a successful attack but rather deals with the resulting damage. *Damage* can be quantified in many different ways including financial (loss/destruction of hardware) or time (re-installation of systems, etc.).

Low *damage* attacks (category 1, *damage* between 0 and 0.1) like a DoS attack only requires a restart of the computer. Higher level *damage* attacks involve danger to sensitive data (i.e. copied, stolen or deleted/unrecoverable).

4.6   Calculation

In table 1 the calculation formulas for parameters is compiled for each particular node type. Since all parameters are normalized to the interval $[0,1]$ a multiplicative linking of the different parameter values is possible. With addition the problem arises that values can possibly exceed 1, in which case the result is set back to the maximum value 1. When multiplying small *probabilities* (category 1, rather improbable) results can become extremely small. In this case the value is reset to a minimum value of 0.05.

Calculating parameter *costs* requires not linear but logarithmic mapping (time) so *costs* C are calculated using a so-called LOG sum:

$$C = \frac{1}{4} \log_{10} \sum_{i=1}^{n} 10^{4a_i},$$

with a_i $(i = 1 \ldots n)$ representing parameter values of the n child nodes.

Table 1. Calculation of the parameter values of an attack tree

	OR	AND
Costs (C)	Minimum	LOG sum
Probability (P)	Maximum	Product
Skills (S)	Minimum	Maximum
Damage (D)	Maximum	Sum

5   Analysis of Attacks on AODV

In the following sections black hole, wormhole, rushing, and Sybil attack on AODV [7] are modelled and analyzed in a consistent way using attack trees and the parameters presented above.

5.1   Black Hole Attack

One problem with route finding in AODV is that not only the destination node can send a route reply message (RREP), it is also possible that a node in the middle knows a valid route and can send an RREP message back to the sender. An attacker who receives a route request message (RREQ) can take advantage of this by sending an RREP package back to the sender, pretending that the destination node is only one or few hops away from the attacking node. The attacker will then be masked as the shortest path and be included within the transmission route.

Instead of pretending to have a shorter route the attacker can fake a higher sequence number in his RREP message. This way the new route overwrites routes transferred by other nodes. A combination of both tactics is also possible.

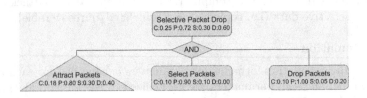

Fig. 4. Black hole attack for the deletion of messages

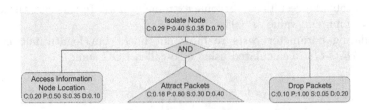

Fig. 5. Black hole attack to isolate a node

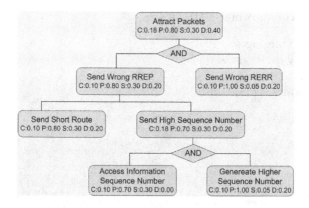

Fig. 6. Attack tree of a black hole attack

Each approach is based on the fact that attackers must await a route discovery process initiated by another node. Attackers can artificially initiate a route finding process by sending a fake route error package (RERR) which pretends that a section of the route between attacker and recipient node is no longer available. This intentionally generated route fault can lead to the initiation of a route request (RREQ) that the attacker can later misuse for his aims.

This procedure creates a "black hole" that collects and discards all arriving data. Possible goals of this attack are the following:

– to selectively delete data (*gray hole*) or
– to isolate a node (DoS)

The related attack trees shown in fig. 4 and fig. 5 both contain the attack tree of fig. 6 (representd by a triangle) which forms the kernel of the black hole attack.

5.2   Wormhole Attack

During a wormhole attack two attackers X and X' work together (as described above) to create an additional channel by an out-of-band connection (see fig. 2). The actions taken by the wormhole attacker first resembles those of a black hole attacker. Upon receipt of a RREQ message both send an RREP message back to the sender which bears fake information intended to attract all data traffic originally intended for another recipient. The difference lies in that a pair of attackers act at two different places of the network, e.g. to control the data traffic between two nodes A and B in a wormhole attack.

As with black hole attacks, wormhole attackers have the possibility to artificially raise the sequence numbers of this RREP message to overwrite route information sent by other nodes with his RREPs.

Possible aims of a wormhole attack are:

– to eavesdrop messages,
– to selectively delete data,
– to manipulate data or
– to isolate nodes (DoS).

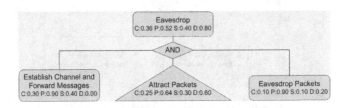

Fig. 7. Wormhole attack to eavesdrop messages

The related attack trees are shown in figures 7 to 10. They all contain the attack tree shown in fig. 11 forming the kernel of the wormhole attack.

5.3 Rushing Attack

The aim of rushing attacks is to become part of as many routes as possible in order to eavesdrop, manipulate data or support other attacks. Node A sends a RREQ package (see fig. 12) to search for a route to C (or to a node behind C). The attacker X receives the RREQ message sent via flooding and will then try to forward this message as soon as possible (possibly violating the rules of lower network layers). If successful node C receives the route request of attacker X before getting the corresponding message from node B. Node C processes this message and subsequently discards the message of B received afterwards since according to AODV specification it must only consider the first route request that bears a certain ID. Node C then returns an RREP message to the sender via X so that X is henceforth on the route between A and C instead of node B that would also have offered this connection.

The attack tree in fig. 13 looks rather simple since the attacker conforms to protocol with the exception of using "hurrying".

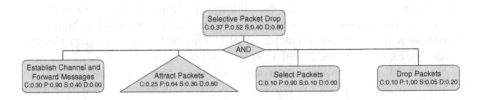

Fig. 8. Wormhole attack for the selective deletion of messages

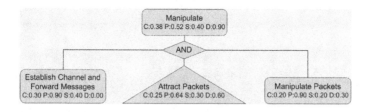

Fig. 9. Wormhole attack for the manipulation of data

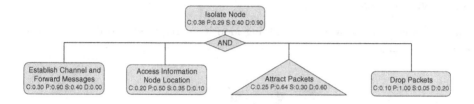

Fig. 10. Wormhole attack to isolate a node

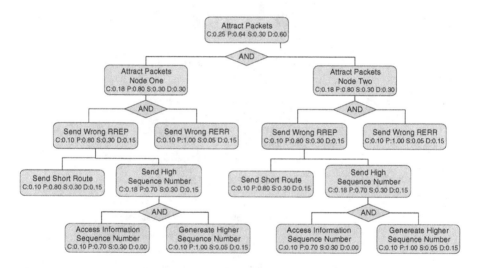

Fig. 11. Attack tree of a wormhole attack

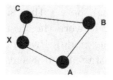

Fig. 12. Node X performs a rushing attack

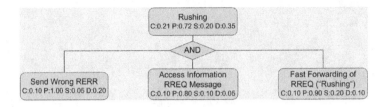

Fig. 13. Attack tree of a rushing attack

5.4 Sybil Attack

Eavesdropping to discover and exsisting identity is required in order to execute a Sybil attack.[2]

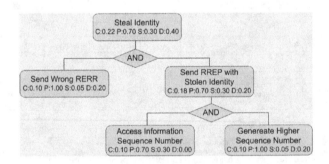

Fig. 14. Node X pretends to be C

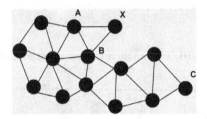

Fig. 15. Attack tree of a Sybil attack

The stealing of an identity can be a rather simple as an attacker must only answer a RREQ message as if it were the destination node (sending a route reply with the identity of another node). Preventative measures include sending a RREP message so early that attacking messages are ignored. As with black hole

---

[2] In principle it is also conceivable that an attacker creates a completely new identity, however this is a less interesting attack to this report and the procedure only differs in few details.

and wormhole attacks, Sybil attacks can benefit from sending RERR messages (pseudo route error) to activate new route requests.

HELLO messages offer another possibility to execute a Sybil. An attacker X could send HELLO messages containing a fake sender address (as shown in fig. 14) for example to node C to spread incorrect neighbourhood information. As HELLO messages are an optional extension of AODV and are usually not used this possibility is not addressed in more detail.

6 Evaluation

The results of all attack trees are now collected to compare and classify them using our defined parameters. This is followed by a discussion and an evaluation.

6.1 Comparison of the Different Attacks

Table 2 displays the results of the attack trees and the related parameter values: *costs*, *probability*, *skills*, and *damage*.

Table 2. Parameter values for the different attacks

Attack target	Costs (C)	Probability (P)	Skills (S)	Damage (D)
Black hole - selectively delete data	0.25	0.72	0.30	0.60
Black hole - isolate node	0.29	0.40	0.35	0.70
Wormhole - eavesdrop	0.36	0.52	0.40	0.80
Wormhole - selectively delete data	0.37	0.52	0.40	0.80
Wormhole - manipulate data	0.38	0.52	0.40	0.90
Wormhole - isolate node	0.38	0.29	0.40	0.90
Rushing	0.21	0.72	0.20	0.35
Sybil	0.22	0.70	0.30	0.40

Costs. The *costs* of each attack range from 0.21 to 0.38. The rushing and Sybil attacks show the lowest *costs* while the wormhole attacks are the most costly.

Probability. The range of *probabilities* of success are between 0.29 and 0.72. The rushing, the Sybil and the simple black hole attack have the greatest *probabilities* of success (successful in most cases) whereas the success of a wormhole attack is the quite unlikely.

Skills. Required technical *skills* fell between 0.20 und 0.40. Higher for a wormhole attack although very similar values to the other attacks (black hole, rushing and Sybil).

Damage. There is a significant gap (ranging from 0.35 to 0.90) between the rushing and Sybil attacks and the black hole and wormhole attacks.

6.2 Evaluation

The greatest *damage* results from a successful black hole or wormhole attack which also requires the greatest effort as all network traffic (or at least part of it) in a MANET region needs to be redirected.

The rushing and the Sybil attacks are easy to perform, relatively inexpensive, and have a high success *probability*. The *damage* caused by these attacks however is also relatively small.

Reactive protocols allow attackers to do considerable *damage* with short-time actions like faking particular packages which in contrast to proactive protocols needs not continuously participate or infiltrate the network to deceive other nodes.

An attacker who attacks an AODV-based network has the possibility to attack directly as he can at any time answer a route request with a wrong route reply or generate a route error to cause a new route request.

He can specifically attack only certain route requests bearing a selected victim as goal but leaves others unaffected. A single faked route in AODV has accordingly only minor impact on the network. In general, it only influences one single node or a small network sector.

7 Summary

In this paper the most important attacks on the MANET routing protocol AODV (Ad-hoc On-demand Distance Vector Routing Protocol) have been modelled and analyzed in a consistent way using attack trees. In contrast to former investigations of single attack possibilities the presented analysis gives a detailed overview of each attack and allows the reader to directly compare and evaluate possible risks using defined criteria. The requirements necessary for a successful attack were analyzed as were the necessary effort, *probability* and *skill* levels. *Damage* resulting from a successful attack was also analysed, completing a full picture of each attack which allowed comparison between the attacks.

This systematic approach proves that the greatest *damage* results from a successful black-hole or wormhole attack, attacks which also require the greatest effort. Conversely, rushing and Sybil attacks are easy to perform have a high success *probability* but cause relatively low levels of *damage*.

References

1. Imad Aad, Jean-Pierre Hubaux, and Edward W. Knightly. Denial of Service Resilience in Ad Hoc Networks. In *Proceedings of the 10th Annual International Conference on Mobile Computing and Networking*, pages 202–215, Philadelphia, PA, USA, September 2004. ACM Press.
2. Mohammad Al-Shurman, Seong-Moo Yoo, and Seungjin Park. Black Hole Attack in Mobile Ad Hoc Networks. In *Proceedings of the 42nd Annual ACM Southeast Regional Conference*, pages 96–97, Huntsville, AL, USA, April 2004. ACM Press.

3. Tobias Bucher. Modellierung und Analyse von Angriffen auf Routingverfahren in mobilen Ad-hoc-Netzen, Diploma Thesis. Master's thesis, Technische Universität Darmstadt, December 2005.
4. J. Douceur. The Sybil Attack. In *Proceedings of the IPTPS02 Workshop*, Cambridge, MA (USA), March 2002.
5. M. Jahnke and J. Toelle. Bedrohungsanalyse taktischer MANETs mittels Attack Trees. Technical report, Forschungsgesellschaft für Angewandte Naturwissenschaften e.V. (FGAN), Wachtberg, Germany, to appear in summer 2006.
6. James Newsome, Elaine Shi, Dawn Song, and Adrian Perrig. The Sybil Attack in Sensor Networks: Analysis & Defenses. In *Proceedings of the Third International Symposium on Information Processing in Sensor Networks*, pages 259–268, Berkeley, CA, USA, April 2004. ACM Press.
7. Charles E. Perkins, Elizabeth M. Belding-Royer, and Samir R. Das. Ad hoc On-Demand Distance Vector (AODV) Routing, July 2003. RFC 3561.
8. Daniele Raffo. *Security Schemes for the OLSR Protocol for Ad Hoc Networks.* PhD thesis, Université Paris 6, September 2005. `http://perso.crans.org/raffo/papers/raffo-phdthesis.pdf`
9. Bruce Schneier. Attack Trees - Modeling Security Threats. *Dr. Dobb's Journal*, 24(12):21–29, December 1999.
10. Flora Rheta Schreiber. *Sybil: The True Story of a Woman Possessed by 16 Separate Personalities.* Henry Regnery Co., Chicago, IL, USA, 1st edition, 1973.
11. Weichao Wang and Bharat Bhargava. Visualization of Wormholes in Sensor Networks. In *Proceedings of the 2004 ACM Workshop on Wireless Security*, pages 51–60, Philadelphia, PA, USA, October 2004. ACM Press.

Initialization for Ad Hoc Radio Networks with Carrier Sensing and Collision Detection*

Jacek Cichoń, Mirosław Kutyłowski, and Marcin Zawada

Institute of Mathematics and Computer Science
Wrocław University of Technology

Abstract. We consider an ad-hoc network consisting of devices that try to gain access for transmission through a shared communication channel. We redesign randomized protocol of Cai, Lu and Wang based on carrier sensing. Our solution improves time efficiency compared to the previous algorithms.

Apart from design of a protocol that behaves well in practical experiments, we provide a rigorous complexity analysis.

Keywords: ad hoc network, radio channel, initialization, carrier sensing, combinatorial class.

1 Introduction

An ad hoc radio network is a collection of processing units, called *stations*. It is temporary in the sense that stations may join and leave the network, so the set of stations in the network is highly unpredictable (in particular, this is the case, when a mobile network is concerned). An ad hoc network should work without aid of any established infrastructure or central control units.

For the problems concerned in this paper we consider single-hop networks in which each station can directly communicate with other stations through a shared communication channel. Therefore, if a single station sends a signal through the communication channel, then it is received by all other stations that have their receivers on.

For the sake of simplicity, we assume that the time is divided into slots and all stations have a local clock that keeps stations synchronous. However, a slot can be further divided into minislots for the sake of resolving conflicts between stations trying to broadcast.

We assume that each station is capable to sense the *carrier*. It means that each station can identify the channel status as either *busy* (some station is sending signals) or *idle* (no station is sending a signal). It is worth to mention that carrier sensing is a standard feature of modern wireless devices [3]. However, the main problem with carrier sensing is that the signal is received with a certain delay, so if a station finds a channel idle, it does not necessarily mean that the channel

---

* Supported by the European Union within the 6th Framework Programme under contract 001907 (DELIS).

T. Kunz and S.S. Ravi (Eds.): ADHOC-NOW 2006, LNCS 4104, pp. 308–320, 2006.

is idle – in the meantime some station might have started sending a signal. This lack of instantaneous propagation of signals makes it hard to design a protocol that coordinates who will transmit within a slot. This is particularly important, since when at least two stations broadcast messages within the same slot, then a *conflict* occurs - interference of signals makes it impossible to decode any of the messages sent.

Initialization Problem. Since we assume that the network is highly dynamic and there are different groups of stations performing different tasks with no coordination between them, it is highly unpredictable who will try to use the communication channel. So we need a level in communication protocol that coordinates between the stations that would like to use the broadcast channel.

We assume that it is impossible or impractical to coordinate the stations by means of serial numbers assigned to them by the manufacturers. Indeed, it is unpredictable which of these numbers are in use in the network – they are more or less random. Therefore, all stations should participate in the competition for temporary ID numbers used to coordinate access to the broadcast channel. The problem of assigning to each of the n stations an ID number in the range $[1, \ldots, n]$ such that no two stations are assigned the same ID is referred to as the *initialization problem*.

Paper organization. In Section 2 we recall previous initialization algorithms. In Section 3.1 we present our initialization algorithm. In Section 3.2 we provide a rigorous complexity analysis of our algorithm.

We assume that the reader familiar with combinatorial classes technique (for details see [2]).

2   Previous Initialization Algorithms

First we recall two initialization algorithms proposed by Nakano and Olariu [7] that work, respectively, for known and unknown number of stations. In the first straightforward protocol the number of stations is known and equal n. Within a time slot each station sends a message with probability $1/n$. If exactly one station has transmitted, then it receives the smallest unused ID. This procedure is repeated by the stations having no ID until no such station is left. The number of time slots required to complete the protocol is $e \cdot n + O(\sqrt{n \log n})$, i.e. with probability exceeding $1 - \frac{1}{n}$ all stations have their ID's.

The second protocol deals with the case when the number of stations is unknown. The stations assign themselves into different groups, so that finally each station is in a different group. At the beginning all stations are in the same group. Within a time slot the stations belonging to a group corresponding to this slot split at random into two subgroups, provided that the group contains at least two stations. The protocol initializes the stations in $\frac{10n}{3} + O(\sqrt{n \log n})$ time slots, i.e. with probability exceeding $1 - \frac{1}{n}$ all stations become their ID's.

In [6], a hybrid initialization method for known n is proposed. It achieves time bound approximately $2.15 \cdot n$.

The algorithms presented in [7,6] assume that the stations that are transmitting can determine whether its transmission was successful (i.e. no conflict has occurred). However, in the practical technical setting a transmitting station cannot simultaneously listen to the communication channel. Therefore, the algorithms may not be suited for practical systems without further modifications.

Cai, Lu and Wang [1] propose initialization algorithm with a proper acknowledgment scheme that utilizes carrier sensing. As shown on Figure 1, they divide the time slots into the following minislots:

– initial period (IP)
– transmission period (TP)
– acknowledgment period (AP)

The idea is that initialization is integrated with transmission: in the initial period one station is chosen (with high probability), then transmission period is used for sending the workload message, and finally during acknowledgement period correctness of transmission is acknowledged to the station that has transmitted.

First, a station that wants to transmit chooses a moment uniformly at random within the minislot IP. At this moment the station checks the status of the communication channel. If it is busy, then the station is not initialized and does not transmit within this slot. Otherwise, it starts to send carrier signal until the end of TP. During TP it sends a workload message.

During the minislot AP, a randomized acknowledgment procedure is executed. Its goal is to inform the station that has succeeded to transmit that no collision has occurred. The acknowledgment is sent at the end of the minislot AP by a station elected previously to a *checker*. If a workload message has been transmitted successfully in a slot, the checker transmits an appropriate acknowledgment packet. Otherwise, no acknowledgment is transmitted by the checker.

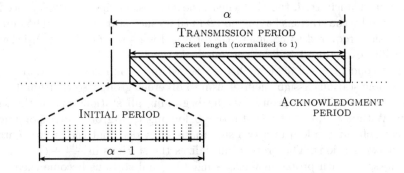

Fig. 1. Slot organization according to the algorithm of Cai, Lu and Wang [1]

Checker selection scheme [1]. For the control purposes, each packet should include time of its transmission. Now, suppose that A is the first station that successfully transmits within a slot. At the end of this slot A is not sure whether its transmission was successful. No station can acknowledge it. Therefore, station A will continue to transmit its initialization packet in the following slots. However, all other stations are aware that A has transmitted successfully and they include the transmitting time of A's successful packet inside their initialization packets. Suppose that B is the second station different from A that transmits successfully. Based on the information included in the packet sent by B, station A can realize that its packet transmission was successful. Then A will serve as the checker during the initialization process, and be responsible for transmitting the acknowledgment packets.

Let us assume that the probability that a slot is successful during the initialization is p. Then the number of slots necessary to elect a checker with probability at least $1 - \frac{1}{n}$ can be determined as

$$k = \frac{1}{p} \left(2 + \log n + \sqrt{4 \log n + (\log n)^2} \right) \tag{1}$$

(apply formula from Appendix A for $d = 2$ and $f = n$).

Time Settings of the Scheme. The duration of a slot and minislots must be set according to technical reality [4,3]. For the sake of simplicity we assume that the initial period length of the initial period plus the transmission period. Let λ be the normalized maximum propagation delay of a signal in the network [1]. The transmission radius for the stations in the single-hop ad hoc network is limited in practical situations, it is usually less then 1000 meters, therefore $\lambda \ll 1$.

Let $\delta = \frac{\lambda}{\alpha - 1}$, that is let δ represents the normalized maximum propagation delay with respect to the amount by which the TP length is larger than the IP length. Since AP is designed to accommodate the transmission of one short acknowledgment packet, its length is negligible compared with the TP length. Hence, the slot length is approximately α.

3   Our Initialization Algorithm

3.1   Description of the Algorithm

The main problem with the previous solution is that it fails if two stations decide to transmit at close moments. Each of them does not sense the other station, due to propagation delay. In our initialization algorithm, we do not use carrier signal to block all other stations to transmit within a slot. So if a station thinks that a channel is idle, but in fact it only looks like that due to propagation delay, then the transmission is not blocked until the end of the slot and another station may succeed. However, it is not obvious, whether this improves success

Procedure Initialization-Algorithm

var l: integer   init 0
       id : integer $\cup \{undef\}$ init $undef$
       δ : float   init 0.0001 //*maximum propagation delay*
       $channel$: $\{idle, busy\}$

For begin of the each slot
    begin
        if $id = undef$ **then**
            $p^* := (\ln(\frac{1}{2\delta^2}) - \ln\ln(\frac{1}{2\delta^2}))/(N - l)$
            if rand$(0,1) \leq p^*$ **then**
                $\gamma := $ rand$(0,1)$
                if $\neg$receive $\langle msg \rangle$ on $(0, (\alpha - 1)\gamma)$ **then**
                    $s := (\alpha - 1)(\gamma + \delta)$
                    $channel := $ sense the channel at $(\alpha - 1)\gamma$
                    if $channel$ *is idle* **then**
                        send $\langle msg \rangle$
                    else
                        $s := (\alpha - 1)\gamma$
                    if $\neg$receive $\langle msg \rangle$ on $(s, \alpha - 1)$ **then**
                        send $\langle workload\ packet \rangle$ at $\alpha - 1$
            end
        end
    end
end

Listen to the AP
    begin
        if received $\langle ack \rangle$ from AP **then**
            $l := l + 1$
            if *my packet has been transmitted in the slot* **then**
                $id := l$
    end

Fig. 2. Pseudo-code of the initialization algorithm

probability in a significant way, while it makes the analysis significantly more complex.

Figure 2 presents procedure executed by our algorithm within a slot in the form of a pseudo-code. Let us describe its general idea. A station that wishes to transmit first generates a value γ uniformly at random from interval $(0, 1)$. The station is monitoring the channel until moment γ. If it receives a transmission (so in particular there is no collision), then the station will not transmit itself.

Otherwise, if it does not sense a carrier signal at moment γ, it transmits a short message of the length δ. This message stops other stations from trying to transmit during this slot. Finally, the station(s) that has succeeded to transmit during initial period sends a workload message. Of course, as before, two stations, say A and B, may start sending at approximately the same time during the initial period. However, the other stations will sense a collision and therefore send their messages. These messages witness that there was a collision and prohibit A and B to make a decision to send during the transmission period.

As in [1], each station that succeeds to transmit leaves the pool of stations waiting to transmit. The algorithm is executed until the pool of these stations becomes empty.

3.2   Properties of the Algorithm

The main problem concerned in the rest of this paper is estimating the number of slots necessary so that every of n stations wishing to send a message finally succeeds. For this purpose we determine the probability that during a slot some station succeeds to transmit. Of course, this occurs if there is a station that chooses a number $\gamma = P$ such that no other station chooses its number γ in the interval $(P-\lambda, P+\lambda)$. For the sake of simplicity of notation we scale the length of the initial period from $\alpha - 1$ to 1. Then obviously the propagation delay is scaled from λ to δ. Assume that there are N stations that wish to transmit the packets during a time slot. So the problem to solve is to estimate the probability that if N numbers in the interval $(0, 1)$ are chosen uniformly at random, then there is a number P chosen such that no other number is contained in the interval $(P - \delta, p + \delta)$.

For analysing properties of the moments of time in which stations start transmissions we use the machinery of combinatorial classes described in [2]. Namely for $D \in \mathbb{N}$ and $a \in \mathbb{N}$ such that $1 \leq a \leq N - 2$ let $A_{N,a,D}$ denotes the combinatorial class

$$Seq(\{\circ\}) \times (\{\bullet\} \times Seq_{<D}(\{\circ\}))^a \times (\{\bullet\} \times Seq_{\geq D}(\{\circ\}))^2 \times (\{\bullet\} \times Seq(\{\circ\}))^b .$$

The class $A_{N,a,D}$ is isomorphic to the class of all sequences

$$(k_1, s_1, \ldots, s_a, l_1, l_2, n_1, \ldots, n_b)$$

of natural numbers, such that $s_1, \ldots, s_a < D$, $l_1, l_2 \geq D$, and $b = N - 2 - a$. Next, we put

$$p_{N,a,\delta} = \lim_{n \to \infty} \frac{[z^n] A_{N,a,\lfloor \delta n \rfloor}(z)}{\binom{n}{N}} \tag{2}$$

Then $p_{N,a,\delta}$ is the probability of the following event: the first interval is of an arbitrary length, then there are a intervals of lengths less than δ, and next we have two intervals of length at least δ.

Lemma 1

$$p_{N,a,\delta} = \sum_{i=1}^{\lfloor 1/\delta \rfloor - 2} \binom{a}{i}(-1)^i(1 - \delta(2 + i))^N \ .$$

Proof. The ordinary generating function of the class $A_{N,a,D}$ is

$$F_a(z) = \frac{1}{1-z}\left(z\frac{1-z^D}{1-z}\right)^a \left(z\frac{z^D}{1-z}\right)^2 \left(z\frac{1}{1-z}\right)^b = \frac{(1-z^D)^a z^{2D} z^N}{(1-z)^{N+1}}.$$

Let us put $G_a(z) = \frac{(1-z^D)^a z^N}{(1-z)^{N+1}}$ and notice that $[z^n]F_a = [z^{n-2D}]G_a$. Since $z^N/(1-z)^{N+1} = \sum_n \binom{n}{N}z^n$ (see [5], page 321), we have

$$G_a(z) = \left(\sum_{i=0}^{a}\binom{a}{i}(-1)^i z^{Di}\right)\left(\sum_{n}\binom{n}{N}z^n\right)$$

$$= \sum_{n} z^n \sum_{Di+j=n} \binom{a}{i}(-1)^i\binom{j}{N} = \sum_{n} z^n \sum_{i=0}^{\lfloor n/D \rfloor} \binom{a}{i}(-1)^i\binom{n - Di}{N}.$$

From this we deduce that

$$[z^n]F_a(z) = \sum_{i=0}^{\lfloor n/D \rfloor - 2}\binom{a}{i}(-1)^i\binom{n - Di - 2D}{N}.$$

Therefore we have

$$p_{N,a,\delta} = \lim_{n\to\infty} \sum_{i=0}^{\lfloor 1/\delta \rfloor - 2} \frac{\binom{a}{i}(-1)^i\binom{n-n\delta(i+2)}{N}}{\binom{n}{N}} = \sum_{i=0}^{\lfloor 1/\delta \rfloor - 2}\binom{a}{i}(-1)^i(1 - \delta(2 + i))^N.$$

This completes the proof.   □

Lemma 2. *If* $\delta \leq \dfrac{1}{N}$, *then*

$$\sum_{a=1}^{N-3} p_{N,a,\delta} = (1 - \delta)^N - (1 - 2\delta)^N + O(N^{\frac{5}{2}}e^{-N}) \ .$$

Proof. Obviously,

$$\sum_{a=1}^{N-3} p_{N,a,\delta} = \sum_{a=1}^{N-2} p_{N,a,\delta} - p_{N,N-2,\delta}.$$

We calculate the first factor on the right side:

$$\sum_{a=1}^{N-2} p_{N,a,\delta} = \sum_{i=0}^{\lfloor 1/\delta \rfloor - 2} \left(\sum_{a=1}^{N-2} \binom{a}{i} \right) (-1)^i (1 - \delta(i+2))^N$$

$$= \sum_{i=0}^{\lfloor 1/\delta \rfloor - 2} \left(\binom{N-1}{i+1} - \binom{0}{i} \right) (-1)^i (1 - \delta(i+2))^N$$

$$= \sum_{i=0}^{\lfloor 1/\delta \rfloor - 2} \binom{N-1}{i+1} (-1)^i (1 - \delta(i+2))^N - (1 - 2\delta)^N .$$

Next we have

$$\sum_{i=0}^{\lfloor 1/\delta \rfloor - 2} \binom{N-1}{i+1} (-1)^i (1 - \delta(i+2))^N =$$

$$= \sum_{i=0}^{\lfloor 1/\delta \rfloor - 2} \sum_{j=0}^{N} \binom{N-1}{i+1} (-1)^i \binom{N}{j} (-1)^j \delta^j (i+2)^j$$

$$= \sum_{j=0}^{N} \binom{N}{j} (-1)^j \delta^j \sum_{i=0}^{\lfloor 1/\delta \rfloor - 2} \binom{N-1}{i+1} (-1)^i (i+2)^j$$

$$= \sum_{j=0}^{N} \binom{N}{j} (-1)^j \delta^j \sum_{i=0}^{\lfloor 1/\delta \rfloor - 2} \frac{i+2}{N} \binom{N}{i+2} (-1)^i (i+2)^j$$

$$= \sum_{j=0}^{N} \binom{N}{j} (-1)^j \delta^j \frac{1}{N} \sum_{i=0}^{\lfloor 1/\delta \rfloor - 2} \binom{N}{i+2} (-1)^{i+2} (i+2)^{j+1}$$

$$= \sum_{j=0}^{N} \binom{N}{j} (-1)^j \delta^j \frac{1}{N} \sum_{i=2}^{\lfloor 1/\delta \rfloor} \binom{N}{i} (-1)^i (i)^{j+1}$$

$$= \sum_{j=0}^{N} \binom{N}{j} (-1)^j \delta^j \frac{1}{N} \left(N! \begin{Bmatrix} j+1 \\ N \end{Bmatrix} - \binom{N}{1} (-1)^1 1^{j+1} \right)$$

$$= \sum_{j=0}^{N} \binom{N}{j} (-1)^j \delta^j \frac{1}{N} N + \sum_{j=N-1}^{N} \binom{N}{j} (-1)^j \delta^j \frac{N!}{N} \begin{Bmatrix} j+1 \\ N \end{Bmatrix}$$

$$= (1-\delta)^N + N(-1)^{N-1} \delta^{N-1} \frac{N!}{N} + (-1)^N \delta^N \frac{N!}{N} \frac{N(N+1)}{2}$$

$$= (1-\delta)^N + O\left(\frac{N!}{N^{N-1}} \right) = (1-\delta)^N + O(N^{\frac{3}{2}} e^{-N}).$$

(Above, $\begin{Bmatrix} j+1 \\ N \end{Bmatrix}$ stands for the second Stirling number [5]. We have used the properties that $\begin{Bmatrix} j+1 \\ N \end{Bmatrix} = 0$ for $j < N-1$, $\begin{Bmatrix} N \\ N \end{Bmatrix} = 1$ and $\begin{Bmatrix} N+1 \\ N \end{Bmatrix} = \binom{N+1}{2}$)

We shall show now that $p_{N,N-2,\delta} = O(N^{\frac{3}{2}}e^{-N})$:

$$p_{N,N-2,\delta} = \sum_{i=0}^{N} \binom{N-2}{i}(-1)^i(1-\delta(i+2))^N$$

$$= \sum_{i=0}^{N} \binom{N-2}{i-2}(-1)^i \sum_{j=0}^{N}\binom{N}{j}(-1)^j\delta^j i^j$$

$$= \frac{1}{N(N-1)}\sum_{j=0}^{N}\binom{N}{j}(-1)^j\delta^j \sum_{i=0}^{N}\binom{N}{i}(-1)^i i^{j+1}(i-1)$$

$$= \frac{N!}{N(N-1)}\sum_{j=0}^{N}\binom{N}{j}(-1)^j\delta^j \left(\left\{{j+2 \atop N}\right\} - \left\{{j+1 \atop N}\right\}\right)$$

$$= \frac{N!}{N(N-1)}\sum_{j=N-2}^{N}\binom{N}{j}(-1)^j\delta^j \left(\left\{{j+2 \atop N}\right\} - \left\{{j+1 \atop N}\right\}\right)$$

$$= \frac{N!\,\delta^{N-2}}{N(N-1)}\left(\frac{1}{2}N(N-1)(-1)^{N-2} + N(-1)^{N-1}\delta\left(\left\{{N+1 \atop N}\right\} - 1\right) + \right.$$
$$\left. +(-1)^N\delta^2\left(\left\{{N+2 \atop N}\right\} - \left\{{N+1 \atop N}\right\}\right)\right)$$

$$= \frac{N!\,\delta^{N-2}}{N(N-1)}\left(\frac{1}{2}N(N-1))(-1)^{N-2} + N(-1)^{N-1}\delta\left(\frac{1}{2}N(N+1) - 1\right) + \right.$$
$$\left. +(-1)^N\delta^2\left(\frac{1}{24}(3N^4 + 10N^3 - 3N^2 - 10N)\right)\right)$$

$$= O\left(\frac{N!}{N^{N-2}}\right) = O(N^{\frac{5}{2}}e^{-N}).$$

So the proof is finished. □

Lemma 3. *Let $\{P_1 < \ldots < P_N\}$ be a random subset of the interval $[0,1]$ and let $x_2 = P_2 - P_1$. Then $\Pr(x_2 > \delta) = (1-\delta)^N$.*

Proof. Let $B_{N,D}$ denotes the combinatorial class

$$Seq(\{\circ\}) \times \{\bullet\} \times Seq_{>D}(\{\circ\}) \times (\{\bullet\} \times Seq(\{\circ\}))^{N-1}$$

The class $B_{N,D}$ is isomorphic to the class of all sequences

$$(k_1, l_1, n_1, n_2, \ldots, n_{N-1})$$

of natural numbers, such that $l_1 > D$. The ordinary generating function of the class $B_{N,D}$ is

$$B_{N,D}(z) = \frac{1}{1-z}z\frac{z^D}{1-z}\left(z\frac{1}{1-z}\right)^{N-1} = z^D\frac{z^N}{(1-z)^{N+1}} = z^D\sum_n\binom{n}{N}z^n.$$

From this we deduce that

$$[z^n]B_{N,D}(z) = [z^n]z^D \sum_n \binom{n}{N} z^n = [z^{n-D}] \sum_n \binom{n}{N} z^n = \binom{n-D}{N}$$

Therefore we have

$$\Pr(x_2 > \delta) = \lim_{n\to\infty} \frac{[z^n]B_{N,\lfloor \delta n \rfloor}(z)}{\binom{n}{N}} = \lim_{n\to\infty} \frac{\binom{n-\delta n}{N}}{\binom{n}{N}} = (1-\delta)^N$$

which completes the proof.   □

Theorem 4. *The probability of successful transmission for N stations and the delay parameter $\delta \le \frac{1}{N}$ is at least $2(1-\delta)^N - (1-2\delta)^N$.*

Proof. We consider two cases for which the algorithm succeeds: the first one occurs when the time distance between the first and the second starting moment of a transmission is larger than δ. By Lemma 3, this event has probability probability $(1-\delta)^N$. The second case is when there are some three consecutive starting moments of a transmission with distance larger than δ. Since we want to avoid double counting the first case, we consider only P_i, P_{i+1}, P_{i+2} such that $i \ge 3$. We may apply Lemma 2 in this case, so we get probability at least $(1-\delta)^N - (1-2\delta)^N$. Putting this together, we obtain the required result.   □

Notice that

$$\lim_{N\to\infty} 2(1-\frac{1}{N})^N - (1-2\frac{1}{N})^N = \frac{2}{e} - \frac{1}{e^2} \approx 0.6004 \,,$$

hence $\delta = \frac{1}{N}$ is slightly too large to ensure optimal chance for a successful execution. However, if we put $\delta = \frac{1}{N \ln N}$, then we have

$$\lim_{N\to\infty} 2(1 - \frac{1}{N \ln N})^N - (1 - 2\frac{1}{N \ln N})^N = 1 \,.$$

Hence the delay $\delta = \frac{1}{N \ln N}$ ensures success for large N.

We assume now that each station chooses to participate in a slot independently with probability $\frac{a}{N}$, where N is a total number of stations. Our goal is to find the best value of the parameter a.

Proposition 5. *Suppose $\delta \le \frac{1}{N}$ and that each station transmits independently with probability $\frac{a}{N}$. Then the probability of a success is at least*

$$2\left(1 - \frac{\delta a}{N}\right)^N - \left(1 - \frac{2\delta a}{N}\right)^N - \left(1 - \frac{a}{N}\right)^N.$$

Proof. The probability of a success can be calculated as

$$\sum_{k=1}^{N} \Pr(\text{success} | k \text{ stations transmits}) \cdot \Pr(k \text{ stations choose to participate})$$

$$\geq \sum_{k=1}^{N} (2(1-\delta)^k - (1-2\delta)^k) \binom{N}{k} \left(\frac{a}{N}\right)^k \left(1-\frac{a}{N}\right)^{N-k}$$

$$= 2 \sum_{k=1}^{N} \binom{N}{k} \left((1-\delta) \cdot \frac{a}{N}\right)^k \left(1-\frac{a}{N}\right)^{N-k}$$

$$- \sum_{k=1}^{N} \binom{N}{k} \left((1-2\delta) \cdot \frac{a}{N}\right)^k \left(1-\frac{a}{N}\right)^{N-k}$$

$$= 2 \left(1-\frac{\delta a}{N}\right)^N - \left(1-\frac{2\delta a}{N}\right)^N - \left(1-\frac{a}{N}\right)^N .$$

This completes the proof. □

Notice that the probability from the last proposition can be approximated by

$$f_\delta(a) = 2e^{-\delta a} - e^{-2\delta a} - e^{-a} .$$

The function f_δ has a unique maximum, but the equation $f_\delta'(a) = 0$ cannot be solved by an analytic formula. However, it can be shown that a good approximation of this equation is given by $a = W(\frac{1}{2\delta^2})$, where W is the Lambert W function. Using the well known approximation $W(x) \simeq \ln(x) - \ln\ln(x)$ we deduce that a good approximation of the maximum of the function f_δ is given by $a_\delta = \ln(\frac{1}{2\delta^2}) - \ln\ln(\frac{1}{2\delta^2})$. For $\delta = 1/100, 1/200, \ldots, 1/1000$ we obtain the following values for optimal value of the parameter a_δ: 6.37511, 7.6106, 8.34283, 8.86588, 9.2734, 9.60745, 9.89059, 10.1364, 10.3535, 10.548.

Theorem 6. *Let* $a_\delta = \ln(\frac{1}{2\delta^2}) - \ln\ln(\frac{1}{2\delta^2})$ *and* $f_\delta(a) = 2e^{-\delta a} - e^{-2\delta a} - e^{-a}$. *Then for each* $\varepsilon > 0$ *we have* $f_\delta(a_\delta) = 1 - o(\delta^{2-\varepsilon})$.

Therefore, if we use that parameter $a_\delta = \ln(\frac{1}{2\delta^2}) - \ln\ln(\frac{1}{2\delta^2})$ for the initialization protocol, then the probability of a success is at least $1 - \delta^2$.

Theorem 7. *To initialize an* n *station ad hoc network, using the proposed algorithm, the initialization will be completed with probability exceeding* $1 - \frac{1}{n}$ *in less then* $\frac{1}{1-\delta^2} n + O(\sqrt{n \log n})$ *time slots.*

Proof. Denote by X the number of successful slots during k sequential slots. Each time slot is successful with probability at least $p = 1 - \delta^2$. Then by Appendix A we have that with probability exceeding $1 - \frac{1}{n}$ in less then $\frac{1}{1-\delta^2} n + O(\sqrt{n \log n})$ time slots the initialization will be completed.

However, up to now, we have considered the length of the slot as $\alpha \geq 1$. Since $\delta = \frac{\lambda}{\alpha - 1}$, then $\alpha = \frac{\lambda}{\delta} + 1$. Hence from Theorem 7 the normalized time cost for initializing the network is

$$\alpha \left(\frac{1}{1 - \delta^2} n + O(\sqrt{n \log n}) \right) = \frac{\lambda + \delta}{\delta(1 - \delta^2)} n + O(\sqrt{n \log n}) \ .$$

Because we want to minimize the normalized time cost we choose δ in the following way

$$\delta^* = \arg \min_{\delta} \frac{\lambda + \delta}{\delta(1 - \delta^2)} \ .$$

Solving this optimization problem we obtain

$$\delta^* = \frac{1}{2} \left(\frac{\lambda^2}{\sqrt[3]{-\lambda^3 + 2\lambda + 2\sqrt{\lambda^2 - \lambda^4}}} + \sqrt[3]{-\lambda^3 + 2\lambda + 2\sqrt{\lambda^2 - \lambda^4}} - \lambda \right)$$

Hence, optimal length $\alpha^* = \frac{\lambda}{\delta^*} + 1$.

Comparison with the previous algorithms. Now let us compare in a table the execution time of our algorithm with the algorithm from [1] for different parameter values:

λ	Algorithm from [1]	Our algorithm
0.00001	$\leq 1.0177n + O\left(\sqrt{n \log n}\right)$	$\leq 1.00088n + O\left(\sqrt{n \log n}\right)$
0.0001	$\leq 1.0520n + O\left(\sqrt{n \log n}\right)$	$\leq 1.00408n + O\left(\sqrt{n \log n}\right)$
0.001	$\leq 1.1532n + O\left(\sqrt{n \log n}\right)$	$\leq 1.01902n + O\left(\sqrt{n \log n}\right)$

Time cost of the algorithm from [7] is $2.7183n + O\left(\sqrt{n \log n}\right)$, which is much higher than for our algorithm and for [1].

4   Conclusions

We developed n-station initialization algorithm which make use of carrier sensing and its computational complexity is very close to a theoretical lower bound. It is worth to mention that the technique of combinatorial classes used by us to estimate a required probability turned out to be a very powerful tool in this setting. It seems that it could be very helpful to improve existing algorithms and to obtain a much better lower bounds in initialization algorithms [7,1] and other closely related problems.

References

1. Z. Cai, M. Lu, X. Wang, *Distributed Initialization Algorithm for Single-Hop Ad Hoc Networks with Minislotted Carrier Sensing*, IEEE, Transactions Parallel and Distributed Systems, vol 14, no. 5, pp. 516-528, May 2003.
2. P. Flajolet, R. Sedgewick, *Analytic Combinatorics*, http://algo.inria.fr/flajolet/Publications/books.html

3. IEEE, *Wireless LAN Medium Access Control (MAC) and Physical Layer (PHY) specifications*, IEEE 802.11 standards, June 1999.
4. K. Jamieson, B. Hull, A. K. Miu, H. Balakrishnan, *Understanding the Real-World Performance of Carrier Sense*, ACM SIGCOMM, Workshop on Experimental Approaches to Wireless Network Design and Analysis (E-WIND), Philadelphia, PA, August 2005.
5. R. Graham, D. Knuth, O. Patashnik, *Concrete Mathematics: A Foundation to Computer Science*, Addison-Wesley, 1994.
6. A. Micic, I. Stojmenovic, *A Hybrid Randomized Initialization Protocol for TDMA in Single-Hop Wireless Networks*, Proceedings of the International Parallel and Distributed Processing Symposium, pp. 147-154, April 2002.
7. K. Nakano, S. Olariu, *Randomized Initialization Protocols for Ad Hoc Networks*, IEEE, Transactions Parallel and Distributed Systems, vol 11, no. 7, pp. 749-759, July 2000.
8. K. Whitehouse, A. Woo, F. Jiang, J. Polastre, D. Culler, *Exploiting the Capture Effect for Collision Detection and Recovery*, The Second IEEE Workshop on Embedded Networked Sensors (EmNetS-II). Sydney, Australia. May 30-31, 2005.

Appendix A

Let X be the random variable following the binomial distribution $\mathbf{B}(k, p)$, which we can interpreted as the number of successes in k independent Bernoulli trials, each succeeding with probability p. To analyze the tail of the binomial distribution we use the Chernoff bound

$$\Pr(X < (1 - \epsilon)\mathrm{E}(X)) \le e^{-\frac{\epsilon}{2}\mathrm{E}(X)} .$$

Our goal is to find k such that $\Pr(X < d) < \frac{1}{f}$. In order to estimate the value of $\Pr(X < d)$ by the properties

$$\Pr(X < d) = \Pr(X < (1 - \epsilon)\mathrm{E}(X)) < e^{-\frac{\epsilon}{2}\mathrm{E}(X)} = \frac{1}{f} ,$$

we have to solve the equations

$$\begin{cases} (1 - \epsilon)\mathrm{E}(X) = \quad d , \\ (\epsilon^2/2)\mathrm{E}(X) = \log f . \end{cases}$$

From these equations we deduce that

$$\mathrm{E}(X) = d + \log f + \sqrt{2d \log f + (\log f)^2} .$$

Since $\mathrm{E}(X) = k \cdot p$, we can determine the number k of Bernoulli trials as

$$k = \frac{1}{p}(d + \log f + \sqrt{2d \log f + (\log f)^2} . \tag{3}$$

Performance Modeling of a Bottleneck Node in an IEEE 802.11 Ad-Hoc Network

Hans van den Berg[1,2], Michel Mandjes[3,4], and Frank Roijers[1,3]

[1] TNO Information and Communication Technology, The Netherlands
[2] Department of Design and Analysis of Communication Systems,
University of Twente, The Netherlands
[3] Centre for Mathematics and Computer Science, The Netherlands
[4] Korteweg-de Vries Institute, University of Amsterdam, The Netherlands

Abstract. The IEEE 802.11 MAC-protocol, often used in ad-hoc networks, has the tendency to share the capacity equally amongst the active nodes, irrespective of their loads. An inherent drawback of this fair-sharing policy is that a node that serves as a relay-node for multiple flows is likely to become a *bottleneck*. This paper proposes a flow-level performance model of such a bottleneck node using fluid-flow analysis. Assuming Poisson initiations of new flow transfers at the bottleneck node, we obtain insightful, robust, and explicit expressions for characteristics related to the overall flow transfer time, the buffer occupancy, and the packet delay at the bottleneck node. The analysis is enabled by a translation of the behavior of the bottleneck node and the source nodes in terms of an M/G/1 queueing model. We conclude the paper by an assessment of the impact of alternative capacity sharing amongst source nodes and the bottleneck in order to improve the performance of the bottleneck.

1 Introduction

Developments in wireless communication technology open up the possibility of wireless ad-hoc networks; these networks can be deployed instantly without a fixed infrastructure or pre-advanced configuration and multi-hop connectivity is one of the key-features. Currently, IEEE 802.11 wireless LAN [12] is the most popular technology used for wireless ad-hoc networks.

In IEEE 802.11 based ad-hoc networks stations have to contend for access to the wireless medium according to the Distributed Coordination Function (DCF). DCF is a distributed random medium access mechanism based on CSMA/CA — for a more elaborate description, see e.g. [2,12]. In the literature it is shown that the DCF tends to share the wireless medium capacity *equally* amongst contending stations, cf. [2,14]. Clearly, DCF is particularly appropriate in the context of ad-hoc networks as it operates in a fully distributed fashion.

DCF has, however, also significant drawbacks. Most notably, it facilitates only equal sharing, i.e., it is not capable of granting stations different shares of the available capacity. The lack of such differentiation options may cause nodes to become bottlenecks, as can be seen as follows. Stations that cannot communicate

T. Kunz and S.S. Ravi (Eds.): ADHOC-NOW 2006, LNCS 4104, pp. 321–336, 2006.

directly with each other use other stations to relay their traffic; in particular, nodes that have a central location in the ad-hoc network are likely to become such a relay node. Let n stations send traffic via the same relay node. Then, due to the sharing policy, the relay-node obtains just the same share of the medium capacity, viz. $1/(n+1)$, as each of the 'sending nodes'. In other words, as soon as $n > 1$, the node's input rate exceeds its output rate, and hence the excess traffic accumulates in the node's buffer; only when $n = 0$ the queue drains. This entails that these relay stations become bottlenecks, and will strongly affect the performance of the flow transfers through the network.

In this paper we study the performance of wireless ad-hoc networks focusing on the Medium Access Control (MAC) layer. In particular, the goal of this paper is to investigate the flow-level performance of a multi-hop flow, e.g. the transfer time of an entire flow consisting of multiple packets. The attractive feature of this performance measure is that it directly relates to the perspective of the users: the quality perceived by a user is primarily based on the time required to transmit the entire flow, rather than the delays or loss probabilities of individual packets. For that purpose we develop and analyze a stochastic model of a simple ad-hoc network scenario capturing the essential characteristics of the resource sharing enforced by the IEEE 802.11 MAC protocol of a 2-hop flow.

In contrast with the analytical modeling approach presented in our paper, almost all earlier studies on the performance of IEEE 802.11 multi-hop ad-hoc networks available in the literature are based on simulation, see e.g. [9,10]. These studies usually capture many details of the ad-hoc network protocols, but, unfortunately, are mostly limited with respect to the considered traffic scenarios and often do not provide much insight into the essentials of the behavior of the system. In particular, in most existing studies it is assumed that there are a fixed number of persistent traffic streams, thus ignoring the dynamics and random nature of the traffic generated by the users in these networks. Analytical studies on ad-hoc networks only consider single-hop flow transfers; the performance of multi-hop flows is not analyzed. Also, they mostly only consider the overall throughput or the node throughput, and not the flow throughput or transfer time. These omissions of the existing literature will be illustrated in the literature overview below.

Literature. This overview covers analytical studies on ad-hoc networks. These studies can roughly be divided into three classes: packet-level in a single-hop network (e.g. WLAN), flow-level in a single-hop network, and packet-level in a multi-hop network. As will become clear, a flow-level analysis of multi-hop ad-hoc networks was not covered yet.

The packet-level behavior of IEEE 802.11 WLAN in a single-hop network has been investigated extensively. A detailed mathematical performance model of the DCF has been developed and analyzed by Bianchi [2]. This paper assumes a constant number of persistently contending stations and rely on a relatively simple Markov chain analysis, neglecting only minor dependencies among the behavior of different stations, and is used to obtain the saturation throughput. Comparison with simulation shows that the analytical results are in general remarkably accurate. Bianchi's model is also used to obtain packet contention delays for saturated [3] and also non-saturated sources [7].

Flow-level behavior in a single-hop network is considered in [8,14,19]. The situation with non-persistent traffic sources is considered, i.e., the number of active stations varies dynamically in time according to the initiation and completion of file transfers at random time instants. These papers propose and analyze simplified analytical models yielding approximations for the expected flow (file) transfer time. In particular, in [14] the analysis is based on the modeling assumption that, from the flow-level point of view, the WLAN can be regarded as a Processor Sharing type of queueing system. The analyses in [8,14,19] ignore the effects of higher layer protocols, in particular TCP, on the traffic behavior. Several papers consider flow transfer times of TCP-flows over WLAN focusing on the impact of the interaction between TCP's feedback control loop and the DCF MAC protocol on TCP throughputs and fairness, see e.g. [16,17].

Analytical models for packet-level performance in ad-hoc networks are presented in e.g. [4,11]. These papers consider situations in which the stations experience different channel-conditions as the number of neighbors and the distances to the neighbors varies per station. The performance measures of interest are the overall aggregate throughput and the throughput per node including the impact of hidden stations; unfortunately they do not consider the performance of multi-hop flows.

Contribution. The present paper analyzes the transfer times of multi-hop flows through a bottleneck of an ad-hoc network. According to a Poisson process inactive source nodes become active and initiate new flow transfers to destinations via the bottleneck node. After a source node has transmitted the entire flow to the bottleneck node, it returns to the inactive state (note that a part of the flow may still be in the buffer of the bottleneck node waiting for service). The model is of a fluid nature: the packet stream is approximated by a continuous stream.

As remarked earlier, the bottleneck node's buffer only drains when there are no source nodes feeding into the node, due to the way DCF shares the medium capacity. It is exactly this property that makes it possible to recast the buffer dynamics in terms of the classical M/G/1 queueing model. This facilitates an explicit and insightful characterization of the mean overall flow transfer time. Interestingly, the resulting formula depends on the flow-size distribution only through its first two moments; in other words, there is a high degree of insensitivity. We also derive expressions for the mean buffer occupancy and the packet delay at the bottleneck node.

Finally, we discuss the possibility of improving the overall flow transfer time by alternative resource sharing between the stations.

Organization. This paper is organized as follows. In Section 2 we introduce the ad-hoc network scenario considered in this paper and develop an analytical performance model for this scenario. Section 3 contains the analysis of this model, whereas the corresponding numerics are presented in Section 4.

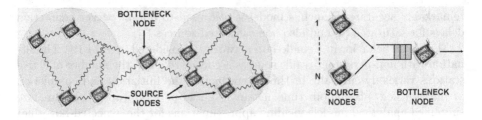

Fig. 1. Left: Bottleneck node connecting two ad-hoc network clusters. Right: Two-hop network with a single (bottleneck) node used as relay node by many sources.

Section 5 contains considerations about the performance improvement that can be obtained by differentiating the capacity between source and bottleneck nodes. Section 6 presents concluding remarks and directions for further research.

2   System and Model Description

This section describes the ad-hoc network scenario and introduces the fluid-flow model that is used to analyze the ad-hoc network in the next section.

2.1   The Ad-Hoc Network Scenario

We consider a wireless ad-hoc network scenario in which a single node is used by many other nodes as a relay node for data transmission to receivers elsewhere in the network. This may, for example, occur when connectivity between two ad-hoc network clusters is (during some time) provided by only one node acting as a kind of 'bridge', see left graph of Fig. 1. It is clear that in such a situation this node forms a potential bottleneck. The aim of the present paper is to investigate the performance of this potential bottleneck in terms of, e.g., the flow transfer times and the queueing delays that are encountered by the traffic generated by other network nodes. In particular, we are interested in the impact of the *interaction* between the traffic generating (neighbor) nodes and the bottleneck node due to sharing common radio transmission resources.

For that purpose we focus on a simple, special case of the above scenario yet capturing the essentials of this interaction; a two-hop ad-hoc network consisting of a number of nodes (sources) that may initiate data transmissions, and a single relay node that forwards the traffic generated by the other nodes to a next-hop destination, see the right graph of Fig. 1. The source nodes and the relay node (bottleneck node) are all within each others transmission range. Data transmissions are controlled by the DCF of the IEEE 802.11B MAC protocol (including the RTS/CTS-access mechanism for collision avoidance), see [12]. Flow transfers by the source nodes are initiated at random time instants; the sizes of the flows transmitted by the source nodes are also random.

The following important observations with respect to the behavior of IEEE 802.11B ad-hoc network nodes sharing common radio transmission resources, obtained in other papers (see e.g. [2,14]), motivate the fluid-flow modeling approach described below in detail:

- Assuming RTS/CTS-access the aggregate throughput C provided by the system is (almost) independent of the number of active nodes (i.e., nodes contending for the radio resources in order to send their data), cf. the model of Bianchi [2]. The aggregate throughput C can be modelled as a fixed rate which is a percentage of the data transmission rate; the actual percentage depends on the used physical layer technology (in case of IEEE 802.11B physical layer with a data transmission rate of 11 Mbit/s rate the aggregate throughput is about 60% of the data transmission rate).
- The aggregate system throughput is more or less equally shared by the active network nodes, cf. see [14].

2.2 Model Description

The ad-hoc network scenario described above is modeled as a fluid-flow queueing system. We assume a large number of source nodes which become active and initiate flow transfers to destinations via the bottleneck node according to a Poisson process with rate λ ('flow arrival rate'). Note, that a source node has at most one flow transfer in progress. The active source nodes and the bottleneck node equally share the system capacity C; i.e., when n source nodes have a flow transfer in progress, any source transmits its traffic (fluid) into the buffer of the bottleneck node at rate $C/(n+1)$, while a rate $C/(n+1)$ is used by the bottleneck node to 'serve' the buffer (i.e., to forward the traffic stored in its buffer to the next node). Recall that for $n > 1$ the bottleneck node's input rate exceeds its output rate, and the excess traffic accumulates in the node's buffer; the queue only drains when $n = 0$. Once a source has completed the flow transmission (i.e., the source node has transmitted the flow's last traffic particle to the bottleneck node), the source node becomes inactive (although the last part of the flow may still be at the buffer of the bottleneck node waiting for service). Flow sizes (in terms of the amount of traffic/fluid) are i.i.d. random variables (denoted by F) with finite mean f and second moment f_2. We define the load of the system by $\rho = \lambda f/C$. All flows have to be served twice, i.e., once by the source node and once by the bottleneck node, resulting in the overall stability condition $\rho < 1/2$.

Our main performance measures of interest are the steady-state buffer occupancy (delay) Q_{BUFFER} (D_{BUFFER}) at the bottleneck node and the overall flow transfer time D_{OVERALL}, i.e., the time required to completely transfer a flow from source to destination. The overall flow transfer time is the sum of two other performance measures: (i) the time (D_{SOURCE}) a source requires to completely transfer a particular flow to the bottleneck node, and (ii) the delay at the bottleneck node (D^*_{BUFFER}) of the last particle of fluid of the flow (the asterisk denotes that the performance measure is considered for the last particle of a flow).

The analysis in the next section aims at the derivation of the mean values of these performance measures; insightful, explicit formulas are obtained, which can easily be evaluated in order to generate numerical results. Derivation of higher-order statistics of the performance measures appears to be also possible, but requires substantially more complicated mathematical analysis. In the complementary paper [15] we consider the same model and derive expressions for the Laplace Transforms of D_{BUFFER} and D_{OVERALL}, and characterize the tail probabilities of these performance measures.

3   Analysis

This section analyzes the model presented in Section 2.2. Exact expressions are derived for the performance measures flow transfer time and the buffer occupancy and approximations are presented for the buffer delay of the last particle and the overall flow transfer time.

3.1   Source Behavior

The dynamics of the source nodes in our model can be described by a Processor Sharing (PS) queueing model with state dependent service rates $cn/(n+1)$, whenever n flows are in progress. This model is a special case of the so called Generalized Processor Sharing (GPS) model extensively studied by Cohen [6]. In the GPS model, whenever there are i customers present in the system, each customer receives service at a rate $r(i)$, where $r(\cdot)$ is some arbitrary positive function (under some weak assumptions). Cohen derives the following result for the joint stationary probability/density function of the number of customers N in the GPS system and their residual service requirements $T := (T(1), ..., T(N))$, cf. formula (7.19) in [6]:

$$\mathbb{P}(N = n, T = \tau) = \frac{\frac{(\lambda\beta)^n}{n!}\varphi(n)}{\sum_{k=0}^{\infty}\frac{(\lambda\beta)^k}{k!}\varphi(k)}\prod_{i=1}^{n}\frac{1 - B(\tau(i))}{\beta}, \quad n = 0, 1, ..., \quad \tau(i) \geq 0,$$
(1)

where $\varphi(0) := 1$ and $\varphi(n) := (\prod_{i=1}^{n} r(i))^{-1}$, for $n = 1, 2, ...$, and where $B(\cdot)$ denotes the customers' service requirement distribution, β is the mean service requirement and λ the customer arrival rate. This general result will be extensively exploited in the remainder of this paper.

In our model $r(i) = c/(i+1)$ and we obtain the following result for the distribution of the number of active source nodes in our model:

$$\pi_n := \mathbb{P}(N = n) = (n+1)(1-\rho)^2\rho^n.$$

Note, cf. equation (1), that the distribution of N, and hence also its mean, is *insensitive* to the flow-size distribution apart from its mean. The mean number of source nodes that is simultaneously active is given by:

$$\mathbb{E}N = 2\frac{\rho}{1-\rho}.$$

Flow Transfer Time D_{SOURCE}. From the above result, using Little's formula, we obtain the mean flow transfer time at the sources:

$$\mathbb{E}D_{\text{SOURCE}} = \frac{\mathbb{E}N}{\lambda} = 2\frac{f/\text{C}}{1-\rho}. \tag{2}$$

$\mathbb{E}D_{\text{source}}$ is also *insensitive* to the flow-size distribution.

The *conditional* flow transfer time $\mathbb{E}D_{\text{source}}(x)$, the time required by a source node to transmit a flow of give size x, is linear in x, see Cohen [6]:

$$\mathbb{E}D_{\text{SOURCE}}(x) = \frac{x}{\text{C}\rho}\mathbb{E}N = \frac{2x/\text{C}}{1-\rho}. \tag{3}$$

3.2   Buffer Occupancy

The mean buffer occupancy is derived for two different epochs: an arbitrary epoch (which relates to the buffer occupancy at the moment that a new flow transfer is initiated by a source node) and the buffer occupancy that is seen by the last particle of a flow that arrives at the bottleneck node.

Buffer Occupancy at an Arbitrary Epoch Q_{BUFFER}. In order to analyze the buffer occupancy we consider the amount of 'work' in the system (i.e., the service time required to serve the buffer occupancy related to the service rate C). The total amount of work in the system W_{TOTAL} consists of two parts: the amount of work W_{SOURCES} present at the source nodes (i.e., fluid that remains to be sent by the actively transmitting source nodes) and the amount of work W_{BUFFER} present in the buffer of the bottleneck node, so

$$\mathbb{E}W_{\text{TOTAL}} = \mathbb{E}W_{\text{SOURCES}} + \mathbb{E}W_{\text{BUFFER}}.$$

$\mathbb{E}W_{\text{TOTAL}}$ is obtained by considering the whole system (i.e., sources plus bottleneck node with buffer) as a single service center. Flows arrive at the service center according to a Poisson process with rate λ and have to be served essentially *twice* (i.e., one service corresponds to a source node transmitting its flow to the bottleneck node and the second service corresponds to the forwarding of the flow by the bottleneck node). Using that the overall system is 'work conserving' (i.e., full capacity C is used whenever work is present in the system), it follows that the (steady-state) distribution of the total amount of work W_{TOTAL} in the overall system equals the distribution of the amount of work $W_{\text{M/G/1}}$ in a 'corresponding' M/G/1 queueing system with flow arrival rate λ, flow size $2F$ and service rate C. Hence, from the Pollaczek-Khintchine formula (see e.g. [18]) we get

$$\mathbb{E}W_{\text{TOTAL}} = \mathbb{E}W_{\text{M/G/1}} = \frac{2\lambda f_2}{1-2\rho}\frac{1}{\text{C}^2}.$$

The amount of work present at the sources W_{SOURCES} consists of the residuals of the flows that are still with the active sources, i.e., not yet transmitted to the bottleneck node.

Formula (1) shows that the residual flow sizes at the sources are all identically distributed, mutually independent and independent of the total number of active sources. In particular, this distribution of the residual flow size (with mean $f_2/(2f)$) is the so-called excess distribution of the initial flow sizes well-known from renewal theory, see e.g. Ch. 1 of [18]. From these observations it follows that the mean total amount of fluid present at the sources is given by $\mathbb{E}N \cdot f_2/(2f)$. Hence, taking into account that the residuals of the flows at the sources have to be 'served' twice (i.e., to be transmitted by the sources and by the bottleneck node), we have:

$$\mathbb{E}W_{\text{SOURCES}} = 2\mathbb{E}N \frac{f_2}{2f\text{C}}.$$

Hence, the mean amount of work in the buffer at the bottleneck node is given by:

$$\mathbb{E}W_{\text{BUFFER}} = \mathbb{E}W_{\text{TOTAL}} - \mathbb{E}W_{\text{SOURCES}} = \frac{2\rho^2 f_2}{f\text{C}} \frac{1}{(1-2\rho)(1-\rho)}. \tag{4}$$

Finally, the expected buffer occupancy in terms of fluid at an arbitrary epoch is obtained from:

$$\mathbb{E}Q_{\text{BUFFER}} = \text{C} \cdot \mathbb{E}W_{\text{BUFFER}}.$$

Buffer Occupancy Seen by the Last Particle of a Flow Q^*_{BUFFER}. The buffer occupancy seen by the last particle of a particular flow is the buffer occupancy at the initiation of the flow plus the buffer increase during its flow transfer time D_{SOURCE} (note that the buffer occupancy cannot decrease during the presence of a flow in our specific sharing policy). By the PASTA-property, the mean buffer occupancy at the epoch of the flow initiation is equal to (the time-average mean) $\mathbb{E}Q_{\text{BUFFER}}$.

The expected growth of the buffer $\mathbb{E}\Delta Q_{\text{BUFFER}}$ during the flow transfer time can be derived easily as the sending rates of the source nodes and the bottleneck node are coupled. In particular, if a source requires time $D_{\text{SOURCE}}(x)$ to transmit a flow of size x to the bottleneck node, the bottleneck node will have served also an amount x. This means that the aggregate input of all sources is $D_{\text{SOURCE}}(x)\text{C} - x$, resulting in a buffer growth of $D_{\text{SOURCE}}(x)\text{C} - 2x$, hence

$$\mathbb{E}\Delta Q_{\text{BUFFER}}(x) = \mathbb{E}D_{\text{SOURCE}}(x)\text{C} - 2x \tag{5}$$

which is linear in x. Hence,

$$\mathbb{E}\Delta Q_{\text{BUFFER}} = \mathbb{E}D_{\text{SOURCE}}\text{C} - 2f = \frac{2f\rho}{1-\rho},$$

and

$$\mathbb{E}Q^*_{\text{BUFFER}} = \mathbb{E}Q_{\text{BUFFER}} + \mathbb{E}\Delta Q_{\text{BUFFER}} = \frac{2\rho^2 f_2/f}{(1-2\rho)(1-\rho)} + \frac{2f\rho}{1-\rho}. \tag{6}$$

Finally, the conditional buffer occupancy, also by the PASTA-property, is

$$\mathbb{E}Q^*_{\text{BUFFER}}(x) = \mathbb{E}Q_{\text{BUFFER}} + \mathbb{E}\Delta Q_{\text{BUFFER}}(x). \tag{7}$$

3.3 Buffer Delay

The mean buffer delay is derived for two different fluid particles: an arbitrary particle and the last particle of a flow that arrives at the bottleneck node.

Buffer Delay of an Arbitrary Particle D_{BUFFER}. The mean delay of an arbitrary fluid particle in the buffer of the bottleneck node can be directly obtained from Little's formula:

$$\mathbb{E}D_{\text{BUFFER}} = \frac{\mathbb{E}Q_{\text{BUFFER}}}{\lambda f} = \frac{2\rho f_2/(f\text{C})}{(1-2\rho)(1-\rho)}. \tag{8}$$

Buffer Delay of the Last Particle of a Flow D^*_{BUFFER}. The buffer delay of the last particle of a flow is equal to the time required by the bottleneck node to serve the buffer content Q^*_{BUFFER} present upon the last particle's arrival. For the analysis of this quantity it is useful to note that, from the point of view of the bottleneck node, the system behaves as a standard Processor Sharing (PS) queueing model; the bottleneck node receives capacity $\text{C}/(n+1)$ if there are n other active sources. So the buffer delay of the last particle corresponds to the time required to serve a given amount of fluid (i.e., the buffer content upon the last particle's arrival) in a PS queueing model.

For the moment we assume exponentially distributed flow sizes; the remark at the end of this subsection generalizes the result to non-exponentially distributed flow sizes. Coffman, Muntz, and Trotter [5] studied the so-called *response time* for jobs in an M/M/1-PS queue. They obtained an explicit expression ((33) of [5]) for the mean response time $\mathbb{E}X_n(\tau)$ of a tagged job requiring an amount τ of service, when there are n other active source nodes upon arrival of the tagged job. In terms of our system this expression reads:

$$\mathbb{E}X_n(\tau) = \tau + \frac{\rho\tau}{1-\rho} + (n(1-\rho)-\rho)\,(f/\text{C})\frac{1-\exp(-(1-\rho)\tau\text{C}/f)}{(1-\rho)^2}. \tag{9}$$

Using this result we can derive an expression for the expected buffer delay of the last particle of a flow in the case that flow sizes are exponentially distributed. Let $w_n(\tau)$ denote the probability density function of the amount of work at the buffer at a departure epoch of a source node leaving n active source nodes behind. Taking into consideration that the distribution of the number of source nodes left behind at a source node departure epoch is distributed according to the stationary distribution ('departure theorem' of product form networks, see also [6]), we obtain

$$\mathbb{E}D^*_{\text{BUFFER}} = \sum_{n=0}^{\infty} \pi_n \int_0^\infty \mathbb{E}X_n(\tau)w_n(\tau)\mathrm{d}\tau.$$

Unfortunately, an exact expression for $w_n(\tau)$ is not available. As an approximation we assume

$$w_n(\tau) \approx w(\tau), \tag{10}$$

where $w(\tau)$ denotes the unconditional probability density function of the amount of work at the buffer at a source node departure epoch. Approximation (10) is expected to work well as the dependency between the number of active source n and the buffer occupancy is not expected to be strong; the buffer occupancy cannot decrease if source nodes are active which means that leaving behind a higher number of active source nodes does not necessarily imply that the buffer occupancy is higher than when a lower number of active source nodes are left behind. Hence,

$$\mathbb{E}D^*_{\text{BUFFER}} \approx \sum_{n=0}^{\infty} \pi_n \int_0^{\infty} \mathbb{E}X_n(\tau)w(\tau)\mathrm{d}\tau.$$

Unfortunately, an expression for $w(\tau)$ is also not available. However, an insightful observation is that if $\mathbb{E}X_n(\tau)$ would be linear in τ (possibly with a non-zero intercept), then $\int \mathbb{E}X_n(\tau)w_n(\tau)\mathrm{d}\tau = \mathbb{E}X_n(\mathbb{E}W^*_{\text{BUFFER}})$ where $W^*_{\text{BUFFER}} = Q^*_{\text{BUFFER}}/\text{C}$. From expression (9) it is seen that only the last part of $\mathbb{E}X_n(\tau)$ does not depend on τ linearly, but for large τ the last part is (almost) constant. This observation justifies our assumption that (9) *is* linear in τ, which entails:

$$\int_0^{\infty} \mathbb{E}X_n(\tau)w(\tau)\mathrm{d}\tau \approx \mathbb{E}X_n(\mathbb{E}W^*_{\text{BUFFER}}). \tag{11}$$

Combining approximations (10) and (11) yields the following approximate expression for the mean delay of the last particle of a flow in the buffer of the bottleneck node:

$$\mathbb{E}D^*_{\text{BUFFER}} \approx \sum_{n=0}^{\infty} \pi_n \mathbb{E}X_n(\mathbb{E}W^*_{\text{BUFFER}}). \tag{12}$$

Remark on Generally Distributed Flow Sizes. Above we considered exponentially distributed flow sizes (note that the assumption of exponentially distributed flow sizes is only used in expression (9)). Now we consider the buffer delay of generally distributed flow sizes. Asare and Foster [1] derived an expression for the expected conditional response time $\mathbb{E}X_n(\tau)$ of an M/G/1-PS queue, which can be considered as the analogue of (9) obtained in [5] for the special case of exponentially distributed flow sizes. The result of Asare and Foster for general flow sizes, however, requires the distribution of the workload in the system, and an explicit formula is not available.

As an approximation we propose to use (6) also for the expected conditional buffer delay for non-exponentially distributed flow sizes. Then, the approximation for the buffer delay of the last particle is approximated by using (6) in (9), which gives approximation (12).

Now, we will argue that using (9) for the buffer delay is expected to work well also for generally distributed flow sizes. An important observation is that rate at which the buffer is drained only depends on the number of active source nodes; recall from Section 3.1 that the steady-state distribution of the number of active source nodes is insensitive to the flow-size distribution apart from its mean. In particular if the buffer occupancy is large, it is expected that the behavior experienced by the last particle will resemble the steady-state behavior of the source nodes. Also for a very small buffer occupancy (smaller than the residuals at the source nodes) the influence of the flow-size distribution is small, as the number of sources will most likely remain constant. For an intermediate buffer occupancy, the influence of the flow-size distribution will be larger; the source nodes do not immediately behave like the steady-state behavior as the expected sizes of the residuals of the flows at the source nodes strongly depend on the flow-size distribution.

The flow-size distribution affects approximation (12) via both (6) and (9). However, as was argued in the previous paragraph, the impact on (9) is only minor. On the contrary, the flow-size distribution has a strong impact on (6); this is captured through the second moment of the flow size.

3.4   Overall Flow Transfer Time D_{OVERALL}

The overall flow transfer time D_{OVERALL} of a flow is the sum of its flow transfer time D_{SOURCE} and the buffer delay of its last particle D^*_{BUFFER}. Hence

$$\mathbb{E}D_{\text{OVERALL}} = \mathbb{E}D_{\text{SOURCE}} + \mathbb{E}D^*_{\text{BUFFER}},$$

where $\mathbb{E}D_{\text{SOURCE}}$ is given by expression (2) and $\mathbb{E}D^*_{\text{BUFFER}}$ is approximated by expression (12). Remark that $\mathbb{E}D_{\text{OVERALL}}$ is linear in f.

Remark on the Overall Flow Transfer Time Conditional on the Flow Size. In the previous section the expectation of the overall flow transfer time was derived for an arbitrary flow. These results can easily be extended to the situation where a flow has a given size x:

$$\mathbb{E}D_{\text{OVERALL}}(x) = \mathbb{E}D_{\text{SOURCE}}(x) + \mathbb{E}D^*_{\text{BUFFER}}(x). \tag{13}$$

The conditional flow transfer time $\mathbb{E}D_{\text{SOURCE}}(x)$ is already presented by (3). The expected conditional buffer delay $\mathbb{E}D^*_{\text{BUFFER}}(x)$ can be obtained also by approximation (12) of the previous section, but now the buffer occupancy is given by $\mathbb{E}W^*_{\text{BUFFER}}(x)$, the expected amount of work in the buffer upon arrival of the last particle of the particular flow with given size x. Note that $\mathbb{E}W^*_{\text{BUFFER}}(x) = \mathbb{E}Q^*_{\text{BUFFER}}(x)/\text{C}$ and $\mathbb{E}Q^*_{\text{BUFFER}}(x)$ is given in (7). Then the conditional buffer delay is obtained by

$$\mathbb{E}D^*_{\text{BUFFER}}(x) \approx \sum_{n=0}^{\infty} \pi_n \mathbb{E}X_n(\mathbb{E}W^*_{\text{BUFFER}}(x)),$$

which completes (13).

4   Numerical Results

This section presents numerical results of the exact and approximate analyses of the previous section. The analytical approximations are validated by simulations of the model as it is described in Section 2.2. The simulation tool was built in Delphi and sufficient replications have been simulated in order to obtain small confidence intervals. The simulations are only performed using $\text{C} = 1$ and $f = 1$ as the performance measures are linear in f, so that the performance measures for different flow sizes f can be directly obtained from $f = 1$. For flow-size distributions we used Deterministic, Erlang-4, Exponential and Hyper-Exponential

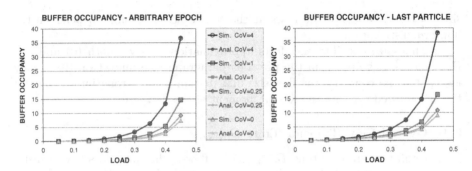

Fig. 2. Mean buffer occupancy. Left: mean buffer occupancy in steady-state ($\mathbb{E}Q_{\text{BUFFER}}$). Right: mean buffer occupancy seen by the last particle ($\mathbb{E}Q_{\text{BUFFER}}^*$).

(with balanced means (see e.g. [18]) and a Coefficient of Variation (CoV) of 4) distributions and the graphs refer to these distributions by their CoVs which are $0, 0.25, 1$, and 4 respectively. The load ρ is varied between 0.05 and 0.45 (by varying the flow arrival rate λ) to observe the system under different load settings. Note that, for stability, $\rho < 0.5$ is required.

Due to spacial limitations we do not present graphs for the mean flow transfer time by the sources nodes. From the formulae in Section 3 it is clear that for moderate and high loads the flow transfer times are small compared to the buffer delay at the bottleneck node.

The graphs of Fig. 2 present the mean buffer occupancy at the bottleneck node for various flow-size distributions. The left graph shows the results for arbitrary fluid particles, whereas the right graph relates to the last particle of a flow. In both graphs it can be seen that the flow-size distribution has a strong impact on the buffer occupancy. Further, it can be observed that the buffer occupancy seen by the last particle is only slightly higher than the buffer occupancy upon flow arrival; As the flow transfer time is relatively short and the number of active sources nodes is also low, resulting in a minor increase of the buffer during the flow transfer time.

Fig. 3 presents the results for the mean buffer delay for both an arbitrary particle (left) and the last particle of a flow (right). The buffer delay of the last particle in the right graph is an approximation, but it captures the behavior very well, not only for exponential flow sizes (for which the approximation is originally designed), but also for other flow-size distributions (cf. remark in Section 3.2). An interesting observation is that buffer delay corresponding to an arbitrary particle is higher than the buffer delay of the last particle. This effect can be explained by the waiting time paradox (see e.g. [18]); with high probability an arbitrary particle belongs to a large flow. Belonging to a large flow has two negative effects: first, before the particle enters the buffer the source node has been transmitting for a long period, so the buffer occupancy will be high, and second, when the particle has entered the buffer, the source node will remain active (in the system) for a long period resulting in a low rate for the bottleneck node.

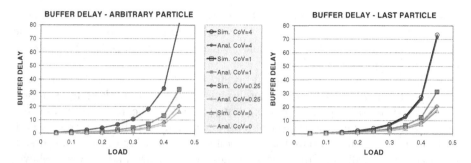

Fig. 3. Mean buffer delay. Left: arbitrary particle ($\mathbb{E}D_{\text{BUFFER}}$). Right: last particle ($\mathbb{E}D^*_{\text{BUFFER}}$).

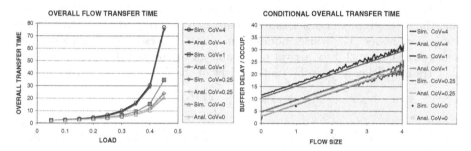

Fig. 4. Overall flow transfer time ($\mathbb{E}D_{\text{OVERALL}}$). Left: unconditional. Right: conditional on the flow size.

The left graph of Fig. 4 presents approximation and simulation results for the mean overall flow transfer time. It can be seen that the mean overall transfer time is largely determined by the buffer delay. There is small deviation which is obviously due to the approximation of the buffer delay of the last particle.

The right graph of Fig. 4 presents the mean overall flow transfer time conditional on the flow size for a load of 0.35, for different flow-size distributions (deterministic flow size ($\text{CoV}=0$) is only a single mark in the graph at $f = 1$). For Erlang-4 and exponentially distributed flow sizes the approximation is accurate as the simulations and analytical results coincide; for hyper-exponential flow sizes it is seen that the analysis underestimates the mean overall flow transfer time, but interestingly the absolute error remains constant for increasing flow sizes.

5   Model Extension: Impact of Resource Allocation Strategy

In the model studied in the previous section the system capacity is equally shared amongst the active source nodes and the bottleneck node. It is interesting to consider the possibility of assigning a larger share of the system capacity to the bottleneck node (meaning a smaller share for each of the active source nodes);

this will reduce the buffer delay at the bottlneck node at the cost of larger flow transfer times by the source nodes. Obviously there is a trade-off between the flow transfer time and the buffer delay, where the main objective is to reduce the *overall* flow transfer time.

With the recently standardized IEEE 802.11E EDCA, a QoS-aware version of the DCF, it is possible to 'assign' more weight to the bottleneck node by appropriately setting EDCA-parameters, e.g. CW_{min}, AIFS or the TXOP-limit. In order to get some feeling for the impact of alternative sharing policies on the network performance, we will compare the situation of equal sharing of the transmission resources among all individual active nodes (as described and analyzed above) with the (extreme) situation in which the bottleneck node is strongly favored and receives exactly as many transmission capacity as *all* active source nodes together. We will denote this latter resource allocation by 'half'.

It is clear that in the 'half' variant, there is no queueing at the bottleneck node and the (overall) flow transfer time is completely determined by the behavior of the source nodes. The source nodes behave as a Processor Sharing model with service capacity $C/2$, flow arrival rate λ and mean flow size f. Hence, the mean flow transfer delay is given by (independent of the flow-size distribution):

$$\mathbb{E}D_{\text{OVERALL}}^{(\text{half})} = \frac{2f/C}{1 - 2\rho}.$$

Under the original 'IEEE 802.11B' resource allocation we have, see Section 3,

$$\mathbb{E}D_{\text{OVERALL}} = \mathbb{E}D_{\text{SOURCE}} + \mathbb{E}D_{\text{BUFFER}}^{*} = 2\frac{f/C}{1 - \rho} + \frac{2\rho f_2/(fC)}{(1 - 2\rho)(1 - \rho)}.$$

These results illustrate the considerable reduction of the overall flow transfer delay that can be achieved by modification of the resource allocation strategy realized by the MAC-protocol. It is expected that resource allocation strategies 'in between' the two strategies considered above will yield mean overall flow transfer times which are in between $\mathbb{E}D_{\text{OVERALL}}$ and $\mathbb{E}D_{\text{OVERALL}}^{(\text{half})}$. From the comparison of $\mathbb{E}D_{\text{OVERALL}}$ and $\mathbb{E}D_{\text{OVERALL}}^{(\text{half})}$ in Fig. 5 it is seen that the scheme 'half' performs significantly better for high loads.

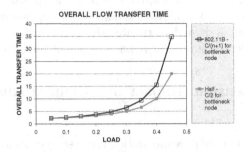

Fig. 5. Mean overall transfer time for a prioritized ('half') and unprioritized (802.11B) bottleneck node

Remark. It is expected that, if the resource allocation at the MAC-layer is according to 'IEEE 802.11B' and, in addition, the traffic rates are end-to-end controlled by TCP, the actually resulting allocation of radio transmission resources to these TCP flows would be according to the 'half' variant described above. The TCP-control loop will achieve that the source nodes can only transmit at a rate at which the bottleneck node can serve the traffic, see e.g. [16].

6   Concluding Remarks and Directions for Further Research

We have developed a mathematical model describing the behavior of a bottleneck node in a wireless ad-hoc network. This model provides useful insights into the (overall) flow transfer time and the buffer delay occurring at the bottleneck node, in particular how it depends on the various system and traffic parameters. The modeling approach presented in this paper may enable refinements and extensions to more general network settings; this is a subject for further research.

Topics for further research also include:

- alternative service disciplines at the bottleneck node. In the above analysis it is assumed that the packet scheduling at the bottleneck node is First Come First Serve. Alternative service disciplines, e.g. round robin, may yield considerably smaller mean overall flow transfer times.
- investigating the influence of higher-layer protocols, such as TCP, on the flow transfer time (cf. the discussion in Section 5).
- investigating the possible performance gain that can be obtained by IEEE 802.11E EDCA (cf. the discussion in Section 5).

Acknowledgements

This work has been carried out partly in the SENTER-NOVEM funded project EASY WIRELESS and the Dutch BSIK/BRICKS project.

References

1. B.K. ASARE and F.G. FOSTER. Conditional response times in the M/G/1 Processor-sharing system. *Journal of Applied Probability*, 20: 910–915, 1983.
2. G. BIANCHI. Performance analysis of the IEEE 802.11 Distributed Coordination Function. *IEEE Journal on Selected Areas in Communications*, 18: 535–547, 2000.
3. M.M. CARVALHO and J.J. GARCIA-LUNA-ACEVES. Delay analysis of IEEE 802.11 single-hop networks. *Proceedings of the 11th IEEE International Conference on Network Protocols*, Atlanta, USA, 2003.
4. M.M. CARVALHO and J.J. GARCIA-LUNA-ACEVES. A scalable model for channel access protocols in multihop ad hoc networks. *Proceedings of MobiCom'04*, Philadelphia, USA, 2004.

5. E.G. COFFMAN JR., R.R. MUNTZ, and H. TROTTER. Waiting time distributions for processor-sharing systems. *Journal of the ACM*, 17: 123–130, 1970.

6. J.W. COHEN. The multiple phase service network with generalized processor sharing. *Acta informatica*, 12: 245–284, 1979.

7. P.E. ENGELSTAD and O.N. ØSTERBØ. Non-saturation and saturation analysis of IEEE 802.11e EDCA with starvation prediction. *Proceedings of MSWiM '05*, Montreal, Canada, 2005.

8. C.H. FOH and M. ZUKERMAN. Performance analysis of the IEEE 802.11 MAC protocol. *Proceedings of European Wireless '02*, Florence, Italy, 2002.

9. Z. FU, P. ZERFOS, H. LUO, S. LU, L. ZHANG, and M. GERLA. The impact of multihop wireless channel on TCP throughput and loss. *Proceedings of INFOCOM '03*, San Francisco, USA, 2003.

10. J. HE and H.K. PUNG. Fairness of medium access control for multi-hop ad hoc networks. Computer Networks, 48: 867–890, 2005.

11. J. HE and H.K. PUNG. Performance modelling and evaluation of IEEE 802.11 Distributed Coordination Function in multihop wireless networks. Computer Communications, 29: 1300–1308, 2006.

12. IEEE P802.11B/D7.0, Supplement: higher speed physical layer extension in the 2.4 GHz band, 1999.

13. IEEE P802.11E-2005, Amendment 8: Medium Access Control (MAC) Quality of Service Enhancements. November 2005.

14. R. LITJENS, F. ROIJERS, J.L. VAN DEN BERG, R.J. BOUCHERIE, and M.J. FLEUREN. Analysis of flow transfer times in IEEE 802.11 wireless LANs. *Annals of Telecommunications*, 59: 1407–1432, 2004.

15. M.R.H. MANDJES and F. ROIJERS. A fluid system with coupled input and output and its application to bottlenecks in ad hoc networks. Available as CWI-report at `http://ftp.cwi.nl/CWIreports/PNA/PNA-R0603.pdf`.

16. F. ROIJERS, J.L. VAN DEN BERG, and X. FAN. Analytical modelling of TCP file transfer times over IEEE 802.11 wireless LANs. *Proceedings of ITC-19*, Beijing, China, 2005.

17. T. SAKURAI and S. HANLY. Modelling TCP flows over an 802.11 wireless LAN. *Proceedings of European Wireless Conference*, Nicosia, Cyprus, 2005.

18. H.C. TIJMS. Stochastic models: an algorithmic approach. Wiley & Sons, 1994.

19. E.M.M. WINANDS, T.J.J. DENTENEER, J.A.C. RESING, and R. RIETMAN. A finite-source feedback queueing network as a model of the IEEE 802.11 Distributed Coordination Function. *Proc. of European Wireless'04*, Barcelona, Spain, 2004.

Circularity-Based Medium Access Control in Mobile Ad Hoc Networks

Mohammad Z. Ahmad[1], Damla Turgut[1], and R. Bhakthavathsalam[2]

[1] School of Electrical Engineering and Computer Science
University of Central Florida, Orlando, Florida 32816
{zubair, turgut}@cs.ucf.edu
[2] Supercomputer Education and Research Center
Indian Institute of Science, Bangalore 560 012, India
bhaktha@serc.iisc.ernet.in

Abstract. The RTS/CTS access scheme, designed to reduce the number of collisions in a IEEE 802.11 network, is known to exhibit problems due to masked nodes, the imbalance between the interference range and the communication range of the nodes, and scenarios in which nodes are unnecessarily silenced, thus preventing parallel transmissions to take place. We present an approach for enhancing the performance of the IEEE 802.11 MAC protocol by selectively discarding or delaying specifically marked RTS and CTS packets. By dropping the circularity-satisfied RTS, we allow certain parallel transmissions to proceed, even if there is a non-zero risk of collision. By delaying the circularity-satisfied CTS, we allow a neighboring parallel transmission to continue. One important feature of the circularity approach is that it is fully compatible with the IEEE 802.11 standard. We implemented the circularity approach in ns-2 simulator. Through a series of experiments, we show that the circularity approach provides a significant improvement in the throughput and end-to-end delay of the network, and contributes to a reduction of the number of collisions in most scenarios.

1   Introduction

A mobile ad hoc network (MANET) is a robust, self-managing and autonomous system of cooperating mobile nodes connected by unreliable wireless links. Devices in communication range can communicate directly with each other while intermediate nodes serve as routers to forward packets from the source to the destination.

Since node communication in a dense network happens at the same frequency band, the problem of packet loss due to collision becomes an area of focus. The IEEE 802.11 standard [1] has been primarily designed for wireless LANs and is responsible for scheduling medium access for multiple stations which are contending for the common channel. It uses a medium access scheme based on the Carrier Sense Multiple Access (CSMA) [2] protocol, where a node transmits only if it finds the medium to be idle for a pre-defined Inter Frame Space (IFS).

T. Kunz and S.S. Ravi (Eds.): ADHOC-NOW 2006, LNCS 4104, pp. 337–348, 2006.

It also uses a Request-To-Send (RTS) and Clear-To-Send (CTS) control packets to coordinate channel access [3] and minimize costly packet collisions in hidden node scenarios. In multi-hop networks, some nodes may not hear control packets from other nodes within the network. This leads to an RTS/CTS exchange with reduced chances of success and increased possibility of packet collisions. The problem is further complicated with packet loss arising due to some transmissions being masked by other on-going transmissions in their neighborhood [4]. The masked node problem is an example of a shortcoming of the IEEE 802.11 when it is used in MANETs.

In [5] the authors propose the selective disengagement of the RTS-CTS hand-shake in IEEE 802.11. They point out that the 802.11 standard was developed keeping in mind that the carrier sensing range is equal to the transmission range whereas it is 1.78 times of the latter [6]. This means that any node can actu-ally hear transmissions going on two hops away thereby resolving the hidden node problem to some extent. This however leads to another problem: due to the basic CSMA/CA scheme, a sender will not transmit neither control nor data packets if it senses the channel is busy. By disengaging the RTS-CTS handshake based on the number of "CTS timeouts", they enable greater fairness and higher network throughput both due to lesser control overhead and enabling parallel transmissions to take place.

In this paper, we use a similar technique of turning off the RTS-CTS hand-shake for particular instances along with delaying the transmission of the CTS packet to enable parallel transmissions to be completed. These changes are made to the IEEE 802.11 MAC protocol and simulations are carried out to compare performance differences. However, revamping the entire MAC standard or the transport layer protocol for use in MANETs is absolutely infeasible due to wide adoption of both. Some modifications need to be incorporated to make the pro-posed and the legacy protocols inter-operable with each other to ensure fairness and optimal resource utilization. Effective changes should be incorporated in an "non-invasive" manner, i.e. the changes should be based on the underlying mechanisms of the current standard, with as few major changes as possible. The modifications proposed in this paper are novice but effective enhancements to the current standard keeping the underlying principles and the workings of the IEEE 802.11 standard almost untouched.

2   Related Work

The IEEE 802.11 standard uses the Distributed Coordination Function (DCF) which includes Carrier Sense Multiple Access with Collision Avoidance (CSMA/CA) as the fundamental access technique. There are two primary access methods in IEEE 802.11: the *basic access* and the *RTS/CTS access* method. The basic access scheme involves only a reliable transfer of the data packets from the source to the destination by using ACK packets. In the RTS/CTS access scheme, the RTS and CTS control packets are first exchanged and the channel is reserved exclusively between the source and destination, which is followed by the DATA/ACK packet transmission. This RTS/CTS dialog helps in the implementation of the virtual carrier sensing mechanism which is also accompanied by physical carrier sensing in IEEE 802.11 DCF. The RTS/CTS frames contain a duration field which is used by the neighbors to set a specific Network Allocation Vector (NAV) during which the nodes are sent to "silenced" state during which the packet exchanges are being carried out between the sender and receiver nodes. On the other hand, physical carrier sensing is implemented using interframe spaces. After a channel is sensed idle for a DCF Interframe Space (DIFS) time interval, the back-off procedure is invoked by the station which has to send the data. A Short Interframe Space (SIFS) is used to separate transmissions belonging to a single session (CTS, DATA and ACK packets). Extensive work has been carried out on the IEEE 802.11 DCF and the usage of the RTS/CTS mechanism [7], [8].

In [6], the authors present mathematical proof that the interference range is typically 1.78 times the communication range. Even though a node may not be within the transmission range to successfully receive a CTS packet, it may still be the cause of interference at the sender. As a simple solution to the above problem, it is suggested that a node should only reply with a CTS when the received RTS is above a certain receiving power threshold, i.e., it is sufficiently close to the transmitter and hence avoid perceptible interference from other nodes. [9] shows that an optimal carrier sensing range along with an appropriate transmission range and an interference model significantly increases network throughput. Another related work [5] tunes the RTS/CTS exchange by selectively disengaging it when there are occurrences of the CTS not being returned.

In [10], the author states with increasing network complexity and/or node mobility, a node which has not heard of a RTS or CTS packet may migrate into the footprint of a receiver and destroy a DATA packet by initiating its own transmission, oblivious to its surroundings. [4] points out another type of nodes in the same class as that of hidden nodes which are termed "masked" nodes. The authors show that the RTS/CTS exchange is not enough under perfect operating conditions since neighbor nodes are masked by other on-going transmissions nearby. Masked nodes cannot decode the RTS/CTS packets correctly and may end up causing Data/ACK packet collisions later on.

Other widely used approaches to resolving medium access include splitting the available channel into separate control and data subchannels. [11] proposes two schemes that attempt to pipeline *contention resolution* with *data transmission* to reduce the idle waiting time and decrease overall delay. For wireless environments, it also proposes a partial pipeline approach to overcome the shortcomings of the total pipelining scheme. The authors show that with proper channel division, a net throughput increase can be obtained. The authors of [12] propose, the Bi-directional Multi-Channel MAC protocol, where the bandwidth is divided into one control channel and several data channels. It is bi-directional because the receiver may also send his own data packet (if any) to the sender using any of the other available channels thereby eliminating the need of another RTS/CTS handshake.

3   Circularity-Based Medium Access Control

3.1   Motivation

The implicit goal of any MAC protocol is to minimize packet collisions and unnecessary retransmissions. The IEEE 802.11 protocol uses a suitable acknowledgement timeout interval to infer a possible collision. The collision detection time is dependent on the packet sizes and the net bandwidth of the channel. Packet collisions may be categorized as follows:

- RTS packets colliding with other RTS packets when two stations start transmission simultaneously.
- DATA packets colliding with RTS/CTS packets from masked/deaf nodes.
- ACK packets colliding with RTS/CTS packets from masked/deaf nodes.

In this work, we define a channel-access scheme for better channel utilization while encouraging parallel transmissions from other non-interfering nodes.

3.2   Circularity

Circularity is defined as a number which enables the identification of specific groups of control packets sent from each node. The total number of packets in each group is equal to its circularity value and the last packet in the group is termed as the *circularity-satisfied* packet. Each node in the network is assigned a circularity value which may be unique to itself or the entire network may have the same value for each of its nodes. For example, if the circularity value is defined as four, then we divide the RTS/CTS packets being created in each node into groups of four and the first such packet in a group is the circularity-satisfied packet. Hence, every fourth packet being created by the node (i.e. every multiple of four) is circularity-satisfied.

Mathematically, a packet is circularity-satisfied if:

$$N \ modulo \ c = 0$$

where N is the current count of the number of packets generated (RTS/CTS) and c is the circularity value for the particular node.

Essential characteristics of applying circularity to RTS and CTS packets include the following:

- By identifying certain RTS and CTS packets as circularity-satisfied, we induce these packets to **behave** differently from the rest. The structure of the packet (size, headers etc) remains the same, so there are no explicit changes which have to be made to the standard IEEE 802.11 protocol.
- There is no absolute grouping of packets taking place to identify the circularity-satisfied packets. The packets are identified through the simple mathematical formula above.
- There are different RTS and CTS circularity values. For simplicity, in this paper, we have considered both of them to be the same. Thus, for the above mentioned example, every fourth RTS packet and fourth CTS packet emanating from one particular node will be the circularity-satisfied packet.

– The RTS-CTS packets identified as circularity-satisfied are independent of the flow for which they are created. Source nodes and any intermediary node mechanism for all the traffic routed through it.

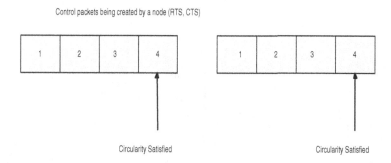

Fig. 1. Identifying Circularity-satisfied control packets for a circularity value of 4. In effect, every fourth packet becomes a Circularity-Satisfied packet.

Essentially, circularity is just a scheme to identify certain packets which behave differently than the rest. Thus, if the RTS circularity value is considered to be four, then a node which sends out a total of 20 RTS packets during the network lifetime will in effect have identified *five* (20/4) of them as circularity-satisfied. Fig 1 refers to such a case where packets are divided into groups of four and the last packet of each group is tagged as circularity-satisfied. Similarly, considering CTS circularity as five, then a node sending out 20 CTS packets would have identified *four* (20/5) of them as circularity-satisfied.

In our scheme, each node is assigned a specific circularity value for both their RTS and CTS packets. These value pairs are used to **drop** circularity-satisfied RTS packets, irrespective of the existing scenario. Similarly, the circularity-satisfied CTS packets are **delayed** to allow parallel DATA or ACK transmissions in the immediate neighborhood to terminate. In our experimental setup we have restricted the delay of the CTS packets to one SIFS time interval.

3.3 RTS Packet Dropping with Circularity

Dropping RTS packets is a technique of selectively disengaging the RTS/CTS dialogue for a particular transmission session. Let us consider the scenario in Figure 2. In this figure we depict the circles as the sensing range (which is roughly 1.78 times the transmission range). Looking at the timeline sequence in the figure, it can be observed that D initiates a transmission to E with an RTS packet. This RTS packet is transmitted to E and C with E sending a corresponding CTS. C then gets blocked from transmitting. This is followed by D starting its data packet transfer to E, but this can also be sensed by B as it is within carrier sensing range of D. At this point, B's neighbor A (who cannot hear beyond C) initiates a transmission by sending an RTS packet to

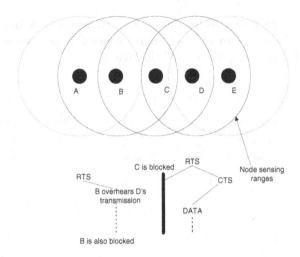

Fig. 2. Hidden Node Scenario with sensing ranges being shown

B. However, as B can sense D sending a long data packet, it will back-off and not reply with a CTS to A. Hence, a parallel transmission is prevented from taking place resulting in decreased network throughput. Another point to be noted is that the flow D to E may capture the channel for a long period of time which results in A giving up re-transmitting RTS packets after the retry count is exceeded and hence reporting a route failure to the routing layer. This will in turn lead to a new process of route discovery and increase network overhead considerably. Disabling the RTS-CTS exchange and simply transmitting the data packet would be of greater effect.

We implement this selective RTS-CTS disengagement by identifying certain RTS packets based on their RTS circularity value and dropping them. We discuss posssible techniques of setting the circularity values in a later section. The authors in [5] take a more conservative approach by waiting for a number of CTS timeouts to occur before sending the data packet directly. We we propose a more aggressive approach of scheduling these packet drops based on the circularity value. The tradeoff is that there may be a higher number of packet collisions occuring but a higher overall network throughput could be obtained due to the parallel transmissions.

3.4   CTS Packet Delay with Circularity

The concept of delaying the CTS packet is also aimed at making possible the occurence of parallel transmission within the network. By delaying the transmission of a CTS packet by a small time interval (one SIFS), we aim to help a neighboring transmission to either continue or complete. Let us consider a sample scenario.

In the Figure 3, we assume that D and C have had a succesful RTS-CTS exchange and D has sent its data packet to C. During the course of this

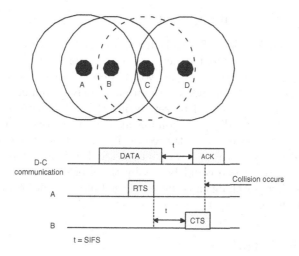

Fig. 3. Scenario depicting CTS collision with ACK packet. The circles denote the transmission range of each node.

transmission, due to the nodes being mobile or if it was earlier a masked node, B comes into transmission range of C. At this point of time, A sends data to B by initiating an RTS-CTS handshake with an RTS packet. Since B is unaware of an ongoing transmission nearby, it responds with its CTS packet which collides with the ACK packet from C. However, if the CTS packet from B was delayed by a small interval, then the ACK packet would be correctly received by D. As only the circularity-satisfied CTS packets are delayed, the impact on the overall delays in the system is reduced. Also it does not guarantee that all CTS-ACK collisions will be avoided but the primary aim is to reduce the probability of these collisions as much as possible. The timeline diagram of CTS delay with circularity is shown in Figure 4.

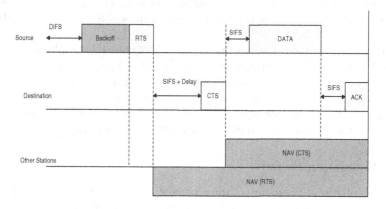

Fig. 4. Timeline Diagram for CTS Delay

3.5   Selecting the Circularity Values

Setting the appropriate pair of circularity values is critical for real-world deployment. In a sparsely populated network, the use of circularity would yield no benefits. In a densely populated networks, we might need to use lower circularity values (which corresponds to a more aggressive dropping and delaying of the packets). This value could be set during the initial network setup time for a device or when the device is joining an existing network. For example, by default the RTS-CTS dialog is turned off in current WLAN cards and can be easily turned on by setting a lower RTS-CTS threshold value. In a similar fashion, the circularity value could be set by the user during or anytime after network bootup. The mechanism could be turned off anytime by setting a very high circularity value.

4   Simulation Study

4.1   Simulation Environment

We carried out simulations using ns-2 [13] with wireless extensions from the CMU Monarch Group. Changes to the MAC source files were carried out to implement the circularity concept and enable the dropping/delaying of circularity-satisfied RTS-CTS packets.

Node movement is modeled by the random waypoint mobility model [14] with nodes moving at a speed between 10 and 100 m/s and a pause time of 20s. Each data point plotted is the average of 10 different scenarios with different initial network configurations.

Each run in our simulation is executed for 900 seconds of simulation time unless otherwise specified, and the nodes move in a 500m x 500m area. Each node in the network generates traffic of fixed size packets of 512 bytes. The propagation model is the two ray ground model and the maximum number of flows in the network is based on the network size. The data rate is 1 Mb/s and FTP is the application generating packets over TCP as the transport layer protocol. AODV is used as the routing layer protocol for all our simulations as it is quite widely used.

Table 1 lists the parameter values used throughout the simulations. The simulation metrics are net throughput, packet delay and packet loss ratio. We define net throughput as the aggregated throughput over all the flows in the network as the foremost performance metric. Along with this, end to end packet delay and packet drop ratio are also considered as reliable performance metrics. The metrics defined above are simulated with different circularity values.

4.2   Simulation Results

In our simulations, the values of the circularity was varied from 10 to 200. Traffic is generated by ns-2's random traffic generator with different random number seeds. For simplicity, we set the circularity of RTS and CTS to the same value.

Table 1. Simulation Parameters

Data Rate	1 Mbps
RTS Size	20 Byte
CTS Size	14 Byte
Data Packet Size	512 byte
ACK Size	14 byte
SIFS	$10\mu s$
DIFS	$50\mu s$
CW Min	31
CW Max	1023
LongRetryLimit	7
Routing Protocol	AODV
Mobility Model	Random waypoint
Propagation Model	Two ray ground

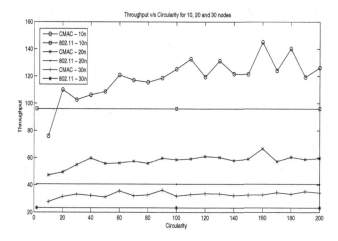

Fig. 5. Throughput versus Circularity for 10, 20 and 30 nodes. Note that for the standard IEEE 802.11, throughput remains constant with increase in circularity.

Figure 5 shows the throughput obtained with respect to circularity for increasing network. We denote our enhanced MAC protocol as Circularity-MAC (C-MAC) and plot the throughput obtained for 10, 20 and 30 node scenario, while the standard IEEE 802.11 protocol naturally remains constant. The graph shows that C-MAC is is obtaining a 15-20% improvement in throughput compared to the 802.11 MAC. Intuitively, this is due to the lower number of packet collisions, which in turn leads to a lower end-to-end delay between source and destination nodes.

Figure 6 confirms this point as it can be observed that with C-MAC, the number of collisions are significantly lower. Packet drop count is an important metric in determining the effectiveness of any MAC protocol, and Figure 6

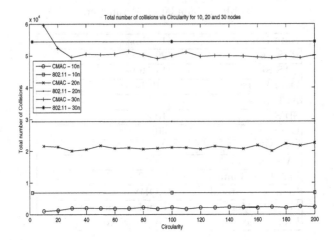

Fig. 6. Total number of collisions versus Circularity for 10, 20 and 30 nodes

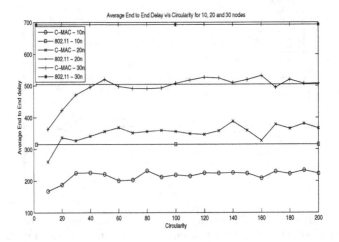

Fig. 7. Average end-to-end delay versus Circularity for 10, 20 and 30 nodes

substantiates the earlier statement that dropping and delaying the circularity-satisfied packets indeed reduces packet collisions.

For both 10 and 20 nodes scenarios, it can be seen that the reduction is almost the tune of 40 percent. Drop in packet collisions leads to lower average end-to-end delay as the network is saved from making successive recoveries from these collisions. The reduced collisions are leading to greater throughput and significantly lower end to end delay (Fig 7). The simulation results show that the improvement can be obtained at all network sizes.

The simulations show that any circularity value greater than 50 results in enhanced performance for the network. Hence, in a real world scenario, any device can have its circularity value set by its user to an arbitrary value within the range of 50-200. Such a value may not result in ideal network performance, but will certainly be an improvement over the IEEE 802.11 standard.

5 Conclusion

In this paper, we have designed and evaluated enhancements to the IEEE 802.11 mechanism for MANETs, using a novel concept of circularity. This concept has been embedded in the existing RTS-CTS handshake and shows better performance, specially in situations where the contention for the medium is high. Simulation results quantify the advantages in terms of both throughput and the number of packet collisions. As a future work, we propose to investigate the effect of greater delays of circularity-satisfied CTS packets, increase traffic loads by increasing packet sizes for arbitrary circularity values. The goal is to identify standard pairs of circularity values for different scenarios, so that they can be effectively deployed according to the various existing network conditions.

References

1. IEEE Std 802.11b-1999. Part 11: Wireless LAN Medium Access Control (MAC) and Physical Layer (PHY) Specifications. IEEE Standard 802.11 (1999)
2. Kleinrock, L., Tobagi, F.: Packet switching in radio channels: Part I - carrier sense multiple-access modes and their throughput-delay characteristics. IEEE Transactions on Communications **COM-23** (1975) 1400–1416
3. Bharghavan, V., Demers, A., Shenker, S., Zhang, L.: MACAW: A media access protocol for wireless LAN's. In: Proceedings of ACM SIGCOMM '94. (1994) 221–225
4. Ray, S., Carruthers, J.B., Starobinski, D.: Evaluation of the masked node problem in ad-hoc wireless lans. IEEE Transactions on Mobile Computing **4** (2005) 430–442
5. Ju, H., Rubin, I., Kuan, Y.: An adaptive RTS/CTS control mechanism for IEEE 802.11 MAC protocol. In: Proceedings of IEEE Vehicular Technology Conference. Volume 2. (2003) 1469 – 1473
6. Xu, K., Gerla, M., Bae, S.: Effectiveness of RTS/CTS handshake in IEEE 802.11 based ad hoc networks. Ad Hoc Networks **1** (2003) 107–123
7. Crow, B.P., Widjaja, I., Kim, J.G., Sakai, P.T.: IEEE 802.11 Wireless local area networks. IEEE Communications Magazine **35** (1997) 116–126
8. Bianchi, G.: Performance analysis of the IEEE 802.11 Distributed Coordination Function. IEEE Journal on Selected Areas in Communications **18** (2000) 535–547
9. Deng, J., Liang, B., Varshney, P.: Tuning the carrier sensing range of IEEE 802.11 MAC. In: Proceedings of IEEE Global Telecommunications Conference. Volume 5. (2004) 2987–2991
10. Haas, Z.: On the performance of a medium access control scheme for the reconfigurable wireless networks. In: Proceedings of the IEEE MILCOM. Volume 3. (1997) 1558–1564
11. Yang, X., Vaidya, N., Ravichandran, P.: Split-channel pipelined packet scheduling for wireless networks. IEEE Transactions on Mobile Computing **5** (2006) 240–257
12. Kuang, T., Williamson, C.: A bidirectional multi-channel MAC protocol for improving TCP performance on multihop wireless ad hoc networks. In: Proceedings of the 7th ACM international symposium on Modeling, analysis and simulation of wireless and mobile systems. (2004) 301–310

13. VINT: The UCB/LBNL/VINT network simulator-ns (version 2). (URL http://
www.isi.edu/nsnam/ns)
14. Broch, J., Maltz, D.A., Johnson, D.B., Hu, Y., Jetcheva, J.: A performance comparison of multi-hop wireless ad hoc network routing protocols. In: Proceedings of
Mobile Computing and Networking. (1998) 85–97

Improving the ID-Based Key Exchange Protocol in Wireless Mobile Ad Hoc Networks

Eun-Jun Yoon and Kee-Young Yoo*

Department of Computer Engineering, Kyungpook National University,
Daegu 702-701, South Korea
ejyoon@infosec.knu.ac.kr, yook@knu.ac.kr

Abstract. In 2005, Liaw et al. proposed an ID-based key exchange protocol without the auxiliary of certification authority that can be applied to wireless mobile Ad hoc networks, which can solve the security problem in the Wireless Mobile Ad hoc Networks (MANET) and is suitable for other wired network structures. Liaw et al.'s protocol, however, still has other serious weaknesses. The current paper will demonstrate the vulnerability of the protocol. Furthermore, we present a method that can avoid attacks.

Keywords: Cryptography, Wireless mobile, Ad hoc networks, Network security, Key exchange, ID-based, Impersonation attack.

1 Introduction

The major problems of Ad hoc networks are providing security for infrastructure less networks and how to manage the cryptographic keys that are needed. Recent research has shown that wireless Ad hoc networks are highly vulnerable to various security threats due to their inherent characteristics [1,2]. This leaves Ad hoc key management and key distribution as a wide open problem. Wireless Mobile Ad hoc Networks (MANET) represent an emerging area of mobile computing. Security issues concerning MANET are attracting a great deal of attention nowadays. MANET is a new wireless networking paradigm without a fixed infrastructure, and it includes base stations or mobile switching centers. The mobile nodes rely on each other to maintain network connections.

In 1984, Shamir [3] first proposed the idea of an Identity Based cryptosystem in order to simplify the conventional public key cryptosystem, and to make managing easier [4]. In 2003, Khalili, et al. [5] proposed a protocol for the management and authentication of Ad hoc networks that is based on an ID-based scheme. In 2005, based on this ID-based scheme, Liaw et al. [6] proposed a key exchange protocol without the auxiliary of certification authority that can be applied to wireless mobile Ad hoc networks, which can solve the security problem in the MANET. It is suitable for other wired network structures. In short, Liaw et al.'s protocol has the following characteristics: (1) it can satisfy network

---

* Corresponding author. Tel.: +82-53-950-5553; Fax: +82-53-957-4846.

T. Kunz and S.S. Ravi (Eds.): ADHOC-NOW 2006, LNCS 4104, pp. 349–354, 2006.

security assumptions without the need for a trusted third party; (2) there is no need to store the public keys of other users; and (3) security is based on factoring a large composite.

Liaw et al.'s protocol, however, does have some serious weaknesses. The current paper will demonstrate the vulnerability of Liaw et al.'s protocol. Using our attacks, we will show that a malicious user can easily obtain a specific legal user's secret signature and impersonate this specific user in order to exchange a common session key with a legal user. Furthermore, we present improvements to the security flaws of the protocol.

This paper is organized as follows: In Section 2, we briefly review Liaw et al.'s protocol. Section 3 shows the security flaws of Liaw et al.'s protocol. In Section 4, we present improvements to the protocol. In Section 5, we analyze the security of our proposed protocol. Finally, conclusions are presented in Section 6.

2   Review of Liaw et al.'s Protocol

This section briefly reviews Liaw et al.'s ID-based key exchange protocol, which is suitable for wireless mobile Ad hoc network [6]. Similar to other ID-based cryptosystems, a trusted key generation center is needed in the protocol in order to verify a user's identity numbers and to generate corresponding private keys. After all users have registered, the key generation center can be closed or off-line. Liaw et al.'s protocol is comprised of four phases: Initialization, user registration, user verification, and key exchange between two users to establish a secure session key. The protocol works as follows:

Initialization Phase: In a RSA [7] scheme, the key generation center calculates the public key (n, e) and private key $(p, q, d, \phi(n))$, where p and q are large and strong primes, n is the product of p and q, $\phi(n) = (p - 1)(q - 1)$, and e is a large prime and $d = e^{-1} \mod \phi(n)$. In addition, the center also determines the primitive element α in both fields $GF(p)$ and $GF(q)$, and a one-way hash function $f(\cdot)$ is chosen. Similarly, $(\alpha, f(\cdot))$ along with (n, e) can be regarded as public information.

Registration Phase: User U_i takes his/her identification number ID_i to the key registration center to obtain the signature g_i for the ID_i. If the center confirms its validity and the relationship between U_i and ID_i, Then, g_i is calculated by using

$$g_i = ID_i^d \mod n \tag{1}$$

and hands g_i to U_i. When all users have been registered and have obtained their respective g_i $(i = 1 \ldots n)$ the center does not need to exist in the Ad hoc network any more. The registration phase of Liaw et al.'s protocol is illustrated in Figure 1.

User Verification Phase: Assume that U_i and U_j are the two users who want to communicate with each other. First, U_i selects a random number r_i and computes two public keys, y_i and t_i as

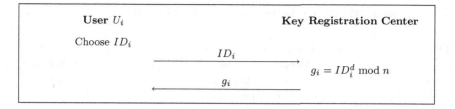

Fig. 1. The Registration Phase of Liaw et al.'s Protocol

$$y_i = g_i \cdot \alpha^{r_i} \bmod n \tag{2}$$

and

$$t_i = r_i^e \bmod n \tag{3}$$

Second, U_i uses a timestamp T_i and the identification number ID_j of user j to perform a one-way function of $f(y_i, t_i, T_i, ID_j)$, then the following is computed:

$$s_i = g_i \cdot r_i^{f(y_i, t_i, T_i, ID_j)} \bmod n \tag{4}$$

Finally, U_i sends $(ID_i, y_i, t_i, s_i, T_i)$ to U_j .

Similarly, U_j selects a random number r_j and a timestamp T_j, then computes

$$y_j = g_j \cdot \alpha^{r_j} \bmod n \tag{5}$$

$$t_j = r_j^e \bmod n \tag{6}$$

$$s_j = g_j \cdot r_j^{f(y_j, t_j, T_j, ID_i)} \bmod n \tag{7}$$

and sends $(ID_j, y_j, t_j, s_j, T_j)$ to U_i.

Before generating the session key, U_i and U_j need to verify whether $(ID_i, y_i, t_i, s_i, T_i)$ and $(ID_j, y_j, t_j, s_j, T_j)$ are sent from user U_i and user U_j, respectively. This can be done by check

$$s_j^e = ID_j \cdot t_j^{f(y_j, t_j, T_j, ID_i)} \bmod n \tag{8}$$

and

$$s_i^e = ID_i \cdot t_i^{f(y_i, t_i, T_i, ID_j)} \bmod n \tag{9}$$

respectively.

Key Exchange Phase: U_i and U_j compute session keys SK_i and SK_j, respectively, as follows:

$$SK_i = (\frac{y_j^e}{ID_j})^{r_i} \bmod n \tag{10}$$

$$SK_j = (\frac{y_i^e}{ID_i})^{r_j} \bmod n \tag{11}$$

SK_i and SK_j are the same, because

$$SK_i = SK_j = \alpha^{e \cdot r_i \cdot r_j} \bmod n \tag{12}$$

3   Cryptanalysis of Liaw et al.'s Protocol

This section shows the security flaws of Liaw et al.'s protocol. In the protocol, an attacker can freely impersonate users (U_i or U_j). This happens because an attacker can obtain the secret signatures (g_i or g_j) of the users (U_i or U_i) after successfully executing the registration phase.

An Impersonation Attack By Malicious U_f Against U_i (or U_j): Suppose that user U_f is an attacker who knows the legal user U_i's ID_i. By using the legal ID_i in the registration phase, U_f can register with the key registration center as follows:

(1) U_f obtains his/her identity ID_f by $ID_f = ID_i^{-1}$ and the ID_f is submitted as a registration request to the key registration center.
(2) The key registration center will compute the secret signature g_f of U_f by $g_f = ID_f^d = ID_i^{-d} = g_i^{-1}(\bmod n)$ and send g_f to U_f with a secure channel. As a result, U_f can obtain the secret signature g_i of the legal user U_i by computing $g_f^{-1} = g_i(\bmod n)$. Then, by using the g_i, so obtained, U_f can freely impersonate U_i in order to exchange a common session key with legal user U_j.

Similarly, U_f can also impersonate U_j by using the aforementioned attack method.

An Impersonation Attack By Malicious U_i (or U_j) Against U_j (or U_i): Regarding another attack on Liaw et al.'s protocol, if a malicious U_i (or U_j), who knows his/her g_i (or g_j), can compute his/her new identity ID_f by $ID_f = ID_i \cdot ID_j$, then the ID_f can be resubmitted as a registration request to the key registration center. The key registration center will compute the secret signature g_f of U_i (or U_j) by $g_f = ID_f^d = (ID_i \cdot ID_j)^d = g_i \cdot g_j(\bmod n)$ and send g_f to U_i (or U_j) via a secure channel. Consequently, a malicious U_i (or U_j) can obtain secret signature g_i of the legal user U_i (or g_j of the legal user U_j), by computing $g_i = g_f \cdot g_j^{-1}(\bmod n)$ or ($g_j = g_f \cdot g_i^{-1}(\bmod n)$), respectively.

4   Countermeasures

This section presents a modification of Liaw et al.'s protocol to correct the security flaws described in Section 3. The proposed protocol employs the concept of hiding identity in order to prevent the aforementioned attacks. We only modify the registration phase which issues a "hashed" identity for every legal user. That is, in the registration phase, the key registration center sends each user U_i (or U_j) a secret signature $g_i = HID_i^d(\bmod n)$ with a secure channel, where $HID_i = f(ID_i)$. The rest of the verification phase is retained except that ID_i is replaced by a "hashed" identity HID_i. Fig. 2 illustrates the proposed registration phase. It can be performed as follows:

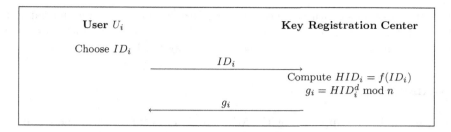

Fig. 2. The Proposed Registration Phase

Registration Phase: User U_i takes his/her identification number ID_i to the key registration center to obtain signature g_i for the ID_i. If the center confirms the validity and the relationship between U_i and the ID_i, the center can compute a "hashed" identity $HID_i = f(ID_i)$. Finally, the center calculates g_i using

$$g_i = HID_i^d \bmod n \tag{13}$$

and hands g_i to the U_i. When all users have registered and obtained his/her g_i ($i = 1 \ldots n$) the center no longer needs to exist in an Ad hoc network.

5   Security Analysis

This section discusses the enhanced security features. The rest are the same as those of the original Liaw et al.'s protocol, as described in literature [6]. Readers are referred to [6] for complete references.

Definition 1. *A one-way hash function assumption [8,9,10]: Let $f(\cdot)$ be a one-way cryptographic hash function; (1) given y, it is computationally unfeasible to find x such that $y = f(x)$; (2) it is computationally unfeasible to find $x_1 \neq x_2$ such that $f(x_1) = f(x_2)$.*

Theorem 1. *In the proposed registration phase, an illegal user cannot obtain a legal user's U_i (or U_j) secret signature g_i (or g_j).*

Proof. The attacks on Liaw et al.'s protocol are effective because a malicious user can successfully register a new ID_f via the ID_i (or ID_j) in the registration phase. In our improved registration phase, since the format of $HID_f^d = f(ID_i^{-1})^d(\bmod n)$ (or $f(ID_i \cdot ID_j)^d(\bmod n)$) is not equal to $ID_f^d = ID_i^{-d}(\bmod n)$ (or $g_i \cdot g_j(\bmod n)$), a malicious user cannot get the legal user U_i (or U_j)'s secret signature g_i (or g_j). Therefore, the proposed protocol can correct the security flaws described in Section 3.

6   Conclusions

The current paper has demonstrated the security flaws of Liaw et al.'s protocol. Using our attacks, we have shown that a malicious user can easily get a specific legal user's secret signature and can impersonate this specific user with legal user to exchange a common session key. For the above mentioned attacks, we have presented improvements to repair the security flaws of Liaw et al.'s protocol.

Acknowledgements

This research was supported by the MIC (Ministry of Information and Communication), Korea, under the ITRC (Information Technology Research Center) support program supervised by the IITA (Institute of Information Technology Assessment).

References

1. Kong, J., Zerfos, P., Luo, H. Lu, S. Zhang, L.: Providing Robust and Ubiquitous Security Support for Mobile Ad-Hoc Networks. Proceedings of the IEEE 9th International Conference on Network Protocols (ICNP '01). IEEE Computer Society. (2001) 251
2. Deng, H., Mukherjee, A., Agrawal, D.P.: Threshold and Identity-based Key Management and Authentication for Wireless Ad Hoc Networks. Proceedings of the International Conference on Information Technology: Coding and Computing (ITCC'04). IEEE Computer Society. Vol. 1. No. 1. (January 2004) 107-111
3. Shamir, A.: Identity-based Cryptosystems and Signature Schemes. In Advances in Cryptology - Crypto '84. Lecture Notes in Computer Science 196. Springer-Verlag. (1984) 47-53
4. Bohio, M., Miri, A.: An Authenticated Broadcasting Scheme for Wireless Ad Hoc Network. Proceedings of the Second Annual Conference on Communication Networks and Services Research (CNSR'04). IEEE Computer Society. (May 19-21 2004) 69-74
5. Khalili, A., Katz, J., Arbaugh, W.: Toward Secure Key Distribution in Truly Ad Hoc Networks. 2003 Symposium on Applications and the Internet Workshop (SAINT 2003). IEEE Computer Society. (2003) 342-346
6. Liaw, S.H., Su, P.C., Chang, H.K.C., Lu, E.H., Pon, S.F.: Secured Key Exchange Protocol In Wireless Mobile Ad Hoc Networks. Security Technology, 2005. CCST '05. 39th Annual 2005 International Carnahan Conference. Vol. 1. No. 1. (Oct. IEEE 2005) 1-3
7. Rivest, R.L., Shamir, A., Adelman, L.: A Method for Obtaining Digital Signatures and Public-Key Cryptosystems. Comm. of ACM. Vol. 21. No. 2. (1978) 120-126
8. Schneier, B.: Applied Cryptography. 2nd ed. John Wiley & Sons. Inc. (1996)
9. Diffie, W., Hellman, M.: New Directions in Cryptography. IEEE Trans Inf Theory. Vol. 22. No. 6. (1976) 644-654
10. Menezes, A.J., Oorschot, P.C., Vanstone, S.A.: Handbook of Applied Cryptograph. CRC Press. New York. (1997)

An Efficient Certificate Management
for Mobile Ad-Hoc Network

Dae-Young Lee[1] and Hyun-Cheol Jeong[2]

[1] Dept. of Computer Science & Statistics, College of Natural Sciences of Chosun University
375 Seosuk-Dong, Dong-Gu, Gwang-ju, 501-759, Korea
Tel.: +82-062-230-7962; Fax: +82-062-234-4326
cssna01@chosun.ac.kr
[2] Dept. of Medical Engineering, Kwangju Health College
683-3 Shinchang-Dong Gwangsan-Gu Gwangju 506-701 Korea
Tel.: +82-62-958-7774; Fax: +82-62-953-4946
hcjeong@kjhc.ac.kr

Abstract. MANET has dynamic composition in which nodes are frequently changing, when a node is damaged, reliability between nodes may be changed and they may join in other administrative domains quickly they can rely on. This study suggests security requirements for MANET which can be a base of ubiquitous system and models that can prevent security threat through application of PKI without trusted certificate authority in MANET. It is also solve excessive loading found in centralized control model by dispersing CA for adjustment to dynamic changes of nodes of MANET quickly and suggests a system model which supports expansion so that existing nodes performing communication within clusters can provide active certificate service without being affected by input of new nodes. In addition, it is to evaluate its stability, effectiveness and strength through simulation of the suggested model.

1 Introduction

Mobile ad-hoc network (MANET) is composed of mobile nodes connected by wireless links. The nodes communicate through a wireless interface to forward packet data. This means that each node acts as a router that moves freely to transmit and receive data [1]. MANET does not require any fixed infrastructure. Since the nodes in the mobile ad-hoc network operate as routers and hosts without base stations or AP, they can forward packets on behalf of other nodes and run user applications [2]. The most appealing aspect of MANET is its independence from the central control, making the communication between mobile devices easier and more flexible.

All nodes are members of MANET. These node members are free to move randomly and form and reorganize even when some members are damaged by nodes dynamically joining and leaving the trusted administrative domains. MANET, however, is more vulnerable than fixed hardwired networks to information and physical security threats. Although there are an array of routing protocols and security mechanisms for mobile ad-hoc networks, wireless vulnerabilities still poses a major threat. Significant effort is underway in many research centers to resolve security problems in mobile ad-hoc networks [3]]4][5].

T. Kunz and S.S. Ravi (Eds.): ADHOC-NOW 2006, LNCS 4104, pp. 355–364, 2006.

This study proposes and evaluates a new public key-based security mechanism for MANET. The mechanism segregates the roles of certification authority to keep with the dynamic mobility of nodes and handle rapid and random topological changes with minimal overhead. That is, this model is characterized by its high expandability that allows the network to perform authentication service without the influence of joining and leaving nodes. The efficiency and security of this concept was evaluated through simulation.

2 System Composition

Once the cluster is formed, the cluster head acts as a certification authority for all its members and is responsible for establishing a new pair of public/secrete keys to be used for certificate authentication. The cluster head will unicast its self-generated public key to other cluster heads through a network backbone. Thus the cluster heads hold shares of public key.

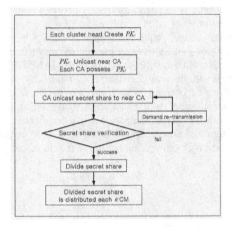

Fig. 1. System Accomplishment process

The certification authority verifies the feasibility of key-share by communicating all active cluster members (k) using the trust relationships in the direct cross-certification. And then the cluster head unicasts its verified key value to other cluster heads for their own verification. If the verification fails, the sensor node requests the requesting cluster head to retransmit the key value. Once random numbers are successfully verified with shared secrete key, the CA segregate secrete key and assign a part of key to its cluster members. This means that the cluster head will request its neighboring cluster members to generate a partial certificate to sign public key certificates at the time of certificate creation. Each partial certificate is transmitted to the requesting node through cluster head. When the number of partial certificates exceeds the threshold number, the requesting node can combine partial certificates together and create a complete certificate.

Given the dynamic topological changes in MANET, it is important for the CA to distribute its public key properly to have the system provide certificate service without being interrupted by nodes joining the cluster. In the proposed concept, the self-initializing mechanism enables the CA to distribute segregated SK certificate signing key to nodes, so the system could perform certificate service without being interfered with nodes joining the cluster.

3 System Design

Cluster heads collaborately generate public key using self-generated random numbers and distribute a partial secrete key to nodes for signing public key certificates. Once the cluster topology is changed rapidly and randomly, the roles of cluster members are changed as well, calling for an autonomic self-configuration mechanism to maintain key agreement between nodes. Dynamic topological changes call for frequent exchange of control information for key agreement and increase network time and overhead. In the proposed concept, control message for key agreement is integrated into a partial secrete key and distributed to each cluster head. A pair of public key/secrete keys is created according to the following process:

3.1 Key Creation Process

The cluster head that assumed the role of CA node generates random numbers x_i and a partial public key value of PK_i and unicasts its self-generated public key on a partial basis to other CA nodes through a network backbone. After a partial public key is delivered to individual CA nodes, each CA node issues domain's public key based on the following equation and keeps the key.

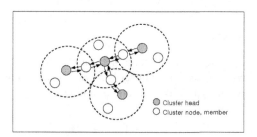

Fig. 2. Transmission of secret share key

If there are 4 clusters in the network, each CA will receive 4 different values of $PK_1, PK_2, \dots, PK_4$ as public keys. CA nodes will generate polynomial f_i with random number of x_i and axis of y when they acknowledge of k number of neighboring nodes To verify the feasibility of secrete key sharing, CA node sends control information F_{ji} of based on polynomial f_i having coefficient of f_{ij} to other CA nodes. CA nodes send secrete key ($SK_{ij} = g^{f_{ij}}$) corresponding to self-generated random numbers to other CA nodes.

Upon receiving secrete key, CA nodes verify whether SK_{ij} value matches public key. If shared secret key is not authenticated, CA node will request the retransmission of SK_{ij} value. When shared secret key authentication is completed, each CA node will generate a partial secrete key using the following equation. Key values generated by each CA node are presented in Table 1.

Table 1. Key values generated by each CA node

S_1	S_{41}	S_{31}	S_{21}	S_{11}
S_2	S_{42}	S_{32}	S_{22}	S_{12}
S_3	S_{43}	S_{33}	S_{23}	S_{13}
S_4	S_{44}	S_{34}	S_{24}	S_{14}

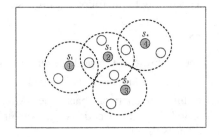

Fig. 3. Creation of partial certification key

3.2 Distribution of Secrete Key

As mentioned earlier, cluster heads assume the role of CA in MANET and coordinating the cluster's behavior. Once each cluster head establishes a pair of public key and secrete key, these keys are partially distributed among the k number of nodes within

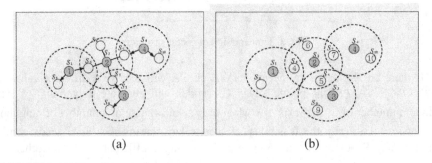

(a)   (b)

Fig. 4. (a) sharing secret share (b) distributed secret share the cluster. Shared public keys will be combined to issue a certificate and secrete keys will be combined to authenticate public key certification.

This partially distributed certificate authority well supports the dynamic nature of the network in which nodes join or leave the trusted administrative domain rapidly and randomly. As a result, the key management scheme is characterized by its mobility and flexibility. Fig. 4. (a) is shown Secrete keys being distributed among all nodes within a cluster and (b) is shown Partially distributed certificate authority.

3.3 Certificate Creation

Although secrete key is partially distributed among N nodes within a cluster, a partial certificates held by k nodes will be combined to issue a certificate that matches the one issued by a CA. In the proposed concept, threshold cryptography allowed CA node to distribute self-generated secrete key to k nodes and create a certificate in collaboration with neighboring nodes within the threshold transmission range. Secrete key shared among all nodes within a cluster will be demanded to sign public key certificate when new nodes join the cluster.

Each cluster member is capable of generating a partial certificate based on key-sharing scheme. A cluster member will broadcast its request for a partial certificate to all its neighbors when public key certificate needs to be verified for certificate creation. Partial certificates will be forwarded to the requesting node through cluster head. Partial certificates are required to combine into a full certificate when the threshold value is met. For instance, CA2 will send its request for secrete key sharing among its neighborhood when it needs to sign public key certificate. When the threshold number of partial certificates are collected, CA2 node can generate a certificate signing key and issue a valid certificate as illustrated in Fig.5.

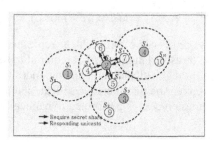

Fig. 5. Creation of complete certificate

A share of secrete key among all nodes within a cluster is based on the polynomial $f(x)$. Security of the ad-hoc network in which public/private keys (SK=<d,n> / PK<e,n> are shared among k nodes can be achieved by constructing a lagrange interpolation polynomial (1) of degree K-1 as follows:

$$f(x) = d + f_1 \cdot x + f_2 x^2 + \cdots + f_{K-1} \cdot x^{K-1}, \ N(> K) \tag{1}$$

However, when the number of nodes is less than k, secrete key information cannot be shared. Shared secrete key is expressed as $f(0) = d$ and a partial key held by node i is expressed as $P_i = f(C_i) \bmod n$.

Each node can sign certificates using shared secrete key. Shared secrete key ($SK=<d,n>$) is obtained by calculating d after collecting k number of partial certificates as defined by the following equation:

$$d \equiv \sum_{i=1}^{K} (P_{C_i} \cdot l_{C_i}(0) \bmod n)$$
$$\equiv \sum_{i=1}^{K} SK_i (\bmod n) \tag{2}$$

Where $l_{C_i}(0)$ represents Lagrange coefficient. $SK=<d,n>$ can be performed by a coalition of k nodes. sk_i can be recovered by Lagrange interpolation from the polynomial (d) (Equation 2) as shown below in equation 3.

$$d \equiv \sum_{i=1}^{K} (P_{C_i} \cdot l_{C_i} \bmod n)$$
$$\equiv t \times n + d \tag{3}$$

Where $(P_{C_i} \cdot l_{C_i} \bmod n)$ represents a module concerning n and will be equal to d in algorithm. However, it would be difficult to create a valid certificate by multiplying k number of partial certificates. A complete certificate can be created using the following K-bounded coalition offsetting algorithm.

$$M^{t \times n + d} \equiv M^{t \times n} \times M^d \equiv M^d (\bmod n) \tag{4}$$

Where M is public key $<e,n>$. The threshold K represents the number of nodes within a cluster. And the threshold number is not required to be large.

A certificate is issued through collaboration among nodes and shared secrete key is used for authentication. Given the distributed certificate authority among every node in the cluster, key matching process is performed dynamically and rapidly achieving more success.

3.4 Certificate Renewal

It is important to have the MANET perform certificate service despite its dynamic nature. As a solution, certificate authority can be partially distributed to all nodes within a cluster as proposed in this study. That is, if a cluster is reformed by joining nodes, existing nodes will incorporate new nodes into the certificate authority by generate secrete key and send it to joining nodes, which in turn generate their own secret key. If the number of nodes within a cluster falls behind the threshold number, it is impossible to obtain the certificate signing key. As a result, network security can be compromised. It is therefore critical to form a cluster maintaining the threshold number of nodes and more and reestablish a new pair of public and secrete keys after assigning the role of CA to a cluster head. It is known that network security increases when the threshold cryptosystem has the ability to generate a new pair of public/private keys on a regular basis [6] [7].

In the proposed concept, cluster head is capable of generating a new public/private key pair regularly regardless the change in the number of nodes within a cluster, making network security stable.

To have MANET perform certificate service even with new nodes joining a cluster, a self-initialization mechanism is used for the distribution of certificate signing key. That is, an existing node will incorporate new nodes into certificate authority by requesting key-sharing to CA node, which will register new nodes and distribute new secrete key to them, leading to the formation of an expanded cluster. Security and scalable expandability of network significantly increase as a result. A cluster is re-formed with joining nodes as follows.

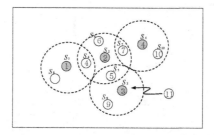

Fig. 6. The joining of new node

As shown in Fig. 6. with node11 joining a cluster, the new node sends its request for registration to CA3 node, which in turn accepts the new node and distributes a partial secrete key to the node. Thus it is possible for the network to offer certificate service without being interfered with joining nodes.

4   Simulation for Testing and Evaluation

Network Simulator 2(NS2) written in C++ with OTcl interpreter, an object-oriented version of TCL, was used to evaluate the performance of the proposed security mechanism. Packet flows were traced using timing and the amount of routing information. And results were compared with those obtained from the use of existing routing protocol. Efficiency, security and robustness of the proposed security mechanism are evaluated as follows.

4.1   Efficiency and Safety

Computation time required to issue a public key certificate is presented in Table 2. PCC represents computation time required to issue a partial certificate using secrete key. Combine represents calculation time required to issue a complete certificate by combining k number of partial certificates.[8]. CPC indicates computation time for a partial certificate based on security mechanism proposed in this study. Sum indicates total time required to obtain secrete key and create a public key certificate based on security mechanism proposed in this study.

The association of computation time with the number of bits was compared for PCC and CPC Fig. 7.

Table 2. Comparison of computation time between CPC and PCC. key: 1024bit, Pentium III/500 laptop.

key(bit)	PCC	Combine	CPC	Sum
512	0.0466	0.0928	0.407	0.301
768	0.1198	0.2416	0.462	0.361
1024	0.2610	0.5280	0.488	0.322
1280	0.4590	0.9742	0.551	0.426
1536	0.7944	1.5598	0.801	0.462
2048	1.7058	3.4410	1.461	0.488

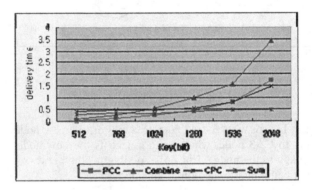

Fig. 7. Comparison of changes in computation time in relation to the number of bits between CPC and PCC. key: 1024bit, Pentium III/500 laptop.

The results revealed that delivery time rose as the number of bits increased when a public key certificate was created using shared secrete key (PCC). Thus the increase in packet size created additional routing overhead and eventually undermines security effectiveness of the ad-hoc network, increasing the risk of losing data or exposing confidential data. Compared with PCC, delivery time slightly rose as the number of bits increased under the security mechanism proposed in this study (CPC). Total computation time (SUM), particularly, maintained a rather steady delivery time despite increase in the number of bits. The efficiency and security of the network is sustained

The evaluation of computation time for a partial certificate and a complete certificate in relation to the number of bits confirmed that the variable k did not have an impact on the efficiency of the proposed concept because individual calculation for a partial certificate is steadily performed despite joining nodes algorithm. Since the node requesting a public certificate authentication initializes other nodes to compute a partial certificate, unnecessary system overhead does not incur.

4.2 Robustness

The robustness of the network was also evaluated by measuring the key agreement and changes in routing overhead in relation to the number of nodes. The match between routing message and secrete key was measured by comparing the changes in the number of nodes and the length of message (packet) sent and received by each node.

The key agreement among 30 and 40 k nodes was illustrated in Fig.8. The average rate of key agreement indicates packet delivery ratio and the length of message indicates packet size.

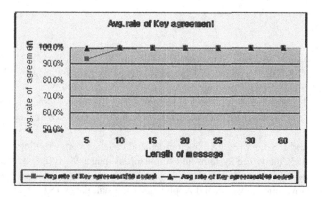

Fig. 8. Avg. rate of Key agreement (node K=30, 40)

The key agreement was maintained at an average rate of almost 100% even after the number of nodes increased to 40 from 30. At the same time, key agreement was not affected by the increase in the length of message. Given the algorithm that allows nodes to forward and receive a packet without incurring unnecessary routing overhead, the robustness of the network is guaranteed.

5 Conclusion

Certificate creation time rose as the number of bits increases under existing certificate-based authentication protocol in which a partial certificate is generated using secrete key shared by nodes and a complete certificate is issued by the combination of partial certificates. The increase in packet size increases routing overhead and eventually undermines data transmission efficiency and security of the network, making routing data more susceptible to misinformation attacks. In the proposed concept, certificate creation time slightly rose as the number of bits increased. But the increased time was much lower than that obtained from the use of existing certificate-based authentication protocol. The proposed concept, however, offered a steady

364   D.-Y. Lee and H.-C. Jeong

delivery time despite the increase in packet size while authenticating a public key
certificate through the combination of shared secrete key. Data transmission efficiency and security can be therefore maintained in the network.

It was also found that the efficiency of the network was not influenced by changes
in the number of nodes (k) because partial certificates are computed by all the nodes
within a cluster without being interfered with topological changes. Since the node
requesting a public certificate authentication initializes other nodes to compute a partial certificate, unnecessary system overhead did not incur. This study suggests an
ideal security mechanism for MANET by eliminating the need for pre-distribution of
certificate signing key and the centralized CA. Also, the proposed concept enhances
the expandability of the network by incorporating joining nodes into certificate authority, so certificate service can be performed despite the dynamic topological
changes.

References

1. Charles E. Perkins, "Ad Hoc Networking," Addison Wesley, 2001.
2. E. M Royer, C-K Toh "A review of Current Routing Protocols for Ad Hoc Mobile Wireless
   Network", IEEE Personal Communication, pp.46-55, April, 1999.
3. D. Johnson, D, Maltz, Y-C. Hu, and J. Jetcheva. The dynamic source routing protocol for
   mobile ad hoc network. IEEE Internet Draft, March 2001. draft-ietf-manet=dsr-05.txt(work
   in progress)
4. C.E Perkins and E. M. Royer. "Ad hoc on-demand distance vector routing". In IEEE Workshop on Mobile Computing Systems and Applications, pages 90-100, Feb. 1999.
5. S. Murthy and J.J. Garcia-Lunca-Aceves. "An efficient routing protocol for wireless networks". ACM Mobile Networks and Applications Journal, pages 183-197, Oct. 1996
6. A. Herzberg, M. jakobsson, S. Jarecki, H. Krawczyk, and M.Yung, "Proactive public-key
   and signature schemes.", In proceedings of the Fourth Annual Conference on computer
   Communications Security", ACM, 1997. pp. 100-120.
7. A. Herzberg, S. Jarecki, H. Krawczyk, and M.Yung, "Proactive secret sharing or : How to
   cope with perpetual leakage", Advances in Cryptolgy-Crypto '95, Santa Barbara, Califonia,
   U.S.A, Aug. 1995, pp.457-469.
8. J. Kong. P.Zerfos, H.Luo. S. Lu, L.Zhang. "Providing Robust and Ubiquitous Security Support for Mobile Ad-Hoc Networks", IEEE Computer Society, Proceedings of the Ninth International Conference on Network Protocols (ICNP'01), PP. 251. 2001.

Performance Improvement of TCP with an Efficient Contention Window Control Mechanism (ECWC) in IEEE 802.11 Based Multi-hop Wireless Networks

Byungjoo Park[1], In Huh[2], and Haniph Latchman[1]

[1] Department of Electrical and Computer Engineering,
University of Florida, Gainesville, USA
{pbj0625, latchman}@ufl.edu
[2] Department of Information Communications Engineering,
Chungnam National University, Daejeon, Republic of Korea
huhin@cnu.ac.kr

Abstract. For multi-hop wireless networks using IEEE 802.11 most TCP performance degradation results from hidden, exposed terminal problems and bandwidth waste caused by exponential backoff of retransmission timer due to node's mobility, not from collisions. However, in normal DCF algorithm, a failed user increases its contention window (CW) exponentially, thus it reduces the success probability of exposed terminal nodes. That is, these problems will cause burst data transmissions frequently in a particular node which already was successful in packet transmission, because the probability of successful packet transmission rate would be increased. To solve these problems, in this paper, we propose an efficient contention window control (*ECWC*) scheme to increase TCP performance in wireless multi-hop network. The proposed *ECWC* scheme is suggested to reduce the hidden and exposed terminal problems of wireless multi-hop network. That is, the proposed scheme increases the number of backoff retransmissions to increase the successful probability rate of MAC transmission, and fixes the contention window at a predetermined value.

1 Introduction

The IEEE 802.11 MAC protocol is the standard for wireless LANs that is currently most widely used in the wireless ad-hoc network. The IEEE 802.11 MAC protocol uses the Distributed Coordination Function (DCF) to transmit packets. IEEE 802.11 DCF applies the method that all stations transmit packets after occupying channels through competition with equal relation, based on the Carrier Sense Multiple Access with Collision Avoidance scheme (CSMA/CA) to avoid collision during packet transmission [1].

The CSMA/CA that is used by the DCF uses a random backoff timer to avoid collision between stations. Each station has minimum contention window

T. Kunz and S.S. Ravi (Eds.): ADHOC-NOW 2006, LNCS 4104, pp. 365–375, 2006.
© Springer-Verlag Berlin Heidelberg 2006

(*CWmin*) and maximum contention window (*CWmax*) to determine the random backoff time. Each station chooses a random contention window (*CW*) value between 0 and *CWmin* before using a channel. The random backoff time is obtained by multiplying this value by the slot time. While each station decreases the backoff time during valid time, the station to first reach 0 gets the right to use the channel. If collision occurs when more than two stations transmit packets simultaneously, they retransmit the packets. In such case, a random value between 0 and *CW-1* is chosen after increasing the range of the backoff contention window to reduce the possibility of collision and then the new backoff time is calculated with this value.

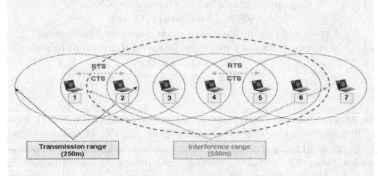

Fig. 1. Hidden terminal problem in mobile Ad-Hoc network

In most cases in which the number of terminals with which the wireless multi-hop network is actually comprised is small, and the size of the currently used *CW* is inefficient. The reason for this is that in the case of the wireless multi-hop network having only a few nodes, despite the small *CW* size, the possibility of collision is very low when more than two nodes transmit packets simultaneously. Since the large size *CW* increases the waiting time of the node for packet transmission, it causes the reduction of the throughput of the wireless multi-hop network [2,3,4]. In Fig. 1, when the No. 4 and No. 5 nodes are engaged in communication, the No. 1 and No. 2 nodes cannot transmit packets due to the hidden-terminal problem by interference range until the No. 4 and No. 5 nodes terminate communication. After that, if the No. 4 and No. 5 nodes transmitted packets successfully, the next contention window has the value of *CWmin*. Conversely, since the No. 1 and No. 2 nodes failed to transmit packets due to the hidden-terminal problem, it postpones current packet transmission and increases the range of their own backoff contention window. Accordingly, the possibility that they will lose in a contention and fail to get channels for packet transmission becomes high. Therefore, a burst of traffic is generated between nodes No. 4 and No. 5. In addition, if the wireless multi-hop network that experiences the hidden-terminal problem frequently fails to receive the Clear-To-Send (CTS) packet after transmitting the Request-To-Send (RTS) packet 7 times, it considers that the corresponding route was cut off and gives up the current packet transmission, resetting the route for the ad-hoc routing protocol. This causes overhead such as re-transmission, thus resulting in the reduction of the TCP performance. In order to solve these problems, in this paper, we propose an efficient contention window mechanism to improve the TCP performance in the wireless multi-hop network by adjusting the size of the maximum contention window and the number of RTS packet retransmission times.

The remainder of this paper is organized as follows. Section 2 provides an overview of IEEE 802.11 Distributed Coordination Function (DCF). In Section 3, we introduce our proposed efficient contention window control ($ECWC$) mechanism. Simulation results and analysis are provided in Section 4. Finally, we will conclude this paper in Section 5.

2   IEEE 802.11 Distributed Coordinate Function (DCF)

The IEEE 802.11 based MAC protocol uses the DCF medium access control method according to CSMA/CA in order to transmit packets successfully without collision. Fig. 2 demonstrates the node movement when the DCF method is used. If a channel idles during DIFS time after completion of the busy medium state, the nodes that have postponed medium access check the medium status and at the same time reduce their own random backoff time. If a medium idles until the backoff time of any node reaches 0, the corresponding node obtains access to the medium. However, if another node uses the medium before the backoff time reaches 0, the node stops using its own backoff time that is decreasing and uses the backoff time remaining after the next DIFS. Therefore, since the possibility that this node will have less backoff time than other nodes that generated random backoff time initially is high, the possibility for this node to obtain access to the medium becomes higher. If the packet transmission is executed successfully, the CW is reduced to the $CWmin$. However, if the packet transmission was failed, the CW value is doubled by exponential backoff. The CW size increases up to the maximum set value, $CWmax$. After that, the packet transmission will be repeated up to the designated number of retry attempts. The frames that

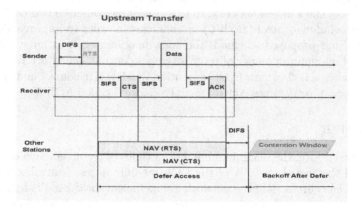

Fig. 2. Basic DCF access method

were not transmitted successfully even after the maximum number of times of retry will be discarded. As such, the node that transmitted packets successfully will begin new transmission from the *CWmin*.

The node that failed to transmit packets sets the contention window by doubling the *CW*. Therefore, the node that succeeded in transmission has permanent priority, thus indicating burst characteristics in traffic transmission. In order to check the unfair transmission between nodes resulting from this problem, the unfairness in the chance of transmission is compared and analyzed in the chain topology.

3   An Efficient Contention Window Control Mechanism (ECWC)

The IEEE 802.11 MAC protocol [5,6] that is most generally used in the wireless network currently uses the DCF as a medium access control scheme. The size of the *CW* used in the current *CW* control algorithm is inefficient for a network with a wireless multi-hop network that is actually comprised with a small number of nodes. The reason for this is that the possibility for the hidden-terminal to cause data transmission failure is higher than that of the collision and also that the waiting backoff time increases if the *CWmax* is set too great. In addition, the node that succeeded in packet transmission has the minimum *CW* value and competes with other nodes waiting for the next transmission to occupy channels. However, the node that failed to transmit packets engages in competition with the *CW* doubled by the exponential backoff to occupy the medium. Therefore, the possibility that the node that succeeded in transmission occupies the medium continuously becomes higher. This forces burst traffic to occur in the wireless multi-hop network. Moreover, when the CTS packet from the recipient is not received after transmitting the RTS packet up to 7 times, performing the route reset procedure for the routing protocol after discarding the corresponding packet for which transmission has been attempted considering that the corresponding route was cutoff is a great overhead in the wireless multi-hop network. This causes a decrease in the throughput of the wireless network. In this section, we describe two new efficient contention window control (*ECWC*) mechanisms in order to improve TCP performance. First proposed scheme is the "new maximum contention window size control and maximum number of retry attempts control" called *MCWR*. Second proposed scheme is the "new fixed contention window size control and the maximum number of retry times control for RTS packet" called *FCWR*.

3.1   *MCWR*

As described above, the maximum number of times of retransmission of the RTS packet and the size of the *CW* of the DCF medium access control scheme that are used in the current wireless network are inefficient for the multi-hop network comprised with a small number of nodes. In order to solve such problem, this paper proposes a method to control the maximum number of retransmission attempts of the RTS packet and the size of the CW of the DCF medium access control scheme. It was found that optimum performance could be achieved by setting the size of the CW to 256 by changing the maximum value to 32, 64, 256, 512, 1024 and 2048 for simulation after setting the minimum CW value to the size that was being used by the current DCF in order to determine the optimum CW size for the wireless network. In addition, this paper proposes a method to increase the number of retry attempts to prevent the execution of the route reset algorithm for the routing protocol by considering that the route was cutoff if the CTS packet was not received after performing 7 attempts at retry for the RTS packet. The reason for this proposal is that the algorithm for frequent route reset of the routing protocol due to the characteristics of a network in which temporary route failure caused by hidden-terminal problem occurs frequently causes unnecessary retransmission in the TCP architecture, which is considered to be big overhead, and this results in decrease in the TCP throughput. It was found that the proposed scheme was superior to the existing DCF in section 4.

3.2 FCWR

In order to improve the TCP performance in the wireless network, this paper proposes a second method that fixes the CW size by setting the maximum and minimum values of the CW to be identical. The reason for fixing the size of the CW regardless of the transmission failure is that doing so will increase the fairness between nodes by preventing the burstiness because the existing DCF method gives priority to the successful node in the case of transmission failure due to hidden-terminal thus causing the burstiness to occur.

Originally, the reason that the CW is doubled each time the transmission fails is to obtain the collision reduction effect when the collision is severe. Both schemes increase the number of retry attempts for RTS packet transmission. This is to reduce the overhead for route reset caused by incorrect judgment of route failure due to the hidden-terminal. Here, attention shall be given to the fact that if the number of retry attempts is too small when the fixed CW value is too low in applying this control method, the number of retry attempts must be greater than the maximum frame transmission time since the transmission may fail once the retry is finished even before the current data frame transmission is completed. In this section, the performance according to the CW size was compared through an ns-2 [7] simulator by controlling the number of retry attempts after setting the maximum and minimum values of the CW to be identical. It was found that the proposed scheme was superior to the existing DCF when fixed CW is maintained after obtaining the optimum CW size.

4   Performance Analysis and Simulation Results

In this section, the throughput, RTT and the number of times of route reset were compared between the existing DCF medium access control scheme and the one proposed by this paper by using the ns-2 simulator. In TCP code, we used TCP-NEWRENO for data traffic.

Table 1. Parameters for simulation

Parameter	Value
Bandwidth	11Mbps
Packet size	1024bytes
Topology size	$5000m \times 1500m$
Traffic	$TCP - NEWRENO$
Simulation time	$300s$
Routing protocol	DSR, AODV
Topology	Chain, Grid topology

In order to obtain a more accurate result through simulation under various environments, the performance was compared between the proposed scheme and the existing DCF in the 8-hop chain topology and 8x8 grid topology by using the DSR and AODV routing protocols. Hereinafter the method that increases the number of retry attempts by changing the $CWmax$ value is referred to as proposed $MCWR$ scheme, and the method that uses the fixed CW value as proposed $FCWR$ scheme. Table 1 indicates the simulation environment and parameters.

Fig. 3 demonstrates the result of the performance simulation according to the number of times of retransmission in the proposed $MCWR$ scheme, which was performed by using the DSR and AODV routing protocols. In the case of the existing DCF, the $CWmax$ was 1024, and the number of retry attempts was 7. The performance based on these values is shown graphically. It could be seen from the simulation result that setting the maximum CW to 256 with more than a certain number of retry attempts resulted in the highest throughput.

This is because the proposed $MCWR$ scheme reduces the delay time of the nodes waiting for transmission as well as the possibility of the burst traffic occurrence by limiting the maximum size of the CW to 256 so that it can obtain improved throughput greater than that of the DCF medium access algorithm. In addition it could be seen that the TCP performance in the wireless multi-hop network could be improved by reducing the overhead due to retransmission and frequent execution of the routing protocol reset procedure caused by the hidden-terminal by raising the number of maximum retry attempts.

Fig. 4 shows the result of the simulation performed for the grid structure in order to evaluate the proposed $MCWR$ scheme by this paper more accurately. The simulation was performed by configuring 8x8 nodes. In the case of the grid structure, unlike the chain topology, the route from the transmitter to the receiver was made to change frequently by causing temporary route failure due to the hidden-terminal. On the other hand, even in a network with such a structure, the number of times of executions of route reset procedure due to the temporary route failure caused by the hidden-terminal was reduced by increasing the number of times of the maximum re-transmission of the RTS packets.

This improves the throughput in the network by reducing its overhead. In addition, it can be seen that in the case of the CW smaller than that of the existing DCF, the performance of the proposed scheme is superior to that of the existing DCF since the proposed scheme reduces the delay time of nodes waiting

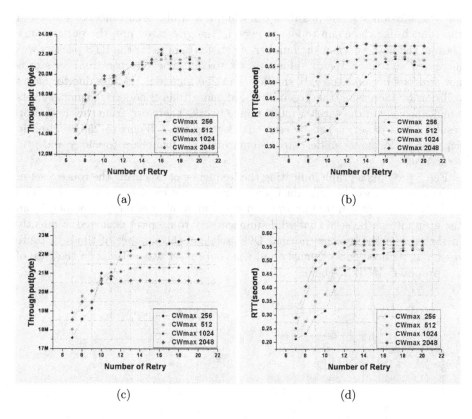

Fig. 3. Performance comparisons between proposed *MCWR* scheme and existing DCF in DSR and AODV networks: (a) DSR Throughput (b) DSR RTT (c) AODV Throughput (d) AODV RTT

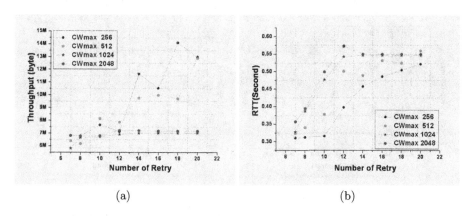

Fig. 4. Performance comparison between proposed *MCWR* scheme and existing DCF for grid structure in DSR: (a) Throughput, (b) RTT

for transmission and the possibility of the burst traffic occurrence. It was found that, much like the chain topology, even in the grid structure, the performance became the highest when the number of retry attempts for the RTS packets was increased and the *CWmax* value was set to 256. The *CW* too small in size is not desirous because the *CW* smaller than 256 increases the loss due to packet collision by increasing the possibility that more than 2 nodes transmit packets simultaneously in the wireless multi-hop network. Therefore, from the simulation result of this paper, it can be seen that setting the *CWmax* to 256 resulted in superior performance under various environments like chain topology and grid structure.

Fig. 5 presents a graph indicating the frequency of executing the route reset of the routing protocol because the CTS packet was not received even after the RTS packets were retransmitted up to the maximum number of retry attempts. From the graph it can be seen that while unnecessary route reset occurred frequently in the case of the existing normal DCF scheme, the number of times of route reset was reduced conspicuously and the route reset was efficient in the case of the proposed *MCWR* scheme.

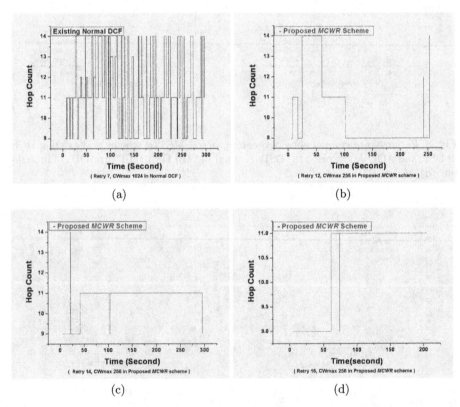

Fig. 5. Number of times of route reset in the grid topology (DSR Case)

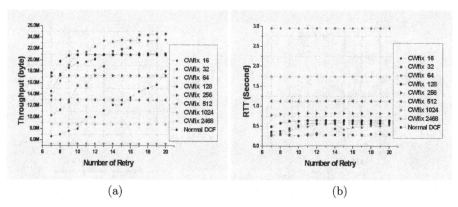

(a)   (b)

Fig. 6. Performance comparison between proposed *FCWR* scheme and existing DCF for chain structure: (a) Throughput, (b) RTT

The second *FCWR* scheme proposed by this paper for the improvement of the TCP performance in the wireless multi-hop network is the scheme that sets the maximum and minimum *CW* value as identical. As in the first proposed *MCWR* scheme, simulation was performed by increasing the number of retransmission times of RTS packets after setting the *CW* value smaller than the existing one.

Fig. 6 shows the performance of the proposed *FCWR* scheme. From the result of the simulation performed to obtain the most suitable CW value by changing the CW size from 16 to 2048, variously it could be seen that the most improved throughput was obtained when the CW size was fixed to 32 or 64.

Fig. 7 demonstrates the number of times of route reset in the chain topology when the proposed *FCWR* scheme was used. From the Fig. 7 it can be seen that while unnecessary route reset occurred frequently in the case of the existing normal DCF scheme, the number of times of route reset was reduced conspicuously and the route reset was efficient in the case of the proposed *FCWR* scheme.

Fig. 8 demonstrates the packet transmission time of each node after obtaining the opportunity for transmission. As can be seen from Fig. 8 (a), IEEE 802.11 DCF shows burst characteristics in the node transmission. This is because the node failed to transmit packets due to a hidden terminal (4 when it is 1). Furthermore, transmission waiting time increased since only *CW* increased and the node that succeeded in transmission has the permanent priority to get the chance of transmission in case of competition.

However, as can be seen from the result indicated in Fig. 8 (b), if the scheme proposed by this paper is used, the network can be used effectively by reducing the waiting time for transmission between nodes by setting the *CW* value small. As well, the fairness of transmission chance can be guaranteed since the *CW* is not increased but it is fixed when transmission failed due to a hidden terminal. As explained above, if the proposed scheme is used, the transmission failed node does not increase the *CW* exponentially and it competes for channel access with other nodes succeeded in the transmission under equivalent conditions. Therefore, the possibility becomes high for the node that failed to transmit the packet

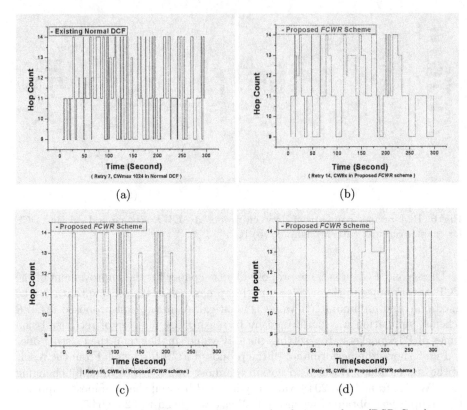

Fig. 7. Number of times of route reset in the chain topology (DSR Case)

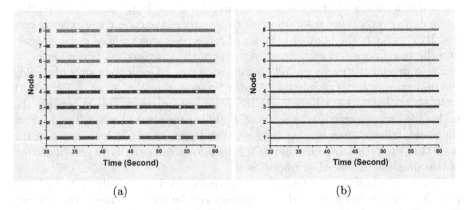

Fig. 8. Comparison of fairness of transmission chance between nodes in IEEE 802.11 DCF (a) and Proposed ECWC scheme (b)

once to occupy the channel for packet transmission. It was confirmed that the proposed *ECWC* scheme through simulation results demonstrated higher performance than the existing DCF.

5   Conclusions

This paper proposed two methods to improve the TCP performance in the wireless network. It was confirmed that the proposed *ECWC* scheme could guarantee the fairness of transmission opportunity between nodes by solving the problem of unfairness of transmission probability through the reduction of the unfairness of transmission chance due to a hidden terminal. Furthermore, since the proposed scheme had smaller *CW* value, the performance could be improved by reducing the overhead for route resetting by reducing the waiting time for transmission and increasing the number of retry attempts. As a subject for further study, the proposed algorithm will be modified and supplemented in preparation for the actual route failure, rather than the temporary route failure due to a hidden-terminal.

References

1. J.P. Macker, M.S. Corson, "Mobile Ad Hoc Networking and the IETF," ACM Mobile Computing and Communications Review, vol. 2, number 1, Jan. 1998.
2. C. E. Perkins. "Ad Hoc Networking," Addison Wesley, 2001.
3. E. M. Royer and C. Toh, "A review of current routing protocols for Ad Hoc mobile wireless networks," IEEE Personal Communications, pp.207-218. April 1999.
4. K. Chandran, S. Raghunathan, S. Venkatesan, P. Prakash, "A Feedback-Based Scheme for improving TCP Performance in Ad Hoc wireless Networks", IEEE Personal Communica-tions, Feb. 2001.
5. Z. Fu, P. Zerfos, H. Luo, S. Li, L. Zhang, M. Geral, "The impact of multihop wireless networks channel on TCP throughput and loss", Proceedings of 22nd Annual Joint Conference of the IEEE Computer and Communications societies '03, vol 4, pp. 1744-1753, March 2003.
6. IEEE Standards Department, "IEEE Standard for Wireless LAN Medium Access Control (MAC) and Physical Layer (PHY) specifications", Institute of Electrical and Electronics Engineers Nov. 1999.
7. The Network Simulator (NS2), http://www.isi.edu/nsnam

New Service Differentiation Model for End-to-End QoS Provisioning in Wireless Ad Hoc Networks

Joo-Sang Youn[1], Seung-Joon Seok[2], and Chul-Hee Kang[1]

[1] Department of Electronics and Computer Engineering, Korea University
5-1ga, Anam-dong, Sungbuk-gu, Seoul, Korea
{ssrman, chkang}@widecomm.korea.ac.kr
[2] Dept. of Computer Engineering, Kyungnam University, Kyungnam, Korea
sjseok@kyungnam.ac.kr

Abstract. In this paper, a new QoS model is presented for end-to-end service provisioning in wireless ad hoc networks. Many previous works focus on the packet scheduling mechanism using multiple service classes implemented for traffic prioritization based service differentiation. However, this paper concentrates on a scheme for dynamically selecting a proper one among several forwarding classes that perform different service rate according to service requirements. There service requirements include low delay, high throughput, and low loss. The proposed solution is a new QoS provisioning model called Dynamic Hop Service Differentiation (DHSD). This model supports soft QoS provisioning to reduce network overhead and wireless PHB (WPHB) to achieve the end-to-end QoS required by applications. The proposed QoS model is evaluated using OPNET simulation. We show that this model outperforms both best-effort and strict priority service models in wireless ad hoc network environments.

1 Introduction

Mobile ad hoc networks (MANETs) are highly dynamic networks formed by a set of mobile nodes connected to others through wireless links. Recently, these networks have been adopted in commercial environments. Moreover, it is important for MANET to support real-time applications such as voice over IP and video conferencing, as well as non real-time applications such as file transfer. Existing challenges including QoS routing, QoS MAC, power management and quality of service (QoS) provisioning model have been explored to obtain realizable QoS provisioning of ad hoc networks. However, this work is not sufficient in supporting these applications because the overall solution is currently too complex. Suitable QoS provisioning for end-to-end delivery of diverse applications is currently significant technical challenge. In this paper, only one situation is investigated, providing end-to-end QoS in MANETs. The focus is specifically on a new service differentiation model to achieve end-to-end QoS.

T. Kunz and S.S. Ravi (Eds.): ADHOC-NOW 2006, LNCS 4104, pp. 376–389, 2006.

Many advanced work [1, 2, 3, 5] for end-to-end QoS provisioning has been reported. There are numerous mechanisms that affect QoS provisioning across the communication protocol stack. However, these mechanisms are very sensitive to exterior environments such as variation in traffic conditions and dynamic topology. Thus, in a good QoS model for MANETs, the fundamental properties of such network; node mobility, shared medium and decentralized access, are taken into account. In addition, an attempt is made to reduce the complexity of QoS management. The goal is to design a service differentiation model adopting the concept of Soft QoS [13].

In this paper, a simple QoS mechanism, called dynamic hop service differentiation (DHSD), is proposed. The proposed solution is a new QoS model based on WPHB (wireless per hop behavior) which can enable soft QoS provisioning in ad hoc environments. In the DHSD model based on WPHB, both N forwarding classes with relatively different service rates for a packet scheduling, which is called 802.11EDCA+, and Per Hop Class Assignment (PHCA) algorithms are implemented. PHCA algorithms are class selection schemes to dynamically select a forwarding class that meets the service requirement per node according to the type of specific end-to-end service. The new four end-to-end services are also defined by considering the characteristics of MANETs. Furthermore, this mechanism assures that, even in the absence of hard resource reservation, accepted flows obtain QoS without requesting end-to-end QoS estimation. The purpose of the proposed QoS model is not to overcome the problems related to legacy EDCA in mobile ad hoc environments, rather to present novel service model using EDCA to obtain better end-to-end QoS performance.

A set of mechanisms are developed to realize the DHSD model in IEEE 802.11e-based ad hoc networks in the presence of traffic variations, and network disturbances due to topology dependent contention (hidden and exposed nodes), fluctuating queuing delays and loss rate including both collision loss rate and buffer loss. To show the effectiveness of the proposed mechanism, the DHSD model is implemented in OPNET. In the simulation, the performance of the DHSD model is evaluated in one-hop and multi-hop networks scenarios. The remainder of the paper is organized as follows. Section 2 presents the existing work related to quality of service of mobile ad hoc networks. Section 3 describes a new QoS model. Section 4 describes the PHCR mechanisms for each end-to-end service. Section 5 summarizes results of simulation studies. This paper concludes in Section 6.

2   Previous QoS Works of MANET

In order to provide end-to-end QoS in MANETs, existing solutions based on the DiffServ and IntServ models are proposed. In the FQMM [2] model, end-to-end QoS is provided by mixing the DiffServ and IntServ model. This method takes advantage of per flow granularity in the IntServ model and aggregation of service in the DiffServ model. Then high priority flows are provided with per flow QoS guarantees, while low priority flows are provided with per class differentiation.

However, FQMM model has the scalability problem of classifying low priority packets into service classes. The INSIGNIA [1] model provides adaptive services. Such model performs an in-band signaling protocol rather than out-of-band signaling protocol as RSVP to decrease reservation overhead. SWAN [3] is based on the reservation-less approach. It assumes a best-effort MAC protocol and uses feedback based control mechanisms to support service differentiation between real-time and best effort in spite of not being able to guarantee QoS requirements of each flow for the entire session. However, this model does not operate well in scenarios where most of the traffic is real-time traffic. Moreover, in the worst case, admitted real-time flows may encounter periodic violations in bandwidth requirement, since intermediate nodes can not provide hard QoS guarantees. QoS provisioning schemes that respect resource reservation have been proposed [6], [7]. These mechanisms, however, require significant overhead to maintain resource reservation and are sensitive to the many changes over the end-to-end route. In order to provide a prioritized service model to guarantee real-time traffic over best-effort traffic, differentiated scheduling and medium access algorithms are proposed [8], [9]. These solutions still face the issue of overhead for QoS guarantees. With regard to media access, some solutions have been proposed in literature trying to improve performance of the 802.11 MAC in multi-hop ad hoc networks. However, no studies have been reported on the feasibility and the scalability of IEEE 802.11e in such networks. The paper addresses the studies that are relevant. Many studies for evaluating EDCA performance are presented to provide service differentiation. Performance results in [17] show that EDCA is capable of providing very good service differentiation. However, other work concentrates on the necessity of adding new features to overcome a few limitations in the performance of the EDCA mechanism. In [16] the authors point out the necessity of local data control and admission control to guarantee a service for real-time traffic under the high traffic load conditions. In this model, each node maps the measured traffic-load condition into backoff parameters locally and dynamically. The solutions of end-to-end QoS provisioning, using the IEEE 802.11e [14], recently investigate in multi-hop environments. In this paper, the 802.11e EDCA mechanism is used to implement several service queues (forwarding classes) with different weighed forwarding packet rates, instead of service queues in network layer.

3   A New QoS Model for Wireless Ad Hoc Networks

In this section, a new QoS model called Dynamic Hop Service Differentiation (DHSD) is described. First, new end-to-end services are defined in terms of QoS of low delay, high throughput and low loss rate. Then, the implementation of the DHSD model is described.

3.1   New End-to-End Services

Four types of new end-to-end service are defined for wireless ad hoc networks; *voice, video, bulk data* and *default service*. First, *Voice service* can achieve lower

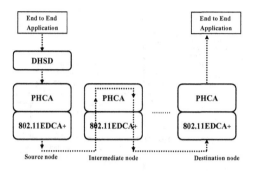

Fig. 1. End-to-end service mechanism

end-to-end delay than any other services. Second, *Video service* can achieve higher end-to-end throughput than any other services. Third, *Bulk data service* can achieve lower end-to-end loss rate than any other services. Finally, *Default service* is one in which the three services are not affected. Hence this service provides no guarantees.

3.2   The DHSD Model

DHSD supports four types of end-to-end services and provides the best service for each service so that QoS requirements of the application can be maintained independent of network state changes along any end-to-end path during its connection duration. The main idea behind the DHSD approach is simply QoS provisioning which is implemented without any state information of end-to-end QoS such as measurement-based end-to-end delay, throughput and loss rate along the route, and signaling protocol for resource reservation. Therefore, the network with DHSD provides no hard guarantees of end-to-end QoS. In addition, DHSD does not need the implementation of several service classes for network-layer service differentiation. Instead, using the 802.11e mechanism, several service queues (forwarding classes) are implemented in MAC layer. Fig. 1 shows end-to-end service mechanism based on the DHSD model with Per Hop Class Assignment (PHCA) and 802.11EDCA+. PHCA and 802.11EDCA+ are defined in the subsection below. Applications with end-to-end flows between arbitrary pairs of nodes have specific QoS requirements. Hence, each flow needs to request different end-to-end QoS provisioning. In order to achieve this demand, in the DHSD model, applications that are only sensitive to end-to-end delay may request low delay service from the network. Therefore, DHSD supports these applications through voice service. In addition, for applications which require both high throughput and low delay such as multimedia streaming, DHSD supports these applications through a video service that provides high throughput. For loss-sensitive applications such as TCP-based FTP, DHSD uses bulk data service. Other applications are selected as default service. As mentioned above, DHSD performs per-hop QoS mechanisms to support low delay for voice service, high throughput for video service, or low loss rate for bulk data service.

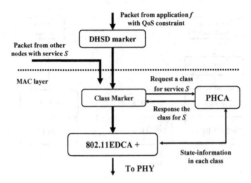

Fig. 2. The operation of DHSD at a node

3.3 Wireless Per-Hop Behavior (WPHB)

WPHB defines the desired behavior that a packet may receive at each hop in wireless ad hoc networks. In this paper, the desired behavior is achieved by the dynamic service class selection scheme of selecting the forwarding class satisfying QoS requirements of each end-to-end service. Therefore, all packets accepting the same service receive the same service treatment per hop. In the DHSD model, the service treatment for each end-to-end service can be constructed by combining four PHCA algorithms with forwarding classes operating by different weight values for packet scheduling called 802.11 EDCA+. 802.11 EDCA+ in the DHSD model is N different forwarding classes ordered in relative forwarding rate. In order to differentiate the forwarding rate at each forwarding class, the different AIFS and the different contention window size (CW_{min}, CW_{max}) is used. At node k, forwarding class i is set at the smaller value of both the AIFS and the contention window size than forwarding class j, for all i < j, i, j$\in$C$_k$, where C_k is the set of service classes for forwarding a packet at node k. The operation of the realized forwarding class is similar to AC operation in 802.11e EDCA. Hence, the DHSD model can construct packet scheduling mechanism by means of the 802.11 EDCA+ at MAC layer. The function of DHSD is implemented in the network and MAC layers, separately, as depicted in fig.2. The DHSD marker in the network layer is responsible for determining the type of end-to-end service according to the characteristics of each flow. Hence, the DHSD marker operates to mark each flow as a specific end-to-end service, which only operate in a source node and the service determined by a source node is fixed at overall nodes along the route during its connection duration. The MAC layer is responsible for forwarding packets to a physical layer, observing all forwarding classes in a node, and selecting a proper one among N forwarding classes. When a packet marked with special type of service arrives at the class marker, the class marker asks PHCA for a forwarding class for this service. At this time, PHCA performs a PHCA algorithm to select the proper class satisfying service-requirements such as low delay, high throughput or low loss rate among N forwarding classes, using measured state information of each forwarding class. After the class selection,

PHCA returns the forwarding class i. Class marker remarks a packet with class i which is returned by PHCA. Consequently, the marked packet is serviced in class i of 802.11EDCA+. This procedure is repeated each hop. The DHSD model is realized with the DHSD marker, PHCA, the class marker and 802.11EDCA+. The following section describes PHCA mechanisms.

4   PHCA Mechanisms

In this section, per hop class assignment (PHCA) mechanisms which can be used to achieve low delay, high throughput and low loss services are described. The state information of each forwarding class used as parameters in PHCA are firs defined. Then, PHCA algorithms to support four end-to-end services are described.

4.1   State Information in 802.11EDCA+

In order to perform WPHB for each end-to-end service, PHCA has to observe the current status of each forwarding class in 802.11EDCA+. PHCA uses the three types of state information of each forwarding class as parameters to select a proper forwarding class according to the type of service. When the class marker requests a class for a specific service from PHCA, PHCA performs the class selection algorithm based on the current state information of each class. Here, the three types of state information are the predicted waiting service delay, mean number of backlogged packets, and mean drop rate. The methods of measuring the three types of state information are discussed.

Predicted waiting service delay estimation. The predicted waiting service delay is defined as the expected time a packet is sent from a forwarding class to the physical layer. This value is then determined with both the current number of backlogged packet in a forwarding class and the mean contention delay consumed for the head-of-line packet to be transmitted to the physical layer at a forwarding class. Here, the contention delay includes the period for successful RTS/CTS exchange, if this exchange is used for that packet. Similarly, if the initial transmission of the packet is delayed due to one or more internal collisions within a node, multiple numbers of backoff periods may also be included.

Let $N_i(t)$ be the current number of backlogged packets of class i and $\overline{ct}_{i,k}$ be the mean contention delay of class i belonging to the kth packet at time t when the class marker requests a forwarding service from the PHCA. The predicted waiting service delay estimation $w_i(t)$ of class i is computed as shown:

$$w_i(t) = N_i(t) \times \overline{ct}_{i,k} \qquad (1)$$

The weighted moving average is used to smooth the estimated value. Therefore the mean contention delay is updated as shown:

$$\overline{ct}_{i,k} = \beta \, \overline{ct}_{i,k-1} + (1 - \beta)n_{i,k} \qquad (2)$$

where β is the weighting factor ($\beta < 1$) and $n_{i,k}$ is the contention delay achieved by kth packet. The initial vale $ct_{i,0}$ is set to the value adding the slot-time of AIFS[i] to the slot-time of middle value between $CW_{min}(i)$ and $CW_{max}(i)$.

Mean number of backlogged packet estimation. The mean number of backlogged packets, N_i, is used to estimate the throughput of a forwarding class. N_i indicates the relationship between the traffic load and transmission rate of a forwarding class because the queue that performs high transmission rate always maintains the minimum number of backlogged packets in saturated condition. Thus N_i is used as the parameter for searching the forwarding class that performs the highest throughput among all forwarding classes.

In order to estimate N_i, the mean number of backlogged packets in each forwarding class is estimated through measuring the number of both queued packets and dequeued packets in each class. The estimator designed for estimating N_i is called the time-sliding window (TSW) estimator [4]. A TSW is used to smooth the estimation value. The design of the estimator is extremely simple. The TSW estimator works as below.

Upon each packet arrival,

$$N_i = \frac{\alpha N_i + 1}{T_{now} - T + \alpha} \tag{3}$$

Upon each packet departure,

$$N_i = \frac{\alpha N_i - 1}{T_{now} - T + \alpha} \tag{4}$$

where α is window length measured in units of time and T_{now} and T are the time for a packet to arrive or depart, and the time when the previous packet arrived or departed, respectively. N_i is updated each time a packet arrives or departs.

Mean drop rate estimation. The mean drop rate estimation is based on a packet drop of class i. Here, packets are dropped at two places in a queue: queue drop and collision drop. In the case of queue drop, a packet in each forwarding class is dropped due to a) a full buffer for transmission, or b) the size of packet in the higher layer, which is greater than the maximum allowed data size defined in the IEEE 802.11 standard. In the case of collision drop, a packet is dropped due to consistently failing retransmissions. This drop includes the number of packets that are discarded because the MAC could not receive any ACKs for the (re)transmissions of those packets or their fragments, and the packets' short or long retry counts reached the MAC's short retry limit or long retry limit, respectively. For convenience, the mean drop rate, D_i is expressed as follows;

$$D_i = \frac{Dr_i^{queue} + Dr_i^{collision}}{L_i} \tag{5}$$

where Dr_i^{queue} and $Dr_i^{collision}$ are the queue drop rate and the collision loss rate of class i , respectively, and $L_i$is the number of transmission of class i from a high layer to LLC layer. Using TWS estimation, each parameter in equation. 5 is expressed as follows;

$$Dr_i^{queue} = \frac{\alpha Dr_i^{queue} + 1}{T_{now} - T + \alpha} \tag{6}$$

$$Dr_i^{collsion} = \frac{\alpha Dr_i^{collsion} + 1}{T_{now} - T + \alpha} \tag{7}$$

$$L_i = \frac{\alpha L_i + 1}{T_{now} - T + \alpha} \tag{8}$$

where α is the window length which is measured in units of time and T_{now} and T are the time when a current packet is dropped or transmitted, and time when previous packet was dropped or transmitted, respectively. D_i is updated when each packet transmits to class i and drops in class i.

4.2   Per Hop Class Assignment (PHCA) Algorithms

In this subsection, PHCA algorithms of four types of end-to-end service are described; VoS (voice service) PHCA, ViS (video service) PHCA, BS (bulk data) service PHCA and DS (default service) PHCA.

VoS (Voice Service) PHCA. VoS PHCA supports the lowest delay service at each node for voice service to achieve low end-to-end delay. Hence, all nodes in the network must assign a forwarding class which performs the lowest predicated waiting service delay at time t when the packet marked with the voice service arrives at a node. VoS PHCA predicates the waiting service time of each class using both the current number of backlogged packet of class i and the mean contention delay of class i. Specifically, let S_B denote the set of forwarding class. VoS PHCA determines class j for voice service such that

$$j = \arg \min_{i \in S_B} w_i(t) \tag{9}$$

where time t is the instantaneous time a packet arrives. After forwarding class j returned by PHCA is determined, the class marker marks a packet with class j. In a steady state, all packets are provided with the lowest service delay and then voice service which needs to service low delay is realized at all nodes in a network.

ViS (Video Service) PHCA. The ViS PHCA algorithm supports video service for achieving high throughput. Hence, PHCA uses the average number of backlogged packet of each forwarding class to search the forwarding class performing the highest throughput among N forwarding classes at all nodes, because the queue that achieves high throughput always maintains the minimum mean number of backlogged packets in saturation conditions. ViS PHCA then assigns a class with the minimum mean number of backlogged packets among N forwarding classes.

Specifically, let S_B denote the set of forwarding class. When packets arrive for this service, ViS PHCA uses the mean number of backlogged packets of each forwarding class measured before time t. ViS PHCA determines forwarding class j which has the lowest mean number of backlogged packets among N forwarding classes such that

$$j = \min_{i \in S_B} N_i(t) \qquad (10)$$

where time t is the time of packet arrival.

BS (Bulk data Service) PHCA. The BS PHCA algorithm supports bulk data service for achieving low end-to-end loss rates. PHCA then assigns a forwarding class with the minimum loss rate among N forwarding classes per hop. Specifically, let S_B denote the set of a forwarding class. When a packet arrives for this service at time t, using the value of loss rate measured before time t, BS PHCA determines forwarding class j which performs the lowest loss rate among N forwarding classes such that

$$j = \arg\min_{i \in S_B} D_i(t) \qquad (11)$$

where time t is the time of packet arrival.

DS (Default Service) PHCA. To select a forwarding class for default service, after calculating the current state in each class, DS PHCA excludes the classes that are selected by three services (voice service, video service and bulk data service) in all forwarding classes not to affect other services. After this procedure, DS PHCA selects the class with the highest throughput among the remainder class. Here, the class selection algorithm selects a class that has the lowest average number of backlogged packets like the class selection algorithm of ViS PHCA.

5   Simulation Studies

In order to illustrate the effectiveness of the DHSD model, a comprehensive performance study is presented and the model is compared with the best-effort service model and strict priority service model in two scenarios: one-hop ad hoc environments and peer-to-peer multi-hop environments. Denoted as the best-effort service, the best-effort service model uses the 802.11 DCF at each node and all traffic in this model is mapped to the same priority. Therefore the performance of the best-effort service represents the QoS perceived in existing IEEE 802.11-based ad hoc networks. The strict priority service represents end-to-end service with a fixed priority based on the 802.11e EDCA scheme. The priority for each flow in the strict priority service model is statically mapped to the MAC priorities as follows: voice traffic $\rightarrow$ forwarding class 0, video traffic $\rightarrow$ forwarding class 1, bulk data traffic $\rightarrow$ forwarding class 2 and default traffic $\rightarrow$ forwarding class 3. Simulations are performed in the OPNET v11.5 [15]. Dynamic

Source Routing (DSR) is employed as a routing protocol. The RTS/CTS mechanism is used. It is assured that forwarding classes in 802.11EDCA+ consist of four queues. Other parameters of 802.11EDCA+ are represented in table 1.

Table 1. Parameters of evaluated service schemes

Scheme	Best-effort	Strict Priority	DHSD
		Voice/video/best/backgroud	Voice/video/best/backgroud
AIFS (us)	N/A	2/2/3/7	2/2/3/7
The number of forwarding class (priorities)	1	4	4
CW_{min}(us)	N/A	7/15/31/31	7/15/31/31
CW_{max}(us)	1023	15/31/1023/1023	15/31/1023/1023
Moving average weight β	N/A	N/A	0.8
Window length α in TSW	N/A	N/A	0.66
802.11 modes	802.11b	802.11EDCA+ over 802.11	802.11EDCA+ over 802.11

A heterogeneous traffic scenario with three types of traffic flows is considered, i.e., voice, video, and data. The traffic flow is characterized by its packet arrival pattern and payload statistics as follows: Voice traffic is characterized as a two state Markov ON/OFF [10, 11]. The ITU-T G.711(silence) speech codec is selected to model good-quality voice calls. Voice traffic is approximately the mean on-time arrival rate of 40kbit/s. The video source rate is modeled by the first-order autoregressive Markov model [12]. It represents the constant bit-rate of a video source during generation of the video frame. It is assumed that both incoming stream frame size and outgoing stream frame size is 1728 byte/pixel, the frame size of video source is 128x120 pixels, and frame inter-arrival time is 10frames/sec. Data packets arrive from the high layer as Poisson sequence, with exponentially distributed packet length. A mean packet length of bulk data traffic is 1024 bytes and the mean packet length of default traffic is 512 bytes. Then, the average data throughput is 40 Kb/s or 20 Kb/s, respectively. The number of flow varies according to the simulation purposes.

5.1 The Performances on One-Hop Ad Hoc Environments

In this scenario, one-hop ad hoc networks are considered using topologies where 20 static nodes are located randomly in 500m x 500m square regions. Each flow

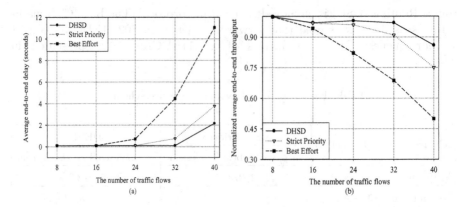

Fig. 3. The performance of each service in the one-hop environments. (a) average end-to-end delay of voice service and (b) normalized average end-to-end throughput of video service.

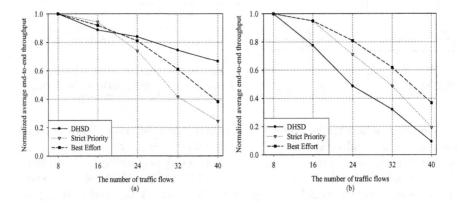

Fig. 4. The performance of each service in the one-hop environments. (a) normalized average end-to-end throughput of bulk data service and (b) normalized average end-to-end throughput of default service.

randomly chooses nodes as sources and destinations in networks. All service flows are increased with the same ratio of number of flows (1:1:1:1) and the number of each flows varies from 2 to 10. The performance of the DHSD model is compared with best-effect service and strict priority service. The metrics used in measuring the performance of voice service and the performance of the other service is end-to-end delay and end-to-end throughput, respectively. 3 different scenarios were randomly selected and the simulation was run for 100s. The average simulation results are as figs. 3-4. It is proven from these results that the proposed DHSD model can outperform both end-to-end delay and normalized average end-to-end throughput. Fig. 3(a) shows average end-to-end delay of voice service and that there is an improvement in end-to-end delay using the DHSD model compared with the other model. As shown in fig. 3(b), the normalized end-to-end throughput of video service is above 0.8 in congested networks. This is because queue drop is decreased using the dynamical forwarding class. Fig 4(a) also shows the there is a great improvement in end-to-end throughput of bulk data service using the DHSD model. Instead, as shown in fig 4(b), the throughput of default service in DHSD model is lower than that of default service in other service models. It is verified that the DHSD model almost provides an optimal end-to-end models.

5.2   The Performances on Multi-hop Ad Hoc Environments

In order to evaluate more realistic performance of the DHSD model, the simulations are run in multi-hop ad hoc environments where 50 static nodes are located randomly in 800m x 800m square regions. In the simulations, three types of service flows, except default traffic, are increased with the same ratio of number of flows (1:1:1) and the number of each flows varies from 2 to 5. The source and destinations of all flows are randomly selected. We randomly choose 3 different scenarios and run the simulation for 400s. The average simulation results are as shown in fig. 5-6. Fig. 5 shows when the number of total flows is greater than 24, the performance of voice service using the DHSD model is decreased more than that of voice service using strict priority service. This is because, as the number of flows of video service and bulk data increase, queue delay of all forwarding classes in a node increase. However, as shown in fig. 6, the normalized end-to-end throughput achieved by video service and bulk data service is a great improvement. This is because, when running concurrently, WPHB schemes for the video and bulk data service will operate well at all nodes.

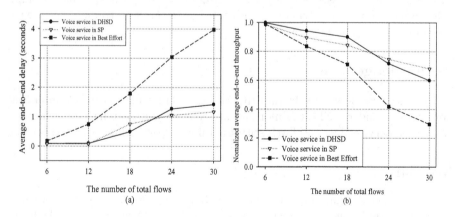

Fig. 5. The performance of each service in the multi-hop environments. (a) average end-to-end delay of voice service and (b) normalized average end-to-end throughput of voice service.

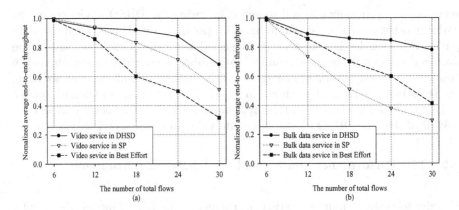

Fig. 6. The performance of each service in the multi-hop environments. (a) normalized average end-to-end throughput of video service and (b) normalized average end-to-end throughput of bulk data service.

6   Conclusion

In this paper, a mechanism to provide end-to-end QoS such as low delay, high throughput and low loss in wireless ad hoc networks, is introduced. Four types of end-to-end service are defined as follows: voice service, video service, bulk data service and default service. The proposed model performs WPHB with the PHCA algorithm, dynamically selecting a service class according to the current service state of each forwarding class at a node. In simulation, the proposed solution for end-to-end QoS provisioning is shown to effectively achieve low end-to-end delay, high end-to-end throughput and low end-to-end loss in wireless ad hoc networks. Consequentially, the DHSD model supports QoS provisioning independent of a node's dynamic bandwidth, drop rate and traffic arrival, with low complexity QoS management.

Acknowledgement

This research was supported by the MIC(Ministry of Information and Communication), Korea, under the ITRC(Information Technology Research Center) support program supervised by the IITA(Institute of Information Technology Assessment).

References

1. S. B. Lee, G. S. Ahn, and A. T. Campbell, "Improving UDP and TCP performance in mobile ad hoc networks with INSIGNIA," IEEE Commun. Mag., vol. 39, no. 6, pp. 156-165, Jun. 2001.
2. A. Lo, H.Xiao and KC Chua, "A Flexible Quality of Service Model for Mobile Ad hoc networks," In IEEE Vehicular Technology Conference Fall 2000, May. 2000, pp. 445-449.

3. A. Veres G.Ahn, A.T Campbell and L.Sun, "SWAN: Service Differentiation in Stateless Wirelass Ad hoc network," In Conference on Computer Communications (IEEE infocom), Jun 2002.

4. D. Clark and W. Fang, "Explicit allocation of best-effort packet delivery service," Networking, IEEE/ACM Transactions on. vol 6, issue 4, Aug. 1998 pp. 362 - 373.

5. K. C. Wang and P. Ramanathan, "QoS assurances through class selection and proportional differentiation in wireless networks," IEEE J. Sel. Areas Commun., vol. 23, no. 3, March. 2005, pp. 573-584.

6. C. H. Yeh, H. T. Mouftah, and H. Hassanein, "Signaling and QoS guarantees in mobile ad hoc networks," in Proc. IEEE Int. Conf. Commun., vol. 5, Apr. 2002, pp. 3284-3290.

7. J. Xue, P. Stuedi, and G. Alonso, "ASAP: an adaptive QoS protocol for mobile ad hoc networks," in Proc. IEEE Int. Symp. Pers., Indoor, Mobile Radio Commun., Sep. 2003, pp. 2616-2620.

8. Z. Ying, A. L. Ananda, and L. Jacob, "A QoS enabled MAC protocol for multi-hop ad hoc wireless networks," in Proc. IEEE Int. Conf. Perform.,Comput., Commun., Apr. 2003, pp. 149-156.

9. J. L. Sobrinho and A. S. Krishnakumar, "Quality-of-service in ad hoc carrier sense multiple access wireless networks," IEEE J. Sel. Areas Commun., vol. 17, no. 8, pp. 1353-1368, Aug. 1999.

10. J. Deng and R. S. Chang, "A priority scheme for IEEE 802.11 DCF access method," IEICE Trans. Commun., vol. E82-B, no. 1, pp. 96-102, Jan. 1999.

11. C. Coutras, S. Gupta, and N. B. Shroff, "Scheduling of real-time traffic in IEEE 802.11 wireless LANs," Wireless Netw., vol. 6, pp. 457-466, 2000.

12. A. Banchs and X. Perez, "Providing throughput guarantees in IEEE 802.11 wireless LAN," in Proc. Wireless Communications Networking Conf., vol. 1, 2002, pp. 130-138.

13. C. Lei and W. B Heinzelman, "QoS-aware routing based on bandwidth estimation for mobile ad hoc networks." IEEE J. Sel. Areas Commun., vol. 23, no. 3, March. 2005, pp. 561 - 572.

14. The IEEE P802.11 Task Group E. The IEEE 802.11e. [Online]. Available: http:// grouper. ieee.org/groups/802/11/.

15. "The OPNET Modeler," http:// www.opnet.com/products/modeler/home.html.

16. Y. Xiao and H. Li. "Local data control and admission control for QoS support in wireless ad hoc networks," Vehicular Technology, IEEE Transaction on. vol. 53, no. 5, Sept. 2004, pp. 1558-1572.

17. A. Iera, G. Ruggeri and D. Tripodi, "An Algorithm for Dynamic Priority Assignment in 802.11e WLAN MAC Protocols," LNCS, vol. 3124/2004, pp.1267 - 1273.

Authenticated In-Network Programming
for Wireless Sensor Networks

Ioannis Krontiris and Tassos Dimitriou

Athens Information Technology,
P.O. Box 68, 19.5 km Markopoulo Ave.,
GR- 19002, Peania, Athens, Greece
{ikro, tdim}@ait.edu.gr

Abstract. Current in-network programming protocols for sensor networks allow an attacker to gain control of the network or disrupt its proper functionality by disseminating malicious code and reprogramming the nodes. We provide a protocol that yields source authentication in the group setting like a public-key signature scheme, only with signature and verification times much closer to those of a MAC. We show how this can be applied to an existing in-network programming scheme, namely Deluge, to authenticate code update broadcasts. Our implementation shows that our scheme imposes only a minimal computation and communication overhead to the existing cost of network programming and uses memory recourses efficiently, making it practical for use in sensor networks.

1 Introduction

The process of programming sensor nodes typically involves the development of the application in a PC and the loading of the program image to the node through the parallel or the serial port. The same process is repeated for all the nodes of the sensor network before deployment. However, after deployment, there is often the need to change the behavior of the nodes in order to adapt to new application requirements or new environmental conditions. This would require the effort of re-programming each individual node with the updated code and relocate it back to the deployment site. Network programming saves this effort by propagating the new code over the wireless link to the entire network, as soon as that code is loaded to only one node. Then, nodes reprogram themselves and start operating with the updated code.

As network programming simplifies things for legitimate users, it also simplifies things for attackers that want to disrupt the normal operation of the network or operate them for their own advantage. In currently deployed networks the nodes do not authenticate the source of the program; therefore an attacker could easily approach the deployment site and disseminate her own malicious/corrupted code in the network.

This possibility makes sensor networks deployments susceptible to outsider attacks. Besides loosing control of the network or getting back altered measurements, it is even possible that the network is reprogrammed with malicious code

T. Kunz and S.S. Ravi (Eds.): ADHOC-NOW 2006, LNCS 4104, pp. 390–403, 2006.

that has the same functionality with the legitimate code but also reports data to the adversary. In such a case legitimate users would never know that something is wrong. Hence, it is important that the sensor nodes can efficiently verify that the new code originates from a trusted source, namely the base station.

2   Problem Definition and Contribution

The goal of this work is to provide an efficient source authentication mechanism for broadcasting a program image from the base station to the sensor network. While the authentication mechanism should still allow efficient dissemination procedures, such as *pipelining*, it should also block malicious updates as early as possible.

By now, what have been studied extensively in sensor networks are point-to-point authentication mechanisms. Using a shared key, two nodes can exchange authenticated messages by appending a message authentication code (MAC) to each packet, computed using that key. Due to its low computational overhead, MACs are an attractive tool for securing communication in sensor networks. However, in order to use it for broadcast authentication, all nodes should share the same key. But then, anyone who could physically capture a node and retrieve that key could impersonate the source. A solution to that problem has been given by Perrig et al. in [1], which is based on delayed disclosure of keys by the sender. The shortcoming of this approach is that it requires time synchronization between the nodes, while current dissemination protocols for in-network programming do not place such bounds.

The most natural solution for authenticated broadcasts is asymmetric cryptography, where messages are signed with a key known only to the sender. Everybody can verify the authenticity of the messages by using the corresponding public key, but no one can produce legitimate signed messages without the secret key. However, public key schemes should be avoided in sensor networks for multiple reasons: long signatures induce high communication overhead of 50 - 1000 bytes per packet, verification time places a lower bound on the computational abilities of the receiver, and so on.

However our goal is not to authenticate just messages, since here we are dealing with *streams*, rather than simple messages. The size of program images that will be sent over the radio is usually between a few hundreds of kilobytes and a few thousands. This fact can allow the use of public key schemes if we manage to reduce the size of the public key and also make signature size to be only a small percentage of the total transmitted stream. Furthermore, if we reduce the verification time down to the order of that of a symmetric scheme, we will have proved that public key cryptography is an attractive solution for such problems.

Therefore, our goal and the contribution of our work is to provide an efficient authentication scheme for a finite stream of data based on *symmetric* cryptography primitives while at the same time having the properties of asymmetric cryptography.

2.1   Design Goals

The solution that we present in this work was designed having the following
requirements in mind:

1. **Low computational cost.** As we mentioned above, asymmetric cryptography involves high computational cost and is not preferable for use in sensor
   networks. Our scheme should impose public-key properties but at the same
   time minimize the computational cost for sign verification at the receivers
   (sensor nodes).
2. **Low verification time.** The rate at which a code segment is transmitted
   to the receiver should not be delayed.
3. **Low communication overhead.** The signature transmitted with data
   should constitute a small percentage of the total bytes, imposing a low communication overhead.
4. **Low storage requirements.** Any cryptographic material that needs to be
   stored in the sensor nodes should be as small as possible, given their limited
   memory resources.

Moreover, since we are providing an authenticated broadcast protocol we need
to assure the following:

1. **Source authentication.** A mote must be able to verify that a code update
   originates from a trusted source, i.e., the base station. This means that
   an attacker should not be able to send malicious code in the network and
   reprogram the nodes.
2. **Node-compromise resilience.** In case an attacker compromises a node
   and read its cryptographic material, she must not be able to reprogram any
   other non-compromised node with malicious code.

Even though we do not address protection against DoS attacks, our protocol
must provide some resilience against such attacks in the following sense: In case
an attacker is trying to transmit malicious code to the network, any receiving
node should be able to realize this as soon as possible and stop receiving it or
forwarding it to other nodes. This means that nodes should not authenticate the
code *after* its reception but rather *during* that process.

3   Related Work

A recent work that proposes a solution for secure dissemination of code updates
in sensor networks is described in [2]. The authors first suggested the use of hash
chains to efficiently authenticate each page of the program image. However, they
make the assumption that there exists a public key scheme to authenticate the
initial commitment of the hash chain, without giving any specific solution.

Another work on the same problem is described in [3], where the authors set
the additional goal of DOS-resilience and therefore they need to authenticate
each packet separately. To do that they construct a signed hash tree scheme

(similar to a Merkle tree) for *every* page in the program image, and they transmit these trees before the actual data. This increases considerably the overhead of packets sent and received by the motes. Moreover, due to memory constrains in the motes, these values need to be stored and loaded from the EEPROM, which is a very energy consuming operation.

In [4] the use of a reverse hash chain computed over the program pages is also used, as in [2] and in our scheme. However the authors use the RSA digital signature scheme for signature verification at the motes, which we have excluded from our design goals. On the other hand, an authentication scheme for broadcasting messages in a sensor network that uses only symmetric primitives is described in [5]. The authors keep the memory and computational overhead of their algorithm efficiently low. However they are concerned about the problem of authenticating broadcasted queries, which are normally less harmful messages with very small size, so their requirements are different than in our case.

4   Overview and Useful Tools

Throughout this paper we are considering Deluge [6] as a paradigm of in-network programming. However other data dissemination protocols like MOAP[7], MNP[8] and INFUSE[9] are following similar principles, and the algorithms presented here should be applicable to those protocols as well.

Deluge propagates a program image by dividing it first into fixed-size pages and then using a demand-response protocol to disseminate them in the network. As soon as a node receives a page, it makes it available to any of its neighbors that also need it. At the same time it sends a request to the sender in order to receive the subsequent pages.

To sign a program image we are following the approach by Gennaro and Rohatgi in [10] for signing digital streams. What they proposed is to divide the stream into blocks and embed some authentication information in each block. In particular, their idea is to embed in each block a hash of the following block. In this way the sender needs to sign just the first block and then the properties of this signature will propagate to the rest of the stream through the "chaining" technique.

So, given a program image divided into N fixed-size pages $P_1, P_2, \ldots, P_N$ and a collision-resistant hash function H, we construct the hash chain

$$h_i = H(P_{i+1}|h_{i+1}), \quad i = 0 \ldots N - 2$$

and we attach each hash value h_i to page P_i, as shown in Figure 1. For the last hash value, $h_{N-1} = H(P_N)$. According to this scheme, we need to authenticate only h_0, which we will sign and release before the transmission of any page. The signing and verification of h_0 constitutes the main overhead of the security protocol, which our goal is to minimize.

Towards this goal, our main design principle is based on the fact that real world software updates in sensor networks do not constitute an every-day operation but rather they are performed occasionally. Therefore, we do not need to

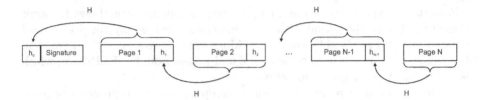

Fig. 1. Applying the hash chaining technique to the pages of a program image. Only h_0 needs to be signed by the sender.

authenticate an unlimited number of broadcasts. We only need to be able to do so for a sufficiently large number of times. This fact allow us to use *one-time signature schemes*, which exhibit fast verification times. Despite their name, there exist one-time signature schemes that can be used r-times instead of just once, r being a design parameter adjustable to our needs.

4.1   One-time Signature Schemes

One-time signatures were first introduced in [11,12]. They are based on the idea of committing a secret key via one-way functions, decreasing dramatically the signing and verification time compared to asymmetric primitives. In the rest of the paper we will describe an efficient one-time signature scheme appropriate for sensor networks and how this can be used for authenticating broadcasts of program images. We believe this technique to be interesting on its own, apart from its usage in devices with limited capabilities.

In one-time signature schemes the signer is generating a set of secrets prior to signing a message along with a set of public commitments to this set which are given to the verifier in an authenticated manner. To sign a message, the signer reveals a subset of these secrets, which is determined by the message content. The verifier authenticates the message by checking the correspondence of these secrets to the commitments that were given earlier. Since a part of the signer's secrets is now revealed, a new key must be generated for the next message.

Although one-time signatures have been known for a relatively long time, they have been considered to be impractical for two main reasons: First, they can be used to sign a message only once and then a new key must be generated; Second, the signature size is relatively long in comparison with common public-key signatures and MACs.

Recently, this area was revisited and some one-time signature schemes were proposed that seem attractive for sensor networks, mainly because they allow the reuse of the same key more than once, but also because they try to reduce the verification time. For example, Reyzin and Reyzin [13] introduced HORS, an r-time signature scheme with efficient signature and verification times. This scheme was further improved by Pieprzyk et al. [14]. Both of these "r-times signatures" can sign several messages with the same key with reasonable security before they can get compromised.

However there are still some drawbacks that prevent us from applying those schemes to sensor networks. The main one is the size of the public/secret key pair and the size of the generated signatures. The public key must be stored on all sensor nodes so its size must be minimized as much as possible. Moreover, the signature is transmitted by the radio and received by nodes, which have to verify it. The larger the signature size, the more energy a node has to spend in order to receive it and verify it.

4.2   Merkle Trees

As we described so far, all verifiers need an authenticated copy of the public commitment to the one-time signature in order to verify the validity of that signature. Merkle [15] introduced a scheme that enables the verification of a large number of public commitments using low storage requirements, i.e. a single hash value. This is done by using a technique called Merkle hash tree. A Merkle hash tree is a complete binary tree where each node is associated with a value, such that the value of each parent node is the hash function on the values of its children:

$$v(parent) = H(v(left)|v(right))$$

where $v()$ here stands for the value of a node and H for a hash function.

If we put the public commitments to the leaves of a Merkle tree, then the root can serve as a short public commitment to all the one-time signatures. Then we only need to give the root to the verifier in a secure and authenticated way. To verify a one-time signature the receiver does not need to know the whole Merkle tree. Instead, the only thing that the signer needs to provide to the verifier is the authentication path, i.e., the values of all the nodes that are siblings of nodes on the path between the leaf that represents the public commitment and the root.

Given that authentication path, a leaf may be authenticated as follows: First apply the one-way hash function to the leaf and its first sibling in the path, then hash the result and the next sibling, etc., until the root is reached. If the computed root value equals the published root value then the signature's commitment is authentic.

4.3   HORS

Our open question so far is the efficient signing of the first value h_0 of the hash chain. This will enable the authentication of the whole chain and therefore the authentication of the program image. To sign h_0 we will modify the HORS scheme so that the sizes of the signature and the public key are reduced to a magnitude proper for use in sensor networks. Here we briefly review the HORS scheme.

First, the signer generates a secret key SK that consists of t random values. The public key PK is computed by applying a one-way function f to each of the values of the secret key and then distributing them to the intended receivers in an authenticated way. A message m is signed according to the following steps:

1. Use a cryptographic hash function H to convert the message to a fixed length output. Split the output into k substrings of length $\log_2 t$ each.
2. Interpret each substring as integer. Use these integers to select a subset σ of k values out of the set SK.
3. This σ is the signature of the message m.

The verifiers recompute the hash value of the message m, re-produce the same indices and pick the corresponding values of the set PK (instead of SK). Then they verify that the hash value of each member of the signature equals to the corresponding member of the public key PK. The signature is accepted if this is true for all k values.

Note that for each message that we sign, a part of the secret key is leaked out. Some typical values for HORS are $l = 80$, $k = 16$ and $t = 1024$. In this case, assuming a hash output of 20 bytes, the public key will be $1024 \times 20 = 20,480$ bytes or 20 KB, which is not suitable for sensor nodes. Moreover, the security of the scheme and the size of the public key are directly related to the number of messages that we can sign. For the above example, we can sign just 4 messages with acceptable security, meaning 4 program image updates in our case. However, to make the scheme practical this number must be higher.

Let r be equal to the number of messages that we allow to be signed with the current instance of the secret key. For an analysis (see also [13]) we assume that the hash function H behaves like a random oracle and that an adversary has obtained the signatures of r messages using the same setting of secret/public key. Then the probability that an adversary can forge a message is simply the probability that after rk values of the secret key have been released, k elements are chosen at random that form a subset of the rk values. The probability of this happening is $(rk/t)^k$. If we denote by Σ the attainable security level in bits, by equating the previous probability to $2^{-\Sigma}$, we see that Σ is given by

$$\Sigma = k(\log_2 t - \log_2 k - \log_2 r). \tag{1}$$

As an example, for $t = 1024$, $k = 16$ and $r = 4$ we get $\Sigma = 64$ bits of security. For $t = 65536$, $k = 8$ and $r = 32$ we get the same level of security but we can sign a lot more messages with the same key. However, since the number t of PK values determines the public/secret keys sizes, it is directly limited by the restrictions imposed by sensor networks capabilities. As a first step we can use equation (1) to solve for t for any desirable security level. Thus we get

$$t = 2^{\Sigma/k} kr. \tag{2}$$

5   Our r-Times Signature Scheme

The reason that makes HORS inappropriate for sensor networks is that the public key grows unacceptably high if we want to sign more messages and keep security at an acceptable level. So we need to effectively reduce the public key size. One way to do this (also proposed in [16]) is to use a Merkle tree, which we discussed in section 4.2. Given the secret key values, we apply a one-way function f to each one and we place the results to the leaves of the Merkle tree. The root of the resulting Merkle tree is the public key.

Even though we have reduced the size of the public key down to a hash output size, we increased the size of the signature to the size of the authentication path. This also results in a corresponding increase to the signature verification time. So, this solution is also not attractive for applications in sensor networks.

Our solution goes one step further and distributes the values of the secret key into *many* Merkle trees, thus achieving a tradeoff between public key size and signature size. First we show how this can be done.

Let f be an $l-$bit one-way function. The generation of the key pair is done by the following algorithm:

Secret Key. Generate t random $l-$bit quantities for the secret key: $SK = (s_1, \ldots, s_t)$.

Public Key. Compute the public key as follows: Generate t hash values $(u_1, \ldots, u_t)$, where $u_1 = f(s_1), \ldots, u_t = f(s_t)$. Separate these values into d groups, each with t/d values. Use these values as leaves to construct d Merkle trees. The roots of the trees are the public key of our scheme.

In this way we have reduced the public key size down to a few hash values that constitute the roots of the Merkle trees. These values need to be passed to all sensor nodes in an authenticated way. This can be done for example during initialization of sensor nodes. Now, a message is signed according to the following steps (see also Figure 2):

1. Use a cryptographic hash function H to convert the message to a fixed length output. Split the output into k substrings of length $\log_2 t$ each.
2. Interpret each substring as integer. Use this integers to select a subset σ of k values out of the set SK.
3. The signature of the message m is made up by the selected secret values along with their corresponding authentication paths.

The verifiers recompute the hash value of the message m, re-produce the same indices and pick the corresponding values of the set PK. Then they evaluate each authentication path of the signature to reproduce the root of the Merkle tree and compare it with the corresponding member of the public key PK. The signature is accepted if this is true for all k values. The detailed description of the algorithm is shown in Figure 3.

As an example, let's apply to our scheme the same values we did for HORS, i.e., $l = 80$, $k = 16$ and $t = 1024$, and assuming a hash output of 20 bytes. If we construct 32 Merkle trees with 32 leaves each (so all 1024 secret values are covered), we will get 32 roots of trees, i.e., 640 bytes that will constitute our public key, compared to 20 KB we got from HORS. These values will provide 64 bits of security for $r = 4$ messages (images).

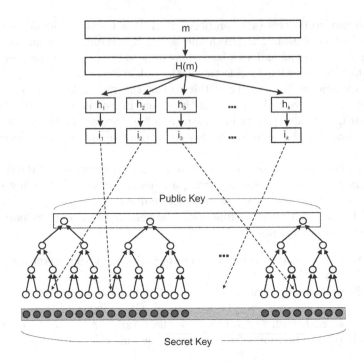

Fig. 2. Signature construction for message m using multiple Merkle trees

If we choose now $l = 80$, $k = 8$ and $t = 65536$, and $r = 32$ we get the same level of security. In this case, by constructing 64 Merkle trees of 1024 leaves each, the public key will become 1024 bytes, which is still an attractive value for use in sensor nodes.

5.1   Tradeoffs

The public key stored in each sensor node is given by the hash values residing at the roots of the trees. The more the number of the trees, the bigger the public key becomes but the smaller the signature size becomes. To see why, notice that signature size depends on the length of the authentication paths, which are ultimately related to the height of the Merkle trees. More trees means less secret values per tree and hence smaller height. To find this tradeoff between public/signature size let T denote the number of trees. Hence the public key size is simply

$$S_{PK} = |h|T, \tag{3}$$

where $|h|$ is the output of the hash function in bits since every root contains a hash value of its children. For example, $|h|$ can be equal to 128 bits in the case of MD5 or 160 bits in the case of SHA-1.

As the number of trees is T, there can be at most t/T values stored at the leaves of each tree. Thus the height of each tree (and the length of each au-

---

Key Generation
  Input: Parameters l,k,t
    Generate t random l-bit strings $s_1, s_2, \ldots, s_t$
    Let $u_i = f(s_i)$ for $1 \leq i \leq t$
    Group t hash values $u_1, u_2, \ldots, u_t$ into d groups of t/d values
    Place each group at the leaves of a Merkle tree, constructing d Merkle trees
    Let $w_1, w_2, \ldots, w_d$ be the roots of the Merkle trees
  Output: $PK = (k, w_1, w_2, \ldots, w_d)$ and $SK = (k, s_1, s_2, \ldots, s_t)$

Signing
  Input: Message m and secret key $SK = (k, s_1, s_2, \ldots, s_t)$
    Let $h = H(m)$
    Split h into k substrings $h_1, h_2, \ldots, h_k$, of length $\log_2 t$ bits each
    Interpret each h_j as an integer i_j for $1 \leq j \leq k$
    Let $\mu_{i_j} = (s_{i_j}, AP(s_{i_j}))$, i.e. the secret ball along with its authentication path
  Output: $\sigma = (\mu_{i_1}, \mu_{i_2}, \ldots, \mu_{i_k})$

Verifying
  Input: Message m, signature $\sigma = (\mu_1', \ldots, \mu_k')$ and public key $PK = (k, w_1, \ldots, w_t)$
    Let $h = H(m)$
    Split into k substrings $h_1, h_2, \ldots, h_k$, of length $\log_2 t$ bits each
    Interpret each h_j as an integer i_j for $1 \leq j \leq k$
    Compute which Merkle tree corresponds to i_j: $M_j = i_j/(t/d)$ for $1 \leq j \leq k$
    Hash the values in each μ_k' to produce the corresponding root w_{M_j}'
  Output: "accept" if for each $j, 1 \leq j \leq k, w_{M_j}' = w_{M_j}$; "reject" otherwise

Fig. 3. Our proposed signature scheme. f is a one-way function and H is a hash function. Both f and H may be implemented using a standard hash function, such as SHA-1 or MD5.

thentication path) is simply $\log_2 t/T$ or $\Sigma/k + \log_2(kr) - \log_2 T$ using equation (2). The signature consists of k such authentication paths, where each path is a sequence of hash values. Thus the signature size is given by

$$S_{sig} = |h|(\Sigma + k \log_2(kr) - k \log_2 T). \qquad (4)$$

From this equation it should be obvious that increasing the number of trees T (and hence the public key size) results in a decrease in the signature size.

This equation can be simplified further if we recall how the k secret values are selected (Figure 2). The message m to be authenticated is first hashed to obtain $H(m)$, a value that is $|h|$ bits long. Then these $|h|$ bits are broken into k parts, where each part references one of the secret values. Thus the number of secrete values t must be equal to $2^{|h|/k}$, or equivalently

$$|h| = k \log_2 t. \qquad (5)$$

Combining with equations (2) and (4), we find that the signature size is given by

$$S_{sig} = |h|(|h| - k \log_2 T). \tag{6}$$

In the same manner, the security level becomes

$$\Sigma = k(|h|/k - log_2 k - log_2 r). \tag{7}$$

In the figures below we tried to keep the public key size equal to approximately 1 KByte so that it fits well in the memory of typical Mica nodes. Assuming $h = 128$, i.e. using MD5 to produce the hash values, we find that the number of trees T should be equal to 64, by equation (3).

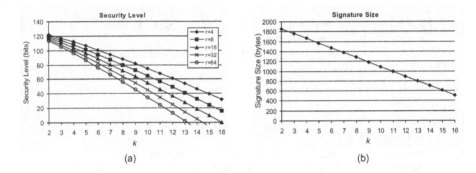

(a)   (b)

Fig. 4. Signature size and security level as a function of k

6   Implementation Details and Evaluation

We implemented our authentication protocol in order to measure its efficiency. Our implementation was built on Deluge 2.0, which we slightly modified in order to include our scheme. At the end we were able to download authenticated images on the sensor nodes, and reprogram them by using Deluge. The whole security mechanisms were kept transparent from the end users, unless they tried to inject a corrupted or malicious program image.

In our implementation design we faced several issues, some of which we address here. First of all, one primary goal was for a mote to be able to authenticate each page separately and stop the downloading of the image as soon as a non-authentic page is received. However, by that time earlier pages that passed the verification procedure will have been propagated to the rest of the network wasting energy of the motes. This may be considered as a kind of DoS attack, if exploited properly by an adversary. However it is a price that must be paid in order to support pipelining. If one is interested in optimizing the protocol from a security point of view, then it must modify it to exclude the pipelining, so that only complete images that have been authenticated can be further forwarded.

Another issue to be considered here is the memory optimization of the protocol. While it is not possible to avoid storing the public key in the mote's memory, we can do so for the signature. This is because the signature is made up by authentication paths and the authentication of each path can be done independently by the others. Referring to Figure 5, the mote first receives the hash value of Image 1. This will provide the indices to the public values. Then, the first authentication path of the signature will be received. The verification of that path evolves only a few hashing operations and a comparison of the result with the corresponding public value. This can be done fast enough by the mote (see Section 6.1) so that the path has been verified before the next path starts coming in. So, only a temporary storing is needed, equal to the size of a path (dependent on the height of the Merkle trees at the base station).

Signature

Fig. 5. The order at which a mote receives the signature, the pages and their hash values. Verification of the signature is possible by storing only one path at the time.

So the only extra memory requirement that we impose is the buffering of the pages, since we want to compute their hash value and compare them with the corresponding commitments in the hash chain. For example, the page size in Deluge is 1104 bytes, which is a large percentage of the available memory in a mote (usually at the order of a few kilobytes). Nevertheless, it is possible to buffer one page of that size at the time, as long as we do not require any more memory of that order, which is true for our scheme.

6.1 Evaluation

Following the discussion of Section 5.1, we used our secure version of Deluge to measure the verification time of the signature attached to the program image. Our implementation was done on the mica2 platform, which exhibits low memory capabilities (4 KB of RAM). For all of our experiments we set a security level equal to 60 bits (although this can be modified accordingly), which is a satisfactory value for most security applications.

For that security level, Figure 6(a) shows the verification time of the signature for different number of Merkle trees T (determining the public key size) and different values of r (number of images that can be signed with the same keys). So, for example, if we take $k = 8$ and want to sign $r = 64$ images using the same secret-public key pair while keeping the public key size down to 1 KB (i.e. $T = 64$), we get a verification time equal to $186.3ms$. This is just the computational time and does not include the time to transmit the stream. So, it defines the computational overhead that our algorithm imposes on Deluge. Notice also that this time uses standard implementations of hash functions, so it can be improved even further using optimized code.

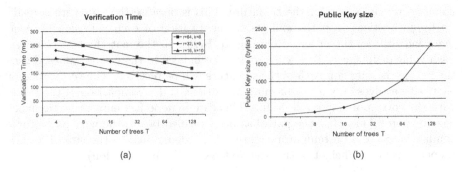

Fig. 6. Verification time and public key size as a function of T

Figure 6(b) shows how the size of the public key changes as a function of the number of Merkle trees T. If the secret values are distributed over more Merkle trees, the public key increases but the verification time decreases accordingly. So, this is a tradeoff that must be decided at design time, depending on the available memory on the sensor nodes, which will determine how big the public key can be.

7   Conclusions

In this paper we presented an efficient and practical scheme for authenticated in-network programming in sensor networks. Our solution imposes asymmetric cryptography properties using *symmetric* cryptography primitives. It minimizes the public key and signature sizes to values that are appropriate for sensor networks. The verification procedure at the motes is also time and computational efficient, since it involves only hashing and comparison operations. Our scheme also provides node compromise resilience, preventing an attacker who captures a node to reprogram any other node in the network. Furthermore, images are authenticated at a per-page basis, which enables a node to stop the downloading of a new image as soon as a page fails the verification procedure.

We implemented our solution and integrated it in Deluge, showing that it can easily adapt to an existing in-network programming protocol. We tested our secure Deluge version and measured the verification time of the signature at the mote's side. This showed that the computational overhead imposed by our scheme is at the order of one to two hundreds milliseconds, which is very efficient for applications running on sensor nodes.

References

1. Perrig, A., Szewczyk, R., Wen, V., Culler, D., Tygar, J.D.: SPINS: Security protocols for sensor networks. Wireless Networks **8**(5) (2002) 521–534
2. Lanigan, P., Gandhi, R., Narasimhan, P.: Secure dissemination of code updates in sensor networks. In: Proceedings of the 3rd international conference on Embedded networked sensor systems (SenSys '05). (2005) 278–279

3. Deng, J., Han, R., Mishra, S.: Secure code distribution in dynamically programmable wireless sensor networks. Technical Report CU-CS-1000-05, Department of Computer Science, University of Colorado, Boulder, CO (2005)
4. Dutta, P., Hui, J., Chu, D., Culler, D.: Securing the deluge network programming system. In: Proceeding of the 5th International Conference on Information Processing in Sensor Networks (IPSN 2006). (2006)
5. Benenson, Z., Pimenidis, L., Hammerschmidt, E., Freiling, F.C., Lucks, S.: Authenticated query flooding in sensor networks. In: Proceedings of the 21st IFIP International Information Security Conference (SEC 2006). (2006)
6. Hui, J.W., Culler, D.: The dynamic behavior of a data dissemination protocol for network programming at scale. In: Proceedings of the 2nd international conference on Embedded networked sensor systems. (2004) 81–94
7. Stathopoulos, T., Heidemann, J., Estrin, D.: A remote code update mechanism for wireless sensor networks. Technical Report CENS-TR-30, University of California, Los Angeles, Center for Embedded Networked Computing (2003)
8. Kulkarni, S.S., Wang, L.: MNP: Multihop network reprogramming service for sensor networks. In: Proceedings of the 25th IEEE International Conference on Distributed Computing Systems (ICDCS'05). (2005) 7–16
9. Arumugam, M.: Infuse: a TDMA based reprogramming service for sensor networks. In: Proceedings of the 2nd international conference on Embedded networked sensor systems (SenSys '04). (2004) 281–282
10. Gennaro, R., Rohatgi, P.: How to sign digital streams. Information and Computation **165**(1) (2001) 100–116
11. Lamport, L.: Constructing digital signatures from a one-way function. Technical Report CSL-98, SRI International Computer Science Laboratory, Palo Alto (1979)
12. Merkle, R.: A digital signature based on a conventional encryption function. In: Advances in Cryptology – CRYPTO '87. Volume 293 of Lecture Notes in Computer Science., London, UK, Springer-Verlag (1988) 369–378
13. Reyzin, L., Reyzin, N.: Better than BiBa: Short one-time signatures with fast signing and verifying. In: Proceedings of the 7th Australian Conference on Information Security and Privacy (ACISP '02), London, UK, Springer-Verlag (2002) 144–153
14. Pieprzyk, J., Wang, H., Xing, C.: Multiple-time signature schemes against adaptive chosen message attacks. In: Selected Areas in Cryptography (SAC 2003), Springer (2003) 88–100
15. Merkle, R.C.: A certified digital signature. In: Proceedings on Advances in cryptology (CRYPTO '89), Springer-Verlag New York, Inc. (1989) 218–238
16. Seys, S., Preneel, B.: Power consumption evaluation of efficient digital signature schemes for low power devices. In: Proceedings of the 2005 IEEE International Conference on Wireless and Mobile Computing, Networking and Communications (IEEE WiMob 2005). Volume 1. (2005) 79–86

A Congestion Window Adjustment Scheme for Improving TCP Performance over Mobile Ad-Hoc Networks

Jung-Hoon Song[1], Kyung-Hwan Ahn[2], Dong-Hoon Cho[1], and Ki-Jun Han[1]

[1] Dept. of Computer Engineering, Kyungpook National University, Korea
{pimpo, firecar}@netopia.knu.ac.kr, kjhan@bh.knu.ac.kr
[2] Mobile Communication Division, SAMSUNG Electronics, Korea
kyunghwan.ahn@samsung.com

Abstract. TCP does not distinguish between congestion and packet losses due to route change and link failures, which are prevalent in mobile ad hoc networks. So, TCP does not show satisfactory performance in ad hoc networks since it assumes that all packet losses are due to network congestions. In particular, when a route is reestablished it needs to be adaptively determined congestion window (CWND) according to the new route features. In this paper, we proposed an adjustable CWND scheme to improve the TCP performance over ad hoc networks. TCP sender effectively adjusts CWND by monitoring the network situation using control packets. Simulation results using NS-2 show that our scheme may increase TCP throughput compared with those of general TCP.

1 Introduction

In a mobile ad hoc network, a group of mobile computing devices communicate among themselves using wireless radios, without the aid of a fixed networking infrastructure. Due to their dynamic properties, mobile ad hoc networks have gained significant attention lately as a way of providing continuous network connectivity to mobile computing devices in various areas such as disaster and military applications. In such applications, reliable packet exchange is an indispensable requirement [8]. TCP, however, does not show a satisfactory performance in mobile ad hoc networks since TCP does not distinguish between congestion and packet losses due to due to a high BER (Bit Error Rate) and node mobility, which are prevalent in mobile ad hoc networks. The node mobility results in frequent route re-computations and occasional network partitions. Discovering new routes takes a significantly longer time than the RTO (Retransmission Time Out) interval at the sender. As a result, the TCP sender times out, retransmits the packet and invokes congestion control [7].

In mobile ad hoc networks, routes are frequently re-computed due to high mobility. At this time, the TCP sender will never get an opportunity to transmit at the maximum negotiated rate because its CWND (Congestion Window) will be much smaller than the receiver's advertised window size. It is also likely that an ad hoc network gets periodically partitioned and the sender and receiver of a connection lie in different partitions. Then, all the sender's packets get dropped which results in the sender

T. Kunz and S.S. Ravi (Eds.): ADHOC-NOW 2006, LNCS 4104, pp. 404–413, 2006.

invoking congestion. If the situation persists for a few seconds, there could be multiple RTO expirations and the RTO may reach its upper bound. Finally, when the receiver and sender get connected, it could take several seconds before there is some kind of transmission [7].

There have been a lot of research works to address the above problems. TCP-Feedback [1], ELFN (Explicit Link Failure Notification) based approach [2] and ATCP (Ad hoc TCP) [3] deal with a route failure due to node mobility. Fixed-RTO [4] and TCP-DOOR (Detection of Out-of-Order and Response) [5] are modified TCP versions which adapt themselves to dynamic ad hoc environments. The main concern is how to avoid invoking unnecessary congestion control in case of packet losses due to node mobility. They, however, did not reflect the dynamic network situation when determining the optimal value of a congestion window, so they could not offer a satisfactory end-to-end throughput. In this paper, we propose a new congestion window scheme which effectively adjusts the TCP's CWND based on network conditions in order to enhance the end-to-end throughout.

The rest of this paper is organized as follows. Section 2 describes previous research works that addressed TCP problems in ad hoc environments. Section 3 describes our CWND adjustment mechanism. Section 4 shows that our scheme enhances TCP throughput via NS-2 simulation. Finally, the conclusion is presented in Section 5.

2   Related Works

In this section, we review some feedback-based schemes that have been proposed to improve TCP performance for wireless ad hoc networks. In feedback-based schemes, the source is informed of a route failure or congestion so that it does not misinterpret a route failure as congestion and gets into the congestion control phase [7].

2.1 TCP-Feedback

TCP-Feedback is a feedback based scheme in which the TCP sender can distinguish between route failure and network congestion by receiving a RFN (Route Failure Notification) message from the intermediate nodes. The idea is to push the TCP into a snooze state when such messages are received. In this state, the TCP stops sending packets and freezes all its variables such as timers and CWND, which makes sense once there is no available route to the destination. Upon receipt of a RRN (Route Reestablishment Notification) message, via routing protocol, which indicates that there is another available path to the destination, the sender leaves the frozen state and resumes transmission using the same variable values prior to the interruption. In addition, a route failure timer is used to prevent an infinite wait for the RRN messages. It is triggered whenever a RFN message is received, and the frozen timers are reset allowing the TCP congestion control to be invoked normally [1] [9].

2.2 ELFN-Based Approach

In this approach, TCP interacts with the routing protocol in order to detect route failure and takes appropriate actions when the failure is detected. This is done via the ELFN messages that are sent back to the sender from the node that detects the failure.

Such messages are carried by the routing protocol that needs to be adapted for this purpose. In fact, the DSR's route failure message was modified to carry a payload similar to the *'host unreachable'* ICMP message. Basically, the ELFN messages contain sender and receiver addresses and ports, as well as the TCP sequence number. In this way, the TCP is able to distinguish losses caused by congestion from the ones caused by node mobility. When the TCP sender receives an ELFN message, it enters a *'stand-by'* mode, which implies that its timers are disabled and the probes packets are sent regularly towards the destination in order to detect route restoration. Upon receiving an ACK packet, the sender leaves the *'stand-by'* mode and resumes transmission using its previous timer values as normal [2] [9].

2.3 ATCP

Unlike the previous approaches, ATCP does not impose changes to the standard TCP itself. Rather, it implements an intermediate layer between network and transport layers to lead TCP to an enhanced performance and still maintain interoperation with non-ATCP machines. In particular, this approach relies on the ICMP protocol and ECN (Explicit Congestion Notification) scheme to detect network partition and congestion, respectively. In this way, the intermediate layer keeps track of the packets to and from the transport layer so that the congestion control is not invoked when it is not really needed, which is done as follows. When three duplicate ACKs are detected, indicating a lossy channel, ATCP puts TCP in *'persist'* mode and quickly retransmits the lost packet from the TCP buffer; after receiving the next ACK, the normal state is resumed. When an ICMP *'destination unreachable'* message arrives, pointing out a network partition, ATCP also puts the TCP in *'persist'* mode which only ends when the connection is reestablished. Finally, when network congestion is detected by the receipt of an ECN message, the ATCP only forwards the packet to TCP so that it can invoke TCP's congestion control normally. In ATCP, the authors argue that it is more appropriate to restart CWND, since the new route might have a lower available bandwidth than the older one. This approach is too conservative [3] [9].

3 CWND Adjustment Mechanisms

As previously described, the TCP assumes that all packet losses are due to network congestion. Previous research works based on feedback messages might distinguish between congestion and packet losses due to route change and link failures, but they could not be applicable to the ad hoc networks since they could not efficiently adjust the congestion window size for dynamic topology change. A new mechanism must be addressed toward an adaptive approach. In this paper, we propose the CWND adjustment scheme to improve the TCP performance over mobile ad hoc networks.

Consider the partial change of an intermediate node due to local repair as shown in Fig. 1. In this case, the sending TCP stops transmission and waits until the route is reconnected through a detour. After this, it re-computes a new CWND based on the information about the buffer status of nodes along the route. The buffer status means how much buffer is being unoccupied at each node, and is measured as the ratio of the amount of the available buffer space to the buffer capacity at each node. If the new route has a higher available bandwidth than the old one, the sending TCP maintains

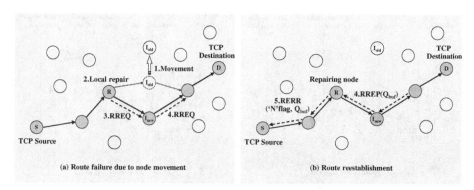

Fig. 1. Local repair at the intermediate nodes

the old CWND value. Otherwise, the sending TCP sets CWND to 1. In particular, if the buffer availability for nodes along the new route, denoted by Q_{buf}, is below some threshold value (Q_{thr}), we consider that the new route has a less available bandwidth than the old one. Here, the buffer availability (Q_{buf}) is defined as the minimum buffer availability at each node along the route.

In AODV routing protocol [6], when there is a link break on an active route, the node upstream of that break chooses to repair the link locally if the destination is not farther than local repair threshold (MAX_REPAIR_TTL) hops away. To repair the link break, the node broadcasts an RREQ message for that destination. The node initiating the repair then waits the discovery period to receive RREPs in response to the RREQ. At this time, in our scheme, information on buffer availability is exchanged among nodes over the route via the RREP. During this process, the minimum buffer availability (Q_{buf}) can be found by comparing the value of Q_{buf} field in the RREP with its own buffer utilization. When the repairing node receives an RREP it issues an RERR message for the destination, with the value of Q_{buf} and 'N' bit set. At this time, the node does not compare the hop count of the new route with the hop count field of the invalid route table entry for the destination, which is not case in the original AODV. Following this, the repairing node sends an RERR to the sending TCP with Q_{buf} and 'N' flag set as shown in Fig. 2(a). When the TCP sender receives an RERR with 'N' flag set indicating that there has been a local repair, it determines the value of CWND using the information on the buffer availability (Q_{buf}).

Now consider the case where a temporary movement of some intermediate node along the route causes link failure or a partition of the network, but the existing route can be reconnected after some time. At this time, if the sender and receiver of a TCP connection temporarily lie in different partition for a short time as shown in Fig. 3, the sender does not have to invoke congestion control. In our scheme, the TCP sender waits until the existing route is reconnected and then resumes transmission with maintaining the old CWND value. If the partition of the network persists for a few seconds, and as a result, the source node receives an RERR message without 'N' flag as shown in Fig. 2(b), a fresh route is established. At this time, the TCP initializes CWND to 1 and resumes transmission.

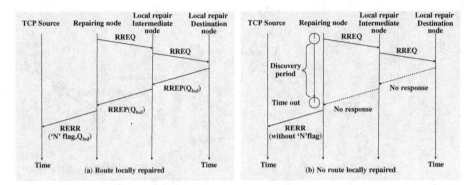

Fig. 2. Signaling for local repair using RREP, RREQ, and RERR messages

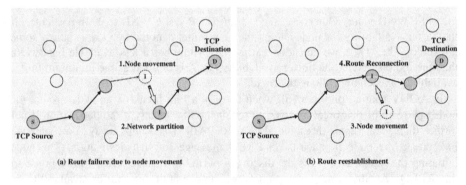

Fig. 3. Route reconnection after temporary movement of intermediate node

As explained so far, in our scheme, the TCP sender monitors the network situation using control packets such as RREP, RERR or ICMP messages, and efficiently adjusts the CWND value based on network conditions. Our scheme is illustrated in Fig. 5 and Table 1 presents a comparison of the existing schemes.

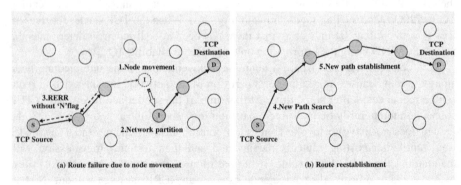

Fig. 4. A new path is established at the source node

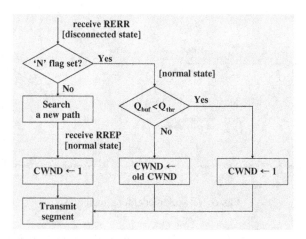

Fig. 5. Our CWND adjustment scheme

Table 1. Comparison of congestion control schemes

Circumstance	General TCP	Feedback based schemes			Our scheme
		TCP feedback	ELFN	ATCP	
The existing route temporarily fails and is reconnected at the intermediate nodes.	Retransmit segments and keep shrinking the CWND until receiving ACK	•Stop transmission and wait until the route is reconnected. •CWND←old CWND	•Stop transmission and wait until the route is reconnected. •CWND←old CWND	•Stop transmission and wait until the route is reconnected. •CWND←1	•Stop transmission and wait until the route is reconnected. •CWND←old CWND
Local repair at the intermediate node	As above	As above	As above	As above	•Stop transmission and wait until the route is reconnected. •Re-compute CWND value based on node buffer status
A fresh route is established at the source node	As above	As above	As above	As above	•Stop transmission and wait until the route is established. •CWND←1

4   Simulations

We evaluated our CWND adjustment scheme comparing with ATCP. ATCP stops transmission until the route is established and sets CWND to 1 after a new route is established. For experiments, we used the NewReno TCP protocol. The simulation study is done by the NS-2 network simulator which is a discrete event simulator developed as part of the VINT project at the Lawrence Berkeley National University [11]. The extensions implemented by the CMU Monarch project enable it to simulate mobile nodes that are connected by wireless network interfaces [12]. We modified the CMU extension of the AODV and TCP code to implement CWND adjustment mechanism.

The network model used in the simulation is shown in Fig. 6. The TCP connection is established between node S and node D. The source-destination pairs of UDP

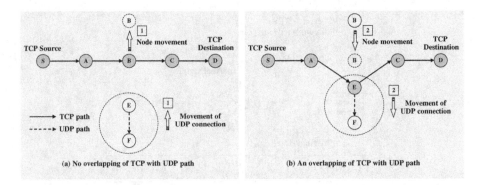

Fig. 6. Network model for simulation

traffic are node E and node F. We made a routing path of the TCP connection change several times by moving node B, E, and F, which are intermediate nodes of the TCP connection.

In the UDP connection, we used a CBR (Constant Bit Rate) as a traffic model. The sending rate was varied between 0 and 500 Kbps at 50Kbps intervals. The load of CBR traffic affects the minimum buffer availability at the intermediate node along the TCP connection route. Table 2 shows the simulation environments in details. In the simulation, the TCP throughput performance was measured.

Table 2. Simulation environments

Application protocol	FTP, CBR
Transport protocol	TCP NewReno
Routing protocol	AODV
Link protocol	IEEE 802.11 MAC
Link bandwidth	2Mbps
Transmission range	250m
Node number	7

The main objective of our simulation study is to prove that CWND adjustment scheme shows a noticeable improvement in the throughput. In the simulation, the buffer availability threshold (Q_{thr}) is given by three quarters of the buffer capacity. Local repair procedure is always executed due to movement of the intermediate node.

Fig. 7 shows the throughput of the proposed. Note that our scheme offers a higher throughput in the CBR traffic rate than the conventional ones. Overall, however, the throughput decreases as the offered load is increased at the intermediate node. In case of low offered load, the sending TCP maintains the old CWND value and

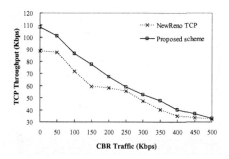

Fig. 7. TCP throughput

results in the higher performance in our scheme. Because the minimum buffer availability of the intermediate node is higher than the threshold value. If the intermediate node is offered high rate of traffic, throughput of proposed scheme is similar to conventional ones. This is shown that the minimum buffer availability of the intermediate nodes is lower than the threshold value and CWND is set to one segment.

Fig. 8 shows the TCP throughput as the movement interval is changed from 10 to 150 seconds without a change in the CBR rate. The movement interval represents how often each node moves. The shorter the interval is, the more frequently the node moves. This figure indicates that our scheme offers a higher throughput than the conventional ones. The more frequently the intermediate node moves, the more sharply performance is shown differences. Because modified NewReno TCP sets CWND to 1 after movement of the intermediate node.

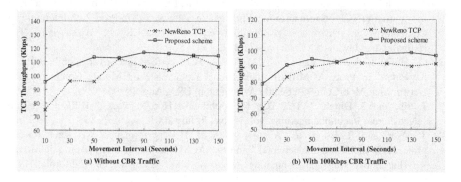

Fig. 8. TCP throughput versus movement interval

Fig. 9 shows the end-to-end delay for transmission of 1 Mbytes as the sending rate was varied between 0 and 300 Kbps with the movement interval of 50Kbps. This figure indicates that our scheme offers a lower delay than the conventional ones.

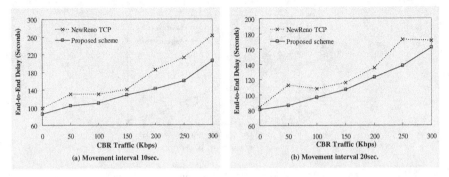

Fig. 9. End-to-end delay for transmission of 1 Mbytes

5 Conclusions

In the existing feedback based schemes, the source could be informed of a route failure so that it could not misinterpret it as congestion and get into the congestion control phase. In case of network topology change, however, they maintained the old value of CWND or just set it to one. This was somewhat extreme approach, and thus degraded the end-to-end throughput. In this paper, we proposed an adaptive CWND adjustable scheme to improve the TCP performance by efficiently coping with dynamic topology changes over mobile ad hoc networks. Simulation results showed that our scheme offered a better performance than the existing schemes.

References

[1] K. Chadran, S. Raghunathan, S. Venkatesan, and R. Prakash, "A feedback based scheme for improving TCP performance in ad-hoc wireless networks," in Proc. 18th International Conference on Distributed Computing Systems (ICDCS'98), Amsterdam, The Netherlands, May 26-29, 1998, pp. 474-479.

[2] G. Holland, and N. Vaidya, "Analysis of TCP performance over mobile ad hoc networks," in Proc. ACM/IEEE International Conference on Mobile Computing and Networking (MobiCom'99), Seattle, Washington, USA, Aug. 1999.

[3] J. Liu, and S. Dingh, "ATCP:TCP for Mobile Ad Hoc Networks," IEEE Journal on selected areas in communications, vol. 19, No. 7, July 2001.

[4] T. Dyer, and R. Boppana, "A comparison of TCP performance over three routing protocols in mobile ad hoc networks," in Proc. ACM International Symposium on Mobile Ad Hoc Networking and Computing (MobiHoc 2001), Long Beach, California, USA, Oct. 2001.

[5] Y. Zhang, and F. Wang, "Improving TCP performance over mobile ad-hoc networks with out-of-order detection and response," in Proc. ACM International Symposium on Mobile Ad Hoc Networking and Computing (MobiHoc 2002), Lausanne, Switzerland, June 2002.

[6] C. Perkins, E. Belding-Royer, and S. Das, "Ad Hoc On-demand Distance Vector (AODV) Routing," IETF RFC 3561, July 2003.

[7] V. Sridhara, "Evaluating Different Techniques to Improve TCP Performance over Wireless Ad Hoc Networks," Research Paper, available from http://www.eecis.udel.edu/~sridhara/ wireless/wireless.html.

[8] J. H. Choi, and C. Yoo, "TCP-aware Source Routing in Mobile Ad Hoc Networks," Eighth IEEE International Symposium on Computers and Communications (ISCC'03), Kemer, Antalya, Turkey, July 2003.

[9] R. De Oliveira, and T. Braun, "TCP in Wireless Mobile Ad Hoc Networks," Technical Report, IAM-02-003, University of Bern, Switzerland, March 2003.

[10] W. Stevens, TCP/IP Illustrated (Vol. 1, The Protocols), Addison-Wesley, 1994.

[11] K. Fall, and K. Varadhan, "NS notes and documentation," the VINT Project, UC Berkeley, LBL USC/ISI, and Xerox PARC, available from http://www.isi.edu/nsnam/ns/, Dec. 2003.

[12] CMU Monarch Group, "CMU Monarch extensions to the NS-2 simulator," Available from http://monarch.cs.cmu.edu/cmu-ns.html, 1999.

Predictive Call Admission Control Algorithm for Power-Controlled Wireless Systems

Choong Ming Chin[1], Moh Lim Sim[2], and Sverrir Olafsson[3]

[1] Asian Research Centre, BT Group, Cyberjaya 63000, Malaysia
eric.chin@bt.com
[2] Multimedia University, Cyberjaya 63100, Malaysia
mlsim@mmu.edu.my
[3] Mobility Research Centre, BT Group, Martlesham Heath, Ipswich, UK
sverrir.olafsson@bt.com

Abstract. In wireless communication systems, a conventional call admission control (CAC) mechanism determines whether a node can be admitted to the network by firstly monitoring the received interference plus noise and estimate the achievable signal-to-interference-plus-noise ratio (SINR). However, in the presence of power control, the SINR may vary over time, thus, rendering the conventional CAC inaccurate. The maximum achievable SINR for a new node in a general wireless system depends on the link gains amongst all the co-channel interfering nodes involved. Thus, one of the challenges of CAC in a power-controlled wireless system is the estimation of maximum achievable SINR when information about global link gains is not available. By ignoring the white noise factor, we present a predictor for the maximum achievable signal-to-interference ratio (SIR) of a new node trying to gain access to the medium. Using the SIR predictor we then calculate an optimal active link protection margin, which together with a SIR threshold would constitute an enhanced threshold value for the new node to attain. By doing so current active communication links would be protected from performance degradation should the maximum achievable SIR value common to all the nodes be lower than the SIR threshold. The accuracy of the predictor is evaluated by means of simulation in terms of mean error and root-mean-square error. Together with finding the corresponding optimal active link protection margin, efficient CAC mechanism to ensure stability of the feasible system can be maintained over a wide range of operating SIR values.

1 Introduction

In wireless ad-hoc networks, CAC algorithm plays an integral part in determining whether a new communication node can gain access to the channel for transmission. The objectives of CAC of a wireless network are to admit as many users as possible so as to maximize the new call admission probability whilst minimizing any performance degradation of existing communication links due to co-channel interference. Generally, the call admission decision is governed

T. Kunz and S.S. Ravi (Eds.): ADHOC-NOW 2006, LNCS 4104, pp. 414–427, 2006.

by several factors amongst others including the available radio resources such as frequency, time channels, transmit power, accessibility to access points and also acceptable link quality in terms of signal-to-interference plus noise ratio (SINR) or signal-to-interference ratio (SIR). In this study, we focus on the SIR requirement. It has been shown that when there are several nodes intending to communicate with each other, power control [2] can improve the number of simultaneous communication pairs through the reduction of unnecessary transmit power or interference and improve the SIR of all communication nodes. In early works on power control [1], balancing the SINRs of all radio links was first proposed via a centralized operation system. The advantage of this approach is that the set of calls that can be admitted while satisfying the required SINR threshold can be determined in advance. However later works on power control tended to shift its focus to a system of distributed SINR-balancing algorithms [2], [3], [4] which do not require global information. In the distributed power control scheme, power level of all the nodes are adjusted iteratively over the time. If the system is *feasible* that is all power constraints are satisfied, transmit powers of all nodes converge to finite levels so that all the user links will meet the required SINR threshold. However if no feasible power assignment is possible for a given SINR threshold, then the power levels will diverge after some iterations of power control and hence required the removal of some existing calls [5]. In the paper by Bambos *et al.* [6] the concept of *active link protection* is first introduced as a means to minimize the degradation of SINR of current active links as new links are accessing the channel. For a newly arrived link it first starts to transmit with a low transmission power and then gradually increases it by a factor of $\delta > 1$ where δ is a pre-assigned parameter. In response to the increased interference caused by the new link, the current active links would then update their powers whilst aiming to satisfy an enhanced SINR threshold target. For the case when new links are not able to be accommodated they would then simply exit from the system without causing any drop of QoS from the existing active links. Extension of this scheme for mobile ad hoc networks is further discussed in [7]. The disadvantage of this scheme [6,7] is that new links are initially admitted and may be rejected by the system after a number of power control iterations.

To summarize, in the presence of power control, the SINR may vary over time, thus, rendering the conventional CAC inaccurate. The maximum achievable SINR for a new node depends on the link gains amongst all the co-channel interfering nodes involved. Thus, one of the challenges of CAC in a power-controlled wireless system is the estimation of maximum achievable SINR when information about global link gains is not available. In the work by [8], the received power upon admission of a call at a cellular CDMA base station is predicted for CAC purpose. The interference from other base stations is assumed to be constant, thus, the SINR would mainly depends on the received signal power. However, in non-cellular type of wireless system, the presence of power control may cause positive feedback among the transmitters and result in changes in the received interference. In this paper, we first present a predictor for the maximum achievable SIR of a new link or node trying to get admitted into the system.

Unlike [9], we assume that the new link does not have any global information concerning all other link gains in the feasible system. Instead, based on some mild assumptions on the relative location of the mobiles to their base stations, it predicts the actual maximum achievable SIR of the network system should it be admitted. In order to achieve a high probability of making accurate prediction (or low decision making error) in relation to the true SIR, a series of simulation studies is first performed to deduce the positions of other existing mobiles with respect to their base stations positions. Based on such information, for each topological scenario, it then constructs a quasi-centralized power control (QCPC) scheme and then calculates the maximum achievable SIR value. For each SIR threshold requirement and based on some probability measures, we then find the corresponding optimal active link protection margin, which together with the SIR threshold would constitute an enhanced threshold value for the new node to attain. Using such an adaptive active link protection strategy, the new link can then make instantaneous decision on whether it should be admitted to the system while current active links would be protected from performance degradation. Furthermore the proposed scheme is also flexible enough to take into account how the power level updates for all existing links can be achieved should the new call be allowed and we envisage the power updating procedure to be based on some distributed and autonomous power control algorithms. The organization of this paper is as follows. In Section 2 we will describe the general framework of the system model and also discuss a few matrix theory results relating to our SIR prediction scheme. In Section 3 we introduce the construction of a quasi-centralized power control scheme using only mild assumptions which we will then use it to predict the maximum achievable SIR value. In this section also based on the SIR threshold and some probability measures we will also discuss how to find the active link protection margin in an adaptive manner. In Section 4 we discuss the statistical results based on extensive simulation studies on a wide range of CAC scenarios. In Section 5 we discuss and extend the SIR prediction as well as the CAC schemes and finally in Section 6 we conclude the paper.

2   System Model

The main objective of utilizing power control is to reduce co-channel interference and ensuring all communication pairs achieve an SINR above a required threshold resulting in efficient power management and better quality of service (QoS). In wireless networks there are many communication channels, but we assume interference generated between links assessing different channels as negligible. Hence only co-channel interference is considered here. As such, the network scenario can be streamlined as a collection of interfering active links accessing a single channel, and the concept of call admission can be rendered equivalent to channel access. In the context of CAC within a distributed power control mechanism, the challenge is to implement a robust admission control scheme to exploit the limited available information like co-channel interference to make admission decisions in a distributed manner. However, in this study CAC is studied from a different angle whereby co-channel interference between links is not featured in our algorithmic scheme. Instead, the maximum achievable common SIR value shared amongst all the interfering active links plays a central role in deciding whether a new call can be successfully admitted to the system.

We assume that in a power-controlled wireless system there are $M - 1$ pairs of communicating links where each pair of communicating links has an SINR above the required threshold. Here such a network is being referred to as a *feasible* network. Assume that a new communicating pair denoted by M wants to access the channel without having to power up gradually and hence disrupt the feasible network. For the communicating pair to be successfully admitted to the system, it can first determine whether the maximum achievable common SIR value for the enlarged network system (M number of interfering links) is above the required threshold. To begin with we denote G_{ij} as the gain of the communication link between the ith mobile station and the jth base station (or frequently called access points in wireless local area networks terminology) where

$$G_{ij} = \frac{S_{ij}}{[(x_{m,i} - x_{b,j})^2 + (y_{m,i} - y_{b,j})^2]^{\nu/2}} \tag{1}$$

such that S_{ij} is the attenuation factor, x and y denote the two dimensional coordinate position, and the subscripts m, b, i and j denote the mobile station, access point, cell i and cell j respectively. The parameter ν is a constant that models the shadowing effect. We assume that $\ln S_{ij} \sim N(0, \sigma^2)$, $1 \leq i, j \leq M$, are independent and identically distributed. The value of σ in the range of $4 - 10$ dB and the propagation constant ν in the range of $3 - 5$ usually provide good models for non-line-of-sight urban propagation [10].

Without loss of generality we assume that the base station i is communicating with the mobile station i. In general, given that there are M interfering links in the enlarged system, we can denote the SINR of the ith mobile station by

$$\bar{\gamma}_i = \frac{G_{ii} P_i}{\sum_{j \neq i} G_{ij} P_j + \eta_i}, \quad 1 \leq i, j \leq M \tag{2}$$

where $P_i > 0$ is the base station transmit power and $\eta_i > 0$ is the white noise of the mobile station i. For each mobile client i there is some common SINR or SIR threshold requirement denoted by $\gamma^\infty > 0$, representing the mobile station i minimal QoS it must support in order to operate successfully. Following the above arguments we then have

$$\bar{\gamma}_i \geq \gamma^\infty, \quad 1 \leq i \leq M. \tag{3}$$

In matrix format, the relationship (2) and (3) can be expressed as

$$(\mathbf{I} - \mathbf{F})\mathbf{P} \geq \mathbf{\Theta}, \quad \mathbf{P} > \mathbf{0} \tag{4}$$

where $\mathbf{\Theta} = \left[\frac{\gamma^\infty \eta_1}{G_{11}}, \frac{\gamma^\infty \eta_2}{G_{22}}, \ldots, \frac{\gamma^\infty \eta_M}{G_{MM}}\right]^T$, $\mathbf{P} = [P_1, P_2, \ldots, P_M]^T > \mathbf{0}$ and $\mathbf{F} = (F_{ij})$ is a matrix having the following entries $F_{ij} = 0$ for $i = j$ and $F_{ij} = \frac{\gamma^\infty G_{ij}}{G_{ii}}$ for

$i \neq j$, $1 \leq i, j \leq M$. Note that $\mathbf{F}$ has nonnegative elements and it can be shown that $\mathbf{F}$ is also irreducible, that is each row of $\mathbf{F}$ has no more than one zero element (see [12]). From the Perron-Frobenius theorem [6] we have the result given below.

Theorem 1. *The following statements are equivalent:*

(i) There exists a power vector $\mathbf{P} > \mathbf{0}$ such that $(\mathbf{I} - \mathbf{F})\mathbf{P} \geq \mathbf{\Theta}$.
(ii) The spectral radius of $\mathbf{F}$, $\rho_F < 1$.
(iii) The matrix $(\mathbf{I} - \mathbf{F})^{-1}$ exists and has positive entries.

Based on Theorem 1, if $\rho_F < 1$ then there exists a unique solution

$$\mathbf{P}^* = (\mathbf{I} - \mathbf{F})^{-1}\mathbf{\Theta} > \mathbf{0} \tag{5}$$

which lies at a vertex of all the linear constraints. However in order to obtain $\mathbf{P}^*$ there is a need to acquire global information on the link gain matrix.

Owing to the complexities associated with acquiring global information on the link gains to assess the matrix properties of $\mathbf{F}$, therefore we propose an alternative strategy of predicting the maximum achievable SIR for all the active links so that the feasibility condition (3) can be ascertained upon arrival of a new link. In an operating environment with many simultaneous transmissions, the system performance is mainly limited by interference rather than noise. Hence $\sum_{j \neq i} G_{ij}P_j \gg \eta_i$ and thus SINR $\approx$ SIR asymptotically. For ease of analysis, by ignoring the white noise factor, the SIR of mobile station i, γ_i is defined as

$$\gamma_i = \frac{G_{ii}P_i}{\sum_{j \neq i} G_{ij}P_j} = \frac{P_i}{\sum_{j \neq i} \frac{G_{ij}}{G_{ii}}P_j} \tag{6}$$

for $1 \leq i \leq M$ where $P_i > 0$ is the base station transmit power. Note also that by comparing (2) and (6) we have the following inequality

$$\bar{\gamma}_i < \gamma_i, \quad 1 \leq i \leq M \tag{7}$$

where the SIR value constitutes an upper bound criteria of SINR for all mobile stations i. In order to find the maximum achievable SIR common for all the links we can first construct an $M \times M$ irreducible, nonnegative matrix $\mathbf{A} = (A_{ij})$ such that $A_{ij} = G_{ij}/G_{ii}$ for $i \neq j$ and $A_{ij} = 0$ for $i = j$. Hence we can write (6) as an eigensystem

$$\mathbf{AP} = \mathbf{\Lambda P} \tag{8}$$

where $\mathbf{\Lambda} = \text{diag}\left(\frac{1}{\gamma_1}, \frac{1}{\gamma_2}, \ldots, \frac{1}{\gamma_M}\right)$. From Grandhi *et al.* [11], we have a matrix theory result given here without proof.

Theorem 2. *Let $\mathbf{A}$ be an $M \times M$ irreducible nonnegative matrix with eigenvalues $\{\lambda_i\}_{i=1}^M$. If we denote $\lambda^* = \max\{|\lambda_i|\}_{i=1}^M$ with a corresponding eigenvector $\mathbf{P}^*$ having strictly positive entries then there exists a unique γ_M^*, which is a maximum achievable or actual SIR value for all the links given as $\gamma_M^* = 1/\lambda^*$ with $\mathbf{P}^*$ being the transmitter power vector to achieve γ_M^* in the system.*

Given a newly arriving link M trying to access the channel and due to lack of global information concerning the positions of active links, our algorithmic scheme therefore tries to approximate the true matrix $\mathbf{A}$ by setting up a quasi-centralized power control (QCPC) scheme and then use Theorem 2 to predict the actual maximum achievable SIR in the system. In this paper we also wish to set up an active link protection mechanism whereby an enhanced SIR threshold is first enforced onto the new node so that current active links are protected prior to the event a new node begins to enter to the network system. To begin with, we define $\delta > 0$ as the active link protection margin and if under some probability measures that the predicted SIR value is greater than the enhanced threshold requirement (threshold plus active link protection margin), and provided the decision error made is small enough then link M is freely admitted to the system and all the active links will then autonomously adjust their power levels according to some distributed power control algorithms.

The procedure described above can lead to two kinds of errors. For instance, if the actual or true SIR value γ_M^* is less than the threshold requirement γ^∞, and the predicted SIR $\hat{\gamma}_M$ concludes that $\hat{\gamma}_M \geq \gamma^\infty + \delta$, then the scheme is committing a *false admission error*. On the other hand, if the actual SIR value, $\gamma_M^* \geq \gamma^\infty$ and the predicted SIR concludes that $\hat{\gamma}_M < \gamma^\infty + \delta$ then the prediction scheme is committing a second kind of error referred to as a *false rejection error*. If in the event we were to commit a false admission error then based on some distributed and autonomous power control algorithms, the power levels for all the links will diverge to infinity. Hence either some of the existing links will drop their calls and exit from the system or the system can negotiate a much lower QoS requirement in response of the new node. Take note that the purpose of this study is to deal with the aspects of SIR prediction as they relate to CAC in wireless systems. Questions on how to update the power control algorithms to take into effect of a new incoming link whilst maintaining the QoS requirements are not considered here.

3   Converged SIR Prediction and Active Link Protection Margin Algorithmic Scheme

For a newly arriving link M, we make the following assumptions so that we can construct a quasi-centralized power control system to estimate the predictor SIR value for any given topological scenario:

- We assume the network structure is quasi-static and the time taken to assess its suitability for admission is less than the relative timescale of mobility of other active links.
- At an instantaneous time, each communication link consists of a base station and a mobile station, that is we exclude handovers in our system model.
- The number of active links, $M - 1$, is known for any new incoming links.

- All the base stations know the positions of their surrounding co-channel base stations positions. i.e., $(x_{b,i}, y_{b,i})$ are known for all i. This is a practical assumption as the positions of base stations in wireless local area networks (LAN) are fixed.
- The link gains between an incoming mobile station M, and all base stations, $\{G_{Mj}\}$ for $j = 1, 2, \ldots, M$, can be measured. Practically this can be achieved by requiring all the base stations to send a constant power pilot tone.
- The maximum radio coverage of each cell, R_o is assumed to be known.

Note that the exact locations of the active mobile stations are unknown to the new link (only cummulative interference can be measured). In order to form the link gain matrix for SIR determination, we assume that the mobiles are placed isotropically around its base station within a common radius R_o. The former (X,Y)-position can be drawn from a uniform distribution within a circle of radius R_o. Based on such information the new link can construct an $M \times M$ irreducible, nonnegative matrix $\mathbf{A} = (A_{ij})$ such that $A_{ij} = G_{ij}/G_{ii}$ for $i \neq j$ and $A_{ij} = 0$ for $i = j$.

Using the results in Theorem 2, for each topological network of size M, we propose a maximum achievable SIR predictor as follows:

$$\hat{\gamma}_M = \frac{1}{n} \sum_{k=1}^{n} \hat{\gamma}^*_{M,k} \tag{9}$$

where n is the sample size, $\hat{\gamma}^*_{M,k}$ is the maximum achievable SIR corresponding to the matrix $\mathbf{A}(k)$ such that

$$A_{ij}(k) = \begin{cases} \hat{G}_{ij}(k)/\hat{G}_{ii}(k), & i \neq j, i \neq M \\ G_{Mj}/G_{MM}, & i = M \\ 0, & i = j \end{cases} \tag{10}$$

where

$$\hat{G}_{ij}(k) = \frac{\hat{S}_{ij}}{[(\hat{x}_{m,i}(k) - x_{b,j})^2 + (\hat{y}_{m,i}(k) - y_{b,j})^2]^{\nu/2}} \tag{11}$$

and the co-ordinates $(\hat{x}_{m,i}(k), \hat{y}_{m,i}(k))$ are randomly positioned within a circle of radius R_o. For $i \neq j$, $i \neq M$ we let $\ln \hat{S}_{ij}(k) \sim N(0, \hat{\sigma}^2)$ where $\hat{\sigma} \sim U(4, 10)$.

The proposed SIR predictor can be used in improving the decision making in CAC. Without the predictor, a call would be first admitted and either it would be served or denied at a later time after undergoing a waiting period. But with the help of an SIR predictor, decision can be made immediately upon arrival of a new node. However in order to protect the existing links to offset the degradation of QoS as the new call is accessing the channel, an active link protection margin $\delta \geq 0$ is introduced. Having this protection margin δ, the probability in making a false admission can be written as

$$P(\hat{\gamma}_M \geq \gamma^\infty + \delta, \gamma^*_M < \gamma^\infty) \tag{12}$$

while the probability in making a false rejection is defined as

$$P(\hat{\gamma}_M < \gamma^\infty + \delta, \gamma_M^* \geq \gamma^\infty) \tag{13}$$

where γ^∞ is the SIR threshold. For a given margin error value $\alpha \in (0, 1)$ and for each SIR threshold value, we can find an optimal δ^* adaptively by solving the following optimization problem

$$\begin{aligned} \underset{\delta \geq 0}{\text{minimize}} \quad & P(\hat{\gamma}_M < \gamma^\infty + \delta, \gamma_M^* \geq \gamma^\infty) \\ \text{subject to} \quad & P(\hat{\gamma}_M \geq \gamma^\infty + \delta, \gamma_M^* < \gamma^\infty) = \alpha. \end{aligned} \tag{14}$$

4   Simulation and Result

In the simulation we use $\nu \sim U(3, 5)$, $\sigma, \hat{\sigma} \sim U(4, 10)$, $R_o = 20$, $n = 30$. The number of pairwise mobile to base station nodes, M varies from 2 to 30 and the base station coordinates, $x_{b,i}, y_{b,i} \sim U(0, 1000)$ for all $i = 1, 2, \ldots, M$. The total simulation runs for each node of size $M = 2, \ldots, 30$ is 1000, and to maintain a minimum QoS, the maximum achievable SIR γ_M^*, and the predictor SIR $\hat{\gamma}_M$ for each simulated nodal topology must be greater or equal than 5 dB.

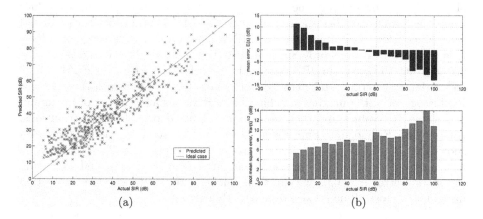

(a)   (b)

Fig. 1. (a) Scatter plot of 500 simulated runs with the straight line indicating perfect matching. (b) Mean error, $E(\Delta)$, and root mean square error, $\sqrt{Var(\Delta)}$ across different range of actual SIR values and topological scenario.

To begin with we first analyze the relationship between the true SIR, γ^* and the predictor SIR, $\hat{\gamma}$ across all topological scenarios. From Figure 1(a), we can see the scatter plot that there is a linear relationship between the predicted SIR and actual SIR values, with a majority of the SIR values seem concentrating between 10 db and 70 dB. This is due to the fact that as the number of active links grow denser, the maximum achievable SIR values for all the links tend to restrict themselves within a more narrower band of values.

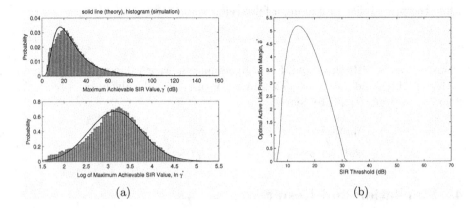

Fig. 2. (a) Histogram of the maximum achievable SIR values and log of maximum achievable SIR values and their corresponding fitted probability distributions. Top plot: fitted a log normal distribution. Bottom plot: fitted a normal distribution. (b) Optimal δ^* at $\alpha = 0.05$ for various SIR threshold values across all nodal topology size.

The statistics of the SIR predictor error, $\Delta = \hat{\gamma}\,(\mathrm{dB}) - \gamma^*\,(\mathrm{dB})$, is expressed in terms of mean and root mean square. In Figure 1(b), we present the histograms of the mean error and root mean square error for a range of actual SIR values where the former plot corresponds closely to the behaviour in which the predicted SIR values tries to estimate the true SIR values across a wide range of values. It is found from Kolmogorov-Smirnov test (test statistic $D = 0.012$, p-value $= 0.7895$) that the error (in dB) is normally distributed with mean $E(\Delta) = 0.17$ and variance, $Var(\Delta) = 67.26$ at 5% significance level. In Figure 2(a) we present the histogram of the maximum achievable SIR values and the log of maximum achievable SIR values and their corresponding fitted probability distributions. Analysis via Kolmogorov-Smirnov goodness of fit test at 5% significance level shows that $\ln \gamma^* \sim N(\mu_{\gamma^*}, \sigma^2_{\gamma^*})$ where $\mu_{\gamma^*} = 3.16$ ($\exp(\mu_{\gamma^*}) = 23.57$) and $\sigma^2_{\gamma^*} = 0.36$ ($\exp(\sigma^2_{\gamma^*}) = 1.43$). Taking note that $\ln \gamma^* \sim N(\mu_{\gamma^*}, \sigma^2_{\gamma^*})$, and by letting the margin error $\alpha = 0.05$ we can then calculate the optimal active protection margin, δ^* across all nodal size topology. By solving

$$\begin{array}{c} \underset{\delta \geq 0}{\text{minimize}} \quad P(\hat{\gamma} < \gamma^\infty + \delta, \gamma^* \geq \gamma^\infty) \\ \text{subject to } P(\hat{\gamma} \geq \gamma^\infty + \delta, \gamma^* < \gamma^\infty) = \alpha. \end{array} \tag{15}$$

where

$$P(\text{false admission}) = P(\hat{\gamma} \geq \gamma^\infty + \delta, \gamma^* < \gamma^\infty)$$

$$= \int_0^{\gamma^\infty} P\left(Z \geq \left. \frac{\gamma^\infty + \delta - \gamma^* - E(\Delta)}{\sqrt{Var(\Delta)}} \right| \gamma^* = y \right) f_{\gamma^*}(y)\,dy$$

$$= \int_0^{\gamma^\infty} \int_{\frac{\gamma^\infty + \delta - y - E(\Delta)}{\sqrt{Var(\Delta)}}}^{\infty} \frac{1}{\sqrt{2\pi}} e^{-\frac{1}{2}z^2}\,dz\, f_{\gamma^*}(y)\,dy \tag{16}$$

and

$$P(\text{false rejection}) = P(\hat{\gamma} < \gamma^\infty + \delta, \gamma^* \geq \gamma^\infty)$$

$$= \int_{\gamma^\infty}^{\infty} P\left(Z < \frac{\gamma^\infty + \delta - \gamma^* - E(\Delta)}{\sqrt{Var(\Delta)}} \middle| \gamma^* = y \right) f_{\gamma^*}(y) \, dy$$

$$= \int_{\gamma^\infty}^{\infty} \int_{-\infty}^{\frac{\gamma^\infty + \delta - y - E(\Delta)}{\sqrt{Var(\Delta)}}} \frac{1}{\sqrt{2\pi}} e^{-\frac{1}{2}z^2} \, dz \, f_{\gamma^*}(y) \, dy. \qquad (17)$$

where $Z \sim N(0,1)$ and $f_{\gamma^*}(\cdot)$ is the pdf of the actual maximum achievable SIR, γ^*, we then obtain the required optimal δ^* for various SIR threshold values (see Figure 2(b)).

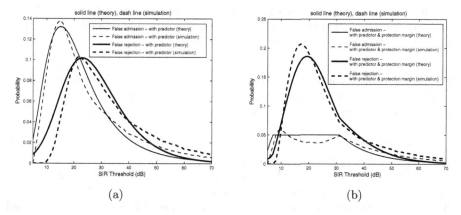

(a)   (b)

Fig. 3. (a) Probability of false admission and false rejection for various SIR threshold values using only the predictor SIR values. The dash lines are from practical simulated work. (b) Probability of false admission and false rejection for various SIR threshold values using predictor SIR values and optimal protection margin found at $\alpha = 0.05$. The dash lines are from practical simulated work.

In Figure 3 we present the probabilities of committing false admission and false rejection with and without the optimal δ^* values. With the utilization of the optimal δ^*, we can see from Figure 3(b) that while the probability of committing false admission is bounded above by the margin error $\alpha = 0.05$, the probability of committing false rejection would tend to increase but bounded above by 0.2 (or 20%). However the reverse is true is we were to utilize the SIR predictor alone. This is due to the fact that as the probability of committing false admission is reduced, even though $\gamma^* \geq \gamma^\infty$, with a non-zero δ^* value, new incoming links will have to overcome a higher enhanced SIR threshold, $\gamma^\infty + \delta^*$ in order to be successfully admitted to the network. Hence, there is a higher chance that new links might not be able to be admitted to the network even though $\gamma^* \geq \gamma^\infty$ which ultimately results in higher probability of committing false rejection. Furthermore if we were to compare the simulated

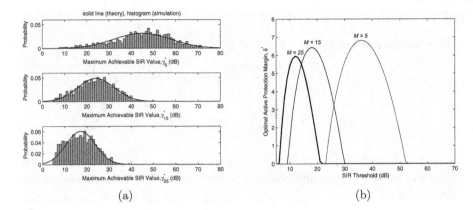

Fig. 4. (a) Histograms of the maximum achievable SIR values for $M = 5, 15, 25$ and their corresponding fitted normal probability distributions. (b) Optimal δ^* at $\alpha = 0.05$ for various SIR threshold values and nodal topology.

results with the theoretical results (solid line) we can see there is a close association between them, and we can therefore deduce that (16) and (17) can be used to provide good probability measures of a power control CAC system.

In the previous discussion we have focussed on analyzing CAC of a new link based on the distribution of SIR values irrespective of nodal topology. We now come to the statistics of the SIR predictor error for each nodal size M, $\Delta_M = \hat{\gamma}_M - \gamma_M^*$ where it is characterized in terms of mean and root mean square. Here, $\hat{\gamma}_M$ and γ_M^* are the predicted and actual achievable SIR for the simulated topology of size M respectively, and it can be verified from Kolmogorov-Smirnov test that $\Delta_M \sim N(E(\Delta_M), Var(\Delta_M))$ and $\gamma_M^* \sim N(E(\gamma_M^*), Var(\gamma_M^*))$ (see Figure 4(a)). In Figure 4(b), we present plots of optimal active link protection margin δ^* for various nodal size M over a wide range of SIR threshold values. The plots are produced based on the optimization solution of problem (14) where in this case $\alpha = 0.05$ with the probability of false admission for nodal topology of size M is expressed by

$$P(\text{false admission}) = P(\hat{\gamma}_M \geq \gamma^\infty + \delta, \gamma_M^* < \gamma^\infty)$$

$$= \int_{-\infty}^{\gamma^\infty} \int_{\frac{\gamma^\infty+\delta-y-E(\Delta_M)}{\sqrt{Var(\Delta_M)}}}^{\infty} \frac{1}{\sqrt{2\pi}} e^{-\frac{1}{2}z^2} \, dz \, f_{\gamma_M^*}(y) \, dy \quad (18)$$

while the corresponding probability of false rejection is defined as

$$P(\text{false rejection}) = P(\hat{\gamma}_M < \gamma^\infty + \delta, \gamma_M^* \geq \gamma^\infty)$$

$$= \int_{\gamma^\infty}^{\infty} \int_{-\infty}^{\frac{\gamma^\infty+\delta-y-E(\Delta_M)}{\sqrt{Var(\Delta_M)}}} \frac{1}{\sqrt{2\pi}} e^{-\frac{1}{2}z^2} \, dz \, f_{\gamma_M^*}(y) \, dy. \quad (19)$$

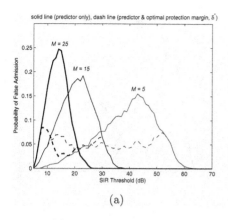 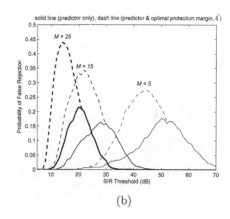

(a)   (b)

Fig. 5. (a) Probability of false admission at $\alpha = 0.05$ for various SIR threshold values and nodal topology. (b) Probability of false rejection at $\alpha = 0.05$ for various SIR threshold values and nodal topology.

where $Z \sim N(0,1)$ and $f_{\gamma_M^*}(\cdot)$ is the pdf of the actual achievable SIR of nodal size M such that $\gamma_M^* \sim N(E(\gamma_M^*), Var(\gamma_M^*))$.

Based on this adaptive active link protection strategy we can see from Figure 5(a) that a majority of a new link committing a false admission to an already feasible system is quite low as compared with the system relying only the predictor. Take note that based on different nodal topology, the probability of committing false admission above the psychological threshold of $\alpha = 0.05$ is centred on different range of SIR threshold values; higher SIR threshold values for low density nodes and vice versa for higher density nodes. In addition, due to the dynamic aspect of selecting the active link protection margin, the probability of committing false rejection of a new incoming node would tend to increase as the nodal size increases (see Figure 5(b)). Hence we can see that if we were to rely on the SIR predictor alone, the false rejection probabilities for various nodal topology are quite low as compared with the CAC scheme exercising the SIR predictor and the optimal active link protection margin. Furthermore due to the distribution of the maximum achievable SIR values, γ_M^*, we can see that for small nodal topology, large false rejection probabilities tend to be centred on higher SIR threshold values, but vice versa for the case of larger nodal topology.

In order to compare both theoretical and simulation studies, Figures 6(a) and 6(b) show that with the usage of a predictor SIR with active link protection margin, the false admission and rejection probabilities from the simulation study do approximate closely with the theoretical one for various nodal topology size. As for the high peaks we encounter in Figure 5(a) which are greater than $\alpha = 0.05$ we feel this is due to lack of simulation samples for each nodal topology and if we were to increase them marginally higher then the simulated and theoretical results would be more conforming. By comparing the close relationship between the theoretical and simulation study results and based on the SIR threshold

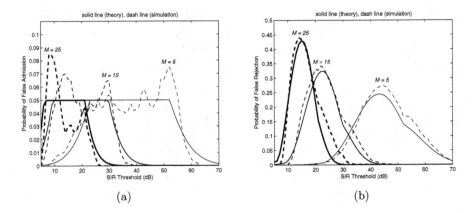

Fig. 6. (a) Theoretical and simulated probability of false admission at $\alpha = 0.05$ for various SIR threshold values and nodal topology. (b) Theoretical and simulated probability of false rejection at $\alpha = 0.05$ for various SIR threshold values and nodal topology.

value, we feel a new link can utilize the corresponding optimal δ^* values to make instantaneous decision whether to admit to the system or not.

5   Discussion

By comparing the theoretical and practical simulated studies, we can deduce that a CAC scheme with an SIR predictor and its corresponding optimal active link protection margin would minimize the probability of committing false rejection whilst maintaining that the false admission probability satisfies the margin error $\alpha \in (0,1)$. Of course based on the analysis, we can utilize two independent approaches to find the appropriate δ^* value as part of our CAC strategy. The first approach is to find the optimal δ^* values across all nodal topology while the second is to isolate each nodal topology according to the number of active links, and then find the appropriate optimal δ^*. Both methods we feel have their own merits where the former approach tends to concentrate solely on the spread of the actual SIR values, and statistically analyze them accordingly while the latter is more into nodal topology. As such new incoming links can have a choice whether should it be admitted to a network based on the size of the network or on the probabilistic modelling of the actual SIR values.

6   Conclusion

In this paper we have studied CAC for power-controlled wireless systems. Such a system usually exhibits high spectrum utilization but in the event of an in-feasible system caused by a new incoming call, the power levels may diverge to infinity. We have therefore proposed an SIR predictor for a power-controlled wireless system and in order to assess the usefulness of the predictor the analysis is performed on a simulated CAC scenario. The performance is then assessed in terms of calculating the error of committing false admission and also false rejection. Ranging from a low density to a high density network topology, the analysis shows that the predictor SIR scheme performs well over the practical operating range of SIR threshold values and is therefore a good estimator to the simulated data. In addition, in order to minimize the probability of false admission of a new link entering to the system, we have also introduced the concept of calculating the optimal active link protection margin in a dynamic fashion. Using the optimal active link protection margin in tandem with the SIR threshold values, we can therefore ensure that current active links are protected from performance degradation since the new link will now have a higher threshold value to attain in order to be successfully admitted to the system. Thus we feel the technique presented here would provide new directions on how CAC issues can be addressed in future.

References

1. Aein, J.M.: Power balancing in systems employing frequency resuse. COMSAT Tech. Rev. (1973) 277 – 299.
2. Zander, Z.: Performance of optimum transmitter power control in cellular radio systems. IEEE Trans. Veh. Technol. **41**(1) (1992) 57 – 62.
3. Grandhi, S.A., Vijayan, R. and Goodman, D.J.: A distributed algorithm for power control in cellular radio systems. Proc. 30th Allterton Conf. Monticello. IL. (1992).
4. Foschini, G.J. and Miljanic, Z.: A simple distributed autonomous power control algorithm and its convergence. IEEE Trans. Veh. Technol. **42**(4) (1993) 641 – 646.
5. Andersin, M., Rosberg, Z. and Zander, J.: Gradual removals in cellular radio networks. Wireless Networks. **2**(1) (1996) 27 – 43.
6. Bambos, N., Chen, S.C. and Pottie, G.J.: Channel access algorithms with active link protection for wireless communications. IEEE/ACM Trans. Networking. **8**(5) (2000) 583 – 597.
7. Holliday, T., Goldsmith A., Glynn, P. and Bambos, N.: Distributed power and admission control for time varying wireless networks. Technical Report. Stanford University (2004).
8. Assa, S., Kuri, J. and Mermelstein, P.: Call admission on the uplink and downlink of a CDMA system based on total received and transmitted powers. IEEE Trans. Wireless Commun. **3**(6) (2004) 2407 – 2416.
9. Andersin, M., Rosberg, Z. and Zander, J.: Soft and safe admission control in cellular networks. IEEE/ACM Trans. Networking **5**(2) (1997) 255 – 265.
10. Lee, W.C.Y.: Elements of Cellular Mobile Radio. IEEE Trans. Veh. Technol. **35** (1986) 48–56.
11. Grandhi, S.A., Vijayan, R., Goodman, D.J. and Zander, J.: Centralized power control in cellular radio systems. IEEE Trans. Veh. Technol. **42**(4) (1993) 466 – 468.
12. Minc, H.: Nonnegative Matrices. John Wiley (1988).

File System Support for Adjustable Resolution Applications in Sensor Networks[*]

Vikram Munishwar[1], Sameer Tilak[2], and Nael B. Abu-Ghazaleh[1]

[1] Department of Computer Science
State University of New York, Binghamton,
Binghamton, NY 13902
{vmunish1, nael}@cs.binghamton.edu
[2] SDSC, UC San Diego, MC 0505, 9500 Gilman Drive
La Jolla, CA 92093-0505
sameer@sdsc.edu

Abstract. Flash memory is often the technology of choice for sensor networks because of its cost-effectiveness and attractive energy properties. In storage-constrained sensor network applications, the monitored data is typically stored in multi-resolution fashion. This allows reclamation of some storage space when needed by reducing the quality of stored data by eliminating some of the precision. Existing sensor network file systems are optimized for sequential logging of the data. However, flash memories have a number of unique properties that require careful consideration in file system design. In this paper, we show that in applications where adjustable resolution occurs, sequential logging file-systems result in an inefficient implementation of adjustable resolution. We propose an alternative implementation of the file system where data components are grouped with each other according to resolution. Thus, reducing resolution is implemented by simply erasing the pages with the excess resolution components. We have implemented the proposed scheme on crossbow MICA2 sensor nodes. In addition, using TOSSIM simulations, we show that as compared to the existing approach, the proposed scheme results in significant savings in read and write operations to the flash (thereby in turn saving energy, and reducing wear). Further, we show that wear leveling can be maintained over time by assigning the most significant data to the most frequently used pages.

1 Introduction

Wireless sensor networks (WSNs) are an important emerging technology that can provide sensing at unprecedented resolution. This capability is of importance to a wide range of scientific, military, industrial and civilian applications. However, many battery-operated sensors have constraints such as limited energy, less computational ability, and small storage capacity, and thus protocols must be designed to deal efficiently with these limited resources. In WSNs, available energy is the primary resource that defines the network's useful lifetime. Thus, concern for energy efficient operation permeates all aspects of sensor network design and operation.

---

[*] This work is partially supported by NSF grant CNS-0454298 and US Army project W911SR-05-C-0014.

T. Kunz and S.S. Ravi (Eds.): ADHOC-NOW 2006, LNCS 4104, pp. 428–443, 2006.

There exists a class of sensor networks where the sensed data (or a subset of it) is not relayed in real-time. In such applications, the data must be stored, at least temporarily, within the network, until it is later collected or queried by an observer. Alternatively, the data may cease to be useful and be discarded or compressed to make room for more important data. The data may also be used in dynamic queries. Two application classes where the storage management problem typically arises are scientific monitoring applications and augmented reality applications. We briefly overview these applications and reader is encouraged to refer to [16] for further details.

Sensors in an unattended network for a remote scientific monitoring application may collect data and store it for extended periods [9,12,18]. Scientists (observers) might periodically visit the network to collect this data. Data collection is not pre-planned: it might be unpredictable and infrequent. However, the data *query* model is simple: data is collected one time. Alternatively, in an augmented reality application, users may dynamically query sensors for the collected data [17]. The queried data can be real-time, recent, or historical data. Therefore, to be able to respond to queries that span temporally long periods, sensors must store data. In these storage-bound networks, the design of the storage system is critical to the overall energy efficiency and performance of the network.

In storage bound sensor networks, it is often the case that the resolution of the collected data will need to be adjusted periodically. For example, in a scenario where a node runs out of storage space, it is desirable to reduce the quality of stored samples to make room for additional data to be collected, rather than discard the new data or overwrite the old data completely due to lack of storage. Similarly, it is possible that data is of interest at the highest resolution only for a period of time; as data ages, low-resolution information may suffice [7]. The data itself may be required at different resolutions in response to different queries. For example, a certain aggregate query may require only the highest resolution component of the data. We call these application *adjustable resolution applications*. Storing data in multi-resolution fashion allows reclamation of some storage space when needed by reducing the quality of stored data by eliminating some of the precision.

For non-volatile storage, flash memory is often the technology of choice for sensor networks and mobile devices because of its cost-effectiveness and attractive energy properties. Section 2 presents some background material related to flash memory organization. A number of research studies have examined various aspects of data storage in sensor networks [11,15,7]. Moreover, file-systems for flash memory devices for use in sensor networks have been developed [8,5], with emphasis on support for sequential data logging (rather than random access). Section 3 overviews these and other related work. Unfortunately, in the case of adjustable resolution applications, the interaction between application and the file-system has received almost no attention. For example, flash memories have properties that require special care from file system designers. Most importantly, flash memory can only be modified (written, or erased) with a page granularity. Thus, if a word needs to be modified in a given page, the whole page has to be flashed (erased) and then rewritten. As a result, random access data modification is a costly operation. In addition, pages can become faulty with usage– a problem called *page wear*. Therefore, it is desired to load balance the number of times each page is written (*page wear leveling*).

Existing sensor network file systems are optimized for sequential logging of the data. We show that in these applications, adjusting resolution under existing file system implementations leads to excessive read and write operations to the flash. These result in high consumption of energy as well as increasing the wear on the flash. Essentially, the data whose resolution is to be adjusted must all be read, then the least significant components are deleted, before the remaining data is stored again. The same argument is true when querying data at different resolutions. In the current implementation, all the data would have to be read and only the relevant resolution data is extracted.

To counter this inefficient operation, we propose an alternative organization where data is organized into pages according to significance. This approach allows reducing resolution by simply erasing the pages with undesired components. Further, querying data by resolution results in only the required data being read. We describe the problem and present the proposed scheme in Sections 5 and 6 respectively. In Section 7 we present an experimental study characterizing the performance of the proposed scheme. Finally, Section 9 presents some concluding remarks.

2 Background

In this section, we first review characteristics of flash memory devices. We then overview two major limitations of existing flash memories, which have significant impact on the design of flash based file systems. Finally, we overview the organization of the flash and EEPROM device on MICA2 sensor nodes, which were used in our implementation and simulation.

2.1 Flash Memory Devices

Flash memory provides non-volatile storage at a lower cost than traditional storage devices such as magnetic disks. Other advantages of flash memory over magnetic disks include: (i) fast read access time; (ii) low-energy dissipation; (iii) light-weight and small form-factor; (iv) better shock-resistance. Because of these properties, flash memories have become a de-facto storage technology in sensor networks and embedded/mobile devices in general.

A flash memory is similar to Electrically-Erasable Programmable Read-Only Memory (EEPROM) in terms of physical architecture; in fact, flash is essentially a block-writable EEPROM (where conventional EEPROM is byte-writable). Flash memory is broadly categorized into two types: NOR flash memory and NAND flash memory. These names are based on the type of logic gate used in its cell. NOR flash memory is mainly used for storing small data, source code, etc. It has a capability of in-place execution of a code. This is because it behaves like ROM memory, mapped to a certain address [4]. NOR flash is characterized by its faster read time, nearly equal to the read times of DRAM [10]. However, it has a disadvantage of slower write and erase operations. On the other hand, NAND flash memory is mainly used for data storage due to its i) higher density; ii) faster erase and write operations; and iii) longer re-write life expectancy [3]. Since NAND memory is accessed in much similar way as block devices, in-place code execution is not possible. To execute code present on NAND flash, the code must be first taken into memory mapped RAM and executed there. Thus, NAND flash always needs to be associated with Memory Management Unit (MMU) for code execution purpose [4].

Table 1. Flash Characteristics

Read cycle	80 - 150 ns
Write cycle for one byte	1 - 10 μs
Erase cycle for one block	1 s
Power consumption	30 - 50 mA in an active state
	20 - 100 μA in a standby state

2.2   Unique Operational Properties of Flash Memory

In comparison with the traditional memories, there are two different operational properties associated with flash memories. These properties have significant implications on the design of a file system for flash memory. These properties are:

1. Limited number of erase/write operations: Flash memory has finite number of erase/write cycles, from 10,000 to 100,000 cycles. So random writes to flash can cause wearing out of some part of the flash while the other part remaining usable. To reduce such problems, it is desirable to reduce the number of erase/write operations (wear) and distribute them evenly across the pages of the flash, thereby achieving *wear leveling*.
2. Lack of in place data overwriting capability: Flash memory doesn't allow writing data to a page before erasing it. Erase is a costly operation because it needs to be done at block granularity, whereas read and write operations are carried out at page granularity. So in effect, to rewrite a single byte in a flash, we need to read all pages in a block, erase that block, modify the requested page, and rewrite all pages back to the flash. Thus, random updates are costly operations.

2.3   Flash Resources on MICA2 Nodes

In our implementation, we used the Atmel AT45DB041B flash [2] chip present on Berkeley MICA2 motes[1] In this section, we describe some important details of this flash. The total storage capacity of the flash is 512KB. The storage area is divided into sectors, blocks, and pages. Each block contains 8 pages, each of 264 bytes. So the maximum number of pages that can be stored on this flash is 2048. Pages need to be erased before being written. All program operations to the flash take place on page-by-page basis; however, the optional erase operations can be done at the block or page level [2]. This complicates the process of in-place modification (or overwriting) of page directly.

---

[1] However, none of the proposed ideas are specific to this flash; they can be generalized to other flash targets directly.

Two on-chip buffers, 264 bytes each, are used during read and write operations on flash. Write operations on flash are performed in two phases, in the first phase, data is written to the on-chip buffer until it gets full, and in the second phase, the data in on-chip buffer is subsequently written to the flash memory.

In addition to the flash, 128K Program Memory, 4K SRAM, and 4K EEPROM are also available on MICA2 mote. The EEPROM provides small, additional storage for storing program related data such as data-structures, variables, etc. EEPROM provides a persistent storage with slow access speeds as compared to SRAM. Byte wise reads and writes are possible with EEPROM. With 1 MHz clock, read and write times for one byte are approximately 4 μs and 8 ms, respectively. Though wear leveling is an issue for EEPROM, it is not as severe as flash because it has an endurance of at least 100,000 writes. We use EEPROM primarily for storing our program specific data structures like bitmaps.

3   Related Work

In this section, we first discuss storage related efforts in sensor networks. We then describe flash based file systems.

3.1   Storage Management in WSN

Recently, storage management in sensor networks has received considerable attention [15,6,11]. Tilak et al. [15] studied the problem of using limited persistent storage of a sensor to store sampled data effectively. They proposed a collaborative storage approach that takes advantage of spatial correlation among the data collected by nearby sensors to significantly reduce the size of the data near the data sources. Their approach provides significant savings in storage space and energy. It also performs load balancing of available storage space. Ganesan et al. [7] proposed a similar data management scheme and motivated use of adjustable resolution scenarios. While these works operate at the level of networked storage, this paper focuses on the local storage (storage organization on flash memory available on a given sensor). We believe that in complement to these works, the proposed scheme provides a lower level service, targeting efficient data organization on flash memory for adapting resolution. The aging policy proposed by Ganesan et al. [7] dictates that resolution of data (across the sensor network) is adapted lower as a function of the age of the data. In our approach we apply a similar aging policy but for data stored on local flash.

3.2   Existing File Systems

We now describe flash based file systems that exist in resource-rich embedded systems (such as PDAs) as well as sensor networks.

YAFFS and JFFS. Existing flash based file systems such as the Journaling Flash File System (JFFS version2) and Yet Another Flash Filing System (YAFFS) are more suitable for resource-rich devices such as PDA. The main reason behind this is their high run-time memory consumption. Therefore, low resources (low computing power, limited memory and storage capability) and different application characteristics demand for design of a new file system for WSNs. We now describe two such efforts in that direction.

Efficient Log-structured Flash file system (ELF). Dai et al. have modified a traditional log structured file system approach to make it adaptable for resource constrained sensor nodes [5]. ELF is an efficient and reliable file system, which also supports general file operations such as create, open, append, delete, truncate, etc. The main idea in ELF is that instead of creating separate log entries for each write-append operation, each log entry is stored in a separate page. This reduces overhead of storing separate log entries even for small write-append operations. In addition they propose using write cache to gather all appends to the same page, to achieve wear leveling. ELF also provides other unique features to the flash file system such as garbage collection, and crash recovery. Like our implementation, ELF makes use of internal EEPROM storage that is available on a sensor node, to store crash recovery data.

Matchbox. Matchbox [8] is a simple filing system designed specifically for sensor nodes. It stores meta-data and files as byte streams in memory. First page of meta-data byte stream is called as *first-meta-data-page*. File data is stored in data pages. There is no fixed root page for the file-system, instead, they maintain a version number for pages of type first-meta-data-page, and choose a page with highest version number as a root page. This helps avoid frequent writes to a single page. Another important data structure they use is a bitmap for free pages. It is constructed during the flash traversal at boot time. Search for a free page starts from the last allocated page, which helps achieve basic wear leveling. Due to sequential reads and appends, matchbox doesn't offer an efficient scheme for handling multi-resolution data. We believe that Matchbox can be extended to incorporate our idea of Resolution-Order Storage organization (ROS) (described in Section 6).

4   Design Goals

In this section, we outline the performance metrics of interest to storage bound applications. The metrics are related to: (1) how efficiently the operations are implemented from an application as well as energy perspective; (2) storage efficiency; and (3) wear and wear leveling. These are described below in more detail.

- **Storage efficiency:** It is important to use the available storage such that even if it gets full, the application should have an option of degrading the quality of existing data and allowing storing new data. Again, this should be achieved with very little or no overhead of extra writes on flash.
- **Flash Access efficiency:** The number of pages that need to be read or written for a given operation depends on how the data is organized and indexed. This metric measures the efficiency in implementing the operations required on the flash.
- **Energy efficiency:** Since the battery life is the most important factor contributing towards the lifetime of a sensor node, additional costs associated to achieve efficient storage management, for example communication overhead, should be avoided.

- **Page Wear:** Since the pages have a limited expected number of writes before they fail, it is desired to reduce the number of writes generated to the flash in the course of application operation. Typical mean writes to failure rating of current generation flash is 100K erase/write cycles.

- **Page Wear leveling:** In addition to reducing wear, it is also desired to load balance the wear (wear-leveling). Wear leveling results in prolonging the time until flash pages start failing, and provide a more graceful failure behavior for the same number of write operations.

Note that these metrics apply to any storage-bound applications. In our case, we specifically target adjustable resolution applications. In this context, we assume that accessing data with variable resolution and deleting data (adjusting resolution) are common operations. Thus, we focus on measuring the metrics above with respect to the implementation of these operations.

5   Adjustable Storage Problem

In this section, we describe the standard *Sequential Logging* (SL) approach and explain how it performs when storage resolution needs to be adjusted. In the SL approach, data is written to the storage device as the sensor node collects it from its transducer or receives it via the network, as shown in figure 1 (a). When resolution adjustment is needed, some portion of the existing data is compressed by discarding lowest precision coefficients and rewriting the remaining coefficients to the flash. Figure 1 (b) shows the flash organization after resolution adjustment phase.

Major disadvantage of this scheme is that the compression or resolution adjustment phase requires additional cost in terms of extra reads and writes on flash. More specifically, in the first pass, the flash is filled sequentially and data is appended to the end of the flash. When the flash is full, data is compressed by discarding least significant coefficients of each data unit, and rewriting the remaining coefficients to the flash. This makes space available for new data.

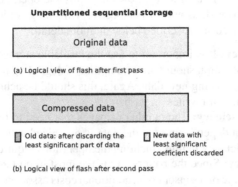

Fig. 1.

6 Proposed Approach

In this section, the proposed Resolution-Order Storage (ROS) organization is described. The discussion is organized into two subsections: the first discusses how the scheme works, while the second discusses wear and wear-leveling issues.

6.1 Resolution-Order Storage (ROS) Organization

The key idea in ROS is to store similar precision coefficients together in the same page of the flash. ROS provides an efficient way to retrieve adjustable multi-resolution data, according to the user's requirements. Further, excess resolution may be eliminated easily by erasing the pages with the unneeded coefficients. In this section we discuss the properties of ROS in more detail.

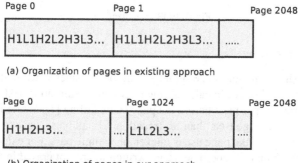

Fig. 2.

Consider an application with two coefficients, High (H) and Low (L). Figure 2 demonstrates ROS comparison between storage organizations of SL and ROS. Figure 2(a) shows how the data is stored sequentially in pages in SL. Figure 2 (b) shows storage organization of ROS where the first half of the flash stores all higher coefficient (H) components and the second half of the flash stores all lower coefficient (L) components.

We assume that the incoming data is in multi-resolution format; more specifically, it can be partitioned into N coefficients, $[C1,C2 \ldots Cn]$, such that the part of data represented by $C1$ is the highest resolution and the most important coefficient while the part expressed by Cn, represents the lowest resolution and the least important coefficient. The examples of such data include wavelet compressed data, and layered compression data used for multimedia formats [14].

ROS facilitates resolution adjustment: we can just directly overwrite a partition that contains data belonging to least significant coefficient with new data. Thus, resolution adjustment has no cost in terms of read and erase/write operations. The logical storage organization of flash for ROS is shown in Figure 3.

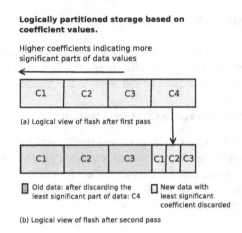

Fig. 3. ROS Organization

6.2   Page Wear leveling

Wear of the flash is a major concern that should be considered while designing storage allocation strategies. A typical flash memory can sustain up to 10,000 to 100,000 erase/write operations per page before the page starts failing [1]. While ROS results in significantly lower average wear than SL (recall, in SL, all pages have to be read, modified and written again), the page use is uneven in a single pass. For example, note that the pages that the initial C_1 coefficients are written to are not erased in the above scenario, while pages where C_4 is initially written to may be erased repeatedly. To avoid early wearing out of individual memory locations, it is important to have a wear leveling mechanism associated with the storage allocation policy for flash.

While wear-leveling within a single scenario is difficult, it is possible to carry out effective wear-leveling across multiple adjustable precision uses of the flash (e.g., every time the base station collects data from the sensor node, data collection starts again with an empty flash). In our implementation, we have used a simple and effective wear leveling policy, which can be applied as pages are allocated. The key idea behind this policy is to track the use count of pages and assign lowest-precision coefficients to the pages that are written least number of times and vice-versa for the highest coefficients. The intuition is that the maximum number of replacements is done in a part of the flash that was assigned to the lowest-precision coefficients and no replacements are done in the part containing highest-precision coefficient values. As we keep on applying this strategy number of times, the flash tends to get better wear leveling. This intuition is shown to hold experimentally in the evaluation section of this paper.

7   Experimental Evaluation

We simulated the proposed file-system organization on TOSSIM simulator. In addition, a preliminary implementation was carried out and tested on MICA2 motes. The proposed approach, Resolution-Order Storage (ROS), can be applied to storing all types of multi-resolution data, but for this paper, without loss of generality, we are using abstract 32 bit data values, which are adjustable. This data can be degraded in terms of precision by dropping least significant bits. For example, this 32-bit quantity can be a floating point number, represented in IEEE 32-bit floating point format, which allows precision loss by erasing lower bits of mantissa. Since many simple quantities measured by sensor networks, such as temperature and pressure generate floating point numbers, this data type may be considered as a representative of such applications. We varied with different breakdowns of the coefficients. For example, a coefficient distribution of [12, 8, 8, 4] means that the original value was divided into four parts of size 12, 8, 8 and 4 bits respectively.

7.1   Implementation Details: Data Structures

Our choice of data structures was heavily influenced by the resource constraints in MICA2 motes. By careful selection of data structures, we were able to implement the proposed approach on a MICA2 mote. The major data structures include two bitmaps, one for storing per page coefficient information and the other for storing age information. The width of each entry in bitmap depends on the total number of coefficients to be stored on flash: the higher the number of coefficients, the more the number of bits required to represent a unique coefficient value. For example, for storing 4 coefficients, each bitmap entry will need 2 bits to represent this information.

7.2   On-Demand Adjustment Scenario

In the first experiment, we assume the on-demand adjustment scenario. Specifically, data is written until the storage is exhausted; at that time, the current lowest coefficient of all data is removed. The new data can be stored on the reclaimed space until it is exhausted, and the process repeats.

To track energy efficiency, we use the total number of read and write operations performed on the flash as a metric to evaluate both the approaches. The operations are in terms of pages, because the flash can only be operated on in units of pages. Figure 4 shows the number of writes required by both the approaches. The number of writes is significantly lower in ROS than it is in the conventional Sequential Logging (SL). To pick a typical point, for the case of three coefficients, [16, 8, 8], total number of writes performed by SL is 6144 pages, which is almost double the number of writes (3200 pages) performed by ROS.

We now describe about the energy consumption required by using EEPROM for storing bitmaps. Since the sizes of bitmaps grow with the increase in total number of coefficients, it is not possible to store these bitmaps on 4K RAM available on MICA2 motes, for an arbitrary number of coefficients. We verified our implementation on MICA2 motes for varying number of coefficients and found that the bitmaps and other variables used in our program can be stored in RAM for only up to 4 coefficients. We addressed this problem by using internal EEPROM of 4K for storing bitmaps when multi-resolution data with more than 4 coefficients needs to be stored. This allows storing bitmaps representing up to 12 coefficient values on EEPROM, leaving RAM to store other program specific variables such as page-buffers, which are used for storing coefficient specific values of incoming data.

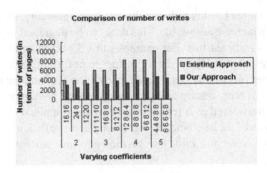

Fig. 4. Write Operations

Access to the EEPROM takes place at the time of compression, which requires reading coefficient and age bitmaps. This time varies based on the total number of coefficients[2]. We argue that this time should be much less than the time required to read all the data $(512KB)$[3] and rewriting the compressed data as in SL. The access time can be further reduced by maintaining a separate cache for storing bitmap values for each coefficient in RAM. The filled buffer can be written to the EEPROM at once during the idle time between sampling schedules of sensor(s). In addition, though wear leveling of EEPROM is not a serious problem, it is automatically taken care by the wear leveling scheme in ROS because writes to the EEPROM constitute of updating bitmap entries of corresponding pages, which ultimately takes place in accordance with the writes to the flash.

Table 2 shows the total number of reads required by both the approaches. For SL, the number of read operations required increases with the number of coefficients. This

Table 2. Comparison of Number of Reads

Number of coefficients	Coefficient distribution	Seq. Logging	ROS
2	16 16	2048	0
	24 8	2048	0
	12 20	2048	0
3	11 11 10	4096	0
	16 8 8	4096	0
	8 12 12	4096	0
4	12 8 8 4	6144	0
	8 8 8 8	6144	0
	6 6 8 12	6144	0
5	4 4 8 8 8	8192	0
	6 6 6 6 8	8192	0

---

[2] The time to read coefficient-bitmap, representing 12 distinct coefficient values, and its associated age-bitmap is approximately 8 to 10 ms (with 1MHz clock).

[3] The time required for reading the whole flash of 512K takes approximately 500ms.

behavior occurs in SL because during the compression activity the whole flash is read in order to discard the lowest resolution coefficient of each data value. In contrast, ROS does not require any reads in the resolution adjustment stage.

Finally, we use a metric called *write-cost* to measure the energy efficiency of the implementation. Write-cost is defined as the ratio of total number of bytes read and moved to another place, to the total number of bytes of new data written to the flash [13]. Ideally, the number of bytes written to the flash should be equal to the number of bytes of new data; however, if data has to be compressed and rewritten, then additional operations are required; this metric measures this overhead. Recall that the new approach simply erases the unneeded resolution components and does not require moving data around. Thus, it achieves ideal performance relative to this metric.

$$WriteCost = \frac{|Bytes\ Read + Bytes\ Written|\ of\ old\ data}{|Bytes\ Written|\ of\ new\ data} \tag{1}$$

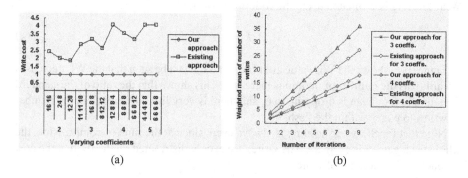

(a)   (b)

Fig. 5. Write-cost and Write distribution study

Figure 5(a) shows that ROS always has a write cost equal to 1.0, as expected. On the other hand, in SL the write cost increases as the number of coefficients increase. This is because, more the number of coefficients, higher the chance of replacing least significant coefficients of old data with new data, and thus higher the overhead for existing approach.

In figure 5(b), we plot values of average wears of all pages in flash, taken after successive iterations of both the approaches, SL and ROS, on flash. Here, one iteration comprises of filling up the whole flash with data, and keep on compressing it every time the flash is full, to make space available for new data. This process continues until the flash contains only the most significant information. By looking at the graph, it can be easily noticed that the wear increases much faster for SL as compared to ROS.

7.3   Wear Leveling

The Figure 6(a) shows that in ROS, the wear leveling of flash improves as the number of iterations increase. During the first iteration, flash memory was statically divided into partitions belonging to different coefficient values. In the successive iterations, highest coefficient value was associated with the pages that were written least number of times and vice versa. To show how the writes are evenly distributed on the flash, we first calculate weighted mean of the number of times a page has been written to the flash. For a given distribution of writes on pages, a1, a2,..., where a1 refers to number of pages written once, a2, number of pages written twice, and so on, the weighted mean W is:

$$W = \frac{\sum_{i=0}^{i=i_{max}} (a_i * i)}{\sum_{i=0}^{i=i_{max}} a_i} \tag{2}$$

Note that the weighted mean indicates the average wear generated to each page in the flash. We then use standard deviation σ in the usual way as:

$$\sigma = \frac{\sqrt{\sum_{i=0}^{i=i_{max}} (a_i * (i - W)^2)}}{\sum_{i=0}^{i=i_{max}} a_i} \tag{3}$$

Wear leveling is estimated using the following metric:

$$WL = \frac{\sigma}{W} \tag{4}$$

Figure 6(a) shows the WL value after every pass. It can be seen that the WL decreases towards 0 as the number of iterations increases. This shows that the ratio of the standard deviation to the mean is dropping to 0, and there is very little difference in the number of writes to pages within the flash.

Note that for SL, wear leveling is always even, since at the time of compression, the whole flash gets written again. The writes are evenly distributed on the flash at the cost of much higher number of writes.

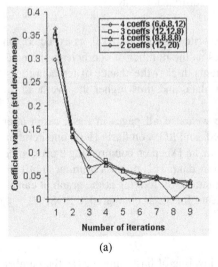

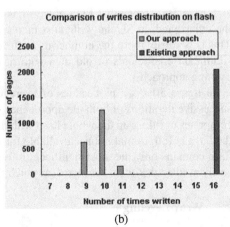

(a)   (b)

Fig. 6. Wear-leveling study of existing and proposed approaches

Figure 6(b) shows a snapshot of write distribution of flash after n iterations, n=4 in the above case. It shows that more than 50% of pages are written 10 times in our approach whereas at the same time, all pages are written 16 times in existing approach.

8   Discussion

In this section we briefly overview the limitations of the proposed approach. The first limitation is that our approach results in higher percentage of reads for random read accesses. This is because, in our approach, each data item is split into multiple components and each of these components are in turn stored on different pages. Therefore, with ROS, reading an entire data item would result in reading N pages, where N is the number of coefficients. In the case of SL, it would simply result in reading of 1 page. However, we argue that random access of data is an uncommon access pattern in an important class of sensor networks applications such as, scientific monitoring applications (described in Section 1). In these applications, the need for random access of data will not arise since the data needs to be read only once before transferring it to the observer.

The second limitation is an artifact of our implementation. For example, we maintain bitmaps to track coefficient values and age values. The size of these bitmaps is linearly proportional to the number of coefficients. Therefore, as the number of coefficients increases, the size of these bitmaps that need to be maintained in RAM increases proportionately. Therefore ROS has higher run time memory consumption that SL. However, with careful selection and management of data structures, we were able to implement ROS on MICA2 nodes with reasonably small memory footprint (for 4 coefficients, run time memory requirement of ROS was only 3667 bytes).

At present, we assume that the importance of data items is inversely proportional to time, i.e. newer data has higher importance. Therefore, in the proposed scheme, we discard least significant coefficient of old data to create space for new data. In future, we would like to extend our scheme to the cases where the importance of data is application specific. In that case, even the least significant parts of important data values of old data should not be deleted. This can be incorporated in our scheme by extending our page allocation policy in a way that would consider importance of data values. In addition, we will extend the Matchbox file-system to incorporate the proposed approach. We will also consider different data structures along with a wide range of data access patterns and their implications on data organization in flash memory.

9   Concluding Remarks

In this paper, we presented a new storage organization (Resolution-Order Storage) that provides better support for adjustable resolution storage applications for sensor networks. In such applications, the resolution of the stored data may need to be adjusted (on-demand, or periodically, for example), to make room for new data when the flash space gets exhausted.

Existing file-systems for sensor networks are optimized to sequential logging. Thus, if adjustable resolution is needed, this requires that the whole data be read, the unneeded coefficients removed, then the remaining data re-written to the flash. In the proposed organization, we group similar data coefficients together on the same page. Thus, adjusting resolution can simply be accomplished by deleting the pages that have unneeded coefficients, greatly enhancing the performance relative to the existing organization. Further, the new organization makes accessing data at different resolutions significantly faster (only the required pages are read). This operation is likely to be common in sensor networks with such data.

References

1. Article on flash memory, http://www.bellevuelinux.org/flash_memory.html.
2. Atmel at45db041b flash specifications: http://www.atmel.com/dyn/resources/prod_documents/doc3443.pdf.
3. Characteristics of nand and nor flash: http://www.linuxdevices.com/articles/at4422361427.html.
4. Wikipedia: Article on flash memory, http://en.wikipedia.org/wiki/flash_memory.
5. DAI, H., NEUFELD, M., AND HAN, R. Elf: An efficient log-structured flash file system for micro sensor nodes. In *Proc. ACM SenSys* (2004).
6. GANESAN, D., ESTRIN, D., AND HEIDEMANN, J. Dimensions: Why do we need a new data handling architecture for sensor networks? *ACM Computer Communication Review 33*, 1 (Jan. 2003).
7. GANESAN, D., GREENSTEIN, B., PERELYUBSKIY, D., ESTRIN, D., AND HEIDEMANN, J. An evaluation of multiresolution storage for sensor networks. In *Proceedings of the First ACM Conference on Embedded Networked Sensor Systems (SenSys)* (2003).
8. GAY, D. Design of matchbox, the simple filing system for motes, 2003. Version 1.0, August 21, 2003, http://www.tinyos.net/tinyos-1.x/doc/matchbox-design.pdf
9. JUANG, P., OKI, H., WANG, Y., MARTONOSI, M., PEH, L. S., AND RUBENSTEIN, D. Energy-efficient computing for wildlife tracking: design tradeoffs and early experiences with zebranet. In *Proceedings of ASPLOS 2002* (2002), ACM Press.
10. KAWAGUCHI, A., NISHIOKA, S., AND MOTODA, H. A flash-memory based file system. In *Proceedings of the 1995 USENIX Annual Technical Conference* (Jan. 1995).
11. RATNASAMY, S., ESTRIN, D., GOVINDAN, R., KARP, B., SHENKER, S., YIN, L., AND YU, F. Data-centric storage in sensornets. In *Proceedings of the First ACM SIGCOMM Workshop on Hot Topics in Networks* (Oct. 2002).
12. A remote ecological micro-sensor network, 2000. (http://www.botany.hawaii.edu/pods/overview.htm)
13. ROSENBLUM, M., AND OUSTERHOUT, J. K. The design and implementation of a log-structured file system. In *Proceedings of the 13th ACM Symposium on Operating Systems Principles* (Feb. 1992).
14. TAUBMAN, AND ZAKHOR. Multi-rate 3-d subband coding of video, 1994.
15. TILAK, S., ABU-GHAZALEH, N., AND HEINZELMAN, W. Collaborative storage management in sensor networks. *International Journal of Ad Hoc and Ubiquitous Computing (IJAHUC) 1*, 1 (2005).
16. TILAK, S., ABU-GHAZALEH, N., AND HEINZELMAN, W. B. Storage management in wireless sensor networks. In *Mobile, Wireless and Sensor Networks* (2005), John Wiley publishers.

17. TILAK, S., MURPHY, A., AND HEINZELMAN, W. Non-uniform information dissemination for sensor networks. In *The 11th IEEE International Conference on Network Protocols (ICNP'03)* (Nov. 2003).

18. VASILESCU, I., KOTAY, K., RUS, D., DUNBABIN, M., AND CORKE, P. Data collection, storage, and retrieval with an underwater sensor network. In *SenSys '05: Proceedings of the 3rd international conference on Embedded networked sensor systems, pp. 154–165, ACM Press, New York, NY, USA, 2005* (2005).

A Classification and Performance Comparison of Mobility Models for Ad Hoc Networks[*]

Emre Atsan and Öznur Özkasap

Koç University, Department of Computer Engineering
Istanbul, Turkey
{eatsan, oozkasap}@ku.edu.tr

Abstract. In mobile ad hoc network research, simulation plays an important role in determining the network characteristics and measuring performance. On the other hand, unrealistic simulation conditions may be misleading, instead of being explanatory. For this reason, constructing simulation models closer to the real circumstances is very significant. Movement behavior of mobile entities is one of the most important concepts for the realistic simulation scenarios in mobile ad hoc networks. In this study, we first provide a survey and a new hybrid classification of existing mobility models in the literature. We implemented the random direction and boundless simulation area models on Scalable Wireless Ad Hoc Network Simulator (SWANS) and conducted simulations of Ad Hoc On-Demand Distance Vector (AODV) protocol for these as well as the random walk and random waypoint models. Our comparative results for the mobility models are discussed on a variety of simulation settings and parameters.

1 Introduction

Wireless and mobile ad hoc networks turn out to be the first option for a wide range of application areas, such as military, environmental, health, home automation and security [1]. Everyday, wireless devices, such as Bluetooth[2]-enabled cell phones, PDAs, and laptops become more popular and indispensable. Wireless Local Area Networks are a necessity even for home users due to the increase in the number of personal wireless electronic devices. In some cases, it turns out to be impossible or too costly to deploy permanent infrastructures (i.e. wireless routers, satellite links, GSM [3] networks) for a wireless network. For several military and civil applications, networking the mobile or static nodes with wireless links in an ad hoc manner can be necessary and/or effective [4].

The idea behind networking wireless nodes in an ad hoc manner dates back to the DARPA packet radio network research [5]. From those days to nowadays, there had been numerous research on mobile ad hoc networks (MANETs) including working groups [6]. Researchers use simulation tools [7,8,9,10] to certify their algorithms and evaluate their performances. These simulation tools offer models that illustrate the

---

[*] This work is supported in part by TUBITAK (The Scientific and Technical Research Council of Turkey) under CAREER Award Grant 104E064.

T. Kunz and S.S. Ravi (Eds.): ADHOC-NOW 2006, LNCS 4104, pp. 444–457, 2006.

characteristics of a mobile ad hoc network. Besides network protocol details, communication channel properties, and spatial properties of simulation area, mobility modeling of nodes plays an important role in the characteristic of a MANET simulation. As argued in [13,14,15], choice of a mobility model may affect the results of simulations significantly. Therefore, a better understanding of the behavior of mobility models and using the appropriate ones give us a chance to achieve realistic conclusions from simulations, which improves the validity of the results.

Several mobility models have been proposed for mobile ad hoc network simulations [11,14,15,16,18,19,24] and various surveys [13,16] have been published about them. In these surveys, authors have offered different classifications of the mobility models by pointing out several aspects. In [13], Camp et.al classified the models based on their *generation* methods (traces and synthetic models) and *social behaviors* (entity and group models). Bettstetter [16] derived a *concept map* for the classification of mobility models. In addition, [16] argues a classification model depending on the *degree of randomness*. Zheng et.al [17] use the same classification criteria founded upon randomness level. Lastly, a recent work [18] makes another classification for mobility models.

In this study, firstly we provide a survey and a new *hybrid* classification of mobility models. We discuss single entity (i.e. a node's movement is independent of other nodes) mobility models with their behavioral analysis studied recently [20,21], and show the resulting node traces for various simulation settings. We implemented the random direction and boundless simulation area models on SWANS (Scalable Wireless Ad Hoc Network Simulator) and conducted simulations of Ad Hoc On-Demand Distance Vector (AODV) protocol for these as well as the random walk and random waypoint models. Our comparative results for the mobility models are discussed on a variety of simulation settings and parameters. The remainder of this paper is organized as follows. Section 2 includes our classification of mobility models, and review of existing single entity mobility models. In Section 3, we present our simulation results of AODV using different mobility models. We aim to convey the importance of choosing the appropriate mobility model to obtain useful conclusions from ad hoc simulations. Section 4 concludes the paper.

2 Mobility Models

There exists different classification criteria proposed by several studies [13,16,17,18] for mobility models in the literature. Table 1 shows a summary of these criteria. We bring these together to form a hybrid classification as depicted in Fig. 1.

2.1 Traces

Traces are pre-determined mobility patterns that are observed in real life. For ad hoc networks, tracing the actual behavior of mobile nodes is a hard process and researchers mostly use synthetic models [16]. Traces hardly let researchers to change simulation parameters, which can be a disadvantage for performance analysis of ad hoc networks.

Table 1. Different criteria proposed for mobility model classifications

Classification Criteria	Classifications
Generation of Mobility	Traces
	Synthetic Models
Social Behavior of	Group Models
Mobile Nodes	Entity Models
Degree of Randomness	Total pseudo-random movement process (Statistical Models)
	Bounded pseudo-random process
	(Constrained Topology based Models)
	Trace-based movement Model
General	Integrated Mobility Models

2.2 Synthetic Mobility Models

a. Random Walk Mobility Model: The mathematical principles of the Random Walk Mobility Model go back to the Einstein's works on Brownian motion [13]. In some sources, this mobility model is called as Brownian Motion Mobility Model [19] or Brownian Walk [18]. In this mobility model, entities (mobile nodes) move randomly choosing a speed and direction from pre-defined ranges ([*minspeed, maxspeed*] and [*0, 2π*], respectively) in constant time intervals (Δt). Border behavior of this model is defined as *bounce-back* [13]. This means that when a mobile node reaches to a simulation boundary, it bounces back to simulation area. We place Random Walk Mobility model into the intersection of statistical and entity concepts in our classification chart (Fig. 1).

Due to its simplicity of implementation, Random Walk is a widely used mobility model in simulations. On the other hand, because of its memoryless behavior (i.e. decision of the next state doesn't depend on previous states) it creates unrealistic mobility patterns with sharp turns and sudden stops. It also exhibits a static behavior when it has a lower velocity range with low continuation. In graphs of Fig.2, it is

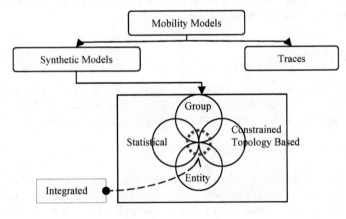

Fig. 1. Hybrid classification structure

shown that lower velocity ranges cause a static behavior. However, increasing the speed range to higher values provides a better spatial distribution. This kind of behavior may lead to an unconnected network if the simulation area is large and number of nodes is small (Fig. 2).

b. Random Waypoint Mobility Model: Random Waypoint Mobility is the most widely used model in simulations by the research community [13]. In this model, a mobile node selects a random position (*x, y*) in the simulation area as a destination point and a velocity (*v*) from a uniformly distributed range [*speedmin, speedmax*]. Then node starts to travel to the chosen destination point with the constant selected speed, *v*. When the node arrives to the destination point, it pauses for a specific time (*pause_time*) defined as a simulation parameter. After this time, node selects a new destination and speed and repeats the process [26]. Random Waypoint Mobility Model is placed into the intersection of Entity and Statistical models in our classification.

Studies on the properties of Random Waypoint Model show that it creates a non-homogenous spatial distribution of nodes if it is used with specific (e.g. delete and bounce-back)

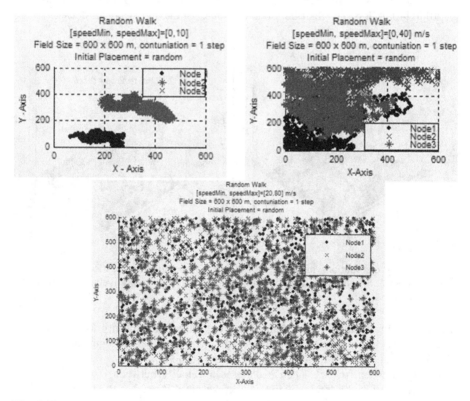

Fig. 2. Traces of Random Walk with speed ranges [0, 10], [0, 40] and [20, 80] m/s in a field size of 600m x 600m. Simulations performed using JIST/SWANS tool's [8] implementation of Random Walk Mobility Model. Continuation value shows the speed and direction change frequency. Continuation = 1 means "change speed and direction in 1 second".

border behavior selections [20]. This means that nodes tend to concentrate on the middle of the field. To avoid this problem, it is suggested to use wrap-around border behavior [16]. Wrap-around border behavior creates a torus-like simulation area by wrapping the edges of area around [24]. Another statistical behavior of Random Waypoint Mobility Model is described in [14]. During their simulations of AODV, Royer et al. discovered a strange characteristic of this model. The average number of neighbors recognized at a given node periodically increases and decreases during the simulation. Frequency of this increase-decrease is relative to the speed of the nodes. This behavior creates a situation in which nodes unite and separate in the center of the simulation area. Authors called this situation *density waves*.

Fig. 3 shows our simulation traces of Random Waypoint model in JIST/SWANS [8] simulation tool. Bounce-back border behavior was implemented in these simulations. Charts were formed by 14000 and 65000 data points, respectively. Each simulation ran for 500 simulation seconds. Our results confirm the previously discussed [20] behaviors of Random Waypoint mobility model. In Fig. 3, it can be observed that nodes generally tend to concentrate in the middle of the simulation area. Furthermore, when the speed range and hence the average speed of mobile nodes is increased from [0, 20] to [40, 80], nodes' distribution in the center of the area becomes denser.

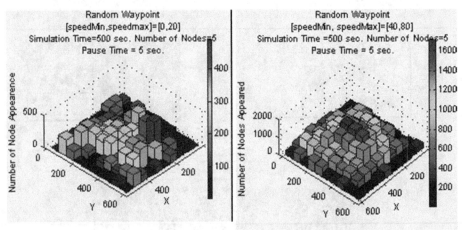

Fig. 3. Density distributions of 5 nodes in 600 x 600 simulation field (Random Waypoint Mobility Model). Speed ranges are [0, 40] and [40, 80] m/sec, respectively. Pause time is 5 simulation seconds.

c. Random Direction Mobility Model: Random Direction Mobility Model was proposed in [14]. It is designed to overcome the *density waves* problem faced in Random Waypoint model. In this mobility model, a node selects a random direction uniformly between [0, 2π]. Then it chooses a velocity between [*speedmin, speedmax*], again uniformly. Node continues to travel until it reaches the boundary of the simulation area. Once it reaches the boundary, it rests for a given pause time and selects a new direction between [0, π]. The direction is limited to π because the node is already on the boundary and cannot exit from the simulation area. We place Random Direction Mobility model into the intersection of statistical and entity models in our classification.

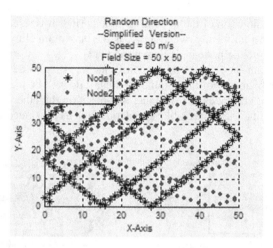

Fig. 4. Movement trace of two nodes. Simplified version of Random Walk Mobility Model used in 50 x 50 simulation area.

There exist several variations of Random Direction Model in the literature. "Modified Random Direction Mobility Model" proposed in [14] doesn't force the mobile node to travel to the boundary for changing speed and direction. A node can select its new direction and speed anywhere on its path. A simplified version of Random Direction model was used by Haas and Pearlman in [12]. In this version, all the nodes have a common constant speed v and move with an initial direction (φ) selected from a uniformly distributed range [0, 2π]. When a node reaches the boundary, it reflects back into the simulation area with an angle (-φ), if it is on a vertical edge; (π- φ), if it is on a horizontal edge. A simulation trace of simplified version for two nodes is shown in Fig. 4.

d. Boundless Simulation Area Mobility Model: In Boundless Simulation Area Mobility Model, different from Random Direction, Waypoint and Walk, there exists a relationship between the previous state of the node and the current state [24]. In other words, the new direction and speed values are dependent on the previous direction and speed values of the node, respectively. In every Δt time period, new values of the direction and speed are assigned according to the given formulas below:

$$v(t +\Delta t) = \min [\max(v(t)+ \Delta v,0), V_{max}]$$
$$\varphi(t +\Delta t) = \varphi(t) + \Delta\varphi$$
$$x(t +\Delta t) = x(t) + v(t) * \cos [\varphi(t)]$$
$$y(t +\Delta t) = y(t) + v(t) * \sin [\varphi(t)]$$

V_{max}: maximum defined velocity of the simulation.
Δv: change in the velocity of the node in Δt time period. It is selected from [-$a_{MAX}*\Delta t$, $a_{MAX}*\Delta t$] range uniformly, where a_{MAX} is the maximum defined acceleration (per Δt^2) parameter for this simulation.

$\Delta\varphi$: change in the direction of the node in Δt time period. It is selected from the range $[-\omega_{MAX}*\Delta t, \omega_{MAX}*\Delta t]$ uniformly, where ω_{MAX} is the maximum allowed angular change in the direction of mobile node in Δt time period [13].

Another important property of this mobility model is its border behavior. When a node reaches to a boundary of the simulation area, it doesn't bounce back into simulation area; instead it continues to travel and reappears on the other side of the simulation area. This is similar to a torus-like simulation area as illustrated in Fig.5.

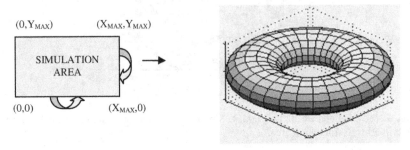

Fig. 5. Construction of a torus-like Boundless Simulation Area from a 2D simulation area

Boundless Simulation Area Mobility Model creates more realistic node movement traces than Random Walk, Waypoint models, because it depends on the previous velocity and direction of the node. Sharp turns and sudden stops can be eliminated in Boundless Simulation Area model by using this state-dependency of the nodes. We place Boundless Simulation Area Mobility Model into the intersection of statistical and entity models in our classification.

In Fig. 6, our simulation results depicting the movement behavior of a node for Boundless Simulation Area Mobility Model are given. Long lines cutting the simulation

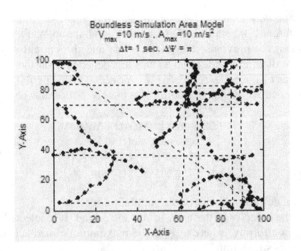

Fig. 6. Simulation trace of Boundless Simulation Area Mobility Model

area from one side to another indicates the reappearance of a node on the other side of the simulation area.

e. City Section Mobility Model: City section mobility model was first proposed by Davies [11]. It uses a simulation area that represents street network of a city. A node is not allowed to choose any point on the graph. It should select a point that is on some street network of the city. After selecting this destination point, the travel path is determined by an algorithm, which calculates the shortest travel time between source and destination points. After reaching to the destination point, node waits there for a defined pause time and randomly chooses another destination on the street network and repeats the process. In addition to this behavior, in City Section Mobility model, a mobile node should obey some pre-defined driving characteristics, such as "speed limit" and "minimum distance allowed between any two nodes". These rules and the use of pre-defined paths (street network) make the movement behavior of the mobile node similar to a vehicle movement in a city central [11]. We place City Section Mobility Model into the intersection of constrained topology based and entity models in our classification.

f. Smooth Random Mobility Model: Smooth Random Mobility Model was originally proposed by Bettstetter [16]. The idea behind the design of this mobility model is to "make the movement of users (pedestrians and vehicles) more smooth and realistic than previously known random models." [16]. Two stochastic processes are used for direction and speed control in which the new values of speed and direction are correlated with the previous ones. This correlation is what makes the movement smoother, because it omits the unrealistic sharp turns and sudden stops in simple random movement models [16]. Another important point that Bettstetter added to his mobility model is the preferred speeds concept. Users select their new target speed from a range $[0, V_{max}]$ uniformly, except V_{pref} has a higher probability than other speed values. By this way, some typical speed patterns can be created, which resembles the movement behaviors of the vehicles and pedestrians. Bettstetter extends his model by adding two new concepts in it. He assumes that in reality generally speed and direction change are not independent from each other. For this purpose, he offers a correlation between speed change and direction change: *"Stop-turn-and-go"* behavior and *"slowdown of turning nodes"* notions. Stop-turn-and-go behavior describes a mathematical model of the reality: a stop of a vehicle often followed by a direction change. So whenever a node comes to a stop $(v=0)$, then node choose a new target direction to move. The selection of the node's new direction is decided by the direction behavior model defined in the same paper [16]. Slowdown of the turning nodes concept was inspired by the physics laws. This means that if a vehicle stops at a given time and point, than it should first slowdown in a given period of time. This concept gives the Smooth Random Mobility Model a more realistic movement behavior than Random Walk and Waypoint models.

Smooth Random Mobility model can be summarized as a (statistical) random mobility model with better movement dynamics of the nodes. It was proposed as a cooperation of simple random mobility models and realistic and complicated mobility models of transportation research [16]. We place Smooth Random Mobility Model into the intersection of statistical and entity models in our classification.

g. Obstacle Mobility Model: Obstacle Mobility Model [15] is designed to model the mobility patterns that may resemble the real world paths on real world topographies. Model provides obstacles inside the simulation area. These obstacles both restrict the movements of the mobile nodes and the wireless transmission range. Nodes are supposed to follow the given provided pathways; instead of the random trajectories. In this model, the simulation area and the pathways are constructed by using the Voronoi diagram of the given obstacles. Obstacle and pathway construction using Voronoi diagrams are described in [15]. Model uses a reachability matrix to represent the transmission possibility between pairs of nodes. Obstacles are assumed opaque and totally block the signal propagation. After defining the pathways from obstacles (using Voronoi graphs) and calculating reachability matrix, each time a node transmits a packet; the model determines whether the intended recipient of the packet is within the reachability matrix of the transmitting node. If so, reception of packet is completed. We place Obstacle Mobility model into the intersection of constrained topology based and entity models in our classification.

3   Simulation Study and Analysis Results

Our simulations have been performed using the SWANS Network Simulator [8]. SWANS is built on Java-based simulation framework JIST (Java in Simulation Time). We investigate the effects of different mobility models on ad hoc network performance using AODV [22] routing protocol as the route establishment choice. We use the available AODV implementation in SWANS, details of which can be found in [25].

For our comparisons of AODV performance with different mobility models, we used Random Walk, Random Waypoint, Random Direction and Boundless Simulation Area Mobility models. First two models are already available in SWANS, and we contributed Random Direction and Boundless Simulation Area models to the simulator. For Random Direction Mobility Model, we implemented the simplified version (constant velocity and reflection) used by Haas and Perlman in [12]. For Boundless Simulation Model, we used the Haas' proposal [24] as a reference for our implementation. Parameters of our simulation settings are listed in Table 2.

In Figures 7 to 9, we present our simulation results. For Figures 7, 8.a and 9.a, the size of the simulation field (1000m x 1000m), pause times of the mobility models, minimum and maximum velocity ranges are same and constant. For Figures 8.b and 9.b,

Table 2. Simulation Default values in SWANS

Routing Protocol	AODV	*Radio Trans. Strength*	15 dBm
Transport Protocol	UDP [23]	*Radio Reception Sensitivity*	-91 dBm
Fading	None	*Radio Reception Threshold*	-81 dBm
Path-loss	Freespace	*Packet Loss*	None
Radio Frequency	2.4 GHz	*Placement (Node Distribution)*	Random
Radio Trans. Bandwidth	1 Mb/s	*Packet send rate per minute for node*	1.0

Table 3. Simulation variables that are kept constant throughout the AODV simulations

Pause time (for related models)	10 sec.	Maximum acceleration (for Boundless S.)	10 m/s^2
[speedmin, speedmax]	[20,80] m/s.	Max. allowed angular change (for Boundless S.)	Π /2

number of nodes in the simulation area is constant (50 nodes) and randomly distributed during the initialization. Some of the simulation variables that are constant in all of the simulations are given in Table 3. In these simulations, we run 10 different simulations for every mobility model and calculated the average of their results.

3.1 Discussion of Simulation Results

Fig.7 illustrates that increasing the number of nodes in the constant sized simulation area causes dissimilar increase in the message activity per node for each mobility model. For the whole simulation, Random Direction mobility model creates a higher message activity. On the other hand, Boundless Simulation Area creates the lowest. We know that message activity is dominated by the sent RREQ messages for AODV implementation which is also discussed in [25]. At this point we investigate the big difference in message activity for different mobility models.

First of all, this difference can result from the border characteristic of the Boundless Simulation Area model. For example, if two nodes are at the two end of the simulation area, they are not within each other's transmission range in Random Direction, Walk and Waypoint models. On the other hand, because of the wrapping of simulation borders,

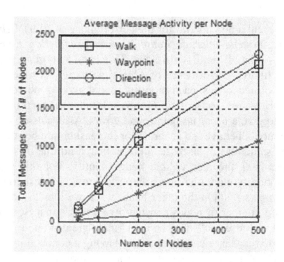

Fig. 7. Average message sent per node for different values of number of nodes in the simulation (all other variables held constant). "Total Messages Sent" is the number of all control messages (RREQ, RREP, RERR, and HELLO) sent by the nodes in AODV algorithm. Data packets are not included in the "Total Messages Sent".

these two nodes became neighbors in Boundless Simulation Area. So, for the same size of the simulation area, nodes near the boundary have two times line-of-sight if they are moving based on the Boundless Simulation Area mobility model. If the line-of-sight (assume that it is equivalent to transmission range in this special case) of the nodes at the boundary increases, than the number of RREQ messages sent decreases. This is because if the destination point is far or not accessible, then sender node starts to transmit large number of RREQ messages in AODV routing. Increasing the line-of-sight for nodes automatically decreases the probability of inaccessible destination nodes.

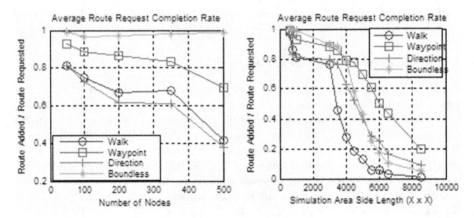

Fig. 8. Average Route Request Completion Rate (for varying number of nodes with constant field size) & (for varying simulation area sizes with constant number of mobile nodes in simulations) gives the rate of routes added (found) to requested routes

In Fig.8, we present a new metric for the performance of AODV simulations under different mobility models. *Average Route Request Completion Rate* is the ratio of completed routes to the all requested routes. This is an important performance criterion for the route finding performance of AODV protocol. As we can easily observe, for small number of node sizes, each mobility model gives good completion rates for requested routes. However, for larger node sizes, Random Direction and Walk mobility models starts to give a lot of no-path found errors. As discussed before, Boundless Simulation Area model behaves so accurate for increasing number of nodes. This may be because of the same reason discussed before: increased line-of-sight at the borders automatically decreases the inaccessible node quantity. For this reason, Boundless Simulation area mobility model has a nearly perfect route request completion ratio for small simulation area sizes. On the other hand, when we increase the simulation area size, while keeping number of nodes constant, we observe that Boundless Simulation Area model starts to loose its "wrapped area" and "greater line-of-sight" advantage. This means that for very large simulation area, having a connection with the nodes on the other boundary of the area doesn't have a big impact on the inaccessible node number. This is because, when the area grows, the probability of nodes' being near the boundary decreases. Therefore, after some very large field sizes, Boundless Simulation Area is expected to behave differently. As we can easily observe for small sizes of simulation area, all of the mobility models give good (nearly perfect) completion ratios. This is expected because of the high node density in the simulation area for small area sizes.

Fig.8 indicates that Random Waypoint model behaves much better than other two random models (random direction and random walk) for varying number of simulation area sizes. This is because of the interesting spatial behavior of the Random Waypoint mobility model, as discussed in Section 2 and originally in [20]. Because of this mathematically proved, behavior of Random Waypoint mobility model, nodes tend to concentrate on the middle of the simulation area, which causes a decrease in the average number of inaccessible nodes for random waypoint model. As a result, Random Waypoint mobility model has a tidier behavior for increasing simulation size, which shows us that Random Waypoint model can be better solution for the simulations that assumes the simulation area size does not a factor for resulting performance values.

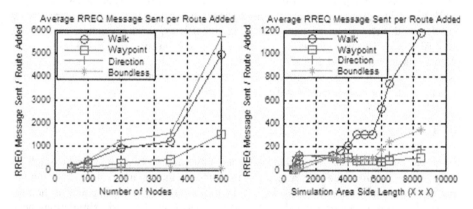

Fig. 9. Average RREQ message sent per route found (for varying number of nodes with constant field size) & (for varying simulation area sizes with constant number of mobile nodes in simulations) is a performance metric that indicates the redundant message creation on the network for a route added. RREQ message is a control message used in AODV routing protocol. Basically, it is sent from the initiator of the route request to its neighbors. Then every neighbor continues this process until the destination is found.

In Fig.9, we present another new metric for the performance of AODV routing: *Average RREQ Message Sent per Route Added*. RREQ is a control message used in AODV. Basically, it is sent from the initiator of the route request to its neighbors. Then every neighbor continues this process until the destination found.   If a source node cannot get a RREP message for a given period of time, it retransmits the RREQ message to the network. Average RREQ message sent per route found (or added) is a measure of the efficient use of network resources for finding a route to the destination node. If too much RREQ message is sent to the network to find a route to the destination, it would not be a very efficient route finding. So, this parameter is an important performance concept for AODV routing protocol which leads to different performance results for different mobility models. From Fig.9, it can be easily observed that Boundless Simulation Area model gives nearly perfect simulation performance for increasing number of nodes. Random Waypoint model has an average performance with high stability in this category. On the other hand, Random Walk and Random Direction mobility models give poor efficiency ratios for large node sizes.

While increasing field size and keeping number of nodes constant in the simulation area, Random Waypoint and Boundless Simulation Area mobility models give efficient performance values for AODV, in contrast to the Random Direction and Random Walk models. However, for very large field sizes, the efficiency of the nodes using Boundless Simulation Area model starts to fall significantly. This fall in the efficiency of AODV routing is again caused by losing "wrapped area" and "greater line-of-sight" advantage of this model for very large field sizes as discussed before. On the other hand, Random Waypoint Model keeps its efficiency even for 8500m x 8500m field. This is mainly due to the nodes' tendency to concentrate on the center of simulation area. This spatial distribution characteristic of nodes was discussed in [20].

These results show us that selection of the mobility model for simulations of ad hoc networks can hugely affect the results. For example, if a researcher wants to use AODV routing protocol for his wireless sensor network structure that nodes move similar to the Brownian motion of molecular particles, for large number of nodes, AODV can lead to poor network resource efficiencies in real life (based on the Random Walk mobility model results in Fig.9). However, this doesn't mean that AODV is an inefficient routing protocol for all cases. The important thing here is to find the most realistic mobility model for the desired case.

4 Conclusions

Everyday mobile ad hoc networking becomes more and more popular and extends its range of application areas. As a result of this, understanding and realistically simulating the behavior of mobile networks is very important for the researchers in the progress of designing better algorithms for ad hoc networks. Realistic simulations need realistic mobility models and better simulation tools. Numerous advantages of using simulation tools for performance analysis of networks [15] put the mobility modeling and its statistical analysis into an important level for the future.

In this paper, we first provide a survey and a new hybrid classification of existing mobility models in the literature. It has been also shown that when the mobility behaviors of nodes change in an ad hoc network, the performance of the network can be vastly affected from this. So that choosing a realistic mobility model for network simulations plays an important role on the validity of the simulation results. For an ad hoc network that propagates data based on AODV routing protocol, when different mobility models applied, so different performance results appear with varying number of nodes or simulation area size. Based on our simulations, we propose that for all of the performance metrics simulated, Random Waypoint Mobility model has not the best performance; however it is the most consistent one for varying simulation values. Using Boundless Simulation Area Mobility Model, for some performance metrics AODV routing protocol performed remarkably well. But, varying factors in the simulation environment enormously affected its performance values in simulations. Two other mobility models, Random Direction and Random Walk, showed the worst performances for AODV routing in simulations. We also demonstrated that Random Walk Mobility is the most affected model from the increase in the simulation size for the performance metrics investigated in this study.

References

1. I. F. Akyildiz, W. Su, Y. Sankarasubramaniam, E. Cayirci. Wireless sensor networks: a survey. *Computer Networks*, Vol. 38, pp. 393 – 422, March 2002.
2. The Official Bluetooth Wireless Info website. http://www.bluetooth.com/.
3. Global System for Mobile Communications (GSM) official website. http://www. Gsm world. com /.
4. M. Frodigh, P. Johansson, P. Larsson. Wireless ad hoc networking—The art of networking without a network. *Ericsson Review No. 4*, 2000.
5. J. Jubin, J. D. Tornow. The DARPA Packet Radio Network Protocols. *In Proceedings of the IEEE*, Vol. 75, 1, pp. 21-32, January, 1987.
6. IETF MANET Group. http://www.ietf.org/html.charters/manet-charter.html.
7. QualNet simulation environment. http://www.scalable-networks.com
8. JIST/ SWANS (Java in Simulation Time / Scalable Wireless Ad hoc Network Simulator) simulation environment. Available at http://jist.ece.cornell.edu/.
9. The Network Simulator - ns-2. Available at http://www.isi.edu/nsnam/ns/.
10. L. Bajaj, M. Takai, R. Ahuja, K. Tang, R. Bagrodia, and M. Gerla. GlomoSim: A Scalable Network Simulation Environment. Technical Report, #990027, UCLA, 1997.
11. V. Davies. Evaluating mobility models within an ad hoc network. Master's thesis, Colorado School of Mines, 2000.
12. Z. J. Haas and M. R. Pearlman. The performance of query control schemes for the Zone Routing Protocol. In Proceedings of *ACM SIGCOMM*, Vancouver, Canada, Sept. 1998.
13. T. Camp, J. Boleng, and V. Davies. A survey of mobility models for ad hoc network research. *Wireless Comm. and Mobile Computing (WCMC)*, 2(5):483–502, 2002.
14. E. Royer, P.M. Melliar-Smith, and L. Moser. An analysis of the optimum node density for ad hoc mobile networks. In *Proc.IEEE International Conference on Communications*, 2001.
15. A. Jardosh, E. Royer, K. Almeroth, and S. Suri. Towards realistic mobility models for mobile ad hoc networks. In *MobiCom'03*, September 2003.
16. C. Bettstetter. Smooth is better than sharp: A random mobility model for simulation of wireless networks. In *Proceedings of MSWiM'01. ACM*, July 2001.
17. Q.Zheng, X. Hong, S. Ray. Recent Advances in Mobility Modeling for Mobile Ad Hoc Network Research. In *ACMSE'04*, April 2004.
18. I. Stepanov, P. J. Marron, K. Rothermal. Mobility Modeling of Outdoor Scenarios for MANETs. In Proceedings of *ANSS'05*, 2005.
19. M. Musolesi, S. Hailes, C. Mascolo. An Ad Hoc Mobility Model Founded on Social Network Theory. In *MSWiM'04*, October, 2004.
20. C. Bettstetter and C. Wagner, The Spatial Node Distribution of the Random Waypoint Model, *Proceedings of the 1st German Workshop Mobile Ad Hoc Networks (WMAN)*, 2002.
21. C. Bettstetter, H. Hartenstein, and X. Perez-Costa, Stochastic Properties of the Random Waypoint Mobility Model, *Wireless Networks* 10, 555–567, 2004.
22. C. E. Perkins, E. M. Royer. The Ad Hoc On-Demand Distance Vector Protocol. In C.E. Perkins, editor, *Ad hoc Networking*, pp. 173-219. Addison-Wesley, 2000.
23. J. Postel. RFC 768: User Datagram Protocol (UDP). http://www.ietf.org/rfc/rfc768.txt.
24. Z. J. Haas. A new routing protocol for reconfigurable wireless networks. *Proc.IEEE International Conference on Universal Personal Communications (ICUPC)*, Oct. 1997.
25. C. Lin. Documentation: AODV Routing Implementation for Scalable Wireless Ad-Hoc Network Simulation (SWANS). JIST/SWANS web site. http://jist.ece.cornell.edu/.
26. J. Broch, D. A. Maltz, D. Johnson, Y.-C. Hu, J. Jetcheva. A Performance Comparison of Multi-Hop Wireless Ad Hoc Network Routing Protocols. In *Proceedings of the 4th Annual ACM/IEEE International Conference on Mobile Computing and Networking (MobiCom)*, pages 85–97, Dallas, Texas, October 1998.

Power-Aware Rate Control for Mobile Multimedia Communications

Hye-Soo Kim, Dinh Trieu Duong, Jae-Yun Jeong,
Byoung-Kyu Dan, and Sung-Jea Ko

Department of Electronics Engineering, Korea University,
Anam-Dong Sungbuk-Ku, Seoul, Korea
Tel.: +82-2-3290-3228
{hyesoo, duongdt, jyjeong, bkdan, sjko}@dali.korea.ac.kr

Abstract. Consumers increasingly demand high quality of service (QoS) for multimedia applications. Rate control scheme is one of the major methods that provide high video quality for mobile multimedia devices over wireless networks. This paper presents an efficient rate-control scheme which estimates the transmission power level of mobile devices, then adaptively adjusts the encoding bit rate according to the estimated power level to minimize overall distortion over sequences of video frames. Experimental results show that the proposed method can efficiently enhance the video quality and provide better peak signal to noise ratio (PSNR) performance than existing TMN8 rate control method.

1 Introduction

With the advantages in small size, utility, and flexible mobility, mobile multimedia devices are prevailing in the market with numerous types such as personal digital assistant (PDA), wideband code division multiple access (WCDMA) devices, and digital multimedia broadcasting (DMB) devices.

This prevalence of mobile multimedia devices has raised the increasing demand in video data that now become one of the most important requirements for the next generation of wireless networks [1]. However, video transmission in these mobile devices has faced many challenges, such as high error rate, bandwidth constraint, time varying channel, especially the strict limitation in transmission power. In addition, maintaining a good video quality and minimizing average power consumption at mobile multimedia devices are in conflict with each other [2]. This means if we want a good video quality, we have to keep the transmission power at a certain high level to avoid negative effects from mobile and wireless environment. Unfortunately, this high level in turn results in the increase of power interference among the users.

Therefore, the problem here is how to guarantee the high QoS for video delivery over wireless networks while the transmission power level of mobile devices is low. To solve this problem, we should develop an efficient rate control scheme, in which the transmission power level of mobile devices is measured, and then the encoding bit rate is adaptively adjusted according to the measured power

T. Kunz and S.S. Ravi (Eds.): ADHOC-NOW 2006, LNCS 4104, pp. 458–471, 2006.

level to minimize the overall distortion over sequences of video frames. These issues are main focus in this paper.

Many rate control schemes have been proposed in [3]-[8]. In general, there are two groups in the rate control techniques. The one is channel based rate control schemes and the other is video coding based rate control schemes. In the first techniques, rate control schemes are performed basing on the supervision and estimation of wireless channel parameters such as channel model, channel rates, or receiving power levels [3]-[4]. In the second one, rate control schemes based on video coding and processing, in which the number of bits and distortion for each image block encoding are controlled by the quantization parameter of the block so that the encoder produces bits at transmission bandwidth and the overall distortion is minimized [5]-[8]. However, applying the transmission power levels for rate control scheme has not been taken into account.

This paper presents an efficient method for rate control scheme, in which both approaches aforementioned are combined. In this proposed rate control method, we first apply two-state Markov modeling for wireless channel. Then, at the streaming server, according to the feedback information of the receiving power level from mobile devices, we estimate key parameters for the channel such as the packet loss ratio (PLR), the channel state and rate, and the frame target bits for current video frame. Based on the estimated channel results, an efficient rate control algorithm is applied to sequence of video frames to minimize the average distortion over an entire sequence as well as variations in distortion between frames. This algorithm uses a non-iterative method with a low computational complexity and low encoding time delay, it is suitable for real-time video communication over wireless networks of mobile multimedia devices [10]. Experiments show that the proposed method not only provides better PSNR performance, but is also less sensitive to the time varying wireless channel condition than existing rate control schemes.

The paper is organized as follows. In section 2, TMN8, which is the conventional rate control algorithm for H.263 standard, is briefly introduced. In section 3, we describe a wireless channel model for video transmission. The proposed frame-layer rate control scheme is presented in section 4. In section 5, presents and discusses the experimental results. Finally, our conclusion is given in section 6.

2   Review on Conventional TMN8 Rate Control

In H.263, the current video frame to be encoded is decomposed into MB of 16×16 pixels per block, and the pixel values for each of the four 8×8 blocks in a MB are transformed into a set of coefficients using the discrete cosine transform (DCT). These coefficients are then quantized and encoded with some type of variable-length coding. The number of bits and distortion for a given MB depend on the MB's quantization parameter used for quantizing the transformed coefficients. For example, in the test model TMN8 [6] of H.263 standard [9], the quantization parameter is denoted by QP whose value corresponds to half the quantization step size. The TMN8 rate control uses a frame-layer rate control to select a target number of bits for the current frame and a MB-layer rate control to select the values of the quantization step-sizes for the MBs. In this paper, the following definitions are used:

B : target number of bits for a frame;
R : target rate in bits per second;
F : frame rate in frames per second;
W : number of bits in the encoder buffer;
M : some maximum value indicating buffer fullness, by default, set R/F;
W_{prev} : previous number of bits in the buffer;
B' : actual number of bits used of encoding the previous frame.

In the frame-layer rate control, the target bit counts for the current frame is estimated as

$$B = \frac{R}{F} - \Delta \tag{1}$$

where Δ is defined below, which is given by TMN8:

$$\Delta = \begin{cases} \frac{W}{F}, & W > Z \cdot M \\ W - Z \cdot M, & \text{otherwise,} \end{cases} \tag{2}$$

$$W = \max\left(W_{prev} + B' - \frac{R}{F}, 0\right) \tag{3}$$

where Z is 0.1 by default. The frame target varies with the nature of the video frame, the buffer fullness, and the channel throughput. To achieve low delay, the algorithm tries to maintain the buffer fullness at about 10% of the maximum M. If W is larger than 10% of the maximum M, the frame target B is slightly decreased. Otherwise, B is slightly increased.

The macroblock-layer rate control selects the values of the quantization step-sizes for all the macroblocks in the frame so that the sum of the bits used in all macroblocks is close to the frame target B in (1). The optimized quantization step size Q_i^* for i^{th} macroblock in a frame can be determined by

$$Q_i^* = \sqrt{\frac{AK}{\beta_i - AN_iC} \frac{\sigma_i}{\alpha_i} \sum_{k=1}^{N} \alpha_i \sigma_i}, \quad i = 1, ..., N, \tag{4}$$

where

K : constant related to the input distribution model;
A : number of pixels in a macroblock;
N_i : number of macroblocks that remain to be encoded in the frame;
σ_i : standard deviation of the ith macroblock;
α_i : distortion weight of the ith macroblock;
C : average rate (in bits per pixel) of encoding the motion vectors and the coder's header and syntax for the frame;
β_i : number of bits left for encoding the frame, where $\beta_1 = B$ at the initialization stage.

3   Wireless Channel Model

In this section, we describe the effects of the wireless channel modeling on video quality. Many different models can be used to model the wireless channel. In this paper, we use a two-state Markov model. This model as described in [8], uses a simplified Gilbert channel at the packet level, which captures the bursty nature of packet errors.

The model has two states, a good state (s_0) and a bad state (s_1). A packet is transmitted correctly when the channel is in the good state and errors occur when the channel is in the bad state as shown in Fig. 1. In Fig. 1, p_{00}, p_{01},

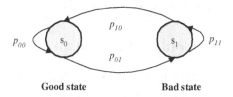

Good state   **Bad state**

Fig. 1. Two-state Markov channel model

p_{10}, and p_{11}, are the state transition probabilities. The transitions between these states occur at each packet instant. The transition probabilities are acquired base on the receiving power levels (P_L) measured in our experimental system. The transition probability matrix for the two-state Markov model can be set up as

$$\mathbf{P} = \begin{bmatrix} p_{00} & p_{01} \\ p_{10} & p_{11} \end{bmatrix} = \begin{bmatrix} 1 - p_{01} & p_{01} \\ p_{10} & 1 - p_{10} \end{bmatrix}. \tag{5}$$

We define the state probability $\pi_n(k|S(t))$ as the probability that the channel is in state s_n at time k given the channel state observation $S(t)$. Base on these values and predefined value of receiving power threshold ($P_{threshold}$), the channel states $S(t)$ at time t are observed before the next step of channel rate estimation.

A vector of state probabilities can be written as (6)

$$\pi(k|S(t)) = [\pi_0(k|S(t)), \ \pi_1(k|S(t))]. \tag{6}$$

The initial state probability $\pi(k|S(t))$ at time t can be set up as

$$\forall n \in \{0,1\}$$
$$\pi(t|S(t)) = \begin{cases} 1, & if \ S(t) = s_n, \\ 0, & otherwise. \end{cases} \tag{7}$$

In the Markov model, the vector of state probabilities $\pi(k|S(t))$ at time k can be derived from the state probabilities $\pi(k-1|S(t))$ at the previous time slot and the transition probability matrix $\mathbf{P}$ in (5) as

$$\pi(k|S(t)) = \pi(k-1|S(t)) \cdot \mathbf{P}. \tag{8}$$

The vector of state probabilities at time k can be obtained by using (8) recursively as

$$\pi(k|S(t)) = \pi(t|S(t)) \cdot \mathbf{P}^{k-t}. \tag{9}$$

In our channel model, packets are transmitted correctly when the channel is in state s_0, while errors occur when the channel is in state s_1. Therefore, $\pi_0(k|S(t))$ is the probability of correct transmission at time k. Let $C(k)$ be the future channel transmission rate where $k > t$. The expected channel rate $\widehat{R}$ given the observation of channel state $S(t)$ can be calculated as

$$\widehat{R} = E\left[C(k)|S(t)\right] = R_{\max} \cdot \pi_0(k|S(t)). \tag{10}$$

where $R_{\max}$ is the maximum channel rates in CDMA or 3G cellular networks.

4   Proposed Rate Control Scheme

The proposed rate control method uses the wireless channel model to estimate the current channel rate, and then adjusts the frame target bits for current frame according to the estimated channel rate. Next, the obtained target bit budget is optimally allocated to each frame to minimize the average distortion over an entire sequence as well as variations in distortion between frames.

Fig. 2 shows overall system block diagram of our experimental system for the proposed rate control method. In this system, the transmission power level of the client is fed into the streaming server through the real-time control protocol (RTCP) feedback. The streaming server first estimates the channel rate, or the available bandwidth using the transmission power level received from the client. Then, the target bit rate for each encoding frame is determined by using the rate control method based on the rate-distortion (R-D) model [3].

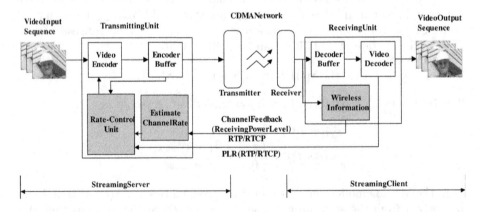

Fig. 2. Block diagram of the proposed rate control system

4.1   Power-Aware Rate Control Algorithm

According to the buffer fullness, the estimated channel bit rate and the channel state feedback information, we can adaptively adjust the frame target bits using the equation in (1). Fig. 3 illustrates the flow chart of the power-aware rate control algorithm. More details of the proposed algorithm is as follows:

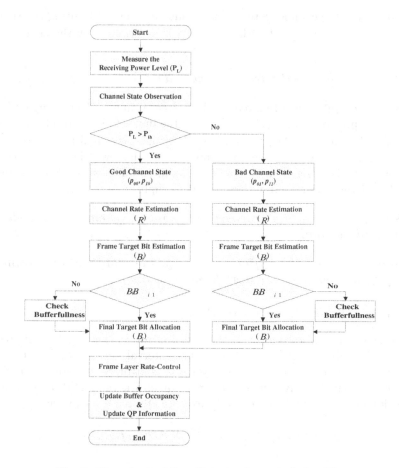

Fig. 3. Flow chart of the efficient rate control algorithm

- *Step 1: Measure the transmission power level and determine the current channel state:*

Before, we start encoding each frame, we first measure power level of the signals, such as receive signal strength indicator (RSSI) and the receiving power level (Rx-power), which have been received from the client through the real-time control protocol (RTCP) (see Fig. 2). The power threshold parameter (P_{th}) is then specified to determine the current channel state. If the receiving power level is greater than the P_{th}, the channel is in the good state at probability of p_{10} and p_{00}, and in the opposite case, the channel is in the bad state at probability of p_{01} and p_{11}.

- **Step 2: Estimate the channel rate and frame target bits for current frame:**

In this step, the channel rate, $\widehat{R}$, is estimated using (10) during the time T. The frame target bits for the ith frame, $\widehat{B}_i$, is determined by the channel rate as shown in (1).

- **Step 3: Compare the frame target bits:**

The frame target bits estimated for the current frame i $(\widehat{B}_i)$ is compared with that for the previous frame (B_{i-1}). Base on the buffer fullness W of the current frame (see in equation (1)), $\widehat{B}_i$ is adjusted again to the final frame target bits B_i to keep the buffer from overflow while remain the low-delay properties. For example, if W is larger than 10% of the maximum buffer size, the frame target bits $\widehat{B}_i$ is slightly decreased. Otherwise, $\widehat{B}_i$ is slightly increased.

- **Step 4: Perform the proposed rate control scheme as described in next subsection.**

- **Step 5: Update the buffer occupancy, the quantization parameter (QP), and prepare for next frame encoding.**

4.2   Frame-Layer Rate Control Scheme

Fig. 4 shows the basic concept of the rate control scheme used in this paper, where the bundle of frames during the time interval is referred to as the temporal frame segment. For the frame-layer rate control, we employ an empirical data-based frame-layer R-D model using the quadratic rate model and the affine distortion model [6] with respect to the average QP in a frame, which is given by

$$\widehat{R}(\overline{q}_i) = (a\overline{q}_i^{-1} + b\overline{q}_i^{-2}) \cdot MAD(\widehat{f}_{ref}, f_{cur}), \tag{11}$$

$$\widehat{D}(\overline{q}_i) = a'\overline{q}_i + b', \tag{12}$$

where a, b, a', and b' are the model coefficients, $\widehat{f}_{ref}$ is the reconstructed reference frame at the previous time instant, f_{cur} is the uncompressed image at the current time instant, $MAD(.)$ is the mean of absolute difference between two frames, $\overline{q}_i$ is the average QP of all MBs in the ith frame, $\widehat{R}(\overline{q}_i)$ and $\widehat{D}(\overline{q}_i)$ are the rate and distortion models of the ith frame, respectively.

We consider a new formulation of frame-layer rate control based on the R-D model as follows: Determine $\overline{q}_i, i = 1, 2, \cdots, N_k^{SEG}$ to minimize

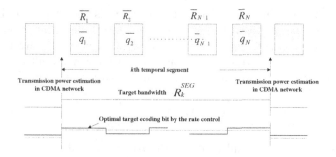

Fig. 4. Target bandwidth and encoding bit rate estimation using channel condition

$$\sum_{i=1}^{N_k^{SEG}} D_i(\overline{q}_i).(\widehat{D_i}(\overline{q}_i) - D_{i-1}), \tag{13}$$

subject to

$$\sum_{i=1}^{N_k^{SEG}} R_i \leq \widehat{R}_k^{SEG} \cdot T_k^{SEG}, \tag{14}$$

where $\widehat{D}_i$ is the estimated distortion of the current frame, D_{i-1} is the actual distortion of the previous frame, T_k^{SEG} is the number of encoding frames in the kth temporal segment, $\widehat{R}_k^{SEG}$ and T_k^{SEG} are the bandwidth and the time intervals of kth temporal segment, respectively. In (15), we introduce a formulation minimizing the average distortion over an entire sequences as well as variations in distortion between frames. The optimization task in (15) and (16) can be elegantly solved using Lagrangian optimization where a distortion term is weighted against a rate term. The Lagrangian formulation of the minimization problem is given by

$$J_i(\overline{q}_i) = \overline{D}_i(\overline{q}_i) \cdot (\overline{D}_i(\overline{q}_i) - D_{i-1}) + \lambda_i \cdot \max\left(\overline{B}_i^{res}, 0\right), \tag{15}$$

$$\overline{B}_i^{res} = \sum_{j=1}^{i} R_j + \overline{R}_i(\overline{q}_i) - \sum_{j=1}^{i} \frac{MAD_k^j}{Ave\_MAD_{k-1}} \cdot \frac{\widehat{R}_k^{SEG} \cdot T_k^{SEG}}{N_k^{SEG}}, \tag{16}$$

where $J_i(\overline{q}_i)$ and λ_i are the cost function and the Lagrange multiplier for the ith frame, R_j is the used bit-rate for the jth frame, MAD_k^j is the MAD between $(j$-$1)$th and jth frames of the kth temporal frame segment, Ave_MAD_{k-1} is the average of $MADs$ of the $(k$-$1)$th temporal frame segment. We use Ave_MAD_{k-1} as a substitute for Ave_MAD_k in real-time encoding, since it exist the correlation of the temporal frame segments. Therefore, the proposed algorithm does not require pre-analysis process. Note that $\overline{B}_i^{res}$ denotes the estimated bit based on the R-D model. It was shown in [7] that $J_i(\overline{q}_i)$ is a convex function generally. Thus, we can get its optimal solution by using the gradient method given by

$$\overline{q}_i^* = arg \ \min J_i(\overline{q}_i) \tag{17}$$

Note that what we finally need is not but which is the target bit budget for the ith frame. The proposed frame-layer rate control algorithm consists of two steps. The first step is to find the optimal bit-rates with the current Lagrange multiplier, and the second step is to adjust the Lagrange multiplier based on residual bit-rates. Thus, we employ the adaptive adjustment rule [7] given by

$$\lambda_{i+1} = \lambda_i + \Delta\lambda, \ \Delta\lambda = \frac{B_i}{B_{target,i}} - 1, \tag{18}$$

where λ_i is the Lagrange multiplier for the ith frame and

$$B_i = \sum_{j=1}^{i} R_j, \tag{19}$$

$$B_{target,i} = \sum_{j=1}^{i} \frac{MAD_k^i}{Ave\_MAD_{k-1}} \frac{\widehat{R}_k^{SEG} \cdot T_k^{SEG}}{N_k^{SEG}}. \tag{20}$$

Therefore, the proposed rate control algorithm produces low encoding time delay.

5 Experimental Results

To prove the effectiveness of proposed power-aware rate control method, we use the CDMA network for our experimental system. Experiments are performed under the same conditions as shown in the architecture in Fig. 5.

This architecture consists of a mobile station (MS), base station (BS), gateway, streaming server, and streaming client, which are typical components in CDMA networks. During the time that an MS randomly moves within one cell area or from one to another cells in the CDMA cellular network, the Rx-Power received from BS is periodically measured to feed into the CDMA channel analyzer as shown in Fig. 6. Then, based on the measured power level, the channel analyzer estimates other desired parameters of channel for our experiments, such as Rx-Powers, maximum channel rate, and PLR.

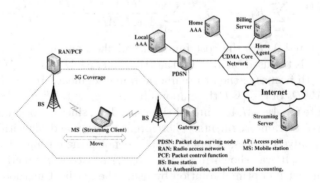

Fig. 5. Experimental rate control architecture

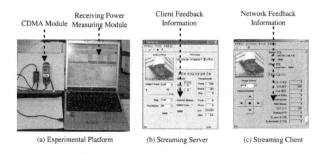

(a) Experimental Platform   (b) Streaming Server   (c) Streaming Client

Fig. 6. CDMA channel condition analyser

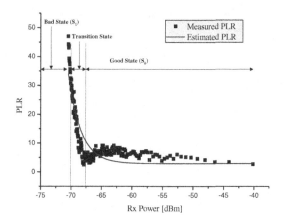

Fig. 7. Relationship between the PLR and Rx-Power

The channel state transition of the proposed channel model is performed by experimental threshold, which is -70(dBm) as shown in Fig. 7. In Fig. 7, the dots are the PLR measured data points while the solid curve is the plot of interpolated one from its measured points.

By using the relationship between PLR and Rx-Power, the transition probability matrix can be found to be p_{00}=0.89314, p_{01}=0.10686, p_{10}=0.26059, p_{11}= 0.73941 in the proposed wireless channel model. Fig. 8 (a) shows the power profile of Rx-Power received in our experiments. It seems that the Rx-Power are mainly varies in the range of [-70,-45] (dBm). If the Rx-Power is lower than -70 dBm, the channel may fall in the bad state and lots of errors occur during this period of time. Once the current channel is in the good state, the channel rate and frame target bits are instantly estimated for the current frame as shown in Fig. 8 (b).

Performance was evaluated by visual judgment since there is no standard measure currently available to evaluate subjective quality. As an objective measure

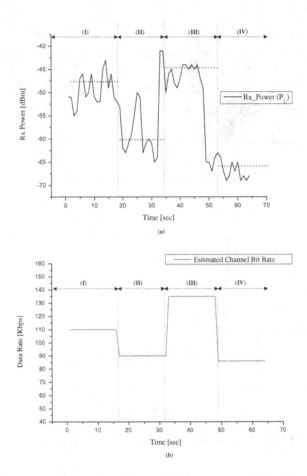

Fig. 8. Expectation target bit rate by channel status: (a) Rx-Power (PL) in CDMA channel and (b) estimated frame target bit

of the distance between an original image and its reconstructed image, PSNR is used. Fig. 9 (a) shows the PSNR values obtained from our experiments. In this figure, the performance of the proposed power aware rate control scheme is compared with that of TMN8.

For the performance comparison, we show the PSNR plot of test sequences "FOREMAN", which is in QCIF format (176 × 144), and the frame rate F is 25 fps. The visual comparisons of the proposed algorithm with the TMN8 are also provided in remain parts of Fig. 9 (a). To make a comparison of the subjective quality more clearly, zoomed images are also presented.

It is observed in Figs. 9 (b), (c), (d), and (e) that video quality obtained from our proposed rate control method is less sensitive to the time varying wireless channel condition. In TMN8 rate control method, even in low channel bit rates, the video quality is dramatically degraded (as shown in Fig. 9 (b), (c), and (d)), but in our proposed method, the high video quality is remained (Fig. 9(e)). It

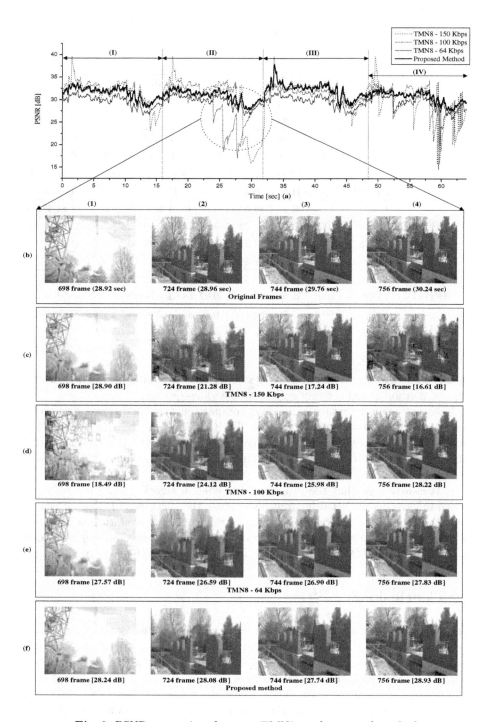

Fig. 9. PSNR comparison between TMN8s and proposed method

Table 1. Performance comparison of the proposed method with the conventional method in CDMA network

Test Sequence	Method	Encoded Bit Rate [Kbps]	PSNR (I) Avg.	Var.	(II) Avg.	Var.	(III) Avg.	Var.	(IV) Avg.	Var.	Total Avg.	Var.
	TMN8	150	31.89	7.74	29.36	19.12	32.27	5.32	28.09	35.20	30.40	18.22
FOREMAN	TMN8	100	31.18	1.76	29.79	7.632	31.18	1.76	28.51	16.56	30.17	8.15
	TMN8	64	29.65	1.62	29.40	2.04	29.65	1.62	29.65	1.62	29.59	1.73
	Proposed Adaptive		31.51	1.76	30.64	1.63	32.24	2.69	30.84	1.70	31.31	2.33
	TMN8	150	34.18	7.50	32.13	6.88	31.14	40.76	29.90	10.24	31.84	17.93
CARPHONE	TMN8	100	33.06	4.69	32.38	4.81	33.06	4.69	32.23	8.49	32.68	5.80
	TMN8	64	31.57	4.44	31.04	5.39	31.57	4.44	31.04	5.39	31.31	4.97
	Proposed Adaptive		33.39	4.37	32.55	4.06	34.54	4.61	32.60	4.90	33.27	5.27
	TMN8	150	39.73	0.49	35.07	8.27	39.73	0.49	28.61	16.11	35.78	35.22
AKIYO	TMN8	100	38.93	0.69	38.93	0.69	38.93	0.69	38.86	0.77	38.91	0.71
	TMN8	64	37.72	0.80	37.72	0.80	37.72	0.80	37.73	0.92	37.72	0.83
	Proposed Adaptive		39.16	0.63	39.73	0.49	39.73	0.49	38.67	0.72	39.82	0.78

means that even in mobile and wireless environment, using our proposed rate control method has effectively enhanced the video quality.

Extensive experimental testing and comparison were performed on several sequences with different characteristics: "FOREMAN" [11], "CARPHONE" [11], and "AKIYO" [11], we show the average (Avg.) PSNR value, the variance (Var.) of PSNR in Table 1. (I), (II), (III), (IV) in Table 1 means territory that Rx-Power changes in Fig. 8.

It is clearly seen that the proposed rate control algorithm can not only improve the average PSNR value, but also reduce the variance of PSNR. It is seen that the proposed rate control can reduce the quality degradation better than TMN8. Now we conclude that the proposed rate control method can be a good improvement of TMN8 in terms of both the PSNR and subjective quality.

6   Conclusions

Once video streams are transmitted over the wireless mobile networks, the compressed video tends to have video quality degradation. In order to overcome this problem, we proposed the efficient rate control scheme based on transmission power level for mobile multimedia devices.

The experimental results indicate that the proposed scheme can effectively enhance the video quality and minimize the average distortion over an entire sequence and variations in distortion between frames, even in time varying wireless channel as CDMA network. Moreover, since our proposed algorithm uses a fast convergence method and does not require pre-analysis, it is suitable for real time video communication. It is expected that our proposed video transmission method can be a useful alternative to existing TMN8 in terms of both the PSNR enhancement and wireless time varying channel consideration.

Acknowledgments. This research was supported by Seoul Future Contents Convergence (SFCC) Cluster established by Seoul Industry-Academy-Research Cooperation Project.

References

1. Hueda, M. R., Marques, C. A.:H.263-based wireless video transmission in multicode CDMA systems. IEEE Vehicular Technology Conference **1** (1997) 433–437.
2. Huang, J., Yao, R. Y., Bai, Y., Wang, S., W.:Performance of a mixed-traffic CDMA2000 wireless network with scalable streaming video. IEEE Trans. Circuits Syst. Video Technol. **13** (2003) 973–981
3. Zhang, Q., Ji, Z., Zhu, W., Zhang, Y., Q.:Power-minimized bit allocation for video communication over wireless channels. IEEE Trans. Circuits Syst. Video Technol. **12** (2002) 398–410
4. Tian, X.:Efficient transmission power allocation for wireless video communications. IEEE Wireless Communications and Networking Conference **4** (2004) 2058–2063
5. Naghshineh, M., Willebeek, M.:End-to-end QoS provisioning in multimedia wireless/mobile networks using an adaptive framework. IEEE Comm. Magazine (1997) 72–81
6. Gardos, T., E.:Video Test Model Number 8 (TMN8). ITU-T SG16/Q15 (1997)
7. Ribas, J., R., Lei, S.:Rate control in DCT video coding for low-delay communications. IEEE Trans. Circuits Syst. Video Technol. **9** (1999) 172–185
8. Zorzi, M., Y., Rao, R., R., Milstein, L., B.:ARQ error control for fading mobile radio channels. IEEE Trans. Veh. Technol. **46** (1997) 445–455
9. ITU-T:Video coding for low bit-rate communication. ITU-T recommendation H.263 Version2 (1998)
10. Kim, Y., Pyun, J., Y., Kim, H., S., Park, S., H., Ko, S., J.: Efficient Real-Time Frame Layer Rate Control Technique for Low Bit Rate Video over WLAN. IEEE Trans. on Consumer Electronics **49** (2003) 621–628
11. http://media.xiph.org/video/derf/

Author Index

Abu-Ghazaleh, Nael B. 46, 428
Ahmad, Mohammad Z. 337
Ahn, Kyung-Hwan 404
Allard, Géraud 170
Atsan, Emre 444

Barbeau, Michel 266
Bhakthavathsalam, R. 337
Blessing, Elijah 197
Bucher, Tobias 294

Chávez, Edgar 101
Chen, Jianer 184
Cheng, Wei-Fang 114
Cheung, Steven 19
Chin, Choong Ming 414
Chin, Kwan-Wu 87
Cho, Dong-Hoon 404
Cho, Yeong-Hun 142
Cichoń, Jacek 308
Clementi, Andrea E.F. 60

Dan, Byoung-Kyu 458
Dimitriou, Tassos 390
Duong, Dinh Trieu 458
Dutertre, Bruno 19

Ebinger, Peter 294
Eom, Doo-seop 32

Fehnker, Ansgar 128

Gadallah, Yasser 5
Gao, Peng 128
Gong, Guang 224

Han, Ki-Jun 404
Hao, Jiang 156
Hoeper, Katrin 224
Hou, Kun Mean 156
Huh, In 365
Hurson, A.R. 238
Hwang, Kwang-il 32

Ianni, Miriam Di 60

Jeong, Hyun-Cheol 355
Jeong, Jae-Yun 458
Jeong, Moon-Sang 142

Kang, Chul-Hee 376
Kim, Hye-Soo 458
Ko, Sung-Jea 458
Krontiris, Ioannis 390
Kutyłowski, Mirosław 308

Latchman, Haniph 365
Laurendeau, Christine 266
Lauria, Massimo 60
Lee, Dae-Young 355
Leung, Victor C.M. 3
Li, Jian-Jin 156
Li, Shan-Shan 114
Liao, Xiang-Ke 114
Lijia, Chen 156
Lindqvist, Ulf 19
Liu, Ke 46
Lowe, Darryn 87

Makri, Effie 211
Mandjes, Michel 321
Minet, Pascale 170
Monti, Angelo 60
Munishwar, Vikram 428

Nguyen, Dang-Quan 170

Olafsson, Sverrir 414
Oliveira, Carlos A.S. 73
Özkasap, Öznur 444

Park, Byungjoo 365
Park, Jong-Tae 142
Peng, Shao-Liang 114
Ploskonka, J.A. 238
Preneel, Bart 252

Roijers, Frank   321
Rossi, Gianluca   60
Ruiz, Pedro M.   101

Sanchez, Juan A.   101
Seok, Seung-Joon   376
Shrestha, Nirisha   170
Silvestri, Riccardo   60
Sim, Moh Lim   414
Singelée, Dave   252
Singhal, Mukesh   280
Song, Jung-Hoon   404
Srivastava, Mani B.   1
Stamatiou, Yannis C.   211
Sukumaran, Sangheethaa   197

Tejeda, Héctor   101
Tilak, Sameer   428
Turgut, Damla   337

van den Berg, Hans   321

Wang, Jianxin   184
Wang, Yongwei   280

Yoo, Kee-Young   349
Yoon, Eun-Jun   349
Youn, Joo-Sang   376

Zawada, Marcin   308
Zhu, Pei-Dong   114
Zhu, Xianman   184

Lecture Notes in Computer Science

For information about Vols. 1–4009

please contact your bookseller or Springer

Vol. 4127: E. Damiani, P. Liu (Eds.), Data and Applications Security XX. X, 319 pages. 2006.

Vol. 4121: A. Biere, C.P. Gomes (Eds.), Theory and Applications of Satisfiability Testing - SAT 2006. XII, 438 pages. 2006.

Vol. 4112: D.Z. Chen, D. T. Lee (Eds.), Computing and Combinatorics. XIV, 528 pages. 2006.

Vol. 4108: J.M. Borwein, W.M. Farmer (Eds.), Mathematical Knowledge Management. VIII, 295 pages. 2006. (Sublibrary LNAI).

Vol. 4106: T.R. Roth-Berghofer, M.H. Göker, H. A. Güvenir (Eds.), Advances in Case-Based Reasoning. XIV, 566 pages. 2006. (Sublibrary LNAI).

Vol. 4104: T. Kunz, S.S. Ravi (Eds.), Ad-Hoc, Mobile, and Wireless Networks. XII, 474 pages. 2006.

Vol. 4099: Q. Yang, G. Webb (Eds.), PRICAI 2006: Trends in Artificial Intelligence. XXVIII, 1263 pages. 2006. (Sublibrary LNAI).

Vol. 4098: F. Pfenning (Ed.), Term Rewriting and Applications. XIII, 415 pages. 2006.

Vol. 4097: X. Zhou, O. Sokolsky, L. Yan, E.-S. Jung, Z. Shao, Y. Mu, D.C. Lee, D. Kim, Y.-S. Jeong, C.-Z. Xu (Eds.), Emerging Directions in Embedded and Ubiquitous Computing. XXVII, 1034 pages. 2006.

Vol. 4096: E. Sha, S.-K. Han, C.-Z. Xu, M.H. Kim, L.T. Yang, B. Xiao (Eds.), Embedded and Ubiquitous Computing. XXIV, 1170 pages. 2006.

Vol. 4094: O. H. Ibarra, H.-C. Yen (Eds.), Implementation and Application of Automata. XIII, 291 pages. 2006.

Vol. 4093: X. Li, O.R. Zaiane, Z. Li (Eds.), Advanced Data Mining and Applications. XXI, 1110 pages. 2006. (Sublibrary LNAI).

Vol. 4092: J. Lang, F. Lin, J. Wang (Eds.), Knowledge Science, Engineering and Management. XV, 664 pages. 2006. (Sublibrary LNAI).

Vol. 4090: S. Spaccapietra, K. Aberer, P. Cudré-Mauroux (Eds.), Journal on Data Semantics VI. XI, 211 pages. 2006.

Vol. 4088: Z.-Z. Shi, R. Sadananda (Eds.), Agent Computing and Multi-Agent Systems. XVII, 827 pages. 2006. (Sublibrary LNAI).

Vol. 4079: S. Etalle, M. Truszczyński (Eds.), Logic Programming. XIV, 474 pages. 2006.

Vol. 4077: M.-S. Kim, K. Shimada (Eds.), Advances in Geometric Modeling and Processing. XVI, 696 pages. 2006.

Vol. 4076: F. Hess, S. Pauli, M. Pohst (Eds.), Algorithmic Number Theory. X, 599 pages. 2006.

Vol. 4075: U. Leser, F. Naumann, B. Eckman (Eds.), Data Integration in the Life Sciences. XI, 298 pages. 2006. (Sublibrary LNBI).

Vol. 4074: M. Burmester, A. Yasinsac (Eds.), Secure Mobile Ad-hoc Networks and Sensors. X, 193 pages. 2006.

Vol. 4073: A. Butz, B. Fisher, A. Krüger, P. Olivier (Eds.), Smart Graphics. XI, 263 pages. 2006.

Vol. 4072: M. Harders, G. Székely (Eds.), Biomedical Simulation. XI, 216 pages. 2006.

Vol. 4071: H. Sundaram, M. Naphade, J.R. Smith, Y. Rui (Eds.), Image and Video Retrieval. XII, 547 pages. 2006.

Vol. 4070: C. Priami, X. Hu, Y. Pan, T.Y. Lin (Eds.), Transactions on Computational Systems Biology V. IX, 129 pages. 2006. (Sublibrary LNBI).

Vol. 4069: F.J. Perales, R.B. Fisher (Eds.), Articulated Motion and Deformable Objects. XV, 526 pages. 2006.

Vol. 4068: H. Schärfe, P. Hitzler, P. Øhrstrøm (Eds.), Conceptual Structures: Inspiration and Application. XI, 455 pages. 2006. (Sublibrary LNAI).

Vol. 4067: D. Thomas (Ed.), ECOOP 2006 – Object-Oriented Programming. XIV, 527 pages. 2006.

Vol. 4066: A. Rensink, J. Warmer (Eds.), Model Driven Architecture – Foundations and Applications. XII, 392 pages. 2006.

Vol. 4065: P. Perner (Ed.), Advances in Data Mining. XI, 592 pages. 2006. (Sublibrary LNAI).

Vol. 4064: R. Büschkes, P. Laskov (Eds.), Detection of Intrusions and Malware & Vulnerability Assessment. X, 195 pages. 2006.

Vol. 4063: I. Gorton, G.T. Heineman, I. Crnkovic, H.W. Schmidt, J.A. Stafford, C.A. Szyperski, K. Wallnau (Eds.), Component-Based Software Engineering. XI, 394 pages. 2006.

Vol. 4062: G. Wang, J.F. Peters, A. Skowron, Y. Yao (Eds.), Rough Sets and Knowledge Technology. XX, 810 pages. 2006. (Sublibrary LNAI).

Vol. 4061: K. Miesenberger, J. Klaus, W. Zagler, A. Karshmer (Eds.), Computers Helping People with Special Needs. XXIX, 1356 pages. 2006.

Vol. 4060: K. Futatsugi, J.-P. Jouannaud, J. Meseguer (Eds.), Algebra, Meaning and Computation. XXXVIII, 643 pages. 2006.

Vol. 4059: L. Arge, R. Freivalds (Eds.), Algorithm Theory – SWAT 2006. XII, 436 pages. 2006.

Vol. 4058: L.M. Batten, R. Safavi-Naini (Eds.), Information Security and Privacy. XII, 446 pages. 2006.

Vol. 4057: J.P. W. Pluim, B. Likar, F.A. Gerritsen (Eds.), Biomedical Image Registration. XII, 324 pages. 2006.

Vol. 4056: P. Flocchini, L. Gąsieniec (Eds.), Structural Information and Communication Complexity. X, 357 pages. 2006.

Vol. 4055: J. Lee, J. Shim, S.-g. Lee, C. Bussler, S. Shim (Eds.), Data Engineering Issues in E-Commerce and Services. IX, 290 pages. 2006.

Vol. 4054: A. Horváth, M. Telek (Eds.), Formal Methods and Stochastic Models for Performance Evaluation. VIII, 239 pages. 2006.

Vol. 4053: M. Ikeda, K.D. Ashley, T.-W. Chan (Eds.), Intelligent Tutoring Systems. XXVI, 821 pages. 2006.

Vol. 4052: M. Bugliesi, B. Preneel, V. Sassone, I. Wegener (Eds.), Automata, Languages and Programming, Part II. XXIV, 603 pages. 2006.

Vol. 4051: M. Bugliesi, B. Preneel, V. Sassone, I. Wegener (Eds.), Automata, Languages and Programming, Part I. XXIII, 729 pages. 2006.

Vol. 4049: S. Parsons, N. Maudet, P. Moraitis, I. Rahwan (Eds.), Argumentation in Multi-Agent Systems. XIV, 313 pages. 2006. (Sublibrary LNAI).

Vol. 4048: L. Goble, J.-J.C.. Meyer (Eds.), Deontic Logic and Artificial Normative Systems. X, 273 pages. 2006. (Sublibrary LNAI).

Vol. 4047: M. Robshaw (Ed.), Fast Software Encryption. XI, 434 pages. 2006.

Vol. 4046: S.M. Astley, M. Brady, C. Rose, R. Zwiggelaar (Eds.), Digital Mammography. XVI, 654 pages. 2006.

Vol. 4045: D. Barker-Plummer, R. Cox, N. Swoboda (Eds.), Diagrammatic Representation and Inference. XII, 301 pages. 2006. (Sublibrary LNAI).

Vol. 4044: P. Abrahamsson, M. Marchesi, G. Succi (Eds.), Extreme Programming and Agile Processes in Software Engineering. XII, 230 pages. 2006.

Vol. 4043: A.S. Atzeni, A. Lioy (Eds.), Public Key Infrastructure. XI, 261 pages. 2006.

Vol. 4042: D. Bell, J. Hong (Eds.), Flexible and Efficient Information Handling. XVI, 296 pages. 2006.

Vol. 4041: S.-W. Cheng, C.K. Poon (Eds.), Algorithmic Aspects in Information and Management. XI, 395 pages. 2006.

Vol. 4040: R. Reulke, U. Eckardt, B. Flach, U. Knauer, K. Polthier (Eds.), Combinatorial Image Analysis. XII, 482 pages. 2006.

Vol. 4039: M. Morisio (Ed.), Reuse of Off-the-Shelf Components. XII, 444 pages. 2006.

Vol. 4038: P. Ciancarini, H. Wiklicky (Eds.), Coordination Models and Languages. VIII, 299 pages. 2006.

Vol. 4037: R. Gorrieri, H. Wehrheim (Eds.), Formal Methods for Open Object-Based Distributed Systems. XVII, 474 pages. 2006.

Vol. 4036: O. H. Ibarra, Z. Dang (Eds.), Developments in Language Theory. XII, 456 pages. 2006.

Vol. 4035: T. Nishita, Q. Peng, H.-P. Seidel (Eds.), Advances in Computer Graphics. XX, 771 pages. 2006.

Vol. 4034: J. Münch, M. Vierimaa (Eds.), Product-Focused Software Process Improvement. XVII, 474 pages. 2006.

Vol. 4033: B. Stiller, P. Reichl, B. Tuffin (Eds.), Performability Has its Price. X, 103 pages. 2006.

Vol. 4032: O. Etzion, T. Kuflik, A. Motro (Eds.), Next Generation Information Technologies and Systems. XIII, 365 pages. 2006.

Vol. 4031: M. Ali, R. Dapoigny (Eds.), Advances in Applied Artificial Intelligence. XXIII, 1353 pages. 2006. (Sublibrary LNAI).

Vol. 4029: L. Rutkowski, R. Tadeusiewicz, L.A. Zadeh, J. Zurada (Eds.), Artificial Intelligence and Soft Computing – ICAISC 2006. XXI, 1235 pages. 2006. (Sublibrary LNAI).

Vol. 4028: J. Kohlas, B. Meyer, A. Schiper (Eds.), Dependable Systems: Software, Computing, Networks. XII, 295 pages. 2006.

Vol. 4027: H.L. Larsen, G. Pasi, D. Ortiz-Arroyo, T. Andreasen, H. Christiansen (Eds.), Flexible Query Answering Systems. XVIII, 714 pages. 2006. (Sublibrary LNAI).

Vol. 4026: P.B. Gibbons, T. Abdelzaher, J. Aspnes, R. Rao (Eds.), Distributed Computing in Sensor Systems. XIV, 566 pages. 2006.

Vol. 4025: F. Eliassen, A. Montresor (Eds.), Distributed Applications and Interoperable Systems. XI, 355 pages. 2006.

Vol. 4024: S. Donatelli, P. S. Thiagarajan (Eds.), Petri Nets and Other Models of Concurrency - ICATPN 2006. XI, 441 pages. 2006.

Vol. 4021: E. André, L. Dybkjær, W. Minker, H. Neumann, M. Weber (Eds.), Perception and Interactive Technologies. XI, 217 pages. 2006. (Sublibrary LNAI).

Vol. 4020: A. Bredenfeld, A. Jacoff, I. Noda, Y. Takahashi (Eds.), RoboCup 2005: Robot Soccer World Cup IX. XVII, 727 pages. 2006. (Sublibrary LNAI).

Vol. 4019: M. Johnson, V. Vene (Eds.), Algebraic Methodology and Software Technology. XI, 389 pages. 2006.

Vol. 4018: V. Wade, H. Ashman, B. Smyth (Eds.), Adaptive Hypermedia and Adaptive Web-Based Systems. XVI, 474 pages. 2006.

Vol. 4017: S. Vassiliadis, S. Wong, T.D. Hämäläinen (Eds.), Embedded Computer Systems: Architectures, Modeling, and Simulation. XV, 492 pages. 2006.

Vol. 4016: J.X. Yu, M. Kitsuregawa, H.V. Leong (Eds.), Advances in Web-Age Information Management. XVII, 606 pages. 2006.

Vol. 4014: T. Uustalu (Ed.), Mathematics of Program Construction. X, 455 pages. 2006.

Vol. 4013: L. Lamontagne, M. Marchand (Eds.), Advances in Artificial Intelligence. XIII, 564 pages. 2006. (Sublibrary LNAI).

Vol. 4012: T. Washio, A. Sakurai, K. Nakajima, H. Takeda, S. Tojo, M. Yokoo (Eds.), New Frontiers in Artificial Intelligence. XIII, 484 pages. 2006. (Sublibrary LNAI).

Vol. 4011: Y. Sure, J. Domingue (Eds.), The Semantic Web: Research and Applications. XIX, 726 pages. 2006.

Vol. 4010: S. Dunne, B. Stoddart (Eds.), Unifying Theories of Programming. VIII, 257 pages. 2006.